SYMPOSIA OF THE
SOCIETY FOR EXPERIMENTAL BIOLOGY

NUMBER XXX

SYMPOSIA OF THE SOCIETY FOR EXPERIMENTAL BIOLOGY

The Journal of Experimental Botany
is published by the Oxford University Press
for the Society for Experimental Biology

SYMPOSIA OF THE
SOCIETY FOR EXPERIMENTAL BIOLOGY

NUMBER XXX

CALCIUM IN BIOLOGICAL SYSTEMS

Published for the Society for Experimental Biology

CAMBRIDGE UNIVERSITY PRESS

CAMBRIDGE
LONDON · NEW YORK · MELBOURNE

574
S67
v.30

Published by the Syndics of the Cambridge University Press
The Pitt Building, Trumpington Street, Cambridge CB2 IRP
Bentley House, 200 Euston Road, London NWI 2DB
32 East 57th Street, New York, NY 10022, USA
296 Beaconsfield Parade, Middle Park, Melbourne 3206, Australia

First published 1976

Printed in Great Britain
at the
University Printing House, Cambridge
(Euan Phillips, University Printer)

Library of Congress Cataloguing in Publication Data
Main entry under title:
Calcium in biological systems.
(Symposia of the Society for Experimental Biology; no. 30)
'Symposium...was held from 9–12 September 1975 in...Royal Holloway College
(University of London) at Englefield Green, Surrey.'
Includes bibliographies and indexes.
1. Calcium in the body–Congresses. 2. Calcium metabolism–Congresses.
I. Duncan, Christopher John. II. Society for Experimental Biology (Gt Brit.)
III. Series: Society for Experimental Biology (Gt Brit.). Symposia; no. 30.
QH302.S622 no. 30 [QP535.C2] 574'.08s [591.1'9'21]
ISBN: 0 521 21236 7   76–2231

CONTENTS

[v]

PREFACE

The thirtieth symposium of the Society for Experimental Biology was held from 9 to 12 September 1975 in the very pleasant surroundings of Royal Holloway College (University of London) at Englefield Green, Surrey, and we are most grateful to the Principal and to the members of the Departments of Zoology and Botany who put their facilities at our disposal and who did so much to make us welcome. In particular, all participants are indebted to Dr E. J. Binyon who acted as local secretary and who worked tirelessly to ensure the smooth running of both the scientific programme and the domestic arrangements. His task was exacerbated because the symposium on 'Calcium in Biological Systems' attracted considerable interest and all available places were fully booked. The people attending the meeting included many non-members of the Society and we were pleased to welcome participants from medical research groups and the pharmaceutical industry and also visitors from the continent of Europe and from the USA.

It was clear that there is currently a great and widespread interest in the role of calcium in the physiology of the cell; the predictions made 35 years ago by Heilbrunn are being fulfilled and calcium, like the cyclic nucleotides, is now seen to occupy an important function in controlling a variety of cellular activities. Only slight changes in the order of the papers in the published volume have been made. The programme began with a consideration of the chemistry of calcium and magnesium and this was followed by papers dealing with the specialized tools that have been developed for the study of calcium, namely the photoprotein aequorin and the divalent cation ionophores. An important feature of calcium is its low intracellular concentration, and a number of papers were devoted to a consideration of the cellular homeostasis of this ion. The major part of the symposium, however, was concerned with studies of calcium as a trigger and as a control mechanism including such topics as membrane permeability, secretion, cell division and interaction with cyclic nucleotides, cilia, cell-to-cell communication, calcium binding proteins and contractility.

It is with very great regret that we record the death of Professor T. Weis-Fogh, some two months after the symposium. Torkel Weis-Fogh had long been a very active member of the Society for Experimental Biology and, in addition to being a joint contributor to the paper on the spasmoneme, acted as one of our chairmen during the symposium.

[vii]

The Symposium Committee are indebted to Dr A. C. Allison, Professor P. F. Baker, Professor P. C. Caldwell, Dr A. W. Cuthbert, Professor R. Miledi, Professor J. W. S. Pringle and Professor K. Simkiss for their advice and guidance in designing the programme and assembling such a distinguished group of workers. I am grateful to many friends for their help in editing and I should like to thank Miss Susan Scott in particular for her assistance in checking manuscripts and references. It is a pleasure to thank two Symposium Secretaries, Professor E. Naylor and Dr J. W. Hannay, for all their work on behalf of the symposia and to acknowledge, once again, the courtesy and assistance that we have received from Cambridge University Press.

C. J. DUNCAN
Editor of the thirtieth
symposium of the Society for
Experimental Biology

CALCIUM CHEMISTRY AND ITS RELATION TO BIOLOGICAL FUNCTION

By R. J. P. WILLIAMS

Inorganic Chemistry Laboratory, South Parks Road,
Oxford OX1 3QR

This introductory paper is concerned with the chemistry of calcium and I shall discuss the available data concerning the structural, thermodynamic and kinetic aspects of the formation of its complex ions and salts. In particular, it will be necessary to refer to competition between Ca^{2+} and the cations H^+, Mg^{2+}, Na^+, K^+, and ammonium bases. The biological action and function of calcium depend on its binding which, in turn, is a function of molecular structure, binding strength and the kinetic parameters that relate a particular complex to its action. Such an interaction may be summarised in the following equation: Action is a function of Concentration × Binding Strength × Structure × Rate Constants. Various points of interest under these four headings that determine the biological action of calcium will be examined, starting from the inorganic chemistry of this divalent cation, but always attempting to emphasise the relevance of these studies to biological systems.

As will be shown in this symposium (see papers by Baker, Carafoli & Crompton, and Borle & Anderson, this volume) the concentration of free Ca^{2+} in the cytosol is of major importance in cellular physiology and consequently we must consider the means by which different cells control the levels of Ca^{2+}, both in the cytosol and within intracellular compartments, such as the mitochondria and sarcoplasmic reticulum.

Finally, I shall describe various probe methods which can be used for studying calcium in biological systems, giving examples (which include peptide, protein and membrane studies) of their use in our own work.

CONCENTRATION OF CALCIUM IONS

All equilibria and rates of reaction depend upon the free calcium ion concentration. Analytical total concentrations in biological systems are likely to exceed greatly this free ion concentration and in general concentrations of free calcium outside cells exceed that inside cells by several orders of magnitude. Some rough values (it is difficult to get accurate values) are given in Table 1. These data are necessary if the thermodynamic

Table 1. *Free calcium ion concentrations* (mM)

	Maximum	Minimum
Plasma (blood)	3–5	2–3
Red blood cell	10^{-2}	$< 10^{-3}$?
Muscle cells	$\geqslant 10^{-3}$	$\leqslant 10^{-4}$
Sea water	~ 10	~ 10
Fresh water	$> 10^{-1}$	$> 10^{-2}$

and kinetic constants discussed in this paper are to be seen in a biological context (Wacker & Williams, 1968).

THE STRUCTURES OF CALCIUM SALTS

Tables 2 and 3 give data from X-ray diffraction studies on single crystals of the coordination sphere around the calcium cation. Looking at the Tables on the structures of small inorganic and organic molecules which form calcium salts or complexes two clear statements can be made.

1. Usually Ca^{2+} has a coordination number of seven or eight.
2. The coordination geometry is irregular in both bond angle and bond length.

In both regards calcium is quite different from magnesium, which maintains six-coordination in a closely regular octahedron, Table 4. The structural

Table 2. *Some typical calcium salt structures*

Salt	Ca (II) coordination no.	Ca–O distances (nm) Min.	Ca–O distances (nm) Max.	Reference
$CaHPO_4 . 2H_2O$	8	0.244	0.282	(a)
$Ca(H_2PO_4)_2 . H_2O$	8	0.230	0.274	(b)
Ca 1,3-disphosphorylimidazole	6	0.226	0.236	(c)
	7	0.227	0.278	(c)
Ca dipicolinate . $3H_2O$	8	0.236	0.257	(d)
Ca $Na(H_2PO_2)_3$	6	0.231	0.233	(e)
Ca tartrate . $4H_2O$	8	0.239	0.254	(f)
$Ca(C_6H_9O_7)_2 . 2H_2O$	8	0.239	0.247	(g)

(a) Beevers, C. A. (1958). *Acta Cryst.*, **11**, 273–277.
(b) MacLennan, G. & Beevers, C. A. (1956). *Acta Cryst.*, **9**, 187–190.
(c) Beard, L. N. & Lenhert, P. G. (1968). *Acta Cryst.*, **24**B, 1529–1539.
(d) Strahs, G. & Dickerson, R. E. (1968). *Acta Cryst.*, **24**B, 571–578.
(e) Matsuzaki, T. & Iitaka, Y. (1969). *Acta Cryst.*, **25**B, 1932–1938.
(f) Ambady, G. K. (1968). *Acta Cryst.*, **24**B, 1548–1557.
(g) Balchin, A. A. & Carlisle, C. H. (1965). *Acta Cryst.*, **19**, 103–111.

N.B. When calcium is six-coordinate the structures may be much more regular.

Table 3. *Some Ca²⁺ structures of small molecules of biological interest*

	Coordination no.	Ca–O distances (nm)		Reference
		Min.	Max.	
Ca²⁺ thymidylate	7	0.230	0.265	(a)
Ca²⁺ diphosphonate	8	0.240	0.260	(b)
Ca²⁺ galactose	8	0.235	0.255	(c)
Ca²⁺ blephavismin	7	0.235	0.245	(d)
Ca²⁺ trehalose	7	0.235	0.255	(e)
Ca²⁺ arabonate	8	0.245	0.250	(f)

(a) Trueblood, K. N., Horn, P. & Luzzati, V. (1961). *Acta Cryst.*, **14**, 965–982.
(b) Uchtman, V. A. (1972). *J. phys. Chem.*, **76**, 1304–1310.
(c) Cook, W. J. & Bugg, C. E. (1973). *J. Am. chem. Soc.*, **95**, 6442–6447.
(d) Kubota, T., Tokoroyama, T., Tsukuda, Y., Koyama, H. & Miyake, A. (1973). *Science, Wash.*, **179**, 400–402.
(e) Cook, W. J. & Bugg, C. E. (1973). *Carbohydrate Res.*, **31**, 265–275.
(f) Furberg, S. & Helland, S. (1962). *Acta chem. scand.*, **16**, 2373–2383.

Table 4. *Some typical magnesium salt structures*

Salt	Mg (II) coordination no.	Mg–O distances (nm)		Reference
		Min.	Max.	
Mg hexa-antipyrine . ClO₄	6	0.206 (Ca, 0.230)		(a)
Mg(C₄H₈O)₄ . Br₂	6	0.216 (Oxygens, 4)		(b)
MgS₂O₃ . 6H₂O	6	0.205	0.212	(c)
MgSO₄ . 4H₂O	6	0.204	0.209	(d)
Mg(HPO₃) . 6H₂O	6	0.200	0.212	(e)
Mg₂P₂O₇	6	0.200	0.211	(f)
Mg(CH₃CO₂)₂ . 4H₂O	6	0.200	0.210	(g)

(a) Vijayan, M. & Viswamitra, M. A. (1968). *Acta Cryst.*, **24B**, 1067–1076.
(b) Pérucaud, M. & Le Bihan, M.-T. (1968). *Acta Cryst.*, **24B**, 1502–1505.
(c) Baggio, S., Amzel, L. M. & Becka, L. N. (1969). *Acta Cryst.*, **25B**, 2650–2653.
(d) Baur, W. H. (1962). *Acta Cryst.*, **15**, 815–826.
(e) Corbridge, D. E. C. (1956). *Acta Cryst.*, **9**, 991–994.
(f) Lukaszewicz, K. (1961). *Roczn. Chem.*, **35**, 31–37.
(g) Shankar, J., Khubchandani, P. G. & Padmanabhan, V. M. (1957). *Proc. Indian Acad. Sci.*, **45A**, 117–119.

NOTE. Several magnesium hexahydrate salts have also been examined but crystal structure data on lower hydrates of magnesium are rare.

role of the two ions in biology is therefore quite different. Magnesium, as it strongly demands a certain geometry, is weakened in its ability to bind to the irregular geometries of coordination sites of biological molecules, e.g. proteins. This same ion does not cross-link structures readily, for cross-linking will usually demand high coordination number and irregular geometry. By way of contrast, the very adaptable calcium coordination sphere makes cross-linking a major feature of its solid-state and solution chemistry. All aspects of the biochemistry of calcium lead to the same conclusion. While —S—S— and sugar–peptide bridges are (almost) irreversible cross-links of proteins and cell-wall structures, the calcium ion provides a *reversible* cross-link, which is therefore rapidly responsive to change of conditions.

Calcium would also act as a cross-link between polydiesterphosphates of RNA and DNA but the very low concentrations of intracellular calcium ($< 10^{-7}$ M) and the poor calcium binding strength of this ligand ($K_{ML} <$ 5 M^{-1}) rule out this possibility.* High calcium would probably prevent the proper functioning of RNA and DNA in cells. By way of contrast magnesium serves as a weak cross-linking agent in tRNA, rRNA and probably in DNA to some degree. Thus magnesium controls cross-linking of one group of polymers inside cells whereas calcium controls cross-linking of a very different set of polymers on the outer surfaces of cells. Exceptions may be provided by some intracellular centres of high calcium binding strength which are used as triggers after an inward calcium concentration pulse, and by magnesium–techoic acid complexes.

In the binding to small molecules and proteins alike, oxygen-atoms and not nitrogen atoms (or sulphur) coordinate the calcium ion. Even the presence of one nitrogen in a coordination sphere leads to the preferential uptake of a transition metal ion or magnesium, e.g. in the second site of concanavalin A. All the known binding sites in proteins, except one (see Table 5) have two or more carboxylate groups and their coordination number is six or greater. The binding sites are again of irregular geometry. These characteristics were all readily predictable from the general chemistry of calcium (Williams, 1970). Moreover it is clear that a high binding strength for calcium usually requires three or more carboxylate residues so that intracellular Ca^{2+} binding proteins generally have at least this array of carboxylate groups.

THE THERMODYNAMICS OF CALCIUM BINDING

The binding of calcium ions by ligands can be treated on an electrostatic model provided that the repulsions between ligands (L) as well as the competitive nature of the equilibria in water are taken into account. The

* Throughout this article K_{ML} in units M^{-1} refers to the *association* constant of ML.

Table 5. Ca^{2+} protein-binding

Protein	Coordination no.	No. of $-CO_2^-$	Reference
Parvalbumin	(a) 6?+H_2O	4	(a)
	(b) 8	4	
Thermolysin	6–8	2–3	(b)
Nuclease	6?+H_2O	3	(c)
Concanavalin A	6?	2	(d)
Lysozyme	(?)	1–2	(e)

(a) Moews, P. C. & Kretsinger, R. H. (1975). *J. molec. Biol.*, **91**, 201–228.

(b) Edelman, G. M., Cunningham, B. A., Reeke, G. N., Becker, J. W., Waxdal, M. J. & Wang, J. L. (1972). *Proc. natn. Acad. Sci. USA*, **69**, 2580–2584.

(c) Cotton, F. A., Brier, C. J., Day, V. W., Hazen, E. E. & Larson, S. (1971). *Cold Spring Harb. Symp. quant. Biol.*, **36**, 243–255.

(d) Matthews, B. W., Weaver, L. H. & Kester, W. R. (1974). *J. biol. Chem.*, **249**, 8030–8044.

(e) Campbell, I. D., Dobson, C. M. & Williams, R. J. P. (1975). *Proc. R. Soc. Lond. A*, **345**, 41–59.

equilibrium equation, ignoring the hydration of L, is

$$Ca(H_2O)_n + L \rightleftharpoons CaL(H_2O)_m + (n-m)H_2O. \tag{1}$$

If we are comparing the affinity of different metal ions (M) for L then we do not have to consider the hydration of the ligand. On the other hand if we wish to know why various ligands bind calcium ions in a definite order the hydration of L is very important. The hydration of $ML(H_2O)_m$ is of the greatest importance in all comparisons for it changes from metal to metal and from one ligand to another.

In the case of metal ion competition the large size of the calcium ion relative to ions such as those of magnesium and beryllium mean that it will bind more weakly to small anions where steric factors are unimportant, as one water molecule in a coordination sphere is almost exactly matched by one oxygen or fluoride anion. The order of stability for the complex ions is then (for F^-, OH^-, RPO_4^{2-}, RCO_2^-)

$$Be^{2+} > Mg^{2+} > Ca^{2+} > Sr^{2+} > Ba^{2+}.$$

On the other hand, a very large anion will displace several water molecules and such anions as SO_4^{2-}, $S_2O_3^{2-}$ compensate less well for the loss of this interaction by the strength of their own interaction for small as opposed to large cations. The binding order for simple large anions is

$$Ba^{2+} > Sr^{2+} > Ca^{2+} > Mg^{2+} > Be^{2+}.$$

Intermediate orders can be obtained as shown by Eisenman (1968), who

Table 6. *Comparison of Ca^{2+} and Mg^{2+} binding constants*

Ligand	Binding constant, $\log K_{ML}$	
	Mg^{2+}	Ca^{2+}
Glycine	3.4	1.4
Imidodiacetate	2.9	2.6
Nitrilotriacetate	5.3	6.4
EDTA	8.9	10.7
EGTA	5.4	10.7
Acetate	0.8	0.7
Malonate	2.8	2.5
Citrate	3.2	4.8

See Williams (1970) for further details.

bases his theoretical discussion on anion field strengths. (For anions of the same charge the larger the anion the smaller its field strength of course.) These considerations are not sufficient in the discussion of ligands of complicated stereochemistry, for now the ligand size and whether it be a flexible or a constrained ring ligand, can dominate selectivity. The idea of competition based upon the ratio of the size of the cation to the size of the best hypothetical or real hole which the ligand can form leads to the generalisation that, no matter what the field strength of the donor atoms, steric factors inherent in the *radius ratio effect* can dominate the order of cation selectivity, Table 6. Ionophores and complicated ligands such as EDTA and EGTA exemplify these facts. (Notional field strengths cannot describe the relative stabilities of complexes of monovalent cations with complicated ligands either.)

Now the above considerations of cation competition for ligands depend upon the equal charge of the cations. Biological systems successfully distinguish mono- from divalent cations and not on the basis of size. Although sodium ions and calcium ions, cations of the same size, compete for ligands, there are circumstances in which one or the other cation is bound much more strongly. Clearly, as well as the above factors, ligand charge and hydration are now important (Dietrich, Lehn & Sauvage, 1973; Williams, 1973). Equation (1) should be re-written

$$M^{p+}(H_2O)_y + L^{q-}(H_2O)_z \rightleftharpoons ML^{(p-q)+}(H_2O)_w + xH_2O, \qquad (2)$$

where $x = z + y - w$.

The higher the charge $(p-q)+$ on the complex, the less stable will be the unit ML in a *low dielectric medium*. Thus a bias toward the monovalent cations can be achieved by building L so that it has very little hydration.

Dietrich *et al.* (1973) proved this point by building ligands L which had very hydrophobic exteriors, showing that the relative affinity of the central core of the ligand L could then be increased in favour of potassium relative to barium. A bias the other way is achieved if $q- > p+$. The simple idea is then that lowly charged (zero or even $+1$) ligands will give selectivity $M^+ > M^{2+}$ and this will be more so the more hydrophobic the ligand; and that highly charged (negative) ligands will give selectivity $M^{2+} > M^+$ and this will be more so the more hydrophilic the ligand. Although the description of this feature of binding can be looked upon as one of the relative importance of the spheres of hydration of M, L and ML, it is unfortunate that the discussion cannot be put on a quantitative basis since in order to do so the relative structural demands of M and L in the states M, L and ML would have to be known, as well as their hydrations. As we have shown above calcium is very flexible in its demands upon the ligand geometry in ML and hence the hydration of ML is also flexible. On the basis of the above considerations the general prediction of factors controlling the stability of calcium binding to ligands such as proteins was a relatively straight-forward affair (Williams, 1970), but its intimate details remain obscure.

The last paragraph is particularly important when extraction of ions from an aqueous to a non-aqueous medium has to be considered, for here residual hydration and the low dielectric constant can have dominating effects.

BINDING SITES

In order to be effective, i.e. taking into account the calcium ion concentration (Table 1), the binding site for calcium ions must have a binding constant of $\geqslant 10^3$ M^{-1} (outside cells) and of $\geqslant 10^6$ M^{-1} (inside cells). If it is to take part in a *permanent* structural feature in a cell, calcium ion must have a binding constant of $> 10^7$ M^{-1}. Such binding strengths can be developed in more than one way. The most obvious ligand of high binding strength is a multicarboxylate donor such as EDTA and, in proteins, such binding centres can be built in subtle ways. (Some other protein must control the free intracellular Ca^{2+} concentration of course.) The need for control of the exact binding strength is partly the requirement for high selectivity $Ca^{2+} \gg Mg^{2+} \simeq Mn^{2+}$, and partly that a very good control on the magnitude of the binding constant must be generated in order for binding to fluctuate with intracellular concentration changes. The way this is managed, using the side-chain carboxylate groups of proteins, can be seen by examining the binding constants of model ligands in which carboxylate groups are attached to a frame $X(CO_2^-)_n$, where X is the frame. The simplest cases are shown in Table 6. Starting from dicarboxylate centres $(CH_2)_n(CO_2^-)_2$ it is seen that stability falls from malonate ($n = 1$) to larger chains. Greater configurational entropy is lost on chelate formation with the larger chains. This can be avoided in three ways:

1. By making a rigid frame of *cis*-conformation, e.g.

$$(CH{=}CH).(CO_2^-)_2.$$

2. By inserting other potential coordinating centres, e.g.

$$(CH(OH))_2.(CO_2^-)_2.$$

3. By increasing the number of carboxylate centres, e.g.

$$(CHCO_2^-).(CO_2^-)_2.$$

Biological systems do not produce multi-carboxylate ligands of rigid frame structure and such structures can be approached in proteins only through folding. An α-helix starting from a glutamate (or aspartate) residue makes a poor chelating agent if the amino acid sequence goes Glu.Glu (adjacent amino acids) because the binding centres are turned away from one another (compare *trans* systems), but it forms a good binding agent if the sequence is Glu.x.x.Glu or Glu.x.x.x.Glu (where x is any amino acid). The helix now generates a *cis* disposition of carboxylate groups. Of course a single helical turn is only one possibility for bringing groups close together in a protein but it is a very common device in metal binding centres, e.g. ferredoxins, zinc proteins, haem proteins and in many calcium proteins.

The second type of centre, i.e. carboxylate centres assisted by hydroxyl, ether or keto functions, is very common in glycolytic intermediates (e.g. citrate) and in sugars themselves, and occurs occasionally in proteins. For small water soluble ligands the additional binding provided by such groups is usually small unless a very rigid frame is provided. Wall polysaccharides provide many such sites, as do ionophores.

Increase in the number of carboxylates centres in a model ligand can be managed in a variety of ways, e.g. EDTA, EGTA, caballylic acid. Biological systems are now known to provide their own centre of this type, γ-carboxylated glutamic acid, containing the unit $-CH(CO_2^-)_2$. This group in itself is still a relatively poor Ca^{2+} chelating agent (log $K_{ML} \leqslant 3$) but if two such groups are juxtaposed, as in a helical turn, γ-Glu.x.x.γ-Glu, then a very potent calcium binding centre can be generated (log $K_{ML} \geqslant 6$).

The fine synthetic work of Dietrich *et al.* (1969) has shown that carboxylate groups are not essential for strong binding to calcium. A rigid disposition of ether oxygens in a cage of the correct dimensions can supply a very good calcium binding site (log $K_{ML} \geqslant 6$). Urry, Cunningham & Ohnishi (1973) have proposed that carbonyl groups of peptide chains could also bind calcium very strongly but I am unaware of a convincing demonstration of this principle *in aqueous media*. A comparison of such chelating devices with the multi-carboxylate centres reveals an important difference. The carboxylate centres are powerful enough chelating ligands to bind to calcium strongly even though they must undergo considerable loss of configurational entropy (conformational change) in doing so. This is not true for the ether and carbonyl residues which must be placed on a very rigid structural frame if high binding is to be achieved. In a biological system it is not just the high binding constant which is required, because usually calcium is triggering the conformational change of an organic moiety and thereby generating biological activity. Again rapid removal of this triggering cation is needed in order for the system to relax rapidly. Thus rapid *on/off* reactions are required as well as high binding constants. Here rigid ligands are not effective, for both the *on* and *off* binding steps are usually very slow. Probably this is the reason for the overwhelming use of multi-carboxylate centres at the binding sites of calcium in biology.

In principle, other anionic centres could be useful and, as a general rule, the higher their pK_a the better the binding: $ROSO_3^- < SO_4^{2-}$ and $(RO)_2PO_2^- < -CO_2^- < ROPO_3^{2-}$, where R is an alkyl or a sugar residue. Clearly, while sulphates are of restricted value, the di-anion of phosphate is very powerful, as in phosphovitin. Immediately we see, however, that these centres are rare in biology. ATP is the most common molecule with such a centre but this has a binding constant of only about 10^4 M^{-1}. In a cell where $[Ca^{2+}] \leqslant 10^{-7}$ M this is not useful. DNA and RNA do not bind calcium in cells, for similar reasons. In the wall of plant cells, calcium can be taken up by sulphated sugars but only with a relatively small binding strength. The uptake is not likely to be well controlled and the process may resemble uptake by an ion exchange bead of sulphonated polystyrene. The combination of sugar and protein can give a much more powerful binding group, as in proteoglycans or glycoproteins. Obviously there are many possible combinations of the different anion and other coordinating centres in cell walls and surfaces.

There may be a need to mobilise calcium through the use of small chelates either in the biological systems *per se* or for medical reasons. Chelating agents such as EDTA, nitriloacetate, and phosphonates $^{2-}O_3P-CH_2-PO_3^{2-}$ are commonly used. (Many such materials are also used or proposed for use in washing detergent materials as water softeners. It must always be remembered that biological systems have evolved in balanced competition so that mobilisation of metals by such reagents is full of potential hazards. It is better to wash with carbonate than polyphosphates and it is better to wash with polyphosphates than with organic chelating agents containing nitrogen. Materials such as EDTA and NTA must mobilise many trace elements. Here we return to the essential selectivity for calcium of the polycarboxylate sites given the availability of the elements in water.)

THE KINETICS OF CALCIUM BINDING

Biological processes are often rate, not equilibrium, controlled. We need to know how fast are the k_{on} and k_{off} rate steps of complex formation as well as the value of their ratio, K.

$$Ca + L \underset{k_{off}}{\overset{k_{on}}{\rightleftharpoons}} CaL' \rightleftharpoons CaL, \tag{3}$$

where L and L′ are initial and final conformational states of this ligand. The main studies in this field are due to Eigen & Hammes (1963). They show that the on-step for calcium binding is diffusion controlled unless there is a slow conformational rearrangement of L before final equilibrium is reached. A protein binding reaction is likely to be divided between a very fast initial binding and a relatively slow conformational change. The off-step is likely to be slow if the binding is strong – as it is for a multidentate ligand such as EDTA, or a protein such as troponin.

We have now introduced the general features of the chemistry of calcium which arise from its specific charge:size ratio, but it must be understood that this ratio cannot give unique chemistry but only highly selective properties. Thus calcium action is antagonised by a range of other cations which may be competitive or non-competitive inhibitors of its actions. The overall influence of a given cation will be seen in the complicated product of concentration.binding constant.structure.kinetic constant. Thus the action of calcium can be modulated by the presence of a range of cations but interest will centre on magnesium, sodium, potassium, the proton and organic cations, especially amines. It would be possible to outline immediately the differences in each of the four different factors (concentration, binding strength, structure and rate constant) for these different cations but I shall take it that many of these differences are well known. It is therefore more important to turn to some problems which are strictly of biological concern, although I shall only illustrate general chemical principles which apply to these systems.

ION EXTRACTION (IONOPHORES) AND PORES

The simple idea of extraction of ions by chelating agents can be extended to the function of those ionophores which carry cations (*not* pore-forming ionophores), and which are soluble in organic solvents, and to the interior of large molecules such as proteins. The extracting chelating agents are made from donor oxygen centres surrounded by hydrophobic groups. The total charge of the chelate is usually zero or small. The calcium ionophores are clearly chelating agents of this kind. There is a binding centre of some five to eight oxygen atoms, one or two of which are charged, and an external set of hydrophobic side-chains from sugar ethers or from amino acids. A ring chelate or horse-shoe chelate forms around the cation; no new principles are involved.

There is a quite different type of transport possible if a molecule forms a connecting pore through a lipid region membrane which separates two aqueous phases. The pore-forming peptides for monovalent cations are the linear gramicidins. By way of contrast F.30 alamethicin (Fig. 1) is a linear pore-forming peptide for the calcium ion (and even such ions as the lanthanides, see below). It has one negative charge. The structure of the pores is not known but they provide a much more rapid transport system for ions than does the ionophore for the latter is made slower by the slow hydration/dehydration steps on forming and dissociating the complex. It could even be that, unlike ionophore transport which requires extensive dehydration of a cation, an ion passes through the pore as a heavily hydrated cation (Martin & Williams, 1975). The control step (gating) of a pore must then be a membrane conformational change not due to ion-binding itself.

Energised calcium ion transport across membranes

The hydration/dehydration step either on formation or dissociation of an ionophore complex or in a pore leads to a possible way to the generation of the selective ion gradients mentioned at the beginning of the article. If the dehydration is differential at one side as opposed to the other side of a membrane, then the uptake at one side has a different energy from that at the other. Suitable positioning of an ATPase could cause the dehydration (Williams, 1966):

$$ML(H_2O) + ATP \rightleftharpoons ML + ADP + P_i. \qquad (4)$$

The reverse of this mechanism of ion movement is related to the mechanism of ATP formation in membranes proposed some 15 years ago (Williams, 1961), and is not to be confused with the chemi-osmotic hypothesis. Evidence that this could be the mechanism is supplied by the work of Schuster & Olson (1974) on energised calcium transport.

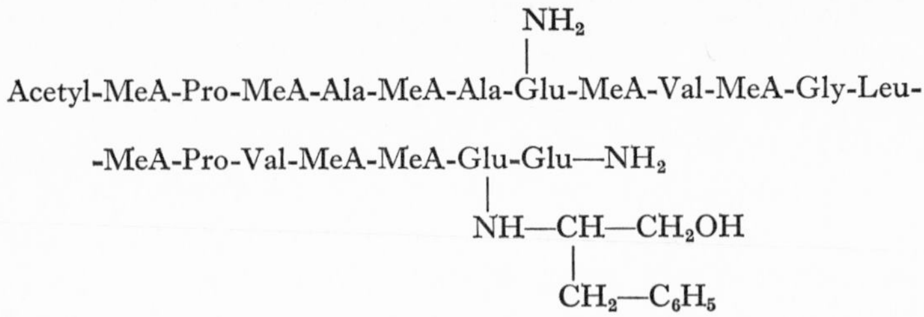

$$NH_2$$
$$|$$
Acetyl-MeA-Pro-MeA-Ala-MeA-Ala-Glu-MeA-Val-MeA-Gly-Leu-

$$-MeA-Pro-Val-MeA-MeA-Glu-Glu—NH_2$$
$$|$$
$$NH—CH—CH_2OH$$
$$|$$
$$CH_2—C_6H_5$$

Fig. 1. F.30 alamethicin. The metal binding site is associated with the free carboxylate terminus, the —OH group of the phenylalaninol, Val 15 and probably MeA 16 or MeA 17. The chain is highly hydrophobic but has considerable helical regions. MeA = methylalanine.

ION PRECIPITATION

The precipitation equilibrium

$$M + L \rightleftharpoons ML \downarrow, \tag{5}$$

where L is now a precipitating anion, is controlled by all the same factors as complex ion formation but in addition there is the cooperative energy of the lattice. Solubility is usually discussed in terms of a solubility product $K_{sp} = [M][L]$. When describing biological systems however this expression is quite misleading. K_{sp} is only a constant for a very large (infinite) crystal. For the small crystals found in biological systems, e.g. $CaCO_3$ in shells and $Ca_2(OH)PO_4$ in bones, K_{sp} varies as much as 10-fold, according to the size of the crystal. Moreover, in biological systems the crystals are maintained as *small* crystals as they are nucleated by proteins in a mesh of protein and polysaccharide. These polymers ensure that the crystals do not grow to any great size and that they do not grow with their crystal axes in any particular direction. The smallness of the crystals means that both the rate of deposition and of re-absorption is very fast so that the precipitated phase of biological systems is in dynamic equilibrium. Nucleation is the source of rate control and ion selectivity is of extreme importance. Magnesium ions can restrict the growth of calcium salt crystals.

Nucleation of precipitation

The nucleation of precipitation of calcium salts is a much studied topic. It would appear that proteins, present in particular regions of an organ, generate special initiation centres but it is quite unclear how this is done or why it goes wrong at times. Such assisted nucleation could imply that a very small crystal which would normally dissolve in the medium surrounding it is stabilised by interaction with a surface, say of a protein. Thus some characteristic structural feature of the initiating protein and the calcium salt structure must correspond. It is easy to envisage a set of calcium binding sites on the protein surface at exactly the repeat distances of the crystal but it has proved to be very difficult to provide hard evidence for this type of relationship.

CALCIUM IONS AND MEMBRANE PHOSPHOLIPIDS

The head groups (or polar ends) of phospholipid molecules present many potential calcium binding sites. For example, phosphatidylserine has a sufficiently high binding constant to bind to the calcium ion at its concentration in blood plasma. However it cannot bind to calcium in cells (10^{-7} M). Thus, in a membrane, phosphatidylserine head groups are more stable pointing outwards (toward the high calcium) than inwards. As pointed out elsewhere energised ion gradients can cause an energised orientation of lipids in membranes (Williams, 1972).

The more general situation is that of the conventional phospholipids such as lecithin. These molecules do not bind calcium ion strongly enough for calcium at 10^{-3} M^{-1}, to affect their behaviour. However, the binding to carboxylate groups arranged in a row, as in the face of a membrane, generates an effective chelating agent and calcium binding strengths can approach 10^4 M^{-1}, see below. These remarks are independent from the dynamics (fluidity) of the membrane but the calcium ion, through binding, does alter fluidity as well as orientation.

LANTHANIDE AND MANGANESE PROBES

The use of lanthanide probes for Ca^{2+} sites was advocated by Vallee & Williams (1968). Independently, Darnall & Birnbaum (1970) started spectroscopic work on lanthanide protein binding and Williams (1970) investigated the binding of lanthanides to proteins and small molecules by a variety of methods. Here it is probably sufficient to give a summary of the procedures, Table 7, and some examples.

SOME EXAMPLES OF THE USE
OF CALCIUM PROBES

Recently we have used probe methods in a study of metal binding to three biological molecules as models for the interaction of ions in channels (pores) of membranes. The first model ligand examined was the pore-forming linear peptide alamethicin (Martin & Williams, 1975). The metal

Table 7. *Probe methods*

Method	Cation	System
UV/visible light	Nd(III)	Albumins
CD	Tb(III)	Ionophores
MCD	Nd(III)	Membranes
NMR shift	Very many reagents	Troponin, lysozyme
NMR relaxation	Mn(II), Gd(III)	Concanavalin, lysozyme
Fluorescence	Tb(III)	Trypsinogen, troponin
X-ray diffraction	Tm(III), Gd(III)	Thermolysin, lysozyme
Electron microscopy stain	La(III)	Membranes
Mössbauer	Eu(III)	

NOTE. The extension of Ln (III) methods to larger bodies such as RNA, DNA and membranes has been initiated.

CD, circular dichroism; NMR, nuclear magnetic resonance; MCD, magnetic circular dichroism.

ions bind in the vicinity of the carboxylate group. It has been possible to show that two different bindings are possible for Ln^{3+} cations in acid ($-CO_2H$) and alkali ($-CO_2^-$) media. The head group of the peptide changes shape on binding. It is significant that the aromatic group of phenylalaninol which is a constitutive part of alamethicin comes quite close to the metal and must control the hydration sphere to some extent (Fig. 1). The potassium ion (and presumably the calcium ion) bind weakly in this same region. Now alamethicin acts indiscriminately as a pore for cations including M^+, M^{2+} and M^{3+} and even anions. Note that weak binding by the carboxylate does not prevent rapid migration through the lipid, despite the fact that NMR experiments show no evidence for mobility of the alamethicin itself in bilayer leaflets. It must not be forgotten that on/off reactions of K^+, Na^+, Ca^{2+} and Ln^{3+} to binding sites of low binding strengths $K < 10^3 M^{-1}$ are very fast (first order rate of dissociation $< 10^{-5}$ sec). *Thus only conformational changes limit the rate.*

The commonly used alamethicin is called F.30 and has a free carboxylate group, but there is another alamethicin, F.50, which does not have this group and yet it is active in ion transport. In place of the carboxylate residue it has an amide. The lanthanide binding is very similar to that detailed in the caption to Fig. 1, i.e. the binding in the acidic form of F.30 alamethicin. Thus, channel formation does not require a charged group. In fact, the gramicidin peptides which are linear and neutral are very similar to alamethicin F.50 in their ability to transport cations.

The second model that we have studied is lysozyme (Campbell, Dobson & Williams, 1975). This protein has several carboxylate metal ion binding sites, Asp 101, Asp 87 and one in the active site between Glu 35 and

Asp 52. Only the last is important here. The binding has again been studied using lanthanide cations as well as calcium. At a pH around 5.5–6.0, Glu 35 ($pK_a > 6.0$), is protonated and lies closely adjacent to the aromatic group Tryp 108. The two carboxylate residues, 35 and 52, are some 0.8–1 nm apart. Adding metal ions causes a movement of the two carboxylates together as Glu 35 leaves the vicinity of Tryp 108. A proton is displaced in the reaction. The relevance of this conformational change to the properties of membranes is as follows. It is easy to picture two very different processes in which two metal atoms, M_1 (of low-binding strength) and M_2 (of high-binding strength), approach the two-carboxylate site of lysozyme. In the case of low-binding strength the cation may not be able to displace the proton from Glu 35 and will bind to it only very weakly and remotely (configuration (1)). Here the retention of the Tryp 108. Glu 35

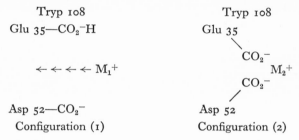

complex could be under the control of groups far away from the site through the mediation of allosteric effectors. A second metal ion M_2 which binds strongly will pull Glu 35 away from Tryp 108 and generate the configuration (2). Now consider the motion of metal ions into the face of the groove of lysozyme near the Glu 35 and Asp 52 region but allow the back of this 'groove' which holds these residues to have an open pore structure. Metal ion M_1 binds very weakly and can pass through the gap between the carboxylate groups. Metal ion M_2 binds more strongly and may close a gate of the carboxylate groups in front of itself. It is frequently observed that lanthanide ions ($K_{(lysozyme)} > 10^3\ \mathrm{M}^{-1}$) do not pass through calcium ($K_{(lysozyme)} < 10^2\ \mathrm{M}^{-1}$) channels in membranes but they do block them. It is worth noting too that measurements of the binding of different metal ions to such calcium channels have given 'binding constant' orders (Hille, Woodhull & Shapiro, 1975):

$$Ln^{3+} > Ni^{2+} > Co^{2+}, Zn^{2+} > Ca^{2+}, Mn^{2+} > Mg^{2+}.$$

Such orders and even the magnitudes of the 'binding constants' are very close to the orders found for lysozyme. Thus the mouth of the calcium channel could be made from two carboxylates held in a particular constraint which is altered as the cation interacts with the site.

The third investigation concerns the reaction of lanthanide metal ions with the polar, usually, anionic groups on the surface of phospholipid membranes. First, we have shown that metal-bound phosphatidylcholine head groups are extended perpendicular to the plane of the membrane but the packing of lipids in the membrane is adjusted. We have then gone on to look at membranes formed from two components, e.g. lecithin and cholesterol, in the presence of metal ions, searching for conformational effects (H. Hauser, B. Levine & R. J. P. Williams, unpublished). The possibility of conformational rearrangement in lipids can be readily illustrated. When lanthanide ions and probably calcium bind to lecithin one metal ion binds between two phosphorus atoms of the phosphatidylcholine head group. Addition of cholesterol apparently prevents this disposition of the two lecithin molecules so that lecithin/cholesterol membranes do not bind metal ions as strongly as lecithin itself. Clearly, in a single phase there will be patches which bind metal ions and patches which do not. Thus metal ions and very hydrophobic molecules can enter into 'allosteric equilibria'. The exact nature of this equilibrium in a membrane phase is hard to judge as yet for it may operate much more closely like a second-order phase change than like a simple molecular equilibrium. The reason for this is that it is known that the lipid medium, the membrane, is highly organised. Thus we must look for cooperative changes which will occur not only in the face of the membrane but also from one side to the other. Addition of reagents to one side of the membrane will then result in effects being transmitted to the other side not through electrostatic field effects but through conformational triggering. Clearly, if this works at different patches in different ways then specific messages can be transmitted by calcium ion binding and the effect can be amplified since only a small degree of binding may cause a phase change. Probe experiments illustrating these effects will be published shortly (J. Krebs, B. Levine & R. J. P. Williams, in preparation). It is likely that cooperative effects associated with calcium binding will be observed in polysaccharide (walls) as well as in membrane and protein systems.

REFERENCES

CAMPBELL, I. D., DOBSON, C. M. & WILLIAMS, R. J. P. (1975). Nuclear magnetic resonance studies of the structure of lysozyme in solution. *Proc. R. Soc. Lond. A*, **345**, 41–59.

DARNALL, D. W. & BIRNBAUM, E. R. (1970). Rare earth metal ions as probes of calcium ion binding sites *J. biol. Chem.*, **245**, 6484–6486.

DIETRICH, B., LEHN, J. M. & SAUVAGE, J. P. (1969). Les Cryptates. *Tetrahedron Lett.*, 2889–2892.

(1973). Cryptates: control over bivalent/monovalent cation selectivity. *J. chem. Soc.* (*Chem. Commun.*), 15–16.

EIGEN, M. & HAMMES, G. G. (1963). Elementary steps in enzyme reactions. *Adv. Enzymol.*, 25, 1–38.

EISENMAN, G. (1968). Ion permeation of cell membranes and its models. *Fedn Proc.*, 27, 1249–1256.

HILLE, B., WOODHULL, A. M. & SHAPIRO, B. I. (1975). Negative surface charge near sodium channels of nerve: divalent ions, monovalent ions, and pH. *Phil. Trans. R. Soc. Ser. B*, 270, 301–318.

MARTIN, D. R. & WILLIAMS, R. J. P. (1975). The nature and structure of ala-methicin. *Biochem. Soc. Trans.*, 3, 166–167.

SCHUSTER, S. M. & OLSON, M. S. (1974). Studies of the energy-dependent uptake of divalent ions by beef heart mitochondria. *J. biol. Chem.*, 249, 7151–7159.

URRY, D. W., CUNNINGHAM, W. D. & OHNISHI, T. (1973). A neutral polypeptide-calcium ion complex. *Biochim. biophys. Acta*, 292, 853–857.

VALLEE, B. L. & WILLIAMS, R. J. P. (1968). Enzyme action: views derived from metallo-enzyme studies. *Chemistry in Britain*, 4, 151–159.

WACKER, W. E. C. & WILLIAMS, R. J. P. (1968). Magnesium/calcium balances of biological systems. *J. theor. Biol.*, 20, 65–78.

WILLIAMS, R. J. P. (1961). Possible functions of chains of catalysts. *J. theor. Biol.*, 1, 1–23.

(1966). The selectivity of metal–protein interactions. In *Protides of the Biologica Fluids*, ed. H. Peeters, vol. 14, pp. 25–36. Elsevier: Amsterdam.

(1970). The biochemistry of sodium, potassium, magnesium, and calcium. *Q. Rev. chem. Soc., Lond.*, 24, 331–365.

(1972). A dynamic view of biological membranes. *Physiol. Chem. Phys.*, 4, 427–439.

(1973). Ion selectivity and ligand design. *Biochem. Soc. Trans.*, 1, 826–828.

CHEMISTRY OF
THE CALCIUM IONOPHORES

By MARY R. TRUTER

Molecular Structures Department, Rothamsted
Experimental Station, Harpenden, Herts. AL5 2JQ

'Ionophore' is used in a biophysical context to mean a compound which facilitates the transport of an ion through a natural or artificial lipid 'membrane' from one aqueous medium to another. This is the definition of B. C. Pressman (Pressman, Harris, Jagger & Johnson, 1967) and dates from his discovery that certain antibiotics stimulated the uptake of potassium by respiring mitochondria. The new feature was the nature of the cation, an alkali metal; compounds which transport transition metals such as iron had been known for much longer, but are not called ionophores.

The characteristics of the organic compounds which will transport alkali or alkaline earth metals into and through lipids are (i) sufficient oxygen (or, rarely, nitrogen) atoms to replace the solvation sphere round the cation, and (ii) sufficient carbon and hydrogen atoms to form the outside of the complex entity so rendering it lipophilic. The complex is described as an entity because it may be neutral, the ionophore being anionic, or positively charged, the ionophore being neutral. The formation constant for the complex has an optimum value for transport, high enough for the formation on one side of the lipid barrier and low enough for there to be release on the other (Ovchinnikov, 1974; M. Kirch & J. M. Lehn, personal communication).

For alkali metals, particularly sodium and potassium, several types of ionophore have been studied. Naturally occurring antibiotics may be cyclic and neutral such as valinomycin (a depsipeptide) and nonactin (a macrotetrolide) or monobasic acids such as monensin or X-537A. Synthetic neutral compounds are the monocyclic polyethers of Pedersen (Pedersen, 1967) called 'crown' compounds and the bi- and tri-cyclic 'cryptates' of Lehn and coworkers (Dietrich, Lehn & Sauvage, 1973a; Cheney, Lehn, Sauvage & Stubbs, 1972). Various aspects of the properties of these compounds have been reviewed in four articles (Lehn, 1973; Truter, 1973; Simon, Morf & Meier, 1973; Izatt, Eatough & Christensen, 1973).

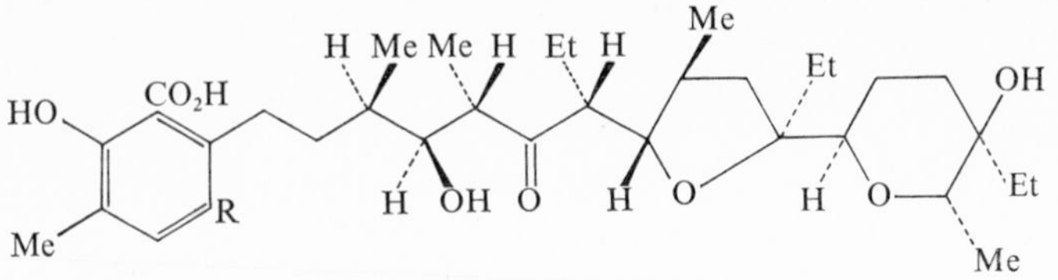

X–537A

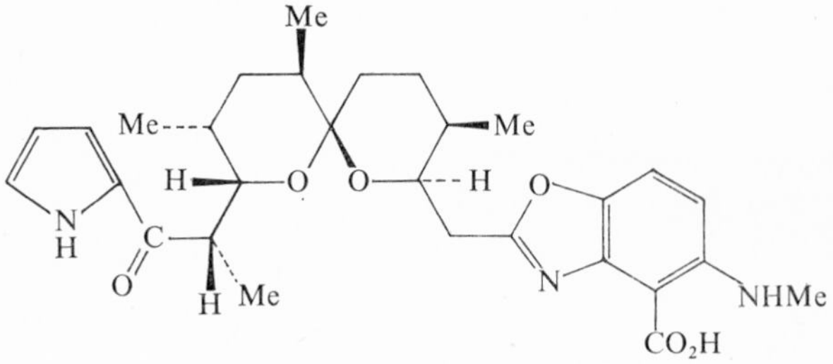

A23187

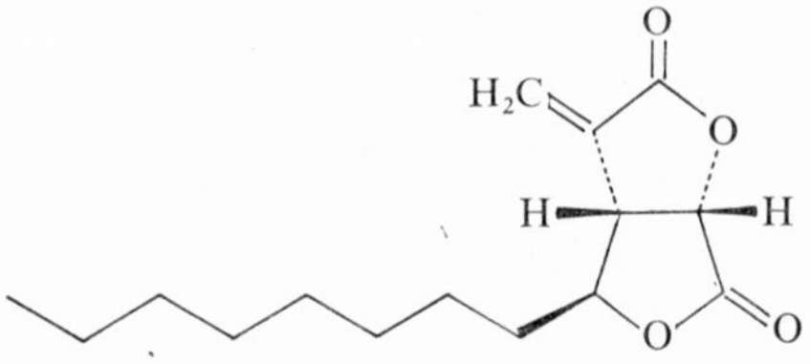

l-Avenaciolide

Fig. 1. Formulae of three naturally occurring ionophores for calcium, X-537A (also called lasalocid-A), A23187, and avenaciolide, as established by chemical methods including X-ray crystal structure analysis for X-537A and A23187. The absolute configurations shown at the many asymmetric carbon atoms were obtained by comparison of fragments of these molecules with others of known configuration.

NATURALLY OCCURRING CALCIUM IONOPHORES

There are three compounds (Fig. 1) which have been used for biophysical experiments in the belief that they have more or less specific calcium transporting activity. X-537A and A23187 have in common that they are monocarboxylic acids, HL, and so give rise to L⁻ anions; neutral compounds with univalent cations, M^+, and divalent cations, M^{2+}, have formulae ML and ML_2 respectively. As the formulae show the carboxylic acid is a substituent on the benzene ring on both compounds. They both have additional carbonyl and ether oxygens which may be used for coordination. They differ, however, quite fundamentally in that A23187 contains two ring nitrogen atoms and there is no nitrogen in X-537A. Avenaciolide is a neutral molecule with two lactone groups providing oxygen atoms and a hydrocarbon tail conferring lipid solubility; to preserve

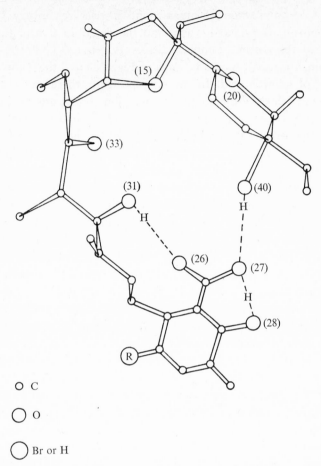

O C

○ O

◯ Br or H

Fig. 2. The molecular structure of the anion of X-537A and its 5-substituted derivatives adapted from Johnson *et al.* (1970*a*, *b*), and Schmidt *et al.* (1974). Broken lines indicate hydrogen bonding and the approximate positions of the hydrogen atoms are shown. The oxygen atoms are designated for comparison with the text and Table 1 shows the hydrogen bonding patterns found in several crystals; the one depicted here is for the anhydrous anion (1) in the barium complex (Johnson *et al.*, 1970*b*).

electroneutrality, anions as well as cations must be transported. Each ionophore is considered in more detail separately.

X-537A

In 1951 an antibiotic was isolated from an unidentified *Streptomyces* sp. and found (Berger *et al.*, 1951) to be an optically active monobasic acid giving sodium and barium salts which are insoluble in water but soluble in non-polar solvents. Nineteen years later the molecular structure was determined by a combination of organic and physical chemistry (Westley, Evans, Williams & Stempel, 1970) and the determination of the crystal structure of the barium complex, $Ba(X-537A^-)_2H_2O$ (Johnson, Herrin, Liu & Paul, 1970a, b). As the formula in Fig. 1 shows, the compound is a derivative of salicylic acid. In the u.v. spectrum, there is an absorption near 310 nm which shifts to the blue region $\sim$ 305 nm on ionization, corresponding to a stronger internal $OH...O{=}C$ hydrogen bond shown in Fig. 2 as between $O(27)$ and $O(28)$. Fig. 2 also shows the typical cyclic form of this anion given by hydrogen bonding from the tertiary alcohol $OH(40)$ at the 'tail' end of the molecule to a carboxylate oxygen at the 'head'. A further internal hydrogen bond reinforces the cyclic form. In the barium complex, two such anions sandwich the barium, one providing six and the other two oxygen atoms (Table 1) which coordinate the cation; a water molecule provides a ninth coordinating atom for the barium. The whole entity $[Ba(X-537A^-)_2H_2O]$ is neutral and the outside consists of carbon and hydrogen atoms so explaining the solubility properties.

The first complex of a univalent cation to be studied by X-ray methods was that of silver (Maier & Paul, 1971); the stoicheiometry is 1:1 but the complex is actually dimeric having each silver ion coordinated on one side by the oxygens of one anion (see Table 1) and on the other by the electron cloud of the benzene ring of the second anion. The second type of co-ordination is characteristic of silver but not of alkali metals or alkaline earth metals in compounds stable in the presence of water so this result did not appear relevant to biologically important cations.

Derivatives of X-537A can be made by substitution in the 5-position of the aromatic ring, shown as R in Figs. 1 and 2. The crystal structure of the acid with R = Br was determined (Bissell & Paul, 1972) and found to be dimeric; each molecule was held in a cyclic form by hydrogen bonding as in the barium and silver complexes with R = H. The main difference between the free molecules and the two metal complexes was that the free molecules were packed 'head'-to-'head', i.e. with the benzene rings near one another and with the two 'tails' or oxacyclohexane rings near one another. The dimer is probably held by hydrogen bonding from one molecule to the other.

An important discovery is that the sodium salt crystallizes in two different forms from solvents of different polarity (Schmidt et al., 1974) and in both forms the stoicheiometry is 2:2, the entities in the crystal being $[Na^+ \ 5\text{-Br-X-}537A^-]_2$. From carbon tetrachloride the dimer has the 'head'-to-'tail' structure with two sodium ions each coordinated by two anions, five oxygens from one and one from another as shown in Table 1. The dimer appears to persist in solution as judged by the osmotic pressure.

Table 1. *Crystalline complexes of X-537A with metals*

R	Metal	Anion	Oxygen atoms coordinated to metal (see Fig. 2 for designations)						Anion	Hydrogen bonding		
			O(26)	O(31)	O(33)	O(15)	O(20)	O(40)				
H	Ba[a]	Anion(1)	O(26)	O(31)	O(33)	O(15)	O(20)	O(40)	Anion(1)	O(40)...O(27)	O(28)...O(27)	O(31)...O(26)
		Anion(2)	O(26)					O(40)	Anion(2)	O(40)...O(27)	O(28)...O(27)	O(31)...O(26)
H	Ag(1)[b]	Anion(1)		O(31)	O(33)	O(15)	O(20)	O(40)	Anion(1)	O(40)...O(26)	O(28)...O(27)	O(31)...O(26)
	Ag(2)	Anion(2)		O(31)	O(33)	O(15)	O(20)	O(40)	Anion(2)	O(40)...O(26)	O(28)...O(27)	O(31)...O(26)
Br	*Head-to-tail form from carbon tetrachloride*											
	Na(1)[c]	Anion(1)		O(31)	O(33)	O(15)	O(20)	O(40)	Anion(1)	O(40)...O(27)	O(28)...O(27)	
		Anion(2)	O(26)						Anion(2)	O(40)...O(27)	O(28)...O(27)	
	Na(2)	Anion(1)	O(26)									
		Anion(2)		O(31)	O(33)	O(15)	O(20)	O(40)				
Br	*Head-to-head form from acetone*											
	Na(1)	Anion(1)		O(31)	O(33)	O(15)	O(20)	O(40)	Anion(1)	O(40)...O(26)	O(28)...O(27)	O(31)...O(26)
		Anion(2)			O(33)				Anion(2)	O(40)...O(26)	O(28)...O(27)	O(31)...O(26)
	Na(2)	Anion(1)			O(33)							
		Anion(2)		O(31)	O(33)	O(15)	O(20)	O(40)				

[a] Johnson *et al.*, 1970.   [b] Maier & Paul, 1971.   [c] Schmidt *et al.*, 1974.

When a more polar solvent is used, such as acetone, a different crystalline form is obtained, the dimeric molecules are now 'head'-to-'head'. The nuclear magnetic resonance spectrum is consistent with a dimer in non-polar solvents very like that of the crystalline salt from carbon tetrachloride. In more polar solvents there appears to be a change corresponding to some 'head'-to-'head' dimer or possibly a monomeric species. Schmidt *et al.* (1974) concluded that there was no evidence for any change in the conformation of the molecular backbone and pointed out that in the 12 different X-537A entities so far found in the solid, the torsion angles in the backbone did not vary by more than 8°.

Pressman (1973) subjected X-537A and several derivatives to a series of physico-chemical measurements to assess their ionophore properties. Acylation of the phenolic group increases the pK_a while substitution in the aromatic ring where R = H (X-537A), Cl, Br, I or NO_2 gives successive reduction in pK_a. The derivatives retain ion-transporting ability consistent with the structure in Fig. 2; loss of the OH(28)...O(27) hydrogen bond on replacement of H by $COCH_3$ in Ac-X-537A, or replacement of R would not affect the shape of the molecule and its coordination. However, just as pK_a is changed by electronic effects, or field strength, on substitution so is the relative affinity for the various cations. For three compounds, Table 2 shows the values obtained by Pressman for pK_a by titration in 90% ethanol with tetramethylammonium hydroxide; these were all lower in the presence of potassium thiocyanate, an indication of complex formation. To obtain biologically relevant values for the complexation constant a two-phase system was used; the aqueous phase was buffered with tricine to pH 8.3 and the organic phase was 70% toluene, 30% butanol. The ionophore concentration was constant at 5×10^{-4} M and the values for the constant K_A were obtained for several (unspecified) cation concentrations from the expressions:

$$K_A{}^I = \frac{[ML_{org}]}{[L_{org}][M_{water}]} \quad \text{or} \quad K_A{}^{II} = \frac{[ML_2]}{[L_{org}]^2[M_{water}^{2+}]},$$

which are derived for the electrically neutral species with L = X-537A⁻. As Pressman pointed out, comparison between $K_A{}^I$ and $K_A{}^{II}$ for mono- and divalent cations respectively is difficult because they do not have the same dimensions, and the values in Table 2 are ratios of $K_A{}^{II}$ for divalent cations taking that for Ca^{2+} with X-537A itself as unity. Comparison of caesium with sodium also showed greater transfer for the larger cation. The numbers quoted by Pressman for $K_A{}^I$ for Na^+ and $K_A{}^{II}$ for Ca^{2+} with X-537A were 36.0 and 0.05×10^3 without dimensions. If the stoicheiometry in the organic layer were 2:2 as suggested by the work of Schmidt

Table 2. *Values of* pK_a *and relative values of* K_A^{II} *for extraction of* M^{2+} *to 30% butanol, 70% toluene (Pressman, 1973) at* 5×10^{-4} M *ionophore*

	Ac-X-537A	X-537A	5-Br-X-537A
pK_a	6.30	5.80	4.95
Mg^{2+}	0.26	0.38	0.54
Ca^{2+}	0.29	1.0	2.8
Sr^{2+}	1.8	8.5	18.0
Ba^{2+}	72.0	2600.0	5600.0

et al. (1974) the relationship would be:

$$K_A^{I} = \frac{[M_2L_{2\,org}]}{[L_{org}]^2[M_{water}]^2},$$

which has different dimensions again and would give larger numerical values.

Confirmation of the effect of substitution was obtained in recent biophysical experiments; 5-Br-X-537A (Nordmann & Dyball, 1975) has been found to be more effective than X-537A (Nordmann & Currell, 1975) in the same system, the release of oxytocin being accompanied by uptake of calcium from intracellular space from rat neurohypophysis. Other biophysical experiments have given less satisfactory results because of the relative affinity of the ionophore for sodium and calcium; Devore & Nastuk (1975) suggest that previous work using X-537A as a purely calcium ionophore might need reinterpretation. In experiments on calcium leakage from retinal lipid liposomes, Hyono, Hendriks, Daemen & Bonting (1975) found that the stoicheiometry of transport for calcium appeared to be 1:1 at low concentrations (less than 10^{-5} M) X-537A and 1:2 at higher concentrations; this is consistent with some of the chemical evidence from spectroscopy.

The molecule has several features which show changes in spectral properties accompanying interaction with protons or metal cations. Because of the asymmetrical environment some of the absorption bands show circular dichroism, for example one at 290 nm due to a transition of an electron in the $C=O(33)$ carbonyl group (see Fig. 2). Such bands are sensitive to changes in the conformation of the molecule and may be used to detect complex formation.

Measurements of circular dichroism in various solvents (Degani & Friedman, 1974) at 10^{-4} to 10^{-6} M show that the free acid conformation is sensitive to the basicity of the solvent, i.e. its tendency to compete with the carboxyl oxygen atoms for the hydroxy hydrogen atoms. The results were interpreted in terms of three possible conformers, I an open chain with only the $O(27)\ldots HO(28)$ hydrogen bond, II a partial ring with an additional hydrogen bond $O(26)\ldots HO(31)$ and III a cyclic form with the $O(27)\ldots HO(40)$ hydrogen bond which results in a hydrophobic exterior. All cation complexes show III as the dominant form which is also the one found for the acid in n-hexane. Alpha & Brady (1973) also observed that the conformation of the free acid in a non-polar solvent, n-heptane, was similar to that of a complex in any solvent (except dimethylsulphoxide) whereas there were changes in more polar solvents. A different technique, measurement of the coupling constants in the nuclear magnetic resonance (NMR) spectrum, showed no effect of solvent on the conformation of the sodium complex of 5-Br-X-537A dissolved at 2×10^{-2} M in deutero-chloroform, deutero-acetone or mixtures of these solvents (Schmidt et al., 1974). These results are consistent with the conformational integrity shown in the crystalline forms described above. If there is a difference in the free acid in dilute solution in different polar solvents, however, there might be differences in its interaction with metals and these effects have been found.

Fluorescence after the absorption at ~ 305 nm has been used (Degani, Friedman, Navon & Kosower, 1973; Degani & Friedman, 1974) to determine the stoicheiometries and association constants of complexes of X-537A with various metal ions in methanol and n-hexane. For a two-phase system with n-hexane as the non-polar medium the results for the reaction,

$$x\mathrm{HL}_{\mathrm{(hexane)}} + \mathrm{M}^{n+}_{\mathrm{(aq.)}} \rightleftharpoons \mathrm{ML}_{n\mathrm{(hexane)}} + n\mathrm{H}^+_{\mathrm{aq.}}$$

are in reasonable agreement with those of Pressman in Table 2, the ratios being $0.13:1.0:5.0:625$ for Mg:Ca (taken as unity):Sr:Ba. A $1:2$ complex for barium is the dominant one. For the alkali metals there is only a small variation, the ratios, with sodium taken as unity, are $0.3:1.0:1.2:0.8:0.5$ for Li:Na:K:Rb:Ca. The transfer to hexane is small, the numerical values being 3.2×10^{-7} for Ca^{2+} and 2×10^{-5} for Na^+. (L is the anion X-537A$^-$.)

With methanol as the solvent and ligand concentrations of 10^{-4} to 10^{-6} M the dominant reaction at high pH is $\mathrm{M}^{n+} + \mathrm{L} \rightleftharpoons \mathrm{ML}^{(n-1)+}$. The concentration dependence established the $1:1$ stoicheiometry and monomeric nature of the complexes which implies that the complexes of the alkaline earth metals are charged, $(\mathrm{ML})^+$. Since the dimensions are the same, mole^{-1}, for alkali and alkaline earth metals the results are strictly comparable, and are shown in Table 3. In this, as in other systems, a change of solvent has a marked effect not only on the absolute values of association constants but also on the relative ones. Cornelius, Gärtner & Haynes

Table 3. *Association constants for X-537A with alkali and alkaline earth metal cations for the reaction* $M^{n+} + L \rightleftharpoons ML^{(n-1)+}$

	Methanol[a]	Ethanol[b]
Na^+	3.7×10^2	1.5×10^6
K^+	3.8×10^3	1.4×10^6
Rb^+	3.7×10^3	1.2×10^6
Cs^+	2.7×10^3	0.95×10^6
Mg^{2+}	6.7×10^3	3.4×10^5
Ca^{2+}	3.7×10^4	2.7×10^5
Sr^{2+}	3.0×10^5	1.4×10^5
Ba^{2+}	2.9×10^6	4.6×10^5

[a] Degani *et al.* (1973) $1-5 \times 10^{-6}$ M HL.
[b] Cornelius *et al.* (1974) 2×10^{-6} M HL.

(1974) confirmed the values for methanol and, with ionophore at 10^{-6} M, determined the association constants in ethanol with the results shown in Table 3, although they did not point out explicitly that their results correspond to the complexed species being $(ML)^+$ for the divalent cations. For the same cations, with thiocyanate as anion, Alpha & Brady (1973) interpreted changes of the circular dichroism at 10^{-5} M X-537A as clear indication for 1:1 complexing with K^+, Rb^+ and Cs^+ and for 1:2 with Ba^{2+}. For Mg^{2+}, Ca^{2+} and Sr^{2+} anomalous results were obtained with an indication that 1:1 was possible in ethanol while Na^+ gave a non-integral stoicheiometry. Unfortunately the ionophore is not soluble in the medium of most biological interest, water (in practice it is solubilized in the minimum amount of ethanol), and addition of water greatly reduces the fluorescence so no measurements for water are available, and the order of selectivity is not known. It seems likely that the discrepancy between the finding of Schmidt *et al.* (1974) that the dimeric forms of 5-Br-X-537A and its sodium complex persist in solution (from chemical shifts in the NMR spectrum) and the monomeric nature of the complexes reported by the other workers lies in the different concentration ranges, down to 10^{-2} M for NMR and less than 10^{-3} M for other measurements.

Identification of the organism producing X-537A as *Streptomyces lasaliensis* has led to the recent name for the antibiotic, lasalocid A. The same organism produces about one tenth as much of an isomer of this compound *iso*-lasalocid A, also a monobasic acid, but much less soluble in non-polar solvents and with a different pattern of antibiotic activity against Gram-positive bacteria. Crystal structure analysis (Westley *et al.*, 1974) of the 5-Br derivative of the free acid showed that the difference was in the terminal ring, 5- instead of 6-membered.

The change from a tertiary alcohol to a secondary alcohol for OH(40) has a profound effect on the hydrogen bonding; a much smaller ring is formed by donation from OH(40) to O(15) and from OH(31) to O(40) leaving the salicylate free. This molecule crystallizes in a monomeric form and would repay further study.

A23187

The first reports of this compound described biophysical and model membrane experiments (Reed, 1972; Reed & Lardy, 1972). A23187 was found to enter the mitochondrial membrane and act as a mobile carrier for Ca^{2+} and Mg^{2+} ions at pH 7.4 allowing them to reach equilibrium. For a given addition of ionophore (0.3 nmole mg^{-1} protein) the loss of Mg^{2+} and Ca^{2+} was 20 and 3 nmole mg^{-1} protein respectively. An effect secondary to the loss of magnesium was leakage of potassium (210 nmole mg^{-1} protein in this example). In model experiments even the chlorides were extracted into 30% 1-butanol, 70% toluene to ratios of M^{2+}:A23187 of approximately 1:2 for Mg^{2+} and 1:3 for Ca^{2+} at both pH 7.4 and pH 9.8. Very little K^+ was extracted at the lower pH but at the higher pH up to 1:8 was achieved. Use of potassium thiocyanate gave slightly better extraction than the chloride. A ratio of 1:2 represents 100% efficiency for the divalent cations on the basis that the compound is a mono-carboxylic acid so two anions are required for neutrality with Mg^{2+} or Ca^{2+}. In the particular system reported there was a 100-fold higher concentration in the aqueous than in the organic layer.

In the following year a paper appeared by the group (Wong et al., 1973) who had extracted the compound. In an investigation of oxidative phosphorylation and Ca^{2+} transport in rat liver mitochondria, they obtained results very dependent on the presence of magnesium which evidently complexes the A23187 more strongly than calcium so addition of magnesium may reverse the effect of the ionophore. There has been no lack of enthusiasm for following the suggestion that A23187 might be useful as a divalent cation-selective compound and several laboratories have produced interesting results without knowing its chemistry.

In 1974 structural formula was established by spectroscopic methods and a crystal structure analysis (Chaney et al., 1974) of the free acid. Ironically the compound is extracted from broths of Streptomyces char-

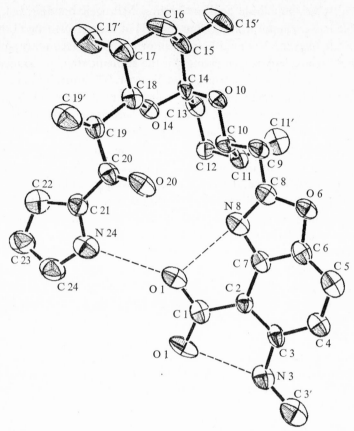

Fig. 3. The molecular structure of the free acid A23187 as found in the crystal (after Chaney, Demarco, Jones & Occolowitz, 1974). In the diagram, broken lines represent hydrogen bonds. In the crystal there are two identical molecules related by symmetry in the unit cell and held to one another only by van der Waal's interactions. The absolute configuration shown was obtained by comparison with that at the two six-membered rings of the spiro group in the anionic polyether antibiotics, monensin (Pinkerton & Steinrauf, 1970) and dianemycin (Czerwinski & Steinrauf, 1971).

treusensis as the magnesium/calcium salt, which would be most interesting structurally but does not form crystals suitable for structure analysis (N. D. Jones, personal communication). The formula of the free acid is shown in Fig. 1 and its molecular structure in Fig. 3. This is consistent with the purely chemical results. It is an optically active acid (pK_a = 6.9 in 90% dimethylsulphoxide), and the i.r. spectrum showed bands in the C=O stretching region at 1640 and 1690 cm^{-1} corresponding to the carboxylate and carbonyl groups. NMR had indicated a 1,2,3,4-tetra-substituted benzene.

As in other monobasic acid antibiotics, hydrogen bonding holds the molecule in a cyclic form, here from the NH of the pyrrole ring to the one carboxylate oxygen. A second bond is shown from the secondary amine to the other carboxylate oxygen giving, for the aromatic ring, an arrangement similar to that of the salicylate group of X-537A. The third hydrogen bond from the carboxylic acid to the nitrogen atom of the benzoxazole would not be present in the anionic form. There is no lack of potential coordinating atoms and it is probable that at least one nitrogen atom, the benzoxazole or the pyrrole acts as a ligand, so, as suggested by Wong *et al.* (1973), accounting in part for the preference for divalent cations. There are four oxygen atoms which could interact with a metal in addition to either or both nitrogen atoms. Spiro oxygen groups have been found in other antibiotics and may or may not both make contact with a cation as shown respectively by dianemycin (Czerwinski & Steinrauf, 1971) or monensin (Pinkerton & Steinrauf, 1970). Intuitively one would expect a carboxylate oxygen atom to be coordinated to the metal but in fact it has been found to do so in some complexes of anionic antibiotics and not in others. Also, as shown in Table 1, the behaviour of one antibiotic is not constant; O(26) is not always coordinated. While it is not possible for the two atoms involved in the N(Me)H...OC hydrogen bond in the aromatic ring both to coordinate the same cation as the other carboxylate oxygen and the benzoxazole nitrogen, they might coordinate a different cation, and possibly the existence of a subsidiary chelating site explains the observation of Hyono *et al.* (1975) that at low concentrations of A23187 the rate of leakage of Ca^{2+} from retinal lipid liposomes indicates a stoicheiometry of $2Ca^{2+}$ to 1 A23187. The recent report (Hovi, Williams & Allison, 1975) that A23187 extracts amino acids into toluene–butanol indicates the need for care in attributing biological effects solely to cation transport.

Avenaciolide

Avenaciolide is a neutral fungicidal molecule. The chemical constitution of the naturally occurring *l*-form shown in Fig. 1 was established by Brookes, Tidd & Turner (1963) by chemical and spectroscopic methods and confirmed by crystal structure analysis (D. L. Hughes, personal communication). The absolute configuration shown was suggested by measurement of the circular dichroism of a fragment (Brookes *et al.*, 1963) but is now uncertain (Anderson & Fraser-Reid, 1975).

The effects of avenaciolide on mitochondria were at first interpreted (McGivan & Chappell, 1970) as indicating competition for the glutamate binding sites but Harris & Wimhurst (1973) pointed out that competition between an anion and a neutral molecule is unlikely. They investigated the effect of avenaciolide on cations, arguing that removal of cations would produce transfer of anions, and showed that the initial effect of avenaciolide on rat liver mitochondria was a rapid release of calcium and continuing release of magnesium. The loss of the divalent ions was, as expected, accompanied by loss of dicarboxylate anions. The behaviour of avenaciolide with divalent cations was similar to that obtained by using A23187 as the ionophore and ruthenium red to inhibit energy-requiring uptake.

The quantitative relations showed (Harris & Wimhurst, 1974) that the ratio of change in calcium concentration to addition of avenaciolide was about 1:40 (nmoles per mg of mitochondrial protein). Some of the avenaciolide was also concerned with the release of magnesium. Additions giving a partial release corresponded to about 10 moles of avenaciolide to one of magnesium; at this stage potassium was taken up in a ratio of several cations per mole of avenaciolide (provided that this cation and a source of energy were available). Addition of avenaciolide to give a magnesium:ionophore ratio of 1:20 resulted in complete release of magnesium and in loss of potassium; as with A23187 the effect on potassium appeared secondary not a simple cation transport.

In model experiments Harris & Wimhurst (1973) showed that avenaciolide facilitated transfer of Mg^{2+}, Ca^{2+} and K^+ cations into a bulk organic phase (30% butanol, 70% toluene) from water provided that the counter ion was reasonably lipophilic, i.e. thiocyanate anions were present. The partition coefficients were increased by a factor of six in the presence of ionophore but were still low, of the order of 10^{-3}. The ratios of metal to avenaciolide in the organic phase were 1:200, 8000 and 43 for Mg^{2+}, Ca^{2+} and K^+ respectively, suggesting that each cation is surrounded by several avenaciolide molecules. The formula in Fig. 1 suggests, and this is confirmed by the use of space-filling models, that it is not possible sterically for all four oxygens to interact with a given metal. Two oxygen atoms, a carbonyl from one ring and an ether oxygen from the other lactone ring are suitably arranged to give a 5-membered chelate ring including the metal. As the usual coordination numbers are 6 for Mg^{2+} and 8–10 for Ca^{2+} and K^+, stoicheiometries of between 1:3 and 1:10 might be expected although the higher ratios implying coordination by only one oxygen from each avenaciolide molecule would give overcrowding. Reaction between magnesium or calcium thiocyanates with avenaciolide (1:2) in methanol gives (D. G. Parsons, personal communication) rapid precipitation of syrups which are probably polymeric, each cation having more than one avenaciolide ligand and in turn each avenaciolide coordinating more than one cation.

SYNTHETIC COMPOUNDS

The ideal ionophore would interact only with one cation and transport this rapidly through a barrier. The naturally occurring ionophores clearly differ from this ideal so there is considerable interest in designing and synthesizing better ionophores. Many factors are involved including (a) the affinity of the ionophore for the cation and its selectivity between similar cations, and (b) the relative solubilities of ionophore and complex in the, usually aqueous, medium and the barrier. The factor (a) depends upon the size and shape of the coordinating assembly of atoms and its field strength; selectivity between cations depends upon the relative affinity of the cation for the solvent and the ionophore so that the sequence of selectivity may change with the solvent for a given ionophore or with changes in the field strength of the ionophore in a given solvent.

Changes in groups not directly attached to the metal (for example the group R in X-537A) may also affect the solubility of the ionophore and its complexes. More detailed discussions of these factors are available (Eisenman, 1962; Morf & Simon, 1971; Williams, 1970); explanation is more advanced than prediction of properties. None of the monobasic acid family of ionophores have been synthesized and this has prevented the study of the effect of systematic variation. Neutral compounds have proved more amenable; they include the cyclic depsipeptide, beauvericin, which was not synthesized as a calcium ionophore, and mono-, bi- and tri-macrocyclic polyethers.

Cyclic depsipeptides

Valinomycin, a dodecadepsipeptide, is a well-known, naturally occurring, highly selective ionophore for potassium. Ovchinnikov and co-workers (Ovchinnikov, 1974) have synthesized it and variants, with more or fewer than three repetitions of the unit and with different substituents in the unit which in valinomycin itself is {L-lactate-L-valyl-D-α-hydroxyisovalerate-D-valyl}$_3$. Most variants had lower, some had higher stability constants for complex formation with potassium, but all were slower at transporting the cation through lipids. This compound, like the hexadepsipeptide, enniatin A, {D-α-hydroxyisovaleryl-N-methyl-L-valyl}$_3$, does not transport calcium ions and beauvericin, {D-α-hydroxyisovaleryl-N-methyl-L-phenyl-analyl}$_3$, seems an unlikely ionophore for calcium. However, beauvericin was found to transport Ca^{2+} as well as alkali metals (all added as chlorides) in liposomes and bacterial chromatophores (carotenoids) (Prince, Crofts & Steinrauf, 1974) and to give the same shape for optical density versus cation concentration as the M$^+$Cl compounds implying transfer of one

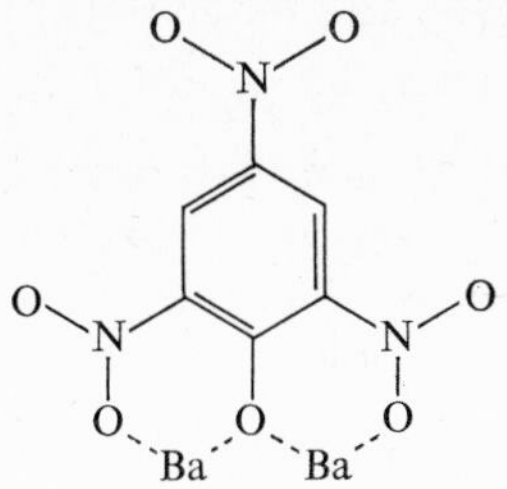

Fig. 4. One picrate anion shared between two barium cations; two more anions form similar bridges between the cations, each of which is coordinated by the three oxygens of the type marked * in beauvericin.

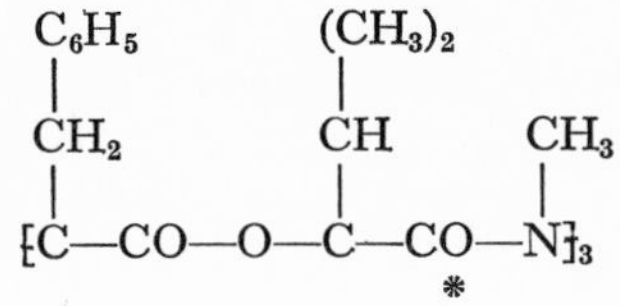

The atoms of *-type are obtained from the figures in Hamilton *et al.* (1975) although the text states that they are oxygen atoms of the α-hydroxyisovaleryl residues. The complex cation is $[Ba_2(picrate)_3(beauvericin)_2]^+$ and the crystal also contains uncomplexed picrate$^-$ ions.

charge per Ca^{2+}. For these systems transfer of calcium was greater than that of barium. Use of picrates in a U-tube experiment, however, gave a greater extraction of barium than of calcium (Roeske, Isaac, King & Steinrauf, 1974) thus indicating the importance of the counter ion with neutral ionophores.

Isolation of a complex of an empirical formula corresponding to barium (picrate)$_2$ beauvericin and its subsequent crystal structure analysis (Hamilton, Steinrauf & Braden, 1975) showed that the compound is actually $[Ba_2(picrate)_3(beauvericin)_2]^+$ picrate$^-$; the cation contains two barium ions 413 pm apart, each coordinated to all three phenolic oxygens and to a nitro group of each picrate ion as shown in Fig. 4. The three bridging picrate ions provide each barium with six oxygen neighbours and three further contacts are made by the carbonyl oxygen atoms of the ester groups of one ionophore. Use of ester not amide oxygen atoms is also shown by valinomycin (Neupert-Laves & Dobler, 1975). Because the phenyl groups of the phenylalanine residues are 'on the outside of the complex cation it is probably lipophilic and the failure of enniatin A to form, or at least to transport a similar complex is explained. The stoicheiometry also explains the observation that this ionophore appears to move as a mono-positive calcium species.

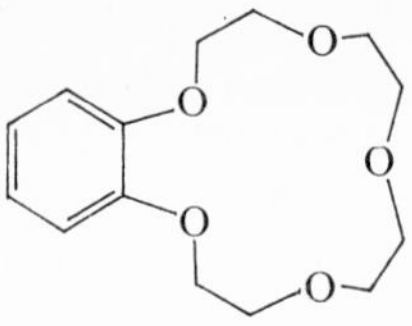

Benzo-15-crown-5 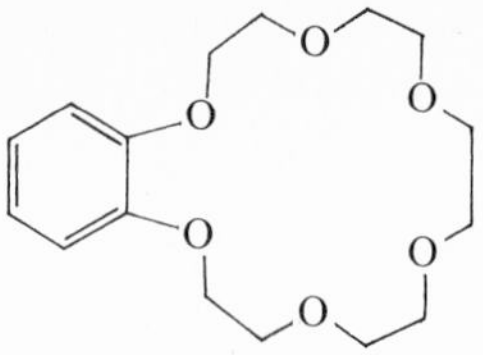　　　　　　　　　　Benzo-18-crown-6

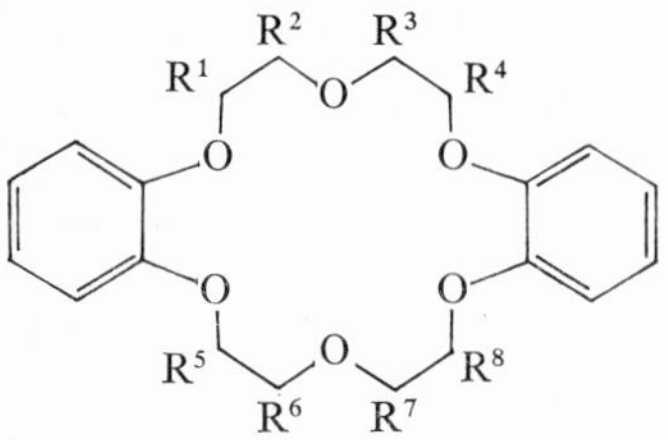

$R.^n$ = H, Pedersen's
dibenzo-18-crown-6 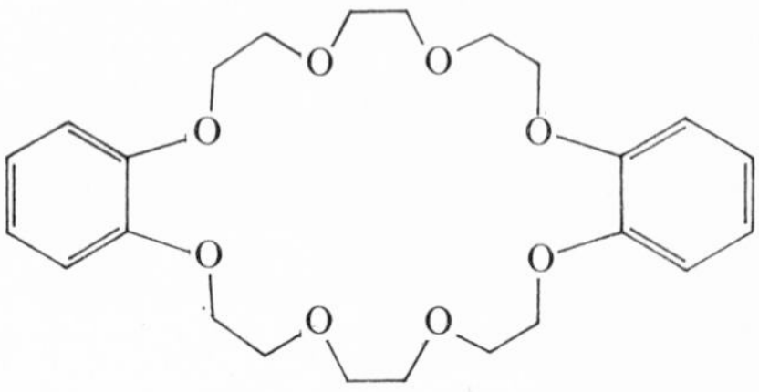　　　　　　　　Dibenzo-24-crown-8

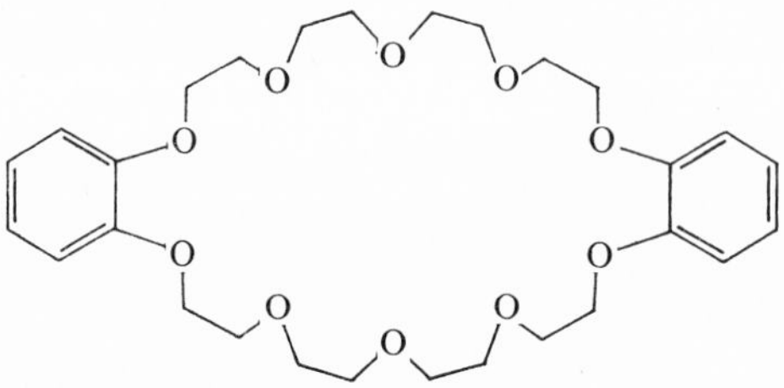

Dibenzo-30-crown-10

Fig. 5. The formula of macrocyclic ethers of 'crown' compounds with the names given by Pedersen (1967). Hydrogen atoms are omitted, > CH_2 groups being shown as ∧. Derivatives of dibenzo-18-crown-6 having > $CH.CH_3$ in some of the eight positions marked R^1 to R^8 (Parsons, 1975) are referred to in the text.

Crown compounds

Monomacrocyclic polyethers were synthesized by Pedersen (1967) and found to form complexes with a range of metal salts. Some of his compounds, named by him 'crowns', and their names in his system are shown in Fig. 5. Calcium parallels the behaviour of sodium as might be expected from the similarity of ionic radii (given in Table 4) with the 15- and 18-membered ring compounds which contain one metal cation per complex; for example 1:1 compounds are formed between calcium salts of small

Table 4. *Log stability constant for the reaction*
$M^{n+} + cryptate \rightleftharpoons [M\ cryptate]^{n+}$ *95 % methanol*

Cryptate (see Fig. 7)	Li+	Na+	K+	Rb+	Cs+	Ca²⁺	Sr²⁺	Ba²⁺	Reference
I [211]	6	6.1	2.3	<2	<2	4.3	2.9	<2	(a)
I [221]	4.2	8.9	7.5	5.8	3.9	9.6	10.7	9.7	(a)
I [222]	1.8	7.2	9.8	8.4	3.6	7.6	11.5	11.8	(a)
I [322]		5.0	7.0		7.0				(d)
II		7.4	9.1					11.1	(b)
III		7.3	8.6					8.5	(b)
IV		3.0	4.4					<2.0	(b)
V		4.5	5.8	6.2	>6.0				(c)
				In water					
I [211]	4.3	2.8	<2	<2	<2	2.8	<2	<2	(a)
I [221]	2.5	5.3	4.0	2.6	<2	7.0	7.4	6.3	(a)
I [222]	<2	3.9	5.3	4.4	<2	4.4	8.0	9.5	(a)
I [322]	<2	1.7	2.1	2.0	1.8		3.4	6.0	(a)
V		1	~1.3	~1.5	1.7				(c)
Radius pm	60	95	133	148	169	99	113	135	

(a) Dietrich *et al.*, 1973*b*.   (b) Dietrich *et al.*, 1973*a*.
(c) Cheney *et al.*, 1972.   (d) M. Kirch & J. M. Lehn, personal communication.

anions and benzo-15-crown-5 but 1:2 compounds if the anion is the non-coordinating tetraphenyl borate (J. N. Wingfield, personal communication; Parsons, Truter & Wingfield, 1975). For the larger crown, dibenzo-24-crown-8, calcium, strontium and barium do not give bridged binuclear complexes comparable with those of sodium and potassium possibly because the metal–metal distances, as found in the alkali metal complexes, ~ 340 pm (Truter, 1973), would give unacceptable repulsion between divalent cations. The largest molecule of Fig. 5, dibenzo-30-crown-10, complexes sodium (1:1 with the tetraphenylborate, 2:1 with small anions) potassium, rubidium; it does not give isolable complexes with calcium but does give the 1:1 complexes to be expected for the larger cations. Synthesis of derivatives of dibenzo-18-crown-6 with methyl groups in the positions R^1, R^4, R^5 and R^8 or R^2, R^3, R^6 and R^7 each having five isomers gave subtle differences in interaction with alkali metals and rapid reaction with calcium (Parsons, 1975).

Replacement of two O atoms by NH groups makes synthesis of three dimensional molecules possible; one of these, Pedersen's 'lantern compound' (Pedersen & Bromels, 1971) shown in Fig. 6 has a higher stability constant for potassium than does dibenzo-18-crown-6. Values of associa-

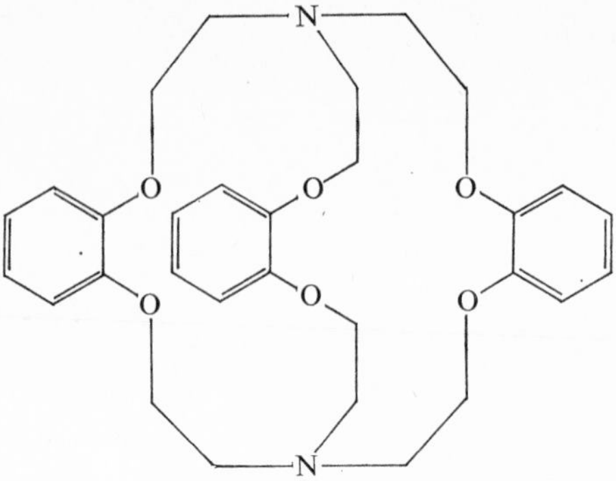

Fig. 6. Pedersen's 'lantern' compound from a derivative of dibenzo-18-crown-6 having —NH— instead of —O— between the aliphatic groups.

tion constants with alkaline earth metals were not quoted but the work on bicyclic polyethers described below suggests that they would be low.

Cryptates

Some of the macrobicyclic compounds, or cryptates, made by Lehn and co-workers (Cheney *et al.*, 1972; Dietrich *et al.*, 1969, 1973a) are shown in Fig. 7; the number of oxygen atoms can be varied to give molecules in which the central cavity is of a controlled size, for example cryptate I [211] has two nitrogen atoms and four oxygen atoms which may surround a central metal, while I [222] has eight ligand atoms. As the number of oxygen atoms increases so the preference for a larger cation goes up and, as the figures in Table 4 show for univalent metals, the maximum stability constant is at Na^+ for I [221] and at K^+ for I [222]. The cryptate I [322] molecule has such a large cavity that the K–O contact distances are larger than the optimum and the stability constant for potassium is less than for I [222] while the Cs–O distances are nearer to the optimum and the stability constants for the two ions are equal. Similar effects are shown for the divalent metals, the ligand I [221] having a maximum for Sr^{2+}, I [211] for Ca^{2+} and I [222] for Ba^{2+}. Because the cryptates are neutral and form 1:1 complexes whatever the charge on the cation, the values of the stability constant for both alkali metals and alkaline earth metals are strictly comparable and at the maximum show large selectivity for the divalent over the univalent cations of the same radius. This selectivity results from the very small number of carbon and hydrogen atoms relative to the oxygen and nitrogen atoms or to put it another way, to the thinness of the shell surrounding the cation. If the shell is thickened by the introduction of benzene rings, as in molecules II, III and IV in Fig. 7, the relative affinity for the alkali metals is considerably increased as shown in Table 4. The

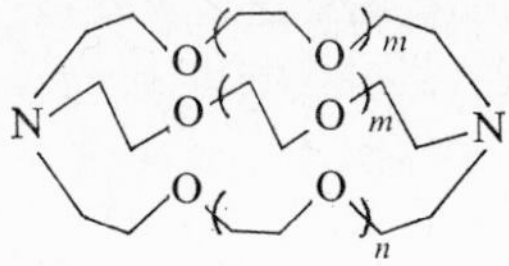

$m=0, n=1$ [211]
$m=1, n=0$ [221]
$m=n=1$ [222]
$m=1, n=2$ [322]

I

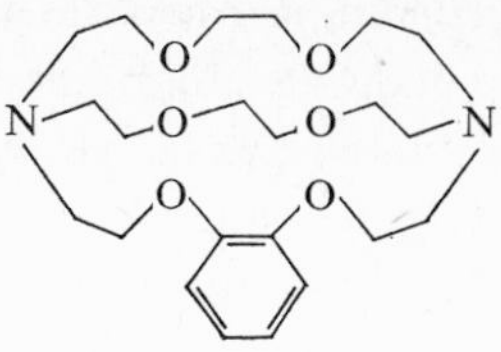

II

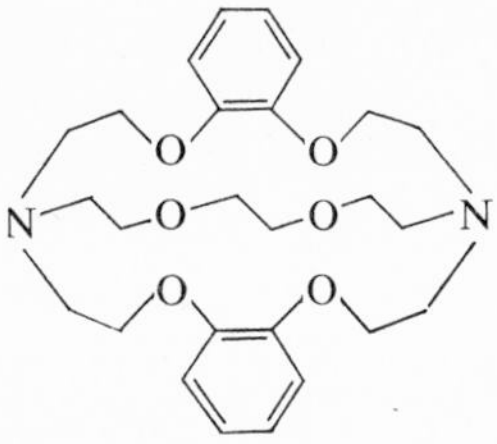

III

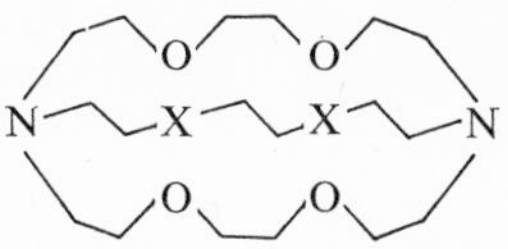

X=CH$_2$

IV

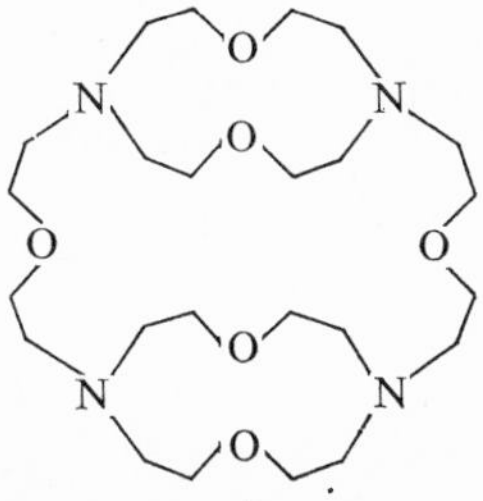

V

Fig. 7. I, Formulae of the original 'cryptate' molecules synthesized by Lehn and co-workers (Dietrich *et al.*, 1969) with the notation indicating the numbers of oxygen atoms in the chains. II, III and IV, derivatives of the cryptates giving controlled divalent/univalent cation selectivities (Dietrich *et al.*, 1973). V, a tri-macrocyclic molecule (Cheney *et al.*, 1972).

change required to confer greater lipid solubility on the cryptate molecule also diminishes the selectivity for divalent cations. Table 4 shows the difference between water and methanol as solvents.

Crystal structures of complexes of univalent and divalent cations have been determined by Weiss and co-workers. For alkali and alkaline earth metals the cation is enclosed by the cryptate, usually without extra contacts for the former, e.g. the sodium ion in NaI [222] (Moras & Weiss, 1973), and with contacts to solvent or anion for the latter, e.g. CaBr$_2$ [222]3H$_2$O in which the calcium is 9-coordinated by the eight atoms of a I [222] cryptate and also by one water molecule which forms weak OH hydrogen bonds to the bromide anion (Metz, Moras & Weiss, 1973).

Experiments on transport through membranes by cryptates of types I and IV have so far been performed on alkali metal picrates with the membrane being a bulk chloroform layer. Results show an optimum stability constant, approximately with log $K \approx 5$, for most rapid transport, although as might be expected the higher the value of K the greater the extraction into chloroform from an aqueous solution (Kirch & Lehn, 1975). These particular compounds are insufficiently lipid-soluble to be used in bilayer experiments and do not appear to have been tested with cell fragments.

The macrotricyclic compound V of Fig. 7 can form complexes enclosing two silver cations. With sodium and the larger alkali metals and with calcium, strontium and barium 1 : 1 complexes only are formed. The stability constants are lower than those for the [222] cryptates and they are more similar to one another, i.e. the selectivity has been reduced (Cheney et al., 1972). This is an example of a more rigid molecule having a lower selectivity.

So far synthetic ionophores require much larger concentrations than naturally occurring ones to produce comparable effects in biological systems. However, the prospects for design and synthesis of ideal compounds seem better than those of lucky finds from culture media.

REFERENCES

ALPHA, S. R. & BRADY, A. H. (1973). Optical activity and conformation of the cation carrier X-537A. *J. Am. chem. Soc.*, **95**, 7043–7049.

ANDERSON, R. C. & FRASER-REID, B. (1975). Synthesis of optically active avenaciolide from D-glucose. Correct stereochemistry of the natural product. *J. Am. chem. Soc.*, **97**, 3870–3871.

BERGER, J., RACHLIN, A. I., SCOTT, W. E., STERNBACH, L. H. & GOLDBERG, M. W. (1951). The isolation of three new crystalline antibiotics from streptomyces. *J. Am. chem. Soc.*, **73**, 5295–5298.

BISSELL, E. C. & PAUL, I. C. (1972). Crystal and molecular structure of a derivative of the free acid of the antibiotic X-537A. *J. chem. Soc. (Chem. Commun.)*, 967–968.

BROOKES, D., TIDD, B. K. & TURNER, W. B. (1963). Avenaciolide, an antifungal lactone from *Aspergillus avenaceus*. *J. chem. Soc.*, 5385–5391.

CHANEY, M. O., DEMARCO, P. V. JONES, N. D. & OCCOLOWITZ, J. L. (1974). The structure of A23187, a divalent cation ionophore. *J. Am. chem. Soc.*, **96**, 1932–1933.

CHENEY, W., LEHN, J.-M., SAUVAGE, J. P. & STUBBS, M. E. (1972). [3]-Cryptates: metal cation inclusion complexes with a macrotricyclic ligand. *J. chem. Soc. (Chem. Commun.)*, 1100–1101.

CORNELIUS, G, GÄRTNER, W. & HAYNES, D. H. (1974). Cation complexation by valinomycin- and nigericin-type ionophores registered by the fluorescence signal of Tl$^+$. *Biochemistry*, **13**, 3052–3057.

CZERWINSKI, E. W. & STEINRAUF, L. K. (1971). Structure of the antibiotic dianemycin. *Biochim. biophys. Res. Commun.*, **45** 1284–1287.

DEGANI, H. & FRIEDMAN, H. L. (1974). Ion binding of X-537A, formulas, formation constants, and spectra of complexes. *Biochemistry*, 13, 5022–5032.

DEGANI, H., FRIEDMAN, H. L., NAVON, G. & KOSOWER, E. M. (1973). Fluorimetric complexing constants and circular dichroism measurements for antibiotic X-537A with univalent and bivalent cations. *J. chem. Soc. (Chem. Commun.)*, 431–432.

DEVORE, D. I. & NASTUK, W. L. (1975). Effects of 'calcium ionophore' X-537A on frog skeletal muscle. *Nature, Lond.*, 253, 644–646.

DIETRICH, B., LEHN, J.-M. & SAUVAGE, J. P. (1969). Diaza-polyoxa-macrocyclic and macrobicyclic compounds. *Tetrahedron Letts.*, 2885–2889.

(1973a). Cryptates: control over bivalent/monovalent cation selectivity. *J. chem. Soc. (Chem. Commun.)*, 15–16.

(1973b). Kryptate: makrocyclische Metallkomplexe. *Chemie in unserer Zeitschrift*, 7, 120–128.

EISENMAN, G. (1962). Cation selective glass electrodes and their mode of operation. *Biophys. J.* (Suppl.) 2, 259–323.

HAMILTON, J. A.. STEINRAUF, L. K. & BRADEN, B. (1975). Beauvericin and divalent cations: crystal structure of the barium complex. *Biochem. biophys. Res. Commun.*, 64, 151–156.

HARRIS, E. J. & WIMHURST, J. M. (1973). Is avenaciolide another ionophore? *Nature New Biol.*, 245, 271–273.

(1974). Ionophoric actions of avenaciolide on mitochondrial cations. *Archs. Biochem. Biophys.*, 162, 426–435.

HOVI, T., WILLIAMS, S. C. & ALLISON, A. C. (1975). Divalent cation ionophore A23187 forms lipid soluble complexes with leucine and other amino acids. *Nature, Lond.*, 256, 70–72.

HYONO, A., HENDRIKS, TH., DAEMEN, F. J. M. & BONTING, S. L. (1975). Movement of calcium through artificial lipid membranes and the effects of ionophores. *Biochim. biophys. Acta*, 389, 34–46.

IZATT, R. M., EATOUGH, D. J. & CHRISTENSEN, J. J. (1973). Thermodynamics of cation-macrocyclic compound interaction. *Structure and Bonding*, 16, 161–189.

JOHNSON, S. M., HERRIN, J., LIU, S. J. & PAUL, I. C. (1970a). Crystal structure of a barium complex of antibiotic X-537A, $Ba(C_{34}H_{53}O_8)_2.H_2O$. *Chem. Comm. (J. chem. Soc. D)*, 72–73.

(1970b). The crystal and molecular structure of the barium salt of an antibiotic containing a high proportion of oxygen. *J. Am. chem. Soc.*, 92, 4428–4435.

LEHN, J. M. (1973). Design of organic complexing agents. Strategies towards properties. *Structure and Bonding*, 16, 1–69.

MAIER, C. A. & PAUL, I. C. (1971). X-ray crystal structure of a silver complex of antibiotic X-537A; a structure enclosing two metal ions. *Chem. Comm. (J. chem. Soc. D)*, 181–182.

McGIVAN, J. D. & CHAPPELL, B. C. (1970). Avenaciolide: a specific inhibitor of glutamate transport in rat liver mitochondria. *Biochem. J.*, 116, 37P.

METZ, B., MORAS, D. & WEISS, R. (1973). Coordination des cations alcalinoterreux dans leurs complexes avec des molécules macrobicycliques. I. Structure cristalline et moléculaire du cryptate de calcium $C_{18}H_{36}N_2O_6.CaBr_2.3H_2O$. *Acta Crystallogr.*, B29, 1377–1381.

MORAS, D. & WEISS, R. (1973). Etude structurale des cryptates. III. Structure cristalline et moléculaire du cryptate de sodium $C_{18}H_{36}N_2O_6.NaI$. *Acta Crystallogr.*, B29, 396–399.

MORF, W. E. & SIMON, W. (1971). Berechnung von freien hydratationsenthalpien und koordinationszahlen für kationen aus leicht zugänglichen parametern. *Helv. chim. Acta*, 54, 794–810.

NEUPERT-LAVES, K. & DOBLER, M. (1975). The crystal structure of a K$^+$ complex of valinomycin. *Helv. chim. Acta*, **58**, 432–442.

NORDMANN, J. J. & CURRELL, G. A. (1975). The mechanism of calcium, ionophore-induced secretion from the rat neurohypophysis. *Nature, Lond.*, **253**, 646–647.

NORDMANN, J. J. & DYBALL, R. E. J. (1975). New calcium-mobilising agent. *Nature, Lond.*, **255**, 414–415.

OVCHINNIKOV, YU. A. (1974). Membrane active complexones. Chemistry and biological function. *FEBS Lett.*, **44**, 1–21.

PARSONS, D. G. (1975). Synthesis of ten isomers of a macrocyclic polyether, tetramethyl-dibenzo-18-crown-6, and their complexes with salts of alkali metals. *J. chem. Soc. (PERKIN I)*, 245–250.

PARSONS, D. G., TRUTER, M. R. & WINGFIELD, J. N. (1975). Alkali metal tetraphenylborate complexes with some macrocyclic, "crown", polyethers. *Inorg. chim. Acta*, **14**, 45–48.

PEDERSEN, C. J. (1967). Cyclic polyethers and their complexes with metal salts. *J. Am. chem. Soc.*, **89**, 7017–7036.

PEDERSEN, C. J. & BROMELS, M. H. (1971). Macrocyclic hetero imine complexing agents. Ger. Patent 2 123 256.

PINKERTON, M. & STEINRAUF, L. K. (1970). Molecular structure of monovalent metal cation complexes of monensin. *J. molec. Biol.*, **49**, 533–546.

PRESSMAN, B. C. (1973). Properties of ionophores with broad range cation selectivity. *Fedn Proc.*, **32**, 1698–1703.

PRESSMAN, B. C., HARRIS, E. J., JAGGER, W. S. & JOHNSON, J. H. (1967). Antibiotic-mediated transport of alkali ions across lipid barriers. *Proc. natn. Acad. Sci. USA*, **58**, 1949.

PRINCE, R. C., CROFTS, A. R. & STEINRAUF, L. K. (1974). A comparison of beauvericin, enniatin and valinomycin as calcium transporting agents in liposomes and chromatophores. *Biochem. biophys. Res. Commun.*, **59**, 697–703.

REED, P. W. (1972). A23187: a divalent cation ionophore. *Fedn Proc.*, **31**, 432 Abs.

REED, P. W. & LARDY, H. A. (1972). A23187: a divalent cation ionophore. *J. biol. Chem.*, **247**, 6970–6977.

ROESKE, R. W., ISAAC, S., KING, T. E. & STEINRAUF, L. K. (1974). The binding of barium and calcium ions by the antibiotic beauvericin. *Biochem. biophys. Res. Commun.*, **57**, 554–561.

SCHMIDT, P. G., WANG, A. H.-J. & PAUL, I. C. (1974). A structural study on the sodium salt of the ionophore, X-537A (lasalocid), by X-ray and nuclear magnetic resonance analysis. *J. Am. chem. Soc.*, **96**, 6189–6191.

SIMON, W., MORF, W. E. & MEIER, P. CH. (1973). Specificity for alkali and alkaline earth cations of synthetic and natural organic complexing agents in membranes. *Structure and Bonding*, **16**, 113–160.

TRUTER, M. R. (1973). Structures of organic complexes with alkali metal ions. *Structure and Bonding*, **16**, 71–111.

WESTLEY, J. W., BLOUNT, J. F., EVANS, R. H., JR, STEMPEL, A. & BERGER, J. (1974). Biosynthesis of lasalocid. II. X-ray analysis of a naturally occurring isomer of lasalocid A. *J. Antibiot.*, **27**, 597–604.

WESTLEY, J. W., EVANS, R. H., JR, WILLIAMS, T. & STEMPEL, A. (1970). Structure of antibiotic X-537A. *Chem. Commun. (J. chem. Soc. D)*, 71–72.

WILLIAMS, R. J. P. (1970). The biochemistry of sodium, potassium, magnesium and calcium. *Q. Rev. chem. Soc.*, **24**, 331–365.

WONG, D. T., WILKINSON, J. R., HAMILL, R. L. & HONG, J.-S. (1973). Effects of antibiotic ionophore, A23187, on oxidative phosphorylation and calcium transport of liver mitochondria. *Archs. Biochem. Biophys.*, **156**, 578–585.

CALCIUM-TRIGGERED LUMINESCENCE OF THE PHOTOPROTEIN AEQUORIN

By O. SHIMOMURA and F. H. JOHNSON

Department of Biology, Princeton University,
Princeton, New Jersey 08540, USA

The light-emitting principle of the bioluminescent jellyfish *Aequorea aequorea* (sometimes called *Aequorea forskalea*; cf. Johnson & Snook, 1927) was isolated more than a decade ago. The purified material was found to consist of a single kind of protein and was given the name 'aequorin' (Shimomura, Johnson & Saiga, 1962).

Aequorin emits light in aqueous solution when Ca^{2+} is added. Neither molecular oxygen nor any other factor is required in this luminescence reaction. The calcium ion or water, however, cannot be a reactant that provides the energy for light-emission, amounting to some 60 kilocalories per einstein at the peak wavelength. Thus, aequorin is evidently an extraordinary protein that stores a large amount of energy which can be readily released in the form of visible radiation by means of an intramolecular chemical reaction triggered by Ca^{2+}. Some progress in understanding the basic mechanism of the intramolecular reaction has been recently made (Shimomura, Johnson & Morise, 1974).

The luminescence reaction of aequorin is extraordinarily sensitive, and under appropriate conditions highly specific to Ca^{2+}. Moreover, aequorin is harmless to many biological systems. Since the pioneering work by Ridgway & Ashley (1967) with single muscle fibres, the use of aequorin in studies of Ca^{2+} in various biological systems is becoming increasingly popular.

PROPERTIES OF AEQUORIN

Molecular weight and other molecular properties

Some confusion has been experienced concerning the molecular weight of aequorin, because of some seemingly anomalous chromatographic behaviour of aequorin in gel filtration, especially on Sephadex gel (Pharmacia).

Thus, gel filtration of a relatively small amount of aequorin through a column of Sephadex suggested that the molecular size of aequorin corresponded to that of a globular protein having a molecular weight of 21000–23000 daltons (Blinks, Mattingly, Jewell & van Leeuwen, 1969; Hastings *et al.*, 1969; Shimomura & Johnson, 1969). Sodium dodecylsulphate gel electrophoresis also indicated a similar molecular size (Kohama, Shimomura & Johnson, 1971). In contrast to these data, however, sedimentation analysis indicated a molecular weight of 30000 daltons (Shimomura & Johnson, 1969; Kohama *et al.*, 1971), and we decided to employ this value as the molecular weight of aequorin on the basis of the methodology. Amino acid analysis and the content of the functional chromophore were consistent with this latter value.

In common buffer solutions, aequorin tends to aggregate, the faster the lower the pH; at a pH higher than 8 aggregation was undetectable (Kohama *et al.*, 1971). A recent study has revealed, however, that even at pH 8 aequorin undergoes strong aggregation if a high concentration of $(NH_4)_2SO_4$ is present in the buffer. For example, when a sample containing more than 0.5 mg ml^{-1} of aequorin was chromatographed on a column of Sephadex G-75 using 10 mM Tris-HCl buffer, pH 7.8, containing 10 mM EDTA and 1 M $(NH_4)_2SO_4$, the protein was eluted at close to the void volume, indicating an apparent molecular weight of over 100000 daltons. This misleading value was clearly due to aggregation. However, the aggregation was completely reversible by removing $(NH_4)_2SO_4$. On the basis of this property, we have designed a new, improved method for purifying aequorin (see under Preparation of aequorin later).

Aequorin is practically non-fluorescent. A concentrated solution appears straw-yellow due to the weak absorption peak at 460 nm (Fig. 1). The value of $E_{1cm}^{1\%}$ at 280 nm was found to be 27.0 (Shimomura & Johnson, 1969). The difference spectrum between aequorin and the protein residue of aequorin, i.e. the spectrum of the functional chromophore, indicated a strong absorption peak at 300 nm in addition to the peak at 460 nm (Shimomura *et al.*, 1974). The isoelectric point was observed at between pH 4 and 5 (Blinks, 1971). The solubility in aqueous buffer solutions is fairly high, and it is necessary to saturate the solution with $(NH_4)_2SO_4$ for complete precipitation. In our latest measurement, 1 mg of pure aequorin emitted 4.3×10^{15} photons ($\lambda_{max} = 470$ nm) when triggered by Ca^{2+} at 25 °C at any pH between 5.5 and 8.5. The quantum yield was computed from this data to be 0.22 at 25 °C.

Reactions of aequorin

Some principal reactions of aequorin are illustrated in Fig. 2. When aequorin is denatured by various methods, such as by addition of acids or urea, or by heating, or by shaking with organic solvents, a blue-fluorescent compound 'coelenteramine' (formerly called AF-350) is split from the protein part. The determination of the structure of this compound (Shimo-

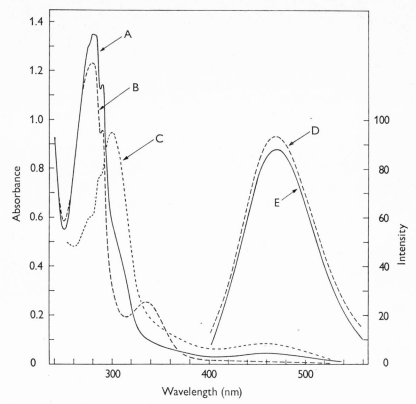

Fig. 1. Absorption spectra of aequorin (A), of blue-fluorescent protein (B) and of the functional chromophore of aequorin (C), fluorescence spectrum of BFP on excitation at 340 nm (D), and luminescence spectrum of aequorin triggered by Ca^{2+} (E). The curve C is the difference spectrum of aequorin and the protein residue (Shimomura *et al.*, 1974); C and D were replotted from Fig. 2 of Shimomura & Johnson, 1970. Protein concentration: 0.5 mg ml^{-1} for A and B, 1.0 mg ml^{-1} for C. Cell path: 1 cm for all.

mura & Johnson, 1972) was a major breakthrough in the study of chemical aspects of aequorin luminescence, and it also helped to elucidate the structure of luciferin of *Renilla*, another coelenterate (Hori & Cormier, 1973).

The luminescence reaction of aequorin triggered by Ca^{2+} results in the formation of 'BFP' (blue-fluorescent protein). BFP can be split by various means to a compound which we have named 'coelenteramide' and the protein part, neither of which is fluorescent in aequeous solution. By simply mixing together in an aqueous solution containing Ca^{2+}, these two components spontaneously combine to form fluorescent BFP identical to the original (Shimomura & Johnson, 1973*a*). The binding constant was

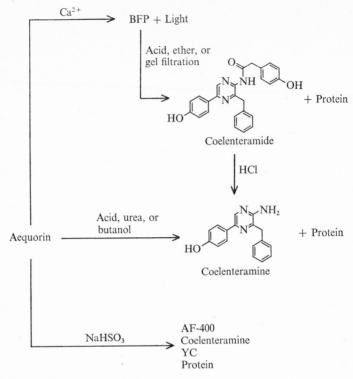

Fig. 2. Main reactions of aequorin. For explanation
of abbreviations see text.

found to be 2×10^5 M^{-1} at pH 7.4 at 25 °C (Morise, Shimomura, Johnson
& Winant, 1974).

The fluorescence spectrum of BFP coincides with the luminescence
spectrum of aequorin (Fig. 1), indicating that the excited state of BFP, or
more specifically the excited state of coelenteramide bound in BFP, is the
light emitter in the luminescence reaction. The absorption peak at 335 nm
in the spectrum of BFP is due to coelenteramide. Coelenteramide can be
converted to coelenteramine by acid hydrolysis.

Treatment of aequorin with NaHSO_3 yields coelenteramine, AF-400
(a blue-fluorescent compound having absorption maximum at 400 nm),
YC (a yellow compound responsible for the yellow colour of native
aequorin), and the residual part of the protein. The amount of AF-400
formed increases with time of treatment, while that of coelenteramine
decreases with time (Shimomura *et al.*, 1974).

Stability and inactivation of aequorin

Purified aequorin solutions containing a sufficient concentration of EDTA are stable enough for most practical purposes at near o °C or when frozen, despite the fact that the material is slowly luminescing and, therefore spontaneously slowly inactivating itself, presumably by contaminant calcium. At room temperature, or in the presence of an insufficient concentration of EDTA, the rate of inactivation varies widely depending on the conditions. The rate of inactivation, however, can be easily estimated by testing the rate of light-emission of the sample. Even in the presence of adequate EDTA, the luminescence activity of aequorin is quickly destroyed at temperatures above 30 °C, at a pH below 5 or above 9, by any denaturation of the protein, by adding $NaHSO_3$, or by adding reagents which react with the SH group of the protein (Shimomura *et al.*, 1962, 1974).

The stability of aequorin is greatly increased by saturating the solution with $(NH_4)_2SO_4$. Thus, no loss in light-emitting activity was detected after three years when an aequorin solution containing 10 mM EDTA was saturated with $(NH_4)_2SO_4$ and then kept at −25 °C. Freeze-drying of aequorin caused a substantial loss of luminescence activity if suitable salts were not present in the sample solution (Shimomura & Johnson, 1969). Moreover, the stability of freeze-dried aequorin at room temperature was found to be no better than that of aequorin in saturated $(NH_4)_2SO_4$ at the same temperature.

CHARACTERISTICS OF Ca^{2+}-TRIGGERED LUMINESCENCE

When a small amount of aequorin is mixed with a large excess of Ca^{2+} at room temperature, the resulting luminescence is complete within a few seconds, and the total light emitted is proportional to the amount of aequorin. The rise of light emission under these conditions is quite fast, with a half-rise time of several milliseconds or less, but the decay is relatively slow with a half-decay time of 0.5–1 sec (van Leeuwen & Blinks, 1969; Hastings *et al.*, 1969; Loschen & Chance, 1971).

The intensity of aequorin luminescence elicited with a low concentration of Ca^{2+} (in the range of less than 10^{-5}–10^{-6} M) has been found to be proportional to the square of Ca^{2+} concentration, and a log–log plot indicates a slope close to 2.0 (Shimomura *et al.*, 1963; Ashley, 1970; Baker, Hodgkin & Ridgway, 1971; see also Fig. 4 of this paper), indicating that one molecule of aequorin reacts with two Ca^{2+} to emit light. In contrast to this, an analysis of BFP labelled with $^{45}Ca^{2+}$ showed that one molecule of BFP is bound with three Ca^{2+}, of which one Ca^{2+} is bound much more tightly than the other two (Shimomura & Johnson, 1970). No substantiated interpretation of this circumstance has yet been made.

The lowest limit of concentration of Ca^{2+} that is detectable by aequorin has not yet been clearly determined, mainly because of the difficulty in preparing test solutions which are sufficiently free from Ca^{2+} as well as EDTA. Nevertheless, the present data (see Fig. 4 and later discussion) indicate that the limit is definitely less than 10^{-8} M Ca^{2+}.

The light emission of aequorin can be elicited to various extents by Sr^{2+}, Pb^{2+}, Co^{2+}, Cu^{2+}, Cd^{2+} and lanthanides, in addition to Ca^{2+}, but no ion other than one of these has been found to be active (Izutsu *et al.*, 1972; Shimomura & Johnson, 1973*b*). In testing biological systems, however, aequorin can be considered sufficiently specific to Ca^{2+} alone, because of the unlikelihood of the occurrence of those other ions in biological systems in significant amounts. For an in-vitro test, all ions mentioned above, except Ca^{2+}, Sr^{2+} and lanthanides, can be completely masked by including 1 mM sodium diethyldithiocarbamate in the test solution (Shimomura & Johnson, 1975*c*).

MECHANISM OF LUMINESCENCE AND REGENERATION OF AEQUORIN

An outline of our present knowledge concerning the mechanism of the Ca^{2+}-elicited luminescence reaction of aequorin is illustrated in the upper part of Fig. 3. It has been postulated that four kinds of functional groups bound to the protein part are involved in the luminescence reaction (Shimomura *et al.*, 1974). The first of the four is the protein-bound enol form of coelenterazine, shown at the bottom of the aequorin molecule. This group is changed to the light-emitter coelenteramide through a Ca^{2+}-triggered oxidation. Coelenterazine itself was first found in the liver of the bioluminescent squid *Watasenia* (Inoue *et al.*, 1975); it is now known to be the 'luciferin' of bioluminescent coelenterates (Shimomura & Johnson, 1975*b*). The second of the functional groups, a hydroperoxide or H_2O_2, is required as the source of oxygen to oxidize the first mentioned group. Thus, we have postulated the presence of a stabilized, bound form of H_2O_2 which easily releases H_2O_2 as needed. In the figure, such a form is indicated as α-hydroxyhydroperoxide, but a peroxy acid or other forms are reasonable possibilities. The third group, designated YC, causes the yellowish colour of native aequorin, but becomes colourless in the luminescence reaction. YC was postulated to be an oxidant for H_2O_2. YC has been isolated and study of its structure is in progress in our laboratory. The fourth group comprises at least one SH group which can be attributed to any of three

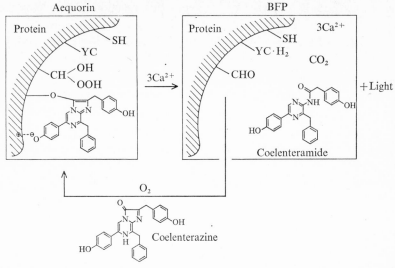

Fig. 3. Postulated mechanism of luminescence and regeneration of aequorin. For explanation of abbreviations see text (from Shimomura & Johnson, 1975c).

cysteine residues known to exist in aequorin (Shimomura & Johnson, 1969). The SH group(s) seems unchanged after the luminescence reaction, but the group(s) is evidently essential for Ca^{2+} to trigger luminescence.

Although each functional group is indicated separately as part of the aequorin molecule in Fig. 3, a possibility remains at present that any of these groups is combined with another group in the actual protein molecule. For example, H_2O_2 might be bound to YC, and, as another possibility, the functions of YC and enolized coelenterazine might be combined by some manner to form a single group. Consequently, more detailed discussion concerning this point must await the determination of the structure of YC.

The mechanism diagrammed below for the chemical reaction of aequorin that is directly involved in light-production appears to be the same as that of the luminescent oxidation of *Cypridina* luciferin with molecular oxygen catalysed by *Cypridina* luciferase (Shimomura & Johnson, 1971). Thus, the coelenterazine moiety of aequorin is first oxidized to hydroperoxide anion by H_2O_2 and YC, then the peroxide forms a dioxetanone (McCapra & Chang, 1967) which immediately decomposes to coelenteramide and CO_2 accompanied by the emission of light:

The mechanism by which Ca^{2+} triggers this reaction is not known at present.

Incubation of the protein residue of BFP, or BFP itself, in a solution containing 5 mM EDTA and 2 mM 2-mercaptoethanol together with an excess of coelenterazine (synthetic) has been found to result in the formation of Ca^{2+}-triggerable photoprotein, and this photoprotein, after isolation, was found to be identical with native aequorin (Shimomura & Johnson, 1975a), indicating that regeneration of aequorin had indeed taken place (Fig. 3, lower part). For regeneration to take place, molecular oxygen is essential and 2-mercaptoethanol seems to protect functional SH group(s) during the regeneration process. The yield in regeneration was 50 % in 20–30 min, and about 90 % within 3 h at 5 °C.

When incubation of the protein part of BFP with coelenterazine was carried out in the presence of Ca^{2+}, the result was, as would now be expected, a spontaneous weak luminescence. In such a circumstance, the protein residue of BFP behaves as an enzyme, catalysing the luminescent oxidation of coelenterazine. It is a very slow-working enzyme, however, having a turnover number of approximately 1–2 h^{-1}.

APPLICATIONAL ASPECTS OF
AEQUORIN LUMINESCENCE

The possible use of aequorin luminescence as a sensitive, specific means for detection and microdetermination of Ca^{2+} was first suggested in 1963 (Shimomura et al., 1963). The first application of aequorin luminescence for such use in research, however, should be credited to Ridgway & Ashley, who reported in 1967 the relationship between tension and intracellular Ca^{2+} of single muscle fibres of a barnacle, monitored by injected aequorin. Since then, a considerable number of studies concerning the role of Ca^{2+} in a wide variety of biological systems based on the aequorin test for Ca^{2+} have been reported, as witnessed by such recent examples as invertebrate photoreceptors (Brown & Blinks, 1974); neurones of marine and land snails (Stinnakre & Tauc, 1973; Chang, Gelperin & Johnson, 1974); squid axons (Baker, Meves & Ridgway, 1973); amphibian twitch muscle (Taylor, Rüdel & Blinks, 1975); coupling conductances (Rose & Loewenstein, 1975).

A brief paper on the use of aequorin was previously published (Johnson & Shimomura, 1972). Thus, the present paper includes original materials on only two aspects of the subject, namely (1) preparation of the aequorin sample, and (2) the leaching of Ca^{2+} into solutions from the surface of clean glassware.

Preparation of aequorin

An average size late-summer specimen of *Aequorea* (8 cm in diameter) collected at Friday Harbor, Washington, usually contains approximately 30 μg of aequorin in the photogenic organs, all of which are located in the margin of the umbrella. To extract the desired material, marginal rings (2–3 mm in width) are carefully cut from the umbrella, and shaken violently in saturated $(NH_4)_2SO_4$ containing 50 mM EDTA to dislodge the photogenic particles, then crude aequorin is extracted from these photogenic particles with 50 mM EDTA, as previously described in detail (Johnson & Shimomura, 1972). It is essential to measure the luminescence activity of the material in solution at each step to make certain that the technique is satisfactory.

Purification of aequorin can be accomplished by five steps of chromatography using three columns, at 0–2 °C. The first step is done on a column of Sephadex G-75 (Pharmacia) with 10 mM Tris-HCl buffer, pH 7–8, containing 10 mM EDTA and 1 M $(NH_4)_2SO_4$. Under these conditions, aequorin is eluted at near the void volume, due to heavy aggregation as mentioned earlier in this paper. The second chromatography is performed on the same column, but involving replacement of $(NH_4)_2SO_4$ by 0.2 M NaCl in the buffer. Aequorin is now eluted mostly as a monomer. In the third step, aequorin is adsorbed on a column of DEAE cellulose prepared with 10 mM Tris-HCl buffer containing 10 mM EDTA, then eluted with a linear gradient concentration of NaCl from 0.05 M to 0.35 M. The fourth and fifth steps are repetitions of the first and second steps, respectively, but with a small size of column. The overall yield of purification is somewhat over 30%, of which about 15% (a half) consists of disc-electrophoretically pure quality and the remainder of less purity.

Purified aequorin can be safely stored in saturated $(NH_4)_2SO_4$ containing some EDTA at 0 °C or preferably in a freezer. Just prior to using, aequorin can be precipitated by centrifugation, and the precipitate dissolved in a desired buffer, at a pH of over 7.5 if avoidance of aggregation is desired. If the concentration of EDTA or $(NH_4)_2SO_4$ in this solution is considered to be undesirably high, then the solvent can be quickly and easily exchanged with a desired buffer by filtration through a small column of Sephadex G-25 or Bio-Gel P-6 (Bio-Rad). If a small volume of concentrated aequorin in a desired buffer is wanted, one drop of concentrated solution of aequorin can be filtered through a micro-column (2.5–3 mm internal diameter, 4–5 cm gel height) equilibrated with that buffer. A suitable micro-column can be conveniently made from plastic tube with a disc of 400 mesh Nylon cloth (Pharmacia) across the opening at the base, held in place by a cylindrical plastic plug of the same OD as the column, firmly juxtaposed with the aid of a short length of tygon tubing surrounding the connection. A satisfactory outlet for the eluate can be provided by a capillary-sized bore in the plastic plug, and small droplets of eluate can be assured by sharpening the exit area of this plug. It is obviously important to avoid any contamination of Ca^{2+} from all sources. Moreover, we recommend including at least 1 to 10 μm EDTA in any aequorin stock solution as the minimum necessary protection.

Addition of chelating resin Chelex 100 (Bio-Rad) offers a convenient way to protect aequorin (Brown & Blinks, 1974). It should be noted, however, that this resin leaches out an easily detectable amount of calcium-chelating compound into the solution, gradually but practically endlessly, even after the resin has been thoroughly washed; thus, an 'EDTA free' aequorin solution containing Chelex 100 beads is meaningless.

Leaching of Ca^{2+} from glass

The composition of many kinds of glass includes calcium, in some instances in an amount as much as several per cent. Thus, in a study of traces of Ca^{2+}, one would quite naturally suspect a possible effect of Ca^{2+} leached from the glassware that is used to contain or measure the solutions. Ordinarily, there would be no simple means to prove the existence or the extent of such effects. The suspected effect due to leached Ca^{2+}, however, was visibly recognized to be actually the case, at least qualitatively, when an aequorin solution containing less than 10^{-4} M EDTA was transferred from a plastic container to a thoroughly cleaned dry Pyrex flask, resulting in an increase in brightness of luminescence of the solution (Shimomura & Johnson, 1969). We now have some quantitative data on the rate of leaching Ca^{2+} from the inside wall of test tubes made of various materials, as described below. These measurements also illustrate another use of aequorin luminescence.

For the experiments involved, all solutions were made in deionized distilled water having a resistance of more than 10 MΩ. They all contained 2 mM sodium acetate (Alfa Inorganics, 'Ultrapure'), and had a pH of 6.5–7.0. The aequorin stock solution contained 10 μM EDTA and approximately 0.7 mg ml^{-1} of aequorin. Light intensity was measured when 2.5 ml of the test solution or the Ca^{2+} standard solution was added into a polycarbonate test tube containing 5 μl of aequorin solution. Only plastics came into contact with solutions throughout the experiment.

The relationship between Ca^{2+} concentration and the initial maximum intensity of luminescence elicited by standard Ca^{2+} solutions was plotted on a log–log scale in Fig. 4; this provided a calibration curve. The curve

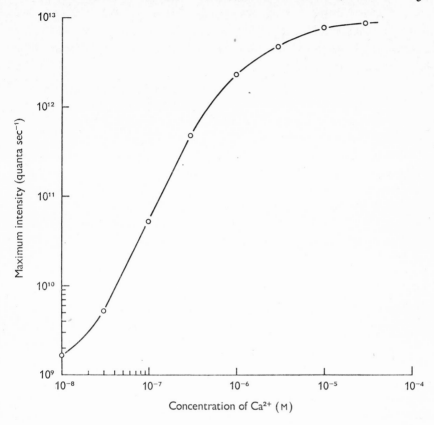

Fig. 4. Relationship between the concentration of Ca^{2+} and the maximum intensity of luminescence when 2.5 ml of 2 mM sodium acetate containing the indicated amount of calcium acetate was added to 5 μl of aequorin stock solution, at 25 °C. The aequorin stock solution contained 0.7 mg of aequorin in 1 ml of 2 mM sodium acetate containing 10^{-5} M EDTA. When no Ca^{2+} was added, the maximum intensity was 1.1×10^{9} quanta sec^{-1}.

is clearly a straight line between Ca^{2+} concentrations of 3×10^{-8} M and 3×10^{-7} M, with a slope of 2.0. The inflexion at high Ca^{2+} concentration is probably due to saturation of the calcium binding site of aequorin, and indicates involvement of more than one rate constant in the luminescence reaction (Hastings *et al.*, 1969; Loschen & Chance, 1971), and the inflexion at low Ca^{2+} concentration seems most likely to be due to the influence of contaminating Ca^{2+}, along with the buffering action of the EDTA present, both in estimated concentrations of the order of 10^{-8} M.

Based on the calibration curve of Fig. 4, the concentrations of Ca^{2+} found in 2 mM sodium acetate solution contained in test tubes made of various kinds of materials are shown in Table 1. The values shown were

Table 1. *Concentration of Ca^{2+} found in 15 ml of 2 mM sodium acetate solution which was placed in various kinds of test tubes having inside diameters of 1.6–1.8 cm^a, at 25 °C*

	Concentration of Ca^{2+} (M)		
Kind of test tube	At 1 min after adding solution to dry tubes^b	At 1 min after adding solution to rinsed tubes	At 24 h after adding solution to rinsed tubes
Polycarbonate (Kimble)	$< 10^{-8}$	$< 10^{-8}$	$< 10^{-8}$
Polypropylene (Nalgene)	$< 10^{-8}$	$< 10^{-8}$	$< 10^{-8}$
Vycor (Silica glass, Corning)	2.6×10^{-8}	2.0×10^{-8}	2.0×10^{-8}
Kimax (Borosilicate glass, Kimble)	3.3×10^{-8}	3.0×10^{-8}	1.3×10^{-7}
Pyrex (Borosilicate glass, Corning)	4.0×10^{-8}	3.6×10^{-8}	1.5×10^{-7}
Boron-free glass (Corning)	5×10^{-8}	$< 10^{-8}$	3.0×10^{-8}
Soda-lime glass (A. H. Thomas)	2×10^{-7}	5.6×10^{-8}	1×10^{-6}

^{a} Brand-new test tubes were washed with detergent, rinsed carefully with deionized distilled water, then oven dried, except the plastics which were air dried. The cleaned test tubes were used in the test within 3 days.

^{b} Test tubes left 1 month after cleaning showed highly irregular results with the average concentration of Ca^{2+} about 10 times over the present results.

obtained on the calibration curve, according to the initial maximum intensities elicited by a portion of the solution inside the test tubes. The data of Table 1 indicates that (1) as expected, plastics did not leach any detectable Ca^{2+}; (2) soft glass (soda-lime glass) is clearly unsuitable in the study of Ca^{2+}, although only one kind was tested; (3) borosilicate glass (Pyrex and Kimax), most common of laboratory glassware, leached Ca^{2+} to an appreciable extent; (4) Vycor (silica glass) and alkali-resistant, boron-free glass leached detectable amounts of Ca^{2+}, but to extents much less than that of borosilicate glass; (5) a simple rinsing of test tubes before use nearly always removed some, clearly detectable amount of Ca^{2+} from the surface of the vessels.

It should be noted that the concentration of leached Ca^{2+} depends on the ratio of the volume of solution to the total wall area of a vessel soaked with the solution, in addition to soaking time, pH of the solution, temperature, etc. Thus, leaching of Ca^{2+} would be greatly amplified in a capillary filled with solution, and sufficient care must be taken in such a case.

We thank the US National Science Foundation for support, in part, of this work.

REFERENCES

ASHLEY, C. C. (1970). An estimate of calcium concentration changes during the contraction of single muscle fibres. *J. Physiol. Lond.*, 210, 133–134P.

BAKER, P. F., HODGKIN, A. L. & RIDGWAY, E. B. (1971). The early phase of calcium entry in giant axons of *Loligo*. *J. Physiol. Lond.*, 214, 33–34.

BAKER, P. F., MEVES, H. & RIDGWAY, E. B. (1973). Effects of manganese and other agents on the calcium uptake that follows depolarization of squid axons. Calcium entry in response to maintained depolarization of squid axons. *J. Physiol. Lond.*, 231, 511–548.

BLINKS, J. R. (1971). Heterogeneous nature of the calcium-sensitive bioluminescent protein aequorin. *Int. Congr. physiol. Sci.*, 9, 68.

BLINKS, L. R., MATTINGLY, P. H., JEWELL, B. R. & VAN LEEUWEN, M. (1969). Preparation of aequorin for use as a calcium indicator in physiological studies. *Fedn Proc.*, 28, 781.

BROWN, J. E. & BLINKS, J. R. (1974). Changes in intracellular free calcium concentration during illumination of invertebrate photoreceptors. *J. gen. Physiol.*, 64, 643–665.

CHANG, J. J., GELPERIN, A. & JOHNSON, F. H. (1974). Intracellularly injected aequorin detects transmembrane calcium flux during action potentials in an identified neuron from the terrestrial slug, *Limax maximus*. *Brain Res.*, 77, 431–442.

HASTINGS, J. W., MITCHELL, G., MATTINGLY, P. H., BLINKS, J. R. & VAN LEEUWEN, M. (1969). Response of aequorin bioluminescence to rapid changes in calcium concentration. *Nature, Lond.*, 222, 1047–1050.

HORI, K. & CORMIER, M. J. (1973). Structure and chemical synthesis of a biologically active form of *Renilla* (sea pansy) luciferin. *Proc. natn. Acad. Sci. USA*, 70, 120–123.

INOUE, S., SUGIURA, S., KAKOI, H., HASIZUME, K., GOTO, T. & IIO, H. (1975). Squid bioluminescence. II. Isolation from *Watasenia scintillans* and synthesis of 2-(p-hydroxybenzyl)-6-(p-hydroxyphenyl)-3,7-dihydroimidazo[1,2-a]pyrazin-3-one. *Chem. Lett.*, 141–144.

IZUTSU, K. T., FELTON, S. P., SIEGEL, I. A., YODA, W. T. & CHEN, A. C. N. (1972). Aequorin: its specificity. *Biochem. biophys. Res. Commun.*, 49, 1034–1039.

JOHNSON, F. H. & SHIMOMURA, O. (1972). Preparation and use of aequorin for rapid microdetermination of Ca^{2+} in biological systems. *Nature New Biol.*, 237, 287–288.

JOHNSON, M. E. & SNOOK, H. J. (1927). *Seashore Animals of the Pacific Coast*, p. 71. Macmillan: New York.

KOHAMA, Y., SHIMOMURA, O. & JOHNSON, F. H. (1971). Molecular weight of the photoprotein aequorin. *Biochemistry*, 10, 4149–4152.

LOSCHEN, G. & CHANCE, B. (1971). Rapid kinetic studies of the light-emitting protein aequorin. *Nature New Biol.*, 233, 273–274.

McCAPRA, F. & CHANG, Y. C. (1967). The chemiluminescence of a *Cypridina* luciferin analogue. *Chem. Commun.*, 1011–1012.

MORISE, H., SHIMOMURA, O., JOHNSON, F. H. & WINANT, J. (1974). Intermolecular energy transfer in the bioluminescent system of *Aequorea*. *Biochemistry*, 13, 2656–2662.

RIDGWAY, E. B. & ASHLEY, C. C. (1967). Calcium transients in single muscle fibers. *Biochem. biophys. Res. Commun.*, 29, 229–234.

ROSE, B. & LOEWENSTEIN, W. R. (1975). Permeability of cell junction depends on local cytoplasmic calcium activity. *Nature, Lond.*, 254, 250–252.

SHIMOMURA, O. & JOHNSON, F. H. (1969). Properties of the bioluminescent protein aequorin. *Biochemistry, N.Y.*, **8**, 3991–3997.

(1970). Calcium binding, quantum yield, and emitting molecule in aequorin bioluminescence. *Nature, Lond.*, **227**, 1356–1357.

(1971). Mechanism of the luminescent oxidation of *Cypridina* luciferin. *Biochem. biophys. Res. Commun.*, **44**, 340–346.

(1972). Structure of the light-emitting moiety of aequorin. *Biochemistry, N.Y.*, **11**, 1602–1608.

(1973a). Chemical nature of the light-emitter in bioluminescence of aequorin. *Tetrahedron Lett.*, 2963–2966.

(1973b). Further data on the specificity of aequorin luminescence to calcium. *Biochem. biophys. Res. Commun.*, **53**, 490–494.

(1975a). Regeneration of the photoprotein aequorin. *Nature, Lond.*, **256**, 236–238.

(1975b). Chemical nature of bioluminescence systems in coelenterates. *Proc. natn. Acad. Sci. USA*, **72**, 1546–1549.

(1975c). Specificity of aequorin luminescence to calcium. In *Analytical Applications of Bioluminescence and Chemiluminescence, NASA SP*-388, pp. 89–94. National Aeronautics and Space Administration: Washington, D.C.

SHIMOMURA, O., JOHNSON, F. H. & MORISE, H. (1974). Mechanism of the luminescent intramolecular reaction of aequorin. *Biochemistry, N.Y.*, **13**, 3278–3286.

SHIMOMURA, O., JOHNSON, F. H. & SAIGA, Y. (1962). Extraction, purification and properties of aequorin, a bioluminescent protein from the luminous hydromedusan, *Aequorea*. *J. cell. comp. Physiol.*, **59**, 223–240.

(1963). Microdetermination of calcium by aequorin luminescence. *Science, Wash.*, **140**, 1139–1140.

STINNAKRE, J. & TAUC, L. (1973). Calcium influx in active *Aplysia* neurones detected by injected aequorin. *Nature New Biol.*, **242**, 113–115.

TAYLOR, S. R., RÜDEL, R. & BLINKS, J. R. (1975). Calcium transients in amphibian muscle. *Fedn Proc.*, **34**, 1379–1381.

VAN LEEUWEN, M. & BLINKS, J. R. (1969). Properties of aequorin relevant to its use as a calcium indicator in biological work. *Fedn Proc.*, **28**, 359.

LASER MEASUREMENT OF
BIOLOGICAL MOVEMENT

By R. W. PIDDINGTON

Department of Physiology, King's College,
University of London, Strand, London WC2R 2LS

When laser light illuminates a system of moving particles the scattered light exhibits fluctuations in intensity when viewed from a given angle in space (Fig. 1). These fluctuations depend on the coherence of the incident light and are related to the particle movements in a known way.

Experiments using laser light-scattering have been done on a wide range of subjects including: diffusion of macromolecules and viruses, acoustic vibration, ultra-sonic air flow, blood flow, streaming plant cells, swimming microorganisms, muscle, adrenal gland and brain. This paper does not attempt to review all the work in this field; some of these topics are included in the bibliography and the reader is especially referred to Cummins & Pike (1974), Chu (1974), and Piddington (1976).

It can be seen from Fig. 1 that a moving system could be analysed either photographically (diffraction pattern) or spectroscopically (photomultiplier). Spectroscopy offers higher resolution of displacement (see Piddington & Ross, 1976) and the particular spectroscopic technique known as photon correlation is emphasised here.

The technique has been schematised in Figs. 1–3. Some of the different kinds of motion detectable are shown in Fig. 4 and samples of the corresponding waveforms are shown in Fig. 5. Two of these systems are described in more detail: (1) cytoplasmic streaming in the giant plant cell (*Nitella*) which, on a fundamental level, occurs at the same speed as muscle contraction (Fig. 6) and, (2) Brownian motion of particles smaller than a wavelength (50 nm vesicles) contained within cells of the bovine adrenal gland (Fig. 7). Streaming in *Nitella* and adrenaline secretion from the adrenal medulla are calcium-dependent processes.

PRINCIPLE OF LASER TECHNIQUE

The precise wavelength of laser light is used as a measuring stick in many applications involving diffraction, holography and spectroscopy. In each, a static or dynamic difference in distance travelled by two or more light rays is registered. If we define this difference as D, then, for diffraction or

[55]

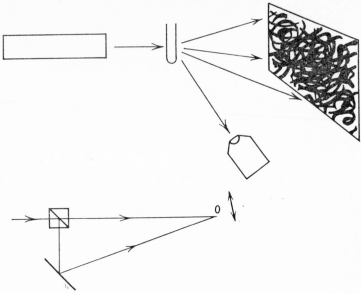

Fig. 1. Two experimental spectroscopic arrangements to measure movement with laser beams. *At top*, a single laser beam illuminates a specimen tube containing (for example) some particles undergoing Brownian motion. Light scattered from the particles forms an interference pattern on a screen (or wall) and this pattern is constantly changing due to the changing positions of the scatterers. For small particles, the pattern changes faster than for large particles and a quantitative measure of the rate of change can be obtained by positioning a photomultiplier at a given point in the scattered field and by analysing the fluctuations in the photo-current output (Fig. 3). *At bottom*, the beam is split in two and the two resulting beams are recombined on the surface of an object. Fringes are set up at the inter-section point (Fig. 2) and motion of the object in a direction across the fringes produces fluctuations in the scattered light which is again detected and analysed by a photomultiplier. A two-beam arrangement such as this is necessary for the analysis of coherent motion such as uniform flow along a tube or uniform oscillation; it is not necessary for systems (such as Brownian motion) in which the component parts move relative to each other.

holography, a fringe is set up every time D changes by one wavelength and in spectroscopy, a complete beat cycle occurs. (See Gabor, Kock & Stroke, 1971; Chu, 1974; Cummins & Pike, 1974; Piddington & Ross, 1976. A fringe spacing here is defined as the distance from light-to-light or dark-to-dark bands.) Fig. 2 illustrates this principle.

The 'D law', as illustrated in Fig. 2, is basically a restatement of an old law in optics, namely Fermat's principle of least time (see Feynman, Leighton & Sands, 1963). For parallel light brought to focus by a para-bolic mirror, the distance travelled from source to focus is constant for any light path. Movement of a reflecting or scattering object along XY in Fig. 2 is equivalent to movement *normal* to the mirror and thus (with a

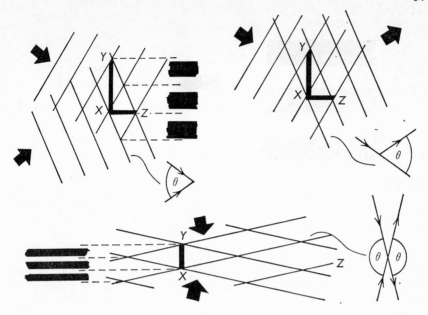

Fig. 2. Geometry for single and double beam spectroscopy. *At top left*, two incident laser beams intersect at an angle θ and produce a series of interference fringes. An object scatters light only when it enters a light band and so movement across the bands (XY) gives fluctuations in the scattered light. Movement along XZ is parallel to the fringes and no changes occur in the scattered light. *At top right*, there is only one incident beam; a second or 'reference' beam from a fixed or differently moving object is also needed but is not shown here. Interference now occurs in the detector which is positioned at the same angle (θ) to the incident beam. The same geometrical laws hold: movement along XY gives a signal but movement along XZ does not. Movement from X to Y would give a signal containing two fluctuations, because the path length difference is two wavelengths. This is the same result as for the two-beam example (*left*) because here two fringes are crossed by movement along XY. Note that θ is the same for right and left diagrams and that XYZ is also the same: the two explanations are equivalent. The bottom diagram is drawn for two beams at large θ but would also work for a single beam. Note that XY is now shorter and the fringes are closer together; motion along XY at a given speed would produce faster fluctuations and, similarly, smaller amplitude movements could be detected. If the diagram were rearranged on its side, we would have the geometry for small θ: wide fringes would result.

reference beam) D changes and we have a signal that we can measure. Changes in D can be measured by causing interference of our two light rays to occur on the detector or on the surface of a scattering object (Figs. 1 and 2).

A MODEL MOTILE SYSTEM, *NITELLA*

Cytoplasmic streaming in the giant algal cells *Nitella* and *Chara* occurs along a helical path inclined at a small angle of around 10–15° to the long

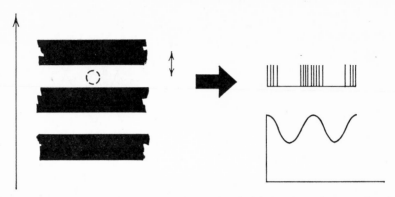

Fig. 3. Photon correlation analysis of two simple kinds of motion; motion in a straight line at constant speed and sinusoidal oscillation at constant frequency. Two intersecting laser beams set up a series of parallel interference fringes and a scattering object moves across the fringes. Photons are scattered every time the object enters a light band and the autocorrelation function for the scattered photon train is a cosine wave of period equal to the period between photon bunches. For motion in a straight line, the speed can be estimated from the period of the correlation function, the fringe spacing and the angle the line makes with the normal to the fringes (XY in Fig. 2). For oscillatory motion, the correlation period equals the oscillation period as long as the oscillation amplitude is a half a fringe spacing or less. At higher amplitudes, harmonics of the oscillation frequency are obtained and the oscillation amplitude can be obtained from the harmonic content (see Piddington & Ross, 1976). The horizontal axis is time; the vertical is correlation.

axis of the cell (see Fig. 3*a* in Piddington, 1976, and Fig. 6). These cells contain actin filaments oriented parallel to the helix and streaming appears to take place along these filaments (Williamson, 1975). The streaming speed is around 50 μm sec^{-1} and this speed equals the contraction speed of fast muscle half-sarcomeres (see Close, 1972; Costantin & Taylor, 1973; Piddington, 1976; Williamson, 1975). Streaming stops at intracellular calcium levels above about 10^{-7} M (Williamson, 1975). If a cell is given an electric shock above a certain threshold, an all-or-none action potential can be elicited and this stops the streaming for about 2 min. The action potential is a calcium-dependent chloride spike lasting about 1 sec and the entry of calcium during the spike shuts off the streaming.

Nitella can be kept in a dish of pond water in the laboratory and a cell can be easily manipulated in a laser scattering apparatus such that the streaming direction makes any desired angle to the incident and scattered directions (Fig. 6; Mustacich & Ware, 1974; Piddington, 1974; Piddington, 1976). In two measurements which take about 1 min to make, the average streaming velocity, distribution of velocities and the relation of these to the Brownian motion can be obtained (Fig. 6). The average velocity

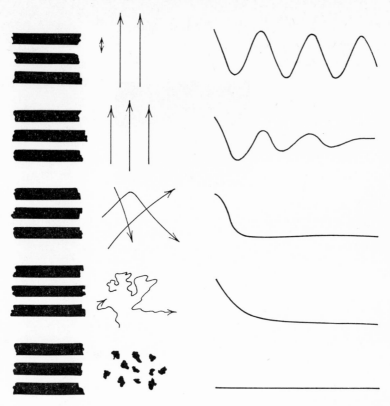

Fig. 4. Photon correlation functions obtained from moving systems of different coherence. The left column shows a set of laser interference fringes, the centre column shows the types of motion through the fringes and the right column shows the resulting photon correlation functions. *At top,* two kinds of coherent motion in a straight line are shown; the correlation function is an undamped cosine curve as explained in Fig. 3. The second correlation, found in streaming plant cells, is a damped cosine wave. This results from particles moving in approximately the same direction with a finite distribution of velocities. The rate of damping gives the distribution of velocities and the period gives the mean velocity. The third example, typical of swimming microorganisms, shows a more highly damped correlation. The particles now move in different directions, but nevertheless travel in fairly straight lines, over distances that are long relative to a fringe spacing, before changing direction. The fourth example shows an exponential correlation. An exponential can be thought of as an extreme form of damping (Feynman *et al.*, 1963); it is typically obtained from particles undergoing Brownian motion. Here each particle executes a random walk and the distance travelled before changing direction is much smaller than a fringe spacing. The last example shows a flat correlation function – this is obtained from 'solid' structures in which movement is minimal. At high amplification a low-amplitude exponential can be obtained. Examples of these different types of correlation function are shown in Fig. 5.

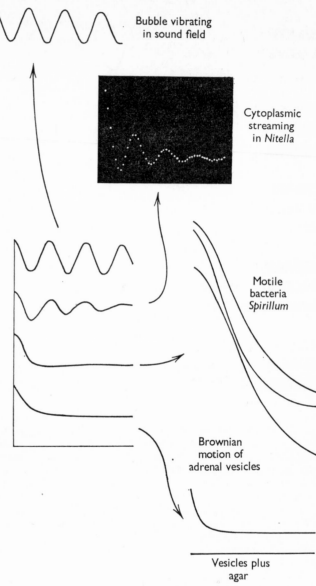

Bubble vibrating
in sound field

Cytoplasmic
streaming
in *Nitella*

Motile
bacteria
Spirillum

Brownian
motion of
adrenal vesicles

Vesicles plus
agar

Fig. 5. Examples of photon correlation functions from several moving systems. The shapes of these correlations can be seen to correspond fairly closely with the shapes diagrammed at left which are the same as in Fig. 4. The correspondence is not exact (see Figs. 6, 7 and text).

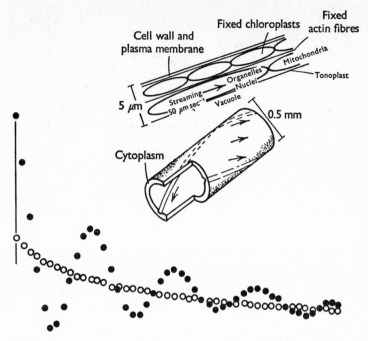

Fig. 6. Photon correlation functions from a streaming giant algal cell (*Nitella*). A cell about 7 cm long and 0.5 mm diameter is placed in a laser beam and the photons scattered at right angles are correlated. The cell is oriented along the bisector of incident and scattered directions (i.e. *XY* in Fig. 2; $\theta = 90°$); the sample time is 0.5 msec per channel. When the cell is streaming, a periodic correlation is obtained (●) but when streaming is temporarily stopped by a supra-threshold shock, a compound exponential (○) results (Brownian motion). This exponential is seen to be present in the streaming cell, added onto the periodic correlation due to the streaming. These measurements require about 1 min of sampling: in this time the average streaming velocity, distribution of velocities and relation of these to the Brownian motion can be obtained.

corresponds closely to that measured microscopically and the half-width at half-height of the distribution is about a tenth of the average velocity (Piddington, 1976). Interestingly, the half-width for the Brownian component has the same value, at least for 90° scattering.

DISTINCTION BETWEEN ACTIVE AND BROWNIAN MOTION

In laser scattering experiments there are three ways of making this distinction:

(1) Active and Brownian systems differ in the *initial curvature* of the correlation function. Motile systems give a correlation which is initially *convex upward* but Brownian systems are *concave upward* (Fig. 4).

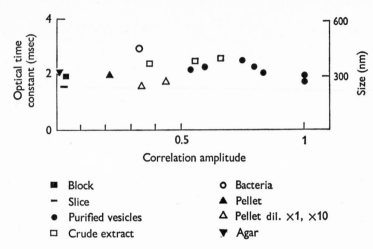

Fig. 7. Plot of correlation time constant against correlation amplitude for adrenal medullary vesicles in different situations. The time constant appears to be independent of the amplitude over a large range of amplitudes. Vesicles isolated by homogenisation and centrifugation are experimentally restricted in their movement by fast centrifugation (pellet) or by suspension in agar. In the intact adrenal gland (block, slice) the vesicles appear to be physically restrained in a similar way (low *amplitude*). However, they appear to vibrate at a speed (*time constant*) equal to the vibration speed of vesicles isolated in water, thus implying a viscosity equal to water in the gland. The effects of large scatterers have been taken into account (see text).

(2) The time constant can be characteristic. For example, in a sample of motile bacteria, the time constant for a fresh active sample is about a third or less of that found with dead cells. A quantitative measure of the distribution of swimming speeds and the proportion of live to dead cells can be easily obtained.

(3) In addition to the shape and time constant of the correlation, the dependence of these on the fringe spacing (Fig. 2) is also characteristic of the kind of motion (see Fig. 2 in Piddington, 1976). It can be shown that the fringe spacing (F) is given by $F = \lambda/[2 \sin(\theta/2)]$. (For θ see Fig. 2.)

For active directed motion, the correlation time constant (or period) is directly proportional to the fringe spacing and hence varies inversely with $\sin(\theta/2)$. This type of angular scaling has been found in the motile systems mentioned at the beginning. For a dead system the situation is different: a Brownian motion particle executing a random walk covers a distance which on average is proportional to the *square root of time* (see Feynman *et al.*, 1963). The time constant for a Brownian motion system is thus found to scale inversely as the *square* of $\sin(\theta/2)$. (See further Cummins & Pike, 1974; Piddington, 1976.)

LASER SCATTERING FROM BRAIN

Nerve cells exist in a large variety of shapes and sizes. Recent discoveries about the structure of the axoplasm in a giant axon from a marine fan worm (*Myxicola*) have given rise to new ideas on how neurone structure is generated and maintained (Gilbert, 1972; Gilbert, Newby & Anderton, 1975). Axoplasm, in the form of a solid gel, can be directly pulled from the worm with forceps in about 10 sec. The fibrous proteins making up the gel are wound in a hierarchical series of at least five helices and the pitches of certain of these helices could act as yardsticks in specifying neurone shape (Gilbert, 1972; Gilbert *et al.*, 1975).

A solid piece of axoplasm can be kept intact for weeks so long as there is no calcium present. The gel breaks up and dissolves in calcium solutions or when calcium is caused to be released from the internal stores (Gilbert *et al.*, 1975). This property of axoplasm was postulated by Shaw & Newby (1972) to be used in the release of synaptic transmitter substances. The vesicles were considered to be trapped in the gel meshwork, but allowed to diffuse to the release site when calcium entry temporarily broke down the gel. This hypothesis has been examined by laser scattering experiments. In accordance with the theory, invertebrate brain and nerve preparations were shown to exhibit a reversible increase in the level of Brownian motion when they were depolarised in potassium solutions. These experiments are discussed in Piddington (1976). The finding of increased particle mobility on depolarisation is quite repeatable but the mechanism is not understood, nor is there an obvious relevance to synaptic release. One of the problems was the impossibility of positively identifying the moving structures in the particular insect brain which was used.

Recently, in collaboration with Professor P. F. Baker, I have examined two simpler systems. One is the isolated axoplasm of *Myxicola* itself and the other is the bovine adrenal gland.

We have found that when a piece of axoplasm is liquefied by calcium there is indeed an increase in the level of Brownian motion. One of these experiments is shown in Piddington (1976). In an improved experiment done on the photon correlator, we have obtained an increase in correlation amplitude of over 10-fold on liquefaction. In an experiment such as this the correlation function changes from an essentially flat form to an exponential form – forms similar to those shown in the bottom two examples of Figs. 4 and 5. The correlation time constant for the dissolved sample was less than half that of the solid sample, indicating the expected presence in the dissolved sample of smaller particles, or lower molecular weight components (see Gilbert *et al.*, 1975). With this laser assay we hope to be able to find the threshold concentration of calcium needed to change the axoplasmic structure.

ADRENAL GLAND

The bovine adrenal medulla is a secretory system particularly suitable for the study of vesicular (or granular) movements. The cells are densely packed with adrenaline-containing vesicles which release their contents on triggering by calcium. Secretion from slices of tissue can be obtained *in vitro* and the vesicles are easily isolated by standard techniques (Fig. 7; Smith & Winkler, 1967).

It is possible to obtain a sufficiently smooth correlation function from a purified sample of vesicles in about 1 min; this is a fairly standard run time for this kind of experiment (Fig. 5: see Pusey, 1974). The optical time constant of the correlation is about 2 msec, and this corresponds to a particle size of about 300 nm diameter (we assume 'homodyning', see Cummins & Pike, 1974; Piddington, 1976). When the correlation is plotted on semi-log paper, a shallow curve deviating slightly from a straight line is obtained and we conclude that there probably exists a finite but narrow distribution of vesicle sizes in our extract (see Pusey, 1974).

We can easily detect the presence of a population of 300 nm particles in our crude extract (Fig. 7). These correlations also contain a slow component due probably to cell debris – we can mimic them by deliberately adding large particles to a pure vesicle extract ('bacteria', Fig. 7). The large particles appear to provide a 'heterodyne' signal and we have taken this into account (see Piddington, 1976).

A standardised measure of the correlation amplitude can be obtained by relating the exponential portion to the 'uncorrelated baseline' (see Cummins & Pike, 1974). (Fig. 3 shows an example of a correlation plus baseline.) The amount of baseline increases in proportion to the amount of fixed or slowly moving material present in a system and so the amplitude measured in this way can be used as a crude measure of the 'amount of motion' present in the system (see Piddington, 1976).

When we examine a slice or block of adrenal medulla we find a component of about 2 msec time constant as before, but this time of very low amplitude (Fig. 7). We can mimic this correlation fairly closely by restricting the movement of *isolated* vesicles by fast centrifugation or by setting them in agar (Fig. 5; 'agar, pellet', Fig. 7). The simplest explanation is that in the gland, the vesicles are restricted in their Brownian movement, by concentration or by a gel meshwork, but that they vibrate at a speed which is governed by the viscosity of water.

If we suspend some isolated vesicles in viscous glycerol–water mixtures we obtain longer time constants (as expected) and so the vesicles are sensitive to viscosity. In the gland, the vesicles could possibly contribute to the slow component, if they were in a viscosity seven times that of water. This is a less likely explanation as the correlation amplitude of this component appears to be too large. We note that a vesicular pellet looks and feels like a 'viscous liquid' on a macroscopic level but that the viscosity the vesicle sees, as measured by the laser, is the viscosity of water! Thus we tentatively conclude that the vesicles are merely *constricted* within the gland.

Our results on the adrenal gland are preliminary and we are checking our particle sizes on the electron microscope. We hope now to observe changes which correlate with the calcium-triggered release process.

Professor P. F. Baker and Drs D. A. Ross, J. P. Robinson, D. B. Sattelle and K. H. Langley collaborated in parts of this work. The photon correlator was supplied by Precision Devices, Malvern, England. Funded by the MRC.

REFERENCES

CHU, B. (1974). *Laser Light Scattering.* Academic Press: New York.

CLOSE, R. I. (1972). Dynamic properties of mammalian skeletal muscles. *Physiol. Rev.*, **52**, 129–197.

COSTANTIN, L. L. & TAYLOR, S. R. (1973). Graded activation in frog muscle fibres. *J. gen. Physiol.*, **61**, 424–443.

CUMMINS, H. Z. & PIKE, E. R. (eds) (1974). *Photon Correlation and Light-beating Spectroscopy* (Proc. NATO Advanced Study Group). Plenum Press: New York.

FEYNMAN, R. P., LEIGHTON, R. B. & SANDS, M. (1963). *The Feynman Lectures in Physics.* Addison Wesley: Reading, Mass.; Menlo Park, Calif.; London; Sydney; Manilla.

GABOR, D., KOCK, W. E. & STROKE, G. W. (1971). Holography. *Science, Wash.*, **173**, 11–23.

GILBERT, D. S. (1972). Helical structure of *Myxicola* axoplasm. *Nature, New Biol.*, **237**, 195–198.

GILBERT, D. S., NEWBY, B. J. & ANDERTON, B. H. (1975). Neurofilament disguise, destruction, and discipline. *Nature, Lond.*, **256**, 586–589.

MUSTACICH, R. V. & WARE, B. R. (1974). Observation of protoplasmic streaming by laser-light scattering. *Phys. Rev. Lett.*, **33**, 617–620.

PIDDINGTON, R. W. (1974). Motion in brain and motile cells. In *Photon Correlation and Light-beating Spectroscopy*, ed. H. Z. Cummins & E. R. Pike, pp. 573–574. Plenum Press: New York.

(1976). Laser light scattering from nerve and motile cells. In *Perspectives in Experimental Biology*, vol. **1**, ed. P. Spencer Davies. Proc. 50th Aniv. meeting Soc. Exp. Biol., pp. 313–322. Pergamon: Oxford.

PIDDINGTON, R. W. & ROSS, D. A. (1976). Laser measurement of auditory displacement. In *Sound Reception in Fish* (A Symposium), Developments in Aquaculture and Fisheries Science Series. Elsevier: Amsterdam, in press.

Pusey, P. N. (1974). Macromolecular diffusion. In *Photon Correlation and Light-beating Spectroscopy*, ed. H. Z. Cummins & E. R. Pike, pp. 387–428. Plenum Press: New York.

Shaw, T. I. & Newby, B. J. (1972). Movement in a ganglion. *Biochim. biophys. Acta*, **255**, 411–412.

Smith, A. D. & Winkler, H. (1967). A simple method for the isolation of adrenal chromaffin granules on a large scale. *Biochem. J.*, **103**, 480–483.

Williamson, R. E. (1975). Cytoplasmic streaming in *Chara*: a cell model activated by ATP and inhibited by cytochalasin B. *J. Cell Sci.*, **17**, 655–668.

THE REGULATION OF INTRACELLULAR CALCIUM

By P. F. BAKER

Department of Physiology, King's College,
Strand, London WC2R 2LS

Those attending this the symposium should not need convincing of the importance of intracellular calcium in the control of many aspects of cell behaviour and function. A knowledge of the various mechanisms involved in the regulation of intracellular calcium is essential for a complete understanding of calcium-dependent processes inside cells and it is my task today to review some of the more important findings in this field.

Much of my paper will be concerned with the regulation of intracellular calcium in the squid giant axon. This preparation has been chosen for intensive study mainly because of its large size – it is up to 1 mm in diameter and many centimetres in length – but also because of its relatively simple internal organization. Although the squid is a cephalopod mollusc, there is growing evidence that results obtained on squid axons may have a general relevance.

This paper will only attempt an overall survey of the regulation of intracellular calcium and for a more extensive discussion of particular topics the reader is referred to a number of recent reviews (Baker, 1972; Reuter, 1973; Blaustein, 1974; Baker & Reuter, 1975; Schatzmann, 1975; Ashley *et al.*, this volume).

THE CALCIUM CONTENT OF AXONS

What is the total concentration of calcium inside axons?

In any discussion of the regulation of intracellular calcium, it is essential to know the total quantity of calcium inside the cell. This is easily measured in squid axons because it is possible to obtain quite large samples of nerve protoplasm, axoplasm, uncontaminated by extracellular fluid. The axon is first dissected from the animal and cleaned of adhering tissue. The central end of the axon is then cut and, by gently squeezing the axon towards the cut end, axoplasm can be extruded rather like toothpaste out of a tube. Freshly extruded axoplasm has the consistency of a stiff gel and is easily picked up with forceps, weighed and analysed. Axoplasm obtained in this way lacks a surface membrane. During the extrusion process the axon membrane, axolemma, remains in the flattened sheath and the sheath can still conduct impulses. If the sheath is refilled with an isotonic solution of potassium salts at pH 7.2 the resulting perfused axon preparation is still excitable (Baker, Hodgkin & Shaw, 1962).

Chemical analysis of the extruded axoplasm using flame photometry and atomic absorption spectrophotometry reveals the following values (mmole kg^{-1} axoplasm): potassium, 400; sodium, 50; calcium, 0.4 and magnesium, 8. These should be compared with the concentration of the same ions in squid blood to which a close approximation is sea water which contains (mM): potassium, 10; sodium, 460; calcium, 11 and magnesium, 55. Insertion of a microelectrode into an axon reveals that the interior of the cell is negative with respect to the exterior by about 60 mV, which means that if purely electrical forces predominate, positively charged ions should be accumulated inside the axon. Of the four ions listed, only the distribution of potassium approximates at all closely to electrochemical equilibrium and even for potassium it is now well established that the high intracellular concentration is maintained by an ATP-consuming process that expels Na^+ in exchange for external K^+: the K^+-gradient so created serving to generate the resting potential.

From the stand point of my topic, the concentration of calcium in axoplasm is far from electrochemical equilibrium and may be much further than the measurements of total calcium suggest if not all the intracellular calcium is ionized.

What fraction of intracellular calcium is ionized?

The observation that a patch of ^{45}Ca injected into an axon broadens very little over a period of many hours and also fails to move in an electric field strongly suggests that the bulk of the intracellular calcium is not ionized (Hodgkin & Keynes, 1956; Baker & Crawford, 1972). This difference is illustrated in Figs. 1 and 2 where the behaviour of injected calcium and magnesium is compared in the same axon. The data suggest that the ionized calcium in axoplasm must be less than 10 μM and more sensitive techniques are needed to make a more accurate estimate.

The photoprotein aequorin provides just such a sensitive probe (see Shimomura & Johnson, this volume) although it is not possible to equate the light emitted by a given amount of aequorin directly with a particular concentration of ionized calcium because a variety of substances found inside cells interfere with the aequorin reaction (Baker, Hodgkin & Ridgway, 1971). The technique used was as follows. Aequorin was micro-injected axially into the centre of an axon and the resting light output monitored by placing the axon in front of a photomultiplier tube in a light-

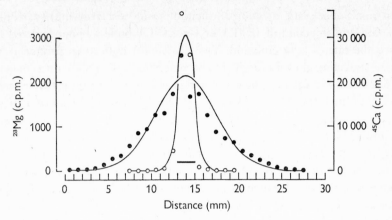

Fig. 1. Self-diffusion of ^{28}Mg (●) and ^{45}Ca (○) in axoplasm. Axon injected with 2 mm of ^{28}Mg, ^{45}Ca and left in artificial sea water at 21 °C for 510 min. The axon was then frozen and cut into 1 mm sections for counting. The cannula was at the left-hand end. The bar marks the nominal site of injection. It is likely that some tracer moved up the track left behind as the injector was withdrawn. The curves are calculated from the equation (Crank, 1956)

$$C = \tfrac{1}{2}C_0 \left(erf \frac{h+x}{2\sqrt{(Dt)}} + erf \frac{h-x}{2\sqrt{(Dt)}} \right),$$

where C is the concentration of tracer at time t and distance x from the middle of the injected patch, and C_0 is its concentration at $t = 0$. h is half the width of the injected patch. The smooth curve for ^{45}Ca is drawn with $D = 10^{-7}$ cm^2 sec^{-1} and for ^{28}Mg with $D = 2 \times 10^{-6}$ cm^2 sec^{-1}. Axon diameter 750 μm (from Baker & Crawford, 1972).

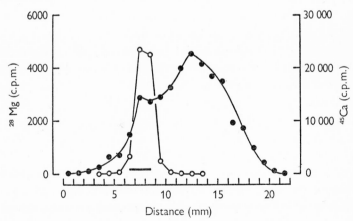

Fig. 2. Mobility of calcium (○) and magnesium (●) in axoplasm. The injector was inserted from the left-hand side and subsequently made the anode. The short horizontal bar marks the site of injection. 1.025 V was applied over 4.3 cm for 240 min. Temperature 19 °C. Axon diameter 790 μm. If magnesium has the same mobility in axoplasm as in free solution the peak should have moved 13.7 mm (from Baker & Crawford, 1972).

tight box. Subsequent injection of calcium produced a transient rise in light emission and injection of EGTA or Ca–EGTA buffers either lowered or raised the rate of light emission. The Ca–EGTA buffer that produced no change in resting glow was argued to be generating inside the axon a concentration of ionized calcium the same as that existing in the undisturbed cell. Assuming an intracellular pH of 7.2 (Caldwell, 1958) and ionized magnesium concentration of 3 mM (Baker & Crawford, 1972), the ionized calcium stabilized by the buffer that produced no change in light emission is close to 10^{-7} M (100 nM). The original estimate of Baker et al. (1971) was 3×10^{-7} M but this was based on the assumption that all the intracellular magnesium is ionized which is now known not to be the case (Baker & Crawford, 1972; Brinley & Scarpa, 1975).

DiPolo et al. (1976) have recently re-examined this question. Using axons obtained from living specimens of Loligo pealii they found values even lower than 10^{-7} M. This does not conflict with the data of Baker et al. (1971) as their axons were obtained from mantles that had been kept in refrigerated sea water for some hours. DiPolo et al. (1976) have also checked the value for ionized calcium obtained with aequorin with that obtained by intracellular injection of the calcium-sensitive dye Arsenazo III. The two values agree closely confirming the conclusion that the ionized-calcium inside squid axons is about 10^{-7} M.

Another estimate of intracellular ionized calcium would be to find the level of ionized calcium that just promotes secretion from the squid giant synapse. Although Miledi (1973) has shown that injection of calcium into the giant synapse promotes transmitter release, a full analysis of the effect of different Ca–EGTA buffers has yet to be made.

To summarize, of a total intracellular calcium in squid axons of 400 μM, only 0.1 μM is ionized. In absolute terms, this amounts to about 60 atoms of ionized Ca μm^{-3} of axoplasmic water.

THE MAINTENANCE OF THE STEADY STATE

General

The results that have just been described raise two important questions:
(1) Why is the total calcium in axoplasm only 400 μM?
(2) Why is only a very small fraction of this total calcium ionized?

One answer to the first question might be that the axon is completely impermeable to calcium. This is clearly not so. Immersion of an axon in sea water containing ^{45}Ca reveals a steady calcium influx of about 0.1 pmole cm^{-2} sec^{-1}. If this represents a net influx of calcium it would be enough to double the total intracellular calcium every 24 h. As squid live for many months, some mechanism or mechanisms must exist for stabilizing the total intracellular calcium at 400 μM and these mechanisms must be capable of pumping calcium out of the axon against a steep electrochemical gradient.

A complete answer to the second question is not possible, although it is of considerable physiological significance that the intracellular binding of calcium maintains such a low intracellular concentration of ionized calcium. As calcium is a potent activator of many intracellular events, very small absolute changes in ionized calcium may have profound effects on cell metabolism and behaviour.

This trigger action of calcium, first envisaged by Heilbrunn (1956), is facilitated by the low level of intracellular ionized calcium. The source of the calcium used for triggering changes in cell function or behaviour can be either intracellular or extracellular. A redistribution of intracellular calcium leading to a rise in ionized calcium happens in the activation of skeletal muscle where calcium is released from the sarcoplasmic reticulum and may possibly also occur in other derivatives of the endoplasmic reticulum and in mitochondria (see papers by Borle & Anderson and Carafoli & Crompton, this volume). The dependence of neurosecretion on external calcium is an example where the source of the triggering calcium is extracellular. Stimulation of the squid giant axon leads to an increased uptake of ^{45}Ca and there is a roughly linear relation between the calcium taken up per pulse and the calcium concentration in the bathing medium. Injection of aequorin into the axon reveals a rise in ionized calcium during stimulation. The use of intracellular aequorin to follow voltage-dependent changes in calcium permeability has recently been reviewed by Baker & Glitsch (1975) and only the main conclusions will be repeated here. These are summarized in Table 1. Depolarization promotes calcium entry in two phases that seem to be pharmacologically distinct: an early phase, blocked by concentrations of tetrodotoxin (TTX) that also block the sodium-channel and a later phase, insensitive to concentrations of TTX and tetraethylammonium ions that block the sodium and potassium channels respectively; but blocked by Mg^{2+}, Mn^{2+}, Co^{2+} and the organic antagonists of calcium, D600 and Iproveratril, at concentrations that have little or no effect on the sodium and potassium channels. The early, TTX-sensitive, phase seems to reflect calcium entering the axon through the sodium channels of the action potential. The properties of the late phase seem different from those of the sodium and potassium channels of the action potential but reveal a striking similarity between these channels and the calcium channels that normally control transmitter release at nerve terminals.

As calcium plays such an important role as a trigger substance for a variety of aspects of cell function and behaviour, control over its entry into

Table 1. *Comparison of some properties of early and late components of depolarization-induced calcium entry in squid axons with those of transmitter release from the squid giant synapse and other nerve terminals (see Baker, 1972)*

	Early calcium entry (via the sodium channel)	Late calcium entry (via the late calcium channel)	Transmitter release
Tetrodotoxin	Blocks	No effect	No effect
Tetraethyl-ammonium ions	No effect	No effect	Increased (action potential lasts longer)
Mg_0^{2+} Mn_0^{2+} Co_0^{2+}	Little or no effect	Blocked	Blocked
La_0^{3+}	Some reduction	Blocked	Blocked (but also prolongs action potential)
D600 Iproveratril	Some reduction	Blocked	Blocked
Maintained depolarization	Inactivates rapidly	Inactivates slowly	Slowly inactivates in some systems. In others not yet certain

the cell may be an important regulatory site. There is now strong evidence that calcium entry into cardiac muscle is affected by cyclic nucleotides (Reuter, 1974) and indirect evidence that a similar mechanism may modulate the late calcium channel in nerve terminals. The evidence to date suggests that a rise in intracellular cyclic AMP promotes calcium entry through the voltage-sensitive late calcium channels.

Intracellular binding

During a period of nervous stimulation the intracellular ionized calcium rises to a steady level, but falls close to its resting value once the stimulation ceases. The rise in ionized calcium is seen only in media containing calcium and seems to reflect calcium entry into the axon. The decline following stimulation is about half complete in 15 sec at 20 °C. The recovery of the intracellular ionized calcium is largely unaffected by the ionic composition of the external media and seems to reflect intracellular binding of calcium. This rapid buffering of calcium in axoplasm is probably the major mechanism involved in terminating transmitter release although it cannot be the only mechanism involved in regulating internal calcium or the cell would accumulate calcium, which seems not to be the case. The long-term maintenance of intracellular calcium seems to be effected by the plasma membrane.

A convenient preparation that enables the mechanisms involved in intra-cellular binding to be studied in isolation from any regulatory functions of the plasma membrane is isolated axoplasm. Axoplasm extruded from a squid axon can be drawn into a tube (pyrex glass or polythene) of roughly the same dimensions as the axon. Subsequent injection of aequorin followed by Ca–EGTA buffers reveals that the preparation can maintain an ionized calcium concentration close to that which normally exists inside intact axons, and in the absence of EGTA, injection of enough calcium to raise the total axoplasmic calcium by 60 μmoles kg^{-1} leads to only a transient increase in light emission indicating that isolated axoplasm is capable of regulating its ionized calcium in response to a calcium challenge.

As mitochondria are known to accumulate calcium, they must be obvious candidates for a regulatory role in axoplasm. Squid axoplasm contains some mitochondria, and the experiments shown in Fig. 3 provide evidence for their involvement in intracellular buffering. Poisoning either an axon or isolated axoplasm with cyanide ultimately results in a rise in ionized calcium which is reversed on removing the poison. Fig. 3a shows that light-emission rises only after a delay and this is explicable in terms of the time taken for intracellular ATP to fall to a low value. The squid axon is part of a very large invertebrate cell and it takes about 1 h after addition of cyanide for the ATP level to decline appreciably (Caldwell, 1960; Baker & Shaw, 1965). Injection of ATP into a poisoned axon produces a prompt reduction in ionized calcium. ATP-dependent uptake of calcium by mito-chondria is blocked by oligomycin and the experiment illustrated in Fig. 3b shows that, in the presence of oligomycin, application of cyanide produces an immediate rise in ionized calcium and subsequent injection of ATP has only a very small effect.

Despite this strong evidence for an important role of mitochondria in intracellular calcium binding, mitochondria seem not to be the only calcium binding system in axoplasm. Thus in both axons and isolated axoplasm poisoning does not liberate all the intracellular calcium in ionized form and in some fully poisoned axons as little as 10 % of the total calcium may be ionized.

In an attempt to obtain more information about calcium binding in axoplasm, Baker & Schlaepfer (1975) have investigated the uptake of ^{45}Ca into isolated axoplasm held in small dialysis sacs and immersed in 'artificial axoplasm' of basic composition: KCl, 350 mM; sucrose, 500 mM; MgCl$_2$, 10 mM; sodium succinate, 10 mM; sodium phosphate, 10 mM; ATP, 5 mM; CaCl$_2$, 5–10 μM; Tris buffer, pH 7.2, 10 mM. Axoplasm accumulated ^{45}Ca from this medium and accumulation ratios (^{45}Ca axoplasm/^{45}Ca medium) often exceeded 30. Removal of ATP, P$_i$ and succinate and addition of

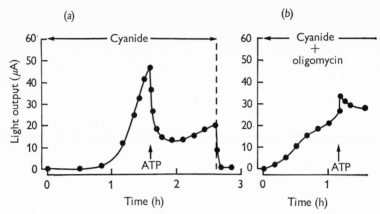

Fig. 3. Effects of cyanide and injected ATP on intracellular ionized calcium in squid axons. Ordinate: light output (μA) from intracellular aequorin. Abscissa: time (h). In (a) the axon was exposed to artificial sea water (ASW) containing 2 mM cyanide at zero time and was returned to cyanide-free ASW at the time marked by a vertical dotted line. In (b), the axon was pretreated for 30 min with oligomycin by injection to give a final concentration of 50 μg ml^{-1} and also by inclusion of 5 μg ml^{-1} in the ASW. At zero time 2 mM cyanide was added to the oligomycin–ASW bathing the axon. In both (a) and (b), ATP was injected at the vertical arrows to give a final concentration of 30 μM. Both axons contained the same amount of aequorin. Temperature 21 °C (from Baker *et al.*, 1971).

cyanide (2 mM) and oligomycin (5 μg ml^{-1}) reduced but failed to abolish accumulation, and provided a convenient means of distinguishing two components of intracellular calcium binding: a component that requires ATP and P_i or succinate and P_i, that is sensitive to cyanide and oligomycin and is identical in properties to a crude preparation of mitochondria isolated from squid axoplasm; and a second component that persists under conditions where mitochondrial uptake has been abolished, and probably reflects binding of calcium to some other axoplasmic constituent.

The properties of these two components of calcium binding are quite different. Calcium uptake by mitochondria both isolated and *in situ* is half-maximal in the calcium concentration range 20–40 μM and uptake continues for many hours. The presumed non-mitochondrial binding equilibrates rapidly with the available calcium and it is possible to distinguish a high-affinity binding component that has a capacity of 20–40 μmole kg^{-1} axoplasm and is half-saturated with calcium at about 0.5 μM, and a low-affinity component that is not saturated at 500 μM Ca (Fig. 4). This latter system may be related to that described by Alema, Calissano, Rusca & Guiditta (1973) in squid axoplasm.

Axoplasmic calcium binding is dependent on the ionic composition of the medium. Thus uptake is highest in potassium and lower in sodium,

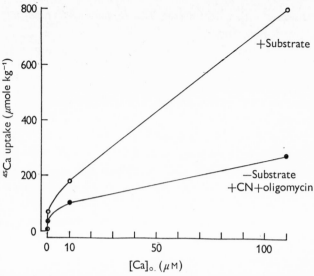

Fig. 4. ^{45}Ca uptake by isolated axoplasm. Axoplasm held in small dialysis bags in artificial axoplasm of basic composition described in text. Calcium or EGTA was added to give the calcium concentrations indicated and incubation was either in the presence of ATP, succinate and inorganic phosphate or in the absence of these substrates together with the addition of cyanide (2 mM) and oligomycin (5 μg ml^{-1}). Incubation was for 4 h in all instances. Temperature 20 °C. (Unpublished results of P. F. Baker & W. Schlaepfer.)

lithium or Tris and uptake is reduced by heavy metals such as La^{3+} and Mn^{2+}. The mitochondria seem most sensitive to the monovalent cation composition of the medium whereas the non-mitochondrial binding system seems most sensitive to lanthanum; but it should be noted that most of the experiments were done at 10 μM Ca, which is close to the apparent K_m for calcium uptake by mitochondria but far in excess of that of the high-affinity binding system and the measurements should be repeated closer to the K_m of this system.

This data is of interest for a number of reasons:

(1) It reveals the existence in isolated axoplasm of a calcium binding system of high affinity. This system has limited capacity but may well be important in rapid buffering of the ionized calcium close to its physiological concentration and as such may play a major role in returning the ionized calcium to its resting concentration after a period of nervous activity. The nature of this high-affinity calcium binding system has not yet been determined. A system of similar properties but much smaller capacity can also be detected in axoplasm from giant axons of the polychaete worm *Myxicola* and it might be significant that the quantity of high affinity calcium binding in squid and polychaete axons roughly parallels the abundance of microtubules in the two preparations. Microtubules are

known to interact with calcium (see *The Biology of Cytoplasmic Tubules*, Ann. N.Y. Acad. Sci., 1975).

(2) It shows that both in intact axoplasm and following isolation from axoplasm, mitochondria have a rather low affinity for calcium when immersed in a medium of composition roughly similar to that existing in an intact axon. Despite this low affinity, mitochondria have a large capacity for calcium and never became saturated during the course of our experiments (up to 4 h). This suggests that, despite their unfavourable K_m, given enough time mitochondria may accumulate considerable amounts of calcium at the concentration of ionized calcium that exists in resting cells. As the apparent K_m of mitochondria for calcium is much greater than the intracellular ionized Ca, any increase in intracellular Ca will lead to a corresponding increase in the rate of mitochondrial calcium uptake. This must contribute, together with the high-affinity binding system, to the termination of transmitter release following nervous activity and may provide a system for protecting the intracellular environment from exposure to high concentrations of ionized calcium.

(3) If the 400 μM Ca in squid axons is distributed between the various binding systems that have been described, at rest the bulk of the calcium must be in the mitochondria. Although poisoning presumably liberates calcium from the mitochondria, the presence of at least two other calcium-binding systems in axoplasm – both of which can operate in the absence of ATP – will ensure that, at most, only about $\frac{1}{4}$ of the total calcium will appear as ionized calcium.

(4) The sensitivity of axoplasmic calcium binding to both monovalent and divalent cations provides examples of mechanisms by which intracellular calcium may be redistributed. Such redistribution may have considerable physiological significance.

It is well known that exposure of the frog neuromuscular junction to either Li–Ringer or Ringer solution containing cardiac glycosides leads to an increase in the spontaneous rate of transmitter release (Birks & Cohen, 1968; Baker & Crawford, 1975; Crawford, 1975). This occurs in the absence of external Ca^{2+} and seems to reflect a redistribution of calcium inside the nerve terminal. Such a redistribution is to be expected from the present observations on isolated axoplasm. In mechanistic terms, it seems possible that the mitochondria may be more permeable to Na^+ and Li^+ than to K^+ and a rise in cytoplasmic sodium or lithium leads to a reduction in the mitochondrial potential. As the mitochondrial potential is thought to provide the energy for calcium accumulation, any reduction in potential will produce a concurrent fall in mitochondrial calcium and a corresponding rise in ionized calcium in the cytosol.

Exposure of a number of neurosecretory tissues to media containing a variety of polyvalent cations such as La^{3+}, Mn^{2+}, Co^{2+} and Pb^{2+} has two effects. It promotes the spontaneous secretion of transmitter and blocks the evoked release. The latter effect probably stems from blockage of the voltage-sensitive calcium channels through which Ca^{2+} enter the nerve terminal and, in view of the present results, it seems possible that the increased spontaneous release of transmitter may reflect a rise in intracellular ionized calcium resulting from both displacement of calcium from high-affinity binding sites and also inhibition of calcium uptake by the mitochondria.

These experiments on calcium-binding in axoplasm will probably find parallels in many other cells once methods become available for their study. Thus the elegant experiments of Rose & Loewenstein (1975) on *Chironomus* salivary glands indicate that in this tissue injected calcium is bound very rapidly resulting in discrete localization of the added calcium. Mitochondria seem to play an important part in sequestering calcium in salivary glands: but in other tissues it is quite probable that derivatives of the endoplasmic reticulum analogous to the sarcoplasmic reticulum of muscle might be involved, as might also be the nucleus and perhaps other organelles in some cells.

Before moving from calcium binding systems in axoplasm to control by the cell membrane of the total calcium available to these binding systems, it seems pertinent to raise a general point. The same general rules that apply to intact cells should also be applicable to membrane-bound intracellular organelles, in that, given enough time, calcium should distribute itself across the membrane of these organelles in equilibrium with the existing electrical potential. According to current views (see Carafoli & Crompton, this volume) respiring mitochondria generate a potential difference (inside negative) and it is this potential that provides the energy for Ca accumulation. It follows that the ratio (ionized Ca in inner mitochondrial space):(ionized Ca in cytosol) should be given by the Nernst equation,

$$E_{(\text{cytosol minus mitoch.})} = 29 \log_{10} \frac{[Ca]_{\text{mitoch.}}}{[Ca]_{\text{cytosol}}}.$$

If mitochondria generate a p.d. of about 150 mV, inside negative, Ca^{2+} inside should be about 10 mM and the total intramitochondrial calcium may be even higher, as, in the presence of phosphate, calcium phosphate precipitates within the mitochondrion. There is no evidence that the intramitochondrial ionized calcium ever reaches this value and probably does not rise above 10 μM (see Carafoli & Crompton, this volume) and the obvious question is why not? Two possibilities are (1) that mitochondria contain an outwardly directed calcium pump and (2) that the mitochondrial potential is not stable. Perhaps as intramitochondrial calcium rises, it inhibits mitochondrial respiration and this in turn causes a collapse of the potential and subsequent release of sequestered calcium. Both possibilities deserve close examination as they might provide important sites of action for agents effecting redistribution of intracellular calcium. Alternatively, the motive force for calcium accumulation within mitochondria may not simply be the mitochondrial potential.

Membrane transport

Although intracellular binding is largely responsible for the rapid buffering of calcium that enters an axon – for instance immediately following a period of nervous activity – over a longer period calcium extrusion from the cell must exactly balance calcium entry otherwise intracellular calcium would not stay constant. As the extra entry of calcium associated with nervous conduction in squid axons is very small (0.01 pmole cm^{-2} pulse), a stable intracellular calcium concentration could be maintained easily by a calcium extrusion mechanism that works at a rather slow and almost constant rate. A more active calcium extrusion mechanism may be required at nerve terminals where a larger calcium entry may be required to evoke secretion (see for instance Katz & Miledi, 1969).

Perhaps the simplest unifying mechanism for calcium extrusion would be to make use of one of the intracellular binding systems already described and arrange every now and then for their bound calcium to be dumped into the extracellular fluid – perhaps by exocytosis. For instance, calcium could be taken up by mitochondria until their calcium content exceeds that in the extracellular fluid. A net extrusion of calcium would be accomplished if these mitochondria underwent something analogous to exocytosis exchanging their bound calcium for extracellular fluid. There is no experimental evidence in favour of this mechanism. Indeed blockage of mitochondrial calcium binding in squid axons leads to an increased rate of calcium extrusion (Blaustein & Hodgkin, 1969).

All the available data favours the existence of calcium pumps located in the plasma membrane. There seem to be two types of calcium pump: those that obtain their energy from the hydrolysis of ATP and those that can operate in the virtual absence of ATP and seem to derive energy for calcium extrusion from the Na$^+$ and perhaps K$^+$ gradients. It is still too early to decide whether these two processes are distinct or merely two extremes of operation of the same basic system. Thus ATP-dependent calcium extrusion may be sensitive to monovalent cations (see review by Schatzmann, 1975) and sodium-activated calcium extrusion is affected by the intracellular level of ATP (Baker & Glitsch, 1973; DiPolo, 1974; Baker & McNaughton, 1976). ATP-dependent calcium extrusion is found in erythrocytes and various cultured cells, whereas the sodium-activated extrusion seems to be present in most excitable cells and also in those epithelial cells that effect calcium transport (see reviews by Baker, 1972; Blaustein, 1974). It may be significant that sodium-activated calcium transport is found in those areas where calcium must be extruded against a very steep electrochemical gradient. For the rest of this paper I shall concentrate on sodium-activated calcium transport and the interested reader should consult the excellent review by Schatzmann (1975) for more information on the ATP-dependent mechanism.

Evidence for the sodium-activated extrusion was obtained first on the squid giant axon (Baker, Blaustein, Hodgkin & Steinhardt, 1967). Examination of the calcium efflux revealed two major components present in roughly equal amounts: one activated by external Ca^{2+} and the other by external Na^+. Both components have a high Q_{10}, are insensitive to concentrations of cardiac glycosides that inhibit the sodium pump and both are inhibited by low concentrations of lanthanum (Blaustein & Hodgkin, 1969; van Breemen & de Weer, 1970). The simplest interpretation of these two components is that calcium-activated calcium efflux probably reflects an exchange of internal for external calcium and sodium-activated calcium efflux may reflect calcium extrusion in exchange for external sodium; but clear proof of these exchanges is still lacking in intact axons (see later).

If these components of the calcium efflux require ATP, poisoning an axon with cyanide ought to inhibit calcium extrusion. This is not the case. As Rojas & Hidalgo (1968) and Blaustein & Hodgkin (1969) showed, poisoning results in a large (10-fold) rise in the calcium efflux. The efflux from poisoned cells still consists of calcium-activated and sodium-activated components. The rise in efflux occurs at a time when the intracellular ionized calcium is elevated because of release of calcium from intracellular (largely mitochondrial) binding sites (Baker et al., 1971); but the observation that a higher rate of calcium extrusion can be maintained in poisoned cells suggests that calcium transport must be rather insensitive to intracellular ATP, and led to the suggestion that the energy for calcium extrusion may be obtained from the Na^+ and possibly K^+ gradients and only indirectly from ATP through the maintenance of these gradients. Thus if the inward movement of sodium down the electrochemical gradient is linked to the outward movement of calcium, a calcium gradient can be established in the opposite direction to that of sodium. The steepness of the gradient will depend on the coupling between Na^+ and Ca^{2+}. The simplest form of coupling would be an electroneutral exchange of 2 Na^+ for 1 Ca^{2+}. At equilibrium the calcium gradient generated would approach

$$\frac{[Ca]_i}{[Ca]_o} = \frac{[Na]_i^2}{[Na]_o^2},$$

where o and i denote external and internal concentrations respectively. As $[Na]_i/[Na]_o$ is roughly $1/10$, $[Ca]_i/[Ca]_o$ would approach $1/100$ which is not enough. In order to achieve the calcium gradient that exists under physiological conditions, the extrusion of 1 Ca^{2+} must be coupled to the entry of at least 3 Na^+. Exchange could be either electrogenic with 1 Na^+ ion entering the cell for each Ca^{2+} extruded ($[Ca]_i/[Ca]_o$ approaching $([Na]_i^3/[Na]_o^3) \exp.^{EF/RT}$) or electroneutral with 1 K^+ leaving the cell together with each Ca^{2+} extruded ($[Ca]_i/[Ca]_o$ approaching $[Na]_i^3.[K]_o/[Na]_o^3.[K]_i$). An electrogenic extrusion of Ca^{2+} is suggested by the recent observation that the sodium-activated, but not the calcium-activated, calcium efflux is inhibited by a reduction in membrane potential, i.e. depolarization (Blaustein, Russell & de Weer, 1974; Mullins & Brinley, 1975; Baker & McNaughton, 1976). The bulk of these measurements made use of extracellular potassium as a depolarizing agent and under these conditions it is not possible to differentiate effects of potassium *per se* from those of potential.

If calcium is extruded in exchange for external sodium, it should be possible to demonstrate an influx of sodium associated with the efflux of calcium. Unfortunately, the expected flux is extremely small (< 0.5 pmole $cm^{-2} sec^{-1}$) and would be difficult to detect in the presence of a resting sodium influx of about 40 pmole $cm^{-2} sec^{-1}$. Blaustein & Russell (1975) have examined this problem in dialysed axons where they could raise $[Ca]_i$ to levels much higher than ever exist under physiological conditions. With 500 μM intracellular calcium, they were able to detect a sodium influx that depended on internal calcium and was roughly three times the size of the $[Na]_o$-activated calcium efflux under the same conditions.

Another approach is to reverse the sodium gradient. If entry of sodium into the cell is associated with calcium efflux, movement of sodium out of the cell ought to be associated with calcium influx and this can be demonstrated under certain conditions. Immersion of axons in sodium-free artificial sea water (ASW) results in a large increase in the influx of calcium and under these same conditions it is possible to detect an external calcium-dependent component of the sodium efflux (Baker *et al.*, 1969). Both internal sodium-dependent calcium influx and external calcium-dependent sodium efflux are increased by raising $[Na]_i$.

The increased calcium influx during exposure to sodium-free solutions

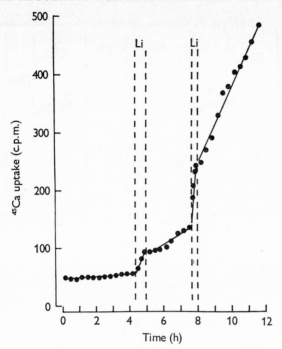

Fig. 5. Continuous measurement of ^{45}Ca uptake into a squid axon. The axon was impaled axially with a rod of scintillator glass (Glass Scintillator Type G.S.F.1, Koch Light Laboratories) and the preparation mounted in the well of a conventional liquid scintillation counter. The axon was immersed in Na–ASW containing (mM); NaCl, 460; KCl, 10; MgCl$_2$, 55; CaCl$_2$, 11 and NaHCO$_3$, 2.5, except for the periods indicated when sodium was replaced isosmotically by lithium. Note the marked increase in ^{45}Ca uptake in lithium solution and its reversibility on return to Na–ASW. Temperature 23 °C. Axon diameter 700 μm. By calibrating the internal scintillator in the external solution and knowing the diameter of the axon in the recording region, it is possible to express the rate of gain of counts as a flux. The first exposure to lithium increased the influx from 0.08 to 3.2 pmole cm^{-2} sec^{-1} and the second from 0.23 to 7.6 pmole cm^{-2} sec^{-1}. In each case on return to Na–ASW the flux was reduced, but not to its value before exposure to lithium (from Baker & Glitsch, 1975).

and its reversal on return to Na–ASW is most clearly demonstrated by use of an internal scintillator (Fig. 5). As exposure to sodium-free solutions leads to an increase in Ca influx, and a decrease in Ca efflux, the intracellular ionized calcium should rise. This can be observed directly in aequorin loaded axons. Fig. 6 provides direct evidence for the importance of the surface membrane in regulating the intracellular calcium and the inability of intracellular binding systems to maintain a constant internal ionized calcium in the face of a maintained alteration in calcium fluxes at the plasma membrane.

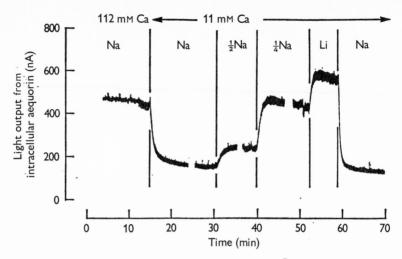

Fig. 6. Changes in intracellular ionized calcium in a squid axon immersed in media of different sodium or calcium contents. The axon was injected with aequorin and mounted in a flow cell. The external solutions were ASWs in which the major cation was either Na⁺, Li⁺ or choline. An increase in light-output (ordinate) reflects a rise in ionized calcium inside the fibre. Squid axon diameter 700 μm. Temperature 20 °C (from Baker *et al.*, 1971).

If the plasma membrane plays a central role in the regulation of intracellular calcium, it must be capable of picking up calcium at the calcium concentrations existing inside the cell. This is clearly the case as sodium-activated calcium efflux is observed in intact unpoisoned axons; but the observation that this efflux can increase 10-fold on poisoning suggests that the system is working well below its maximum rate. Experiments on intact axons suggest that the apparent K_m for internal calcium is probably about 1 μM (Baker & McNaughton, 1976) whereas a somewhat higher value of 3–10 μM is found in poisoned axons and axons dialysed in the absence of ATP (DiPolo, 1973; Blaustein & Russell, 1975; Brinley, Spangler & Mullins, 1975; Baker & McNaughton, 1976).

All the observations discussed so far are consistent with exchange of one intracellular Ca^{2+} for three external Na^+, probably in an electrogenic fashion, in a reaction that is not very sensitive to the level of ATP inside the cell. This conclusion is supported by experiments on dialysed axons (DiPolo, 1973) where sodium-activated calcium efflux persists in axons dialysed for many hours with solutions lacking ATP or other substrate.

Nevertheless, there is growing evidence for an involvement of ATP, or some derivative of it, in calcium transport in squid axons (Baker, 1972; Baker & Glitsch, 1973; DiPolo, 1974; Baker & McNaughton, 1976). The evidence can be summarized:

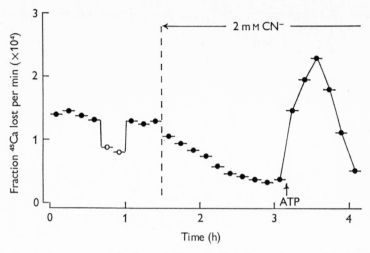

Fig. 7. Dependence on metabolism of calcium efflux from intact squid axon. Axon injected with a mixture containing 100 mM CaCl₂ and 200 mM K–EGTA, pH 7.2 and subsequently exposed to cyanide (2 mM). ●, Na–ASW; ○, Ca-free Li–ASW. ATP was injected at the time shown to give a final concentration of 2 mM. Temperature 20 °C.

(1). Cyanide completely inhibits the external calcium-dependent sodium efflux (Caldwell, Hodgkin, Keynes & Shaw, 1960; Baker *et al.*, 1969).

(2). Application of cyanide sometimes produces a fall in calcium efflux before the large rise.

(3). Injection of the ATP-destroying enzyme apyrase causes a fall in calcium efflux.

(4). If intracellular calcium is stabilized at its physiological value by injection of a suitable Ca–EGTA buffer, subsequent exposure to cyanide always results in a maintained inhibition of the calcium efflux. Both sodium-activated and calcium-activated components fall to about 30% of their initial value and recover on removing the cyanide. The efflux from cyanide-poisoned axons can be reactivated by injection of ATP (Fig. 7) but not AMP or cAMP.

(5). Inclusion of ATP in dialysed axons increases both calcium- and sodium-activated components of the calcium efflux. Under these conditions, the apparent K_m for internal ATP is about 600 μM.

(6). Detailed examination of the kinetics of calcium extrusion from intact axons reveal striking differences between unpoisoned and poisoned axons. In both unpoisoned and poisoned axons, activation of calcium efflux by external calcium occurs along a section of a rectangular hyperbola; but the calcium concentration required for half-maximal activation is about 2 μM in unpoisoned axons and 5 mM in fully poisoned

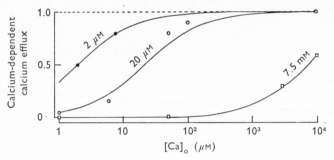

Fig. 8. Dependence on metabolism of the apparent affinity for external calcium of external calcium-activated calcium efflux in intact squid axon. Normalized calcium-dependent calcium efflux plotted against the log of the external calcium concentration. The apparent $K_m^{Ca_o}$ is shown on each curve. All data from the same axon. The top curve was obtained in the unpoisoned state, the middle curve 2–3 h after application of cyanide (2 mM), and the bottom curve 5 h later. Axon diameter 800 μm. Temperature 20 °C (from Baker & McNaughton, 1976).

cells (Fig. 8). The activation of calcium efflux by external sodium also occurs along a section of a rectangular hyperbola in unpoisoned axons but changes to a markedly sigmoidal curve in poisoned axons, the sodium concentration required for half-maximal activation increasing from about 60 mM in unpoisoned axons to 300 mM in poisoned cells. These changes in kinetics occur in a progressive manner, are seen when $[Ca]_i$ is held constant and can be reversed either by removing the poison or by injection of ATP into a poisoned axon.

Taken together, this data strongly suggests that ATP, or some derivative of it, must participate in calcium extrusion and the observed differences in kinetics can be explained on this basis. The argument is clearest in the case of calcium-activated calcium efflux where activation occurs in a Michaelis–Menten fashion. The dependence of the apparent K_m for external calcium in the metabolic state of the cell can be explained quantitatively if it is assumed that for an external Ca^{2+} to activate calcium efflux, an ATP or some derivative of it must first bind to the transport system – presumably at its inner face. Thus if the transport site is designated M,

$$M + ATP \underset{K_m^{ATP}}{\rightleftharpoons} M.ATP \underset{K_m^{Ca_o}}{\overset{Ca_o}{\rightleftharpoons}} M.ATP.Ca_o \longrightarrow Ca \text{ efflux},$$

assuming that the transport step is rate-limiting,

$$\frac{v}{V_{max}} = \frac{1}{1 + K_m^{Ca_o}(1 + K_m^{ATP}/[ATP])/[Ca]_o}$$

and the apparent K_m for external calcium,

$$K_m^{Ca_o}(\text{apparent}) = K_m^{Ca_o}(1 + K_m^{ATP}/[ATP]).$$

Thus as $[ATP]_i$ falls to vanishingly low levels, the apparent K_m for $[Ca]_o$ rises to very high levels. Taking

$$K_m^{ATP} = 0.5 \text{ mM}, \quad [ATP]_i = 3 \text{ mM}, \quad K_m^{Ca_o} = 2 \ \mu\text{M},$$

the apparent $K_m^{Ca_o}$ is $2.3 \ \mu\text{M}$. If poisoning reduces ATP, to $1 \ \mu\text{M}$, the apparent $K_m^{Ca_o}$ will increase to 1 mM which is the order of change observed. In an intact cell, poisoning increases the concentration of breakdown products of ATP and if these, in turn, compete with ATP for its binding site, the effective concentration of ATP in poisoned axons may be lower than its measured value.

It seems possible to apply a similar argument to the change in apparent affinity for external sodium on poisoning, although here the situation is complicated by the change in the form of the activation curve from a rectangular hyperbola to a sigmoidal relation despite the presumed involvement of 3 Na^+ – as evidenced by sensitivity to potassium depolarization – under both sets of conditions. One way round the problem is to assume that, in the unpoisoned axon, two of the three external sodium binding sites have a very much higher affinity for sodium than the third and it is possibly the affinity of these two sites that is mainly reduced in the poisoned cell. Further support for the suggestion that sodium activation of calcium efflux from squid axons may involve more than one sodium binding site comes from the finding that in a number of other tissues, including muscle, synaptosomes and adrenal medulla, the activation of calcium efflux by external sodium is sigmoidal even in unpoisoned cells.

The essential feature of the kinetic argument mentioned above is that ATP – or some derivative of it – must bind to the transport site *before* external calcium can activate calcium efflux. It is immaterial to the argument whether or not external calcium is transported or ATP is hydrolysed; but the observation that ATP, or some derivative of it, participates in calcium extrusion reopens the question whether ATP hydrolysis could be necessary for sodium-activated calcium transport because traces of ATP remain even in fully poisoned and dialysed axons. It is perhaps worth noting that under suitable conditions both the Na–K-ATPase of plasma membranes and the Ca-ATPase of sarcoplasmic reticulum can be activated by micromolar concentrations of ATP.

One way to examine the possible involvement of ATP hydrolysis is to use analogues of ATP that cannot be hydrolysed. Two analogues, AMP–PCP and AMP–PNP, both increase the calcium efflux from cyanide-poisoned axons with constant internal calcium concentrations. Unfortunately, both analogues contain traces of ATP and their injection may block the breakdown of endogenous ATP and in so doing lead to a rise in intracellular

ATP. Further work, especially on dialysed axons, is needed to decide whether or not ATP is hydrolysed and it would seem worth re-examining whether nerve cell membranes contain an ATPase that requires both calcium and sodium for activation.

THE RELATIVE IMPORTANCE OF INTRACELLULAR BINDING AND MEMBRANE TRANSPORT

The theme of this paper has been to stress that both intracellular binding and membrane transport contribute to the maintenance of a stable intracellular level of ionized calcium. As the bulk of the intracellular calcium is bound, any alteration in the various rate constants governing binding should lead to a redistribution of intracellular calcium with concomitant and perhaps dramatic changes in ionized calcium. Release of 1% of the bound calcium in the axoplasm would elevate the cytosolic ionized calcium 40-fold. But important as these mechanisms must be in the short-term control of intracellular ionized calcium and in the regulation of calcium-dependent aspects of cell function and behaviour, some other mechanism must exist for controlling the total calcium available for intracellular binding and this mechanism seems to be the calcium pumps located in the plasma membrane. Energetically, these pumps are capable of maintaining an intracellular ionized calcium of about $0.1\ \mu M$ and measurement of their affinity for intracellular calcium shows that they can respond with an increased rate of pumping to an elevation in intracellular ionized calcium of up to 10 times the physiological. It follows that they have the required characteristics to provide long-term regulation of the total calcium available to the various intracellular binding systems.

This work was supported by a grant to P.F.B. from the Medical Research Council.

REFERENCES

ALEMA, S., CALISSANO, P., RUSCA, G. & GUIDITTA, A. (1973). Identification of a calcium-binding, brain specific protein in the axoplasm of squid giant axons. *J. Neurochem.*, **20**, 681.

BAKER, P. F. (1972). Transport and metabolism of calcium ions in nerve. *Prog. Biophys. molec. Biol.*, **24**, 177–223.

BAKER, P. F., BLAUSTEIN, M. P., HODGKIN, A. L. & STEINHARDT, R. A. (1967). The effect of sodium concentration on calcium movements in giant axons of *Loligo forbesi. J. Physiol., Lond.*, **192**, 43P.

(1969). The influence of calcium on sodium efflux in squid axons. *J. Physiol., Lond.*, **200**, 431–458.

BAKER, P. F. & CRAWFORD, A. C. (1972). Mobility and transport of magnesium in squid giant axons. *J. Physiol., Lond.*, **227**, 855–874.

(1975). A note on the mechanism by which inhibitors of the sodium pump accelerate spontaneous release of transmitter from motor nerve terminals. *J. Physiol., Lond.*, **247**, 209–226.

BAKER, P. F. & GLITSCH, H. G. (1973). Does metabolic energy participate directly in the Na^+-dependent extrusion of Ca^{2+} ions from squid giant axons? *J. Physiol., Lond.*, **233**, 44–46P.

(1975). Voltage-dependent changes in the permeability of nerve membrane to calcium and other divalent cations. *Phil. Trans. R. Soc. B.*, **270**, 389–409.

BAKER, P. F., HODGKIN, A. L. & RIDGWAY, E. B. (1971). Depolarization and calcium entry in squid giant axons. *J. Physiol., Lond.*, **218**, 709–755.

BAKER, P. F., HODGKIN, A. L. & SHAW, T. I. (1962). Replacement of the axoplasm of giant nerve fibres with artificial solutions. *J. Physiol., Lond.*, **164**, 330–354.

BAKER, P. F. & McNAUGHTON, P. A. (1976). Kinetics and energetics of calcium efflux from intact squid giant axons. *J. Physiol., Lond.*, in press.

BAKER, P. F. & REUTER, H. (1975). *Calcium Movement in Excitable Cells*. Pergamon Press: Oxford.

BAKER, P. F. & SCHLAEPFER, W. (1975). Calcium uptake by axoplasm extruded from giant axons of *Loligo*. *J. Physiol., Lond.*, **249**, 37–38P.

BAKER, P. F. & SHAW, T. I. (1965). A comparison of the phosphorus metabolism of intact squid nerve with that of the isolated axoplasm and sheath. *J. Physiol., Lond.*, **180**, 424–438.

BIRKS, R. I. & COHEN, M. W. (1968). The action of sodium pump inhibitors on neuromuscular transmission. *Proc. R. Soc. Lond. B*, **170**, 381–399.

BLAUSTEIN, M. P. (1974). The interrelationship between sodium and calcium fluxes across cell membranes. *Rev. Physiol. Biochem. exp. Pharmacol.*, **70**, 33.

BLAUSTEIN, M. P. & HODGKIN, A. L. (1969). The effect of cyanide on the efflux of calcium from squid axons. *J. Physiol., Lond.*, **200**, 497–527.

BLAUSTEIN, M. P. & RUSSELL, J. M. (1975). Sodium–calcium exchange and calcium–calcium exchange in internally dialysed squid giant axons. *J. Memb. Biol.*, **22**, 285–312.

BLAUSTEIN, M. P., RUSSELL, J. M. & DE WEER, P. (1974). Calcium efflux from internally dialysed squid axons: the influence of external and internal cations. *J. supramolec. Struct.*, **2**, 558–581.

BRINLEY, F. J. & SCARPA, A. (1975). Ionized magnesium concentration in axoplasm of dialysed squid axons. *FEBS Lett.*, **50**, 82.

BRINLEY, F. J., SPANGLER, S. G. & MULLINS, L. J. (1975). Calcium and EDTA fluxes in dialysed squid axons. *J. gen. Physiol.*, **66**, 223–250.

CALDWELL, P. C. (1958). Studies on the internal pH of large muscle and nerve fibres. *J. Physiol., Lond.*, **142**, 22–62.

(1960). The phosphorus metabolism of squid axons and its relationship to the active transport of sodium. *J. Physiol., Lond.*, **152**, 545–560.

CALDWELL, P. C., HODGKIN, A. L., KEYNES, R. D. & SHAW, T. I. (1960). Partial inhibition of the active transport of cations in the giant axons of *Loligo*. *J. Physiol., Lond.*, **152**, 591–600.

CRANK, J. (1956). *The Mathematics of Diffusion*. Clarendon Press: Oxford.

CRAWFORD, A. C. (1975). Lithium ions and the release of transmitter at the frog neuromuscular junction. *J. Physiol., Lond.*, **246**, 109–142.

DiPOLO, R. (1973). Calcium efflux from internally dialysed squid giant axons. *J. gen. Physiol.*, **63**, 5–36.

(1974). Effect of ATP on the calcium efflux in dialysed squid giant axons. *J. gen. Physiol.*, **64**, 503–517.

DiPOLO, R., REQUENA, J., MULLINS, L. J., BRINLEY, F. J., SCARPA, A. & TIFFERT, T. (1976). Ionized calcium concentrations in squid axons. *J.gen.Physiol.*, in press.

HEILBRUNN, L. V. (1956). *The Dynamics of Living Protoplasm*. Academic Press: New York.

HODGKIN, A. L. & KEYNES, R. D. (1956). Movements of labelled calcium in squid giant axons. *J. Physiol., Lond.*, **138**, 253–281.

KATZ, B. & MILEDI, R. (1969). Tetrodotoxin-resistant electric activity in presynaptic terminals. *J. Physiol., Lond.*, **203**, 459–487.

MILEDI, R. (1973). Transmitter release induced by injection of calcium ions into nerve terminals. *Proc. R. Soc. Lond. B*, **183**, 421–425.

MULLINS, L. J. & BRINLEY, F. J. (1975). Sensitivity of calcium efflux from squid axons to change in membrane potential. *J. gen. Physiol.*, **65**, 135–152.

REUTER, H. (1973). Divalent cations as charge carriers in excitable membranes. *Prog. Biophys. molec. Biol.*, **26**, 1–43.

(1974). Localization of *beta* adrenergic receptors, and effects of noradrenaline and cyclic nucleotides on action potentials, ionic currents and tension in mammalian cardiac muscle. *J. Physiol., Lond.*, **242**, 429–451.

ROJAS, E. & HIDALGO, C. (1968). Effect of temperature and metabolic inhibitors on ^{45}Ca outflow from squid giant axons. *Biochim. biophys. Acta*, **163**, 550–556.

ROSE, B. & LOEWENSTEIN, W. R. (1975). Permeability of cell junctions depends on local cytoplasmic calcium activity. *Nature, Lond.*, **254**, 250–252.

SCHATZMANN, H. J. (1975). Active calcium transport and Ca^{2+}-activated ATPase in human red cells. *Curr. Topics Membr. Transport*, **6**, 125–168.

VAN BREEMEN, C. & DE WEER, P. (1970). Lanthanum inhibition of ^{45}Ca efflux from the squid giant axon. *Nature, Lond.*, **226**, 760–761.

CALCIUM IONS AND MITOCHONDRIA

By E. CARAFOLI and M. CROMPTON

Laboratory of Biochemistry, Swiss Federal Institute
of Technology (ETH), Zurich, Switzerland

HISTORICAL ASPECTS, AND GENERAL PROPERTIES

The striking ability to extract Ca^{2+} from the ambient medium is a property common to mitochondria from nearly all sources so far examined (Lehninger, Carafoli & Rossi, 1967). A list of the tissues with mitochondria known to possess transport activity is very extensive, and includes heart, liver, kidney, brain, adrenal, spleen and thyroid from mammals, birds and reptiles, and some plant tissues, e.g. sweet and white potato, corn seedlings and *Neurospora crassa* (Carafoli & Lehninger, 1971; Chen & Lehninger, 1973; Bygrave, Daday & Doy, 1975). The ability to take up Ca^{2+} is notably absent or of low activity in mitochondria from yeasts (Carafoli, Balcavage, Lehninger & Mattoon, 1970) and the blowfly at certain stages in the life of the adult insect (Carafoli, Hansford, Sacktor & Lehninger, 1971; Bygrave *et al.*, 1975).

In the absence of metabolic energy, mitochondria bind 50–70 nmoles Ca^{2+} mg^{-1} protein with rather low affinity (K_d about 100 μM, Rossi, Azzi & Azzone, 1967; Carafoli & Lehninger, 1971). About half of this energy-independent binding capacity in liver mitochondria is located in the outer membrane and, therefore, is not related to calcium transport (Carafoli & Gazzotti, 1973). In fact it seems unlikely that the remaining energy-independent binding sites, associated with the inner membrane, are part of the translocation process, since they greatly outnumber the quantity of 'translocase' units as judged by the amount of specific inhibitors required to block the translocation process (Carafoli, 1973). It has been proposed that a small amount of Ca^{2+} (1–5 nmole mg^{-1} protein) may also be bound by the inner membrane with high affinity ($K_d = 0.1$–1.0 μM, Reynafarje & Lehninger, 1969; Carafoli & Lehninger, 1971), although the existence of such high-affinity sites in the fully de-energised state has recently been questioned (Åkerman, Saris & Järvisalo, 1974; Southard & Green, 1974).

In the presence of metabolic energy, and in the absence of permeant anions, mitochondria are able to bind 100–150 nmole Ca^{2+} mg^{-1} protein and, in so doing, to reduce the external calcium concentration to the μM range or less (Drahota *et al.*, 1965). The energy-dependent uptake of Ca^{2+} is obviously inhibited completely by uncouplers of oxidative phosphoryla-

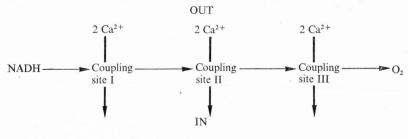

Fig. 1. The stoichiometry between Ca^{2+} uptake and
electron transport in mitochondria.

tion and, depending on the source of energy, by either oligomycin or
respiration inhibitors. It is also inhibited completely by La^{3+} (Mela, 1968)
and ruthenium red (Moore, 1971; Vasington, Gazzotti, Tiozzo & Carafoli,
1972) added at concentrations as low as 2–5 nmole mg^{-1} mitochondrial
protein. Energy from either substrate oxidation or ATP hydrolysis is
expended stoichiometrically with the amount of calcium bound, 1.7–2.00
Ca^{2+} being accumulated per pair of electrons traversing each energy-
conserving site of the respiratory chain (Fig. 1) (Chappell, Cohn &
Greville, 1963; Rossi & Lehninger, 1964; Chance, 1965; Bielawski &
Lehninger, 1966). Under these conditions, i.e. in the absence of parallel
anion fluxes, most of the calcium taken up probably remains bound at the
inner membrane (Chappell et al., 1963; Chance, 1965; Chance & Mela, 1966a;
Gear, Rossi, Reynafarje & Lehninger, 1967; Gunter & Puskin, 1972) to
lipid and protein components, although the relative contribution of these
in binding calcium is subject to some speculation (Chappell et al., 1963;
Gunter & Puskin, 1972; Gunter, Puskin & Russell, 1975). There is some
evidence that the energisation of the inner mitochondrial membrane
results in a higher density of negative charges, probably on the outer
surface (Azzi, Gherardini & Santato, 1971; Schäfer & Bojanowski, 1972).
Such changes may be responsible for the energy-induced binding of a
variety of organic cations (Gitler, Rubalcava & Caswell, 1969; Schäfer &
Bojanowski, 1972; Colonna, Massari & Azzone, 1972; Davidoff, 1974) and
the decreased binding of anions (Azzi, 1969; Jasaitis, Kuliene & Skulachev,
1971). Whether a higher electrostatic attraction is the only force responsible
for the energised binding of Ca^{2+} is doubtful, since the process is not only
energy-dependent, but is energy-consuming, which implies that an
energy-utilising translocation step also takes place. From their studies
using the paramagnetic ion Mn^{2+} as an analogue of Ca^{2+}, Gunter and co-
workers (1975) have suggested that as much as 35–40 % of the surface of

the inner membrane might be occupied by the divalent cation; such areas are considered to arise by surface binding, providing high local concentrations, and may be stabilised by phosphate.

The uptake of Ca^{2+} is associated with a stoichiometric ejection of protons (H^+:Ca^{2+}, o.8–1.4; Chappell et al., 1963; Chance, 1965; Rossi, Bielawski & Lehninger, 1966; Schäfer & Bojanowski, 1972) and a net increase in the titratable alkalinity of the mitochondria (OH^-:Ca^{2+}, about 1; Chance & Mela, 1966a, b; Gear et al., 1967; Addanki, Cahill & Sotos, 1968; Gunter & Puskin, 1972). It is not established where, within the mitochondria, alkalinisation might occur (Chance & Mela, 1966a, b; Gear et al., 1967; Gunter et al., 1975). It is possible that the external pH changes accompanying calcium binding reflect displacement of protons from anionic sites in the membrane, so that alkalinisation becomes real only during the subsequent manipulations necessary for the titrations (Gear et al., 1967). However, the potential capacity of energy-dependent Ca^{2+} uptake to facilitate the release of OH^- in the matrix compartment is in little doubt. This is most evident from the fact that calcium uptake under energised conditions may be accompanied by a parallel influx of those anions which permeate electroneutrally and cause acidification of the matrix (e.g. phosphate, acetate, bicarbonate, Chappell et al., 1963; Lehninger, 1974), but not of those anions which permeate electrogenically without net transfer of protons into the inner compartment (e.g. nitrate, thiocyanate, Lehninger, 1974). The uptake of about 100 nmole Ca^{2+} mg^{-1} protein accompanied by phosphate produces little change in the structure or phosphorylating efficiency of the mitochondria, but further uptake tends to cause the matrix space to swell, and the mitochondria to become uncoupled (Rossi & Lehninger, 1964; Hackenbrock & Caplan, 1969), probably because of membrane damage. As a result, the accumulated Ca^{2+} and phosphate are lost to the medium. The maximal figure of 100 nmole mg^{-1} protein, however, may underestimate the capacity of mitochondria to accumulate calcium phosphate without deleterious results, since the initial integrity of the mitochondria is also important (Chappell & Crofts, 1965).

For reasons that are not understood, the presence, specifically, of ADP (or ATP) and Mg^{2+} in the external medium prevents mitochondrial swelling, following calcium phosphate uptake. Under these conditions, Ca^{2+} is not released, but precipitation of calcium phosphate instead occurs within the matrix. The precipitates are often seen in the electron microscope near the periphery following the outline of the cristae, but may occupy the majority of the matrix space when very high levels of loading are reached (Brierley & Slautterback, 1964; Greenawalt, Rossi & Lehninger, 1964; Peachey, 1964; Geenawalt & Carafoli, 1966). The electron

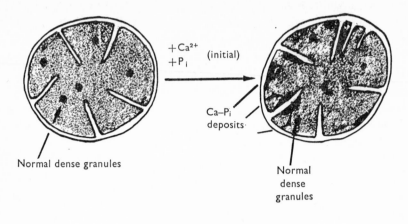

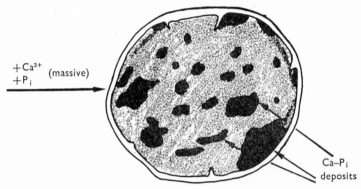

Fig. 2. Stages in the massive accumulation of
calcium and phosphate by mitochondria.

microscopic appearance of the massive loading with Ca^{2+} and phosphate is
schematically represented in Fig. 2. The scheme stresses the fact, to be
further discussed below, that the 'normal' dense granules of the matrix do
not participate in the deposition of the precipitates. Although the crystalline form of the precipitate is mainly hydroxyapatite $[Ca_3(PO_4)_2]_3 . Ca(OH)_2$
(Weinbach & von Brand, 1965), it appears that the granules are deposited
initially in the amorphous form (Greenawalt et al., 1964; Thomas &
Greenawalt, 1968), and assume the typical X-ray diffraction pattern of
hydroxyapatite only after certain manipulations of the samples. In the
presence of ATP and inorganic phosphate, the maximum level of calcium
accumulated by mitochondria increases by about one order of magnitude,
to about 2–3 μmole mg^{-1} protein. Concomitant with the massive loading of
the matrix, is a net uptake of adenine nucleotides (Carafoli, Rossi &
Lehninger, 1965; Pfaff, Heldt & Klingenberg, 1969; Gunter et al., 1975)

which is not in accord with the fact that, under normal conditions at least, the adenine nucleotide translocase responsible for adenine nucleotide transport across the inner membrane catalyses an obligatory 1:1 exchange. However, *net* movement of adenine nucleotides through the inner mitochondrial membrane has also been observed in liver mitochondria in the presence of Mg^{2+} and phosphate (Meissner & Klingenberg, 1968), a somewhat similar situation, and furthermore the penetration of adenine nucleotides in the presence of large amounts of Ca^{2+} and phosphate is only partially sensitive to the inhibitor of the adenine nucleotide translocase atractyloside (Carafoli, Rossi & Lehninger, 1965). Approximately one molecule of adenine nucleotide is accumulated per 12 Ca^{2+} (Carafoli, Rossi & Lehninger, 1965), a finding that may be related to the observation by Krane & Glimcher (1962) that, during the growth of synthetic hydroxyapatite crystals, ATP is bound in approximately the same ratio (1:12). It looks very likely that adenine nucleotides stabilise the calcium phosphate deposits, or perhaps 'prime' their precipitation in the matrix. In fact, Weinbach & von Brand (1965) have found adenine nucleotides in the granules isolated from calcium phosphate-loaded mitochondria.

Although the formation of calcium phosphate granules is a reversible process, since much of the calcium can leave the mitochondria under appropriate conditions (Greenawalt *et al.*, 1964), the release of the Ca^{2+} accumulated under these conditions is very slow, and it is thus unlikely that these deposits can participate in any rapid calcium fluxes between mitochondria and cytoplasm. The 'massive' loading of mitochondria with Ca^{2+} and phosphate may however have a role in cell physiology and pathology, as will be discussed below. At this point it can be mentioned that one of the purposes of massive uptake of calcium phosphate in the matrix may be to sequester calcium and phosphate in the form of an insoluble deposit, thus providing, perhaps, the initial step in the process of biological calcification. Such a role for mitochondria has been championed in particular by Lehninger (1970) and is supported by many observations of calcium phosphate deposits within the mitochondria of chondrocytes, osteoclasts and the cells of the calciferous gland of the earthworm (Gonzales & Karnovsky, 1961; Crang, Holsen & Hitt, 1968; Martin & Matthews, 1969). Calcium phosphate precipitation is the unavoidable end-result of uptake *in vitro* when adenine nucleotides are included in the medium and, therefore, must be considered a property inherent in most mitochondria *in situ*, since both adenine nucleotides and inorganic phosphate are normally available in the cytosol. However, the presence of calcium in the dense granules commonly observed in mitochondria *in situ* has been questioned (Pasquali-Ronchetti, Greenawalt & Carafoli, 1969) and the morphological (electron microscopic) equivalent of the 'normal' deposits of calcium and phosphate *in vivo* is still an open question. In addition, it has been shown (Vasington & Greenawalt, 1964) that the precipitation of calcium and phosphate during massive loading does *not* occur on the normal dense granules of the mitochondrial matrix.

The nature of the immediate driving force for calcium uptake is of considerable importance in understanding the transport mechanism, and also for the conclusions that may be drawn regarding the reversibility of the uptake process per se. Early statements (Rasmussen, Chance & Ogata, 1965; Chance & Mela, 1966b; Rasmussen & Ogata, 1966) on the mechanism of active calcium uptake represent determined attempts to formulate reaction schemes within the chemical coupling hypothesis, and embracing the $Ca^{2+}:H^+$ ratios observed during respiration-supported calcium uptake. According to these mechanisms, the initial step of calcium uptake consists of its binding to a hypothetical, non-phosphorylated, high-energy inter-mediate, with the displacement of one proton per calcium bound. Translocation of the complex to the inner side of the inner membrane is accompanied by dissipation of the energised state. Whilst these schemes do account for the overall $Ca^{2+}:H^+$ stoichiometry, both in the presence and absence of permeant, proton-carrying anions, the nature of the translocation mechanism is thereby expressed in a form that has proved to be un-amenable to experimental verification.

The observation (Selwyn, Dawson & Dunnett, 1970) that calcium influx may be driven in the absence of respiratory or ATPase activity by a diffusion potential set up by a gradient of H^+ (in the presence of uncoupler to permit electrogenic proton permeation) or SCN^-, which is permeant (and also K^+ in the presence of valinomycin, Scarpa & Azzone, 1970) clearly indicated that the movement of Ca^{2+} is an electrophoretic process, and placed the nature of the link between calcium transport and energy production within chemiosmotic principles, as originally proposed by Mitchell (1966). This is confirmed by the recent quantitative measurements of Rottenberg & Scarpa (1974), in which the distribution of Ca^{2+} between mitochondria and medium has been compared to that of Rb^+. In the presence of valinomycin, to permit its electrophoretic movement, Rb^+ is known to distribute according to the membrane potential. Rottenberg & Scarpa (1974) have demonstrated that the distribution of Ca^{2+} (in the presence of acetate to minimise the binding of Ca^{2+} to membrane sites) between the inner and outer compartments is also determined by the membrane potential and occurs with a net charge transfer of 2 (Fig. 3). Thus, in summary, the driving force for calcium uptake in the presence of permeant anions is seen to be the electrical component of the electro-

$$E = RT/ZF \ln (Rb^+_i)/(Rb^+_o) \quad \text{(Valinomycin)}$$

$$^{86}Rb^+_i/^{86}Rb^+_o = 13.3 \quad \text{Log ratio} = 1.24$$

$$^{45}Ca^{2+}_i/^{45}Ca^{2+}_o = 230 \quad \text{Log ratio} = 2.36$$

$$\text{Log} (Ca^{2+}_i/Ca^{2+}_o) = 2 \log (Rb^+_i/Rb^+_o)$$

Fig. 3. The distribution of Rb^+ and Ca^{2+} across the inner membrane of energised mitochondria (Rottenberg & Scarpa, 1974, and see text). It should be noted that the electrical potential across the inner membrane according to these data is substantially lower than the values reported from other laboratories (see text).

chemical proton gradient derived from respiration or ATP hydrolysis (Mitchell, 1966). The values of the electrical potential across the inner membrane of liver mitochondria, estimated in various laboratories from the distribution of K^+ or Rb^+ between the intra- and extramitochondrial compartments, are in reasonable agreement, being in the range 125–145 mV during oxidative phosphorylation (Mitchell & Moyle, 1969; Padan & Rottenberg, 1973; Nicholls, 1974). In the absence of phosphate acceptor (state 4 respiration) the values are somewhat higher (150–170 mV) (Mitchell & Moyle, 1969; Nicholls, 1974). On these grounds, electrophoretic influx of Ca^{2+} would be predicted to lead to a distribution ratio of $[Ca_i/Ca_o]$ equal to 10^4–10^5 at equilibrium during state 3 respiration, and a ratio of 10^5–10^6 during state 4 respiration. These considerations underline the remarkable potential of the electrophoretic transporting system of mitochondria to accumulate Ca^{2+}, and to reduce the cytoplasmic Ca^{2+} concentrations to extremely low levels.

The utility of the Ca^{2+} pump in intracellular Ca^{2+} physiology demands that it functions rapidly within the range of free calcium concentrations found in the cell (10^{-7}–10^{-5} M). The kinetic parameters available for calcium uptake into mitochondria from different tissues are given in Table 1. The V_{max} values at 25 °C for the energy-dependent Ca^{2+} uptake by mitochondria from mammalian heart and liver are high, being within the range 3–13 nmoles Ca^{2+} mg^{-1} mitochondrial protein sec^{-1}; at physiological temperatures, these rates are approximately doubled. Unfortunately, there is considerable disagreement in the published values for the affinities of the transport systems for Ca^{2+}, particularly in heart mitochondria. Techniques using murexide as a Ca^{2+}-sensitive chromogen consistently give values for the free Ca^{2+} concentration needed for half maximal velocity that are much higher than the probable physiological limit of cytoplasmic calcium. It must be pointed out that the spectrophotometric murexide technique prevents the use of Ca^{2+} buffers, a limitation which requires the inclusion of excess Mg^{2+} in the reaction medium, to eliminate

Table 1. *Kinetic properties of mitochondrial calcium transport*

Tissue	[Ca] required for $\frac{1}{2}V_{max}$ (μM)	V_{max} (nmole mg^{-1} sec^{-1})	Sigmoidicity	Method	Reference
Liver	50–70	8–13	Yes	Direct (murexide) Mg^{2+} present	Vinogradov & Scarpa, 1973
Liver	2	—	Yes	Direct (Ca–EGTA buffers)	Bygrave et al., 1971
Liver	2–3	—	—	Indirect (redox shift of cyt. b)	Carafoli & Azzi, 1972
Liver	4	—	—	Direct (EGTA buffers, RR quenching)	Reed & Bygrave, 1975
Liver	8	—	—	Indirect (O$_2$ consumption)	Reynafarje & Lehninger, 1974
Lucilia flight muscle	5	1	Yes	Direct (EGTA buffers, RR quenching)	Bygrave et al., 1975
Heart	2	—	—	Indirect (redox shift of cyt. b)	Chance & Schoerner, 1966
Heart	55–56	4–5	No	Direct (murexide) Mg^{2+} absent	Sordahl, 1974
Heart	85–105	8–14	Yes	Direct (murexide) Mg^{2+} present	Scarpa & Graziotti, 1973
Heart	8–10	3	Yes	Direct (inhibitor stop, EGTA buffers)	E. Sigel, M. Crompton & E. Carafoli, unpublished
Heart	0.1–1.0	-	—	Indirect (redox shift of cyt. b)	Jacobus et al., 1975

Abbreviations used: RR = Ruthenium red; EGTA = ethyleneglycol-*bis*-(β-aminoethyl ether)N,N'-tetraacetic acid; cyt. = cytochrome.

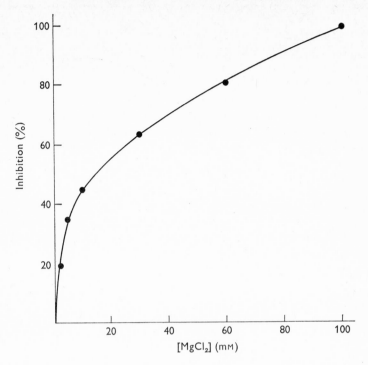

Fig. 4. Inhibition of Ca^{2+} transport in heart mitochondria by Mg^{2+}. Rat heart mitochondria were prepared essentially with the Polytron method of Sordahl & Schwartz (1967), and incubated in a medium containing, in a final volume of 5 ml at 15 °C (the temperature was kept low to slow the rate of uptake) 0.13 M NaCl, 0.02 M Tris-Cl, pH 7.0, 70 μM $^{45}CaCl_2$, 0.002 M Na-succinate, 5 μg ml^{-1} rotenone, 0.5 mg ml^{-1} bovine serum albumin, the amounts of $MgCl_2$ indicated, and 1 mg mitochondrial protein. Samples of mitochondria were removed at different time-intervals, and the uptake of Ca^{2+} was determined isotopically.

the complexation of Ca^{2+} with components of the medium and its non-specific binding to mitochondria. Recently, however, it has been shown that Ca^{2+} uptake by heart mitochondria, but not by liver mitochondria, is inhibited by Mg^{2+} (Sordahl, 1974; Jacobus et al., 1975). The amounts of Mg^{2+} necessary to inhibit the uptake of Ca^{2+} completely depend on the concentration of Ca^{2+} employed, but could be as low as 5 mM at 1–2 μM external Ca^{2+}. Using 70 μM Ca^{2+} (Fig. 4), complete inhibition is obtained using 100 mM Mg^{2+}. The inhibition by Mg^{2+} has been exploited in our laboratory to determine the affinity of the heart system for Ca^{2+} by an inhibitor-stop method, using Ca^{2+} buffers to maintain accurately the free Ca^{2+} concentration at determined levels, and, after a certain influx has occurred, simultaneously stopping the reaction and removing non-specific Ca^{2+} binding by high concentrations of Mg^{2+}. According to this technique,

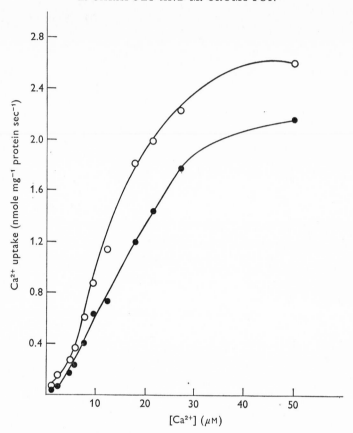

Fig. 5. Initial rate of Ca^{2+} uptake by heart mitochondria, measured with an inhibitor-stop method. Rat heart mitochondria were prepared essentially with the Polytron method of Sordahl & Schwartz (1967) and preincubated for 2 min at 25 °C in a medium containing 0.13 M NaCl (or KCl), 0.02 M Tris-Cl, pH 7.0, 0.002 Na- (or K-) succinate, 5 μg ml^{-1} rotenone, 0.5 mg m^{-1} bovine serum albumin and 0.52 mg of mitochondrial protein in a final volume of 1 ml. The reaction was initiated with the addition of a ^{45}Ca–HEDTA (N'-(2-hydroxyethyl)-ethylenedi-amine-N,-N-N'-triacetic acid) buffer, equal to the free calcium concentrations indicated. 10 sec after the addition of the Ca^{2+} buffer, the uptake was stopped by the addition of 0.1 M $MgCl_2$, and the uptake of Ca^{2+} was measured by difference from a control experiment where the Ca^{2+} buffer and $MgCl_2$ were added simul-taneously. Mitochondria were removed from the reaction medium by rapid centrifugation, and the uptake of Ca^{2+} was determined isotopically. (●, NaCl; ○, KCl.)

half-maximal velocity is attained at about 12 μM Ca^{2+} when about 2.6 nmole Ca^{2+} are transported per mg protein per sec at 38 °C (Fig. 5). (This rate would, in principle, be adequate to meet the requirements of heart relaxation, as discussed in more detail below.) In common with other data (Bygrave, Reed & Spencer, 1971; Vinogradov & Scarpa, 1973), the initial rate of uptake displays a sigmoidal dependence on the free calcium concentration (Hill coefficient, $n = 2$) implying that the hypothetical carrier transports 2 Ca^{2+} per cycle across the inner membrane. However, the data have also revealed that the measured uptake is in reality a composite of two processes, one that shows cooperativity, as mentioned previously, and another that obeys Michaelis–Menten kinetics. Preliminary evidence indicates that the $n = 1$ component has a K_m value of about 30 μM and a V_{max} of about 1 nmole mg^{-1} protein sec^{-1} at 25 °C. It is also interesting that the initial rate of uptake seems to be higher in media containing K^+, rather than Na^+, as the main monovalent cation.

Concerning the molecular components of the Ca^{2+} translocation process, not very much can be said at the moment. The kinetic parameters discussed above (saturation kinetics, competitive inhibition by Sr^{2+}, specific inhibition by lanthanum and ruthenium red) have in general been taken to indicate the existence of a Ca^{2+} carrier (for a discussion, see Lehninger & Carafoli, 1969). The specificity of ruthenium red for carbohydrates has directed the search for a Ca^{2+} carrier towards glycoproteins, and indeed several glycoproteins capable of binding Ca^{2+} have been isolated from mitochondria (see Carafoli & Crompton, 1976, for a review). It is well to remember, however, that the kinetic parameters mentioned above could equally well be accounted for by a specific superficial receptor *not* functioning as a Ca^{2+} carrier (Carafoli, 1975*a*), and, therefore, the existence of a transmembrane, mobile or immobile, Ca^{2+} carrier remains for the moment hypothetical.

A number of other divalent cations, e.g. Sr^{2+}, Mn^{2+}, Ba^{2+}, and Mg^{2+}, can also be accumulated by mitochondria. Of these, Sr^{2+} is transported most rapidly, at rates approaching those for calcium transport (Chappell & Greville, 1963; Carafoli, 1965*a*). Strangely, continued uptake of Sr^{2+} and phosphate is obtained in the absence of added adenine nucleotides; the mitochondria remain structurally intact and functionally coupled, unlike after calcium phosphate uptake (Carafoli, Weiland & Lehninger, 1965). Mn^{2+} and Ba^{2+} are also accumulated by energy-dependent processes, although much more slowly that Ca^{2+} (Chappell *et al.*, 1963; Chance & Mela, 1966*c*; Grabske, 1968; Drahota, Gazzotti, Carafoli & Rossi, 1969). In addition, Mg^{2+} accumulation together with phosphate in heart mitochondria has been reported (Brierley, Murer, Bachmann & Green, 1963) although hardly any influx into liver mitochondria was observed under the same experimental conditions (Carafoli *et al.*, 1964; Drahota *et al.*, 1969). The question of whether all these cations employ the same transport system is not settled. Mutual competition has been observed between Ca^{2+} and Sr^{2+} uptake (Carafoli, 1965*b*), but it is important to remember that competition may exist at levels other than combination with the transport system, e.g. for the energy supply. The use of refined techniques for kinetic measurements has suggested that Ca^{2+} and Mn^{2+} may share a common transport system in liver mitochondria (Vinogradov & Scarpa, 1973). It will be recalled that the rate of Ca^{2+} uptake by liver mitochondria displays a sigmoidal dependence on the Ca^{2+} concentration, and such is true also for Mn^{2+} uptake, albeit with lower affinity (Vinogradov & Scarpa, 1973). It was found that low concentrations of Ca^{2+} induce a hyperbolic relation between the rate of Mn^{2+} influx and the Mn^{2+} concentrations, and this may be interpreted to indicate the existence of a mobile ternary complex of the form, Ca–carrier–Mn.

Thus, mitochondria possess a capacity for Ca^{2+} transport which, by virtue of its complexity and activity, one may reasonably suppose to be an important process in the regulation of intracellular Ca^{2+} (see below). Conversely, Ca^{2+} transport by mitochondria *per se* must be subject to rigorous control so that uptake is limited. This requirement is emphasised by the fact that, in liver mitochondria at least (Rossi & Lehninger, 1964), Ca^{2+} uptake takes primacy over oxidative phosphorylation in utilising respiratory energy (although in heart mitochondria, perhaps significantly, such is not the case, Jacobus *et al.*, 1975). Whether the role of the Ca^{2+} uptake process is to contribute to rapid fluctuations in the Ca^{2+} concentration in the cytosol, or whether the process endows the mitochondria with the capacity to serve as a Ca^{2+} sink, maintaining calcium levels in the cytosol within certain 'background' limits, is subject to speculation and will be discussed in more detail below.

REVERSIBILITY OF THE TRANSPORT PROCESS

An evaluation of the true role of the process of mitochondrial Ca^{2+} transport requires that much more be known of the conditions which not only limit Ca^{2+} uptake, but reverse it, in mitochondria from different tissues. At this point, it is pertinent to consider whether, given the magnitude of the membrane potential in respiring mitochondria, a system catalysing Ca^{2+} transport with a net charge movement of 2 might reach equilibrium *in vivo*. As stated earlier, such a process would be predicted to lead to a transmembrane gradient of ionised Ca^{2+} of 10^4–10^6, assuming that the electrical potential across the inner membrane *in vivo* is comparable to that measured in isolated mitochondria. This in turn would lead to either an intolerably high concentration of ionised Ca^{2+} in mitochondria or to a lowering of the ionised Ca^{2+} concentration in the cytosol below levels compatible with the functioning of the cell. A lower transmembrane gradient of ionised Ca^{2+}

could, in principle, be achieved by equilibrium with a substantially lower electrical potential. However, it is now appreciated that the membrane potential is a central element of energy transduction in mitochondria, and any large change in this potential would be predicted to have diverse consequences on mitochondrial function. Thus, the possibility must be considered that this system may not be permitted to attain equilibrium *in vivo* and, therefore, be incapable of catalysing net efflux of calcium; in other words, it would imply that the electrophoretic influx of Ca^{2+} operates as a one-way pump, providing for Ca^{2+} influx only. Naturally, these considerations would not apply so forcibly if the net charge transfer per Ca^{2+} transported is less than 2. Nevertheless, it seems logical to keep in mind the possibility that the efflux of Ca^{2+} from mitochondria is catalysed by a system (carrier?) different from that responsible for Ca^{2+} uptake. In fact, that such could be the case receives some support from the observation that, although ruthenium red completely inhibits the energy-dependent uptake of Ca^{2+}, it does not inhibit the release of accumulated Ca^{2+} induced by uncouplers (Vasington et al., 1972). This release is about as rapid as the uptake when conditions are chosen to prevent intramito-chondrial precipitation of calcium phosphate. Other conditions which dissipate the energised state (e.g. respiratory inhibitors, anaerobiosis) also discharge Ca^{2+}. However, such unphysiological means of inducing release can have little relevance to the in-vivo state, and enable one to speculate only how quickly the release might occur. Certainly, if the energised state is maintained, the rate of efflux (i.e. recycling) of calcium is very low (about 0.05 nmole mg^{-1} protein sec^{-1}; Stucki & Ineichen, 1974). In the search for possible physiological release-inducing agents, attention has been directed towards the changes in the extramitochondrial milieu that might be expected to apertain *in vivo* and, in the case of heart mitochondria, the possible consequences of a sudden influx of Na^+ into the muscle cell have been studied. It has been shown (Carafoli et al., 1974) that the addition of NaCl to energised heart mitochondria after limited calcium uptake (10 nmole mg^{-1} protein) causes a rapid release of the accumulated calcium (Fig. 6); with the partial exception of Li^+, no release is elicited by K^+ or other monovalent cations. When ruthenium red is added to inhibit the re-uptake of the released calcium, the effect of Na^+ is evident at concentrations as low as 5 mM, and is not inhibited by K^+, phosphate, or Mg^{2+}. Interestingly, Na^+ has little effect on the ability of liver mitochondria to retain Ca^{2+}. The interaction between Ca^{2+} and Na^+ with heart mitochondria clearly merits further study to evaluate its physiological implications. A further mechanism of Ca^{2+} release not restricted to heart mitochondria is that effected by prostaglandins (Kirtland & Baum, 1972; Carafoli &

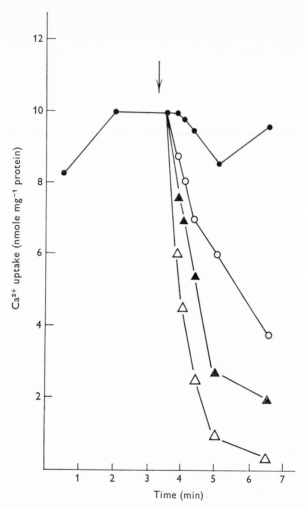

Fig. 6. Release of Ca^{2+} from heart mitochondria by Na^+. Rat heart mitochondria were prepared by the method of Pande & Blanchaer (1971) slightly modified, and incubated in a medium containing 0.21 M mannitol, 0.07 M sucrose, 0.01 M Tris-Cl buffer, pH 7.4, 0.01 M Tris-succinate, 5 mg of mitochondrial protein, and 50 nmole $^{45}CaCl_2$. The final volume was 4 ml, the temperature 25 °C. 15 sec before the time indicated by the arrow, 0.0025 mM ruthenium red was added, followed, at the time indicated by the arrow, by the addition of NaCl (○, 0.005 M; ▲, 0.01 M; △, 0.05 M; ●, medium). Aliquots of mitochondria were separated from the medium by rapid Millipore filtration, and the movements of Ca^{2+} were estimated isotopically.

Crovetti, 1973; Malmström & Carafoli, 1975). These hormones are able to discharge accumulated Ca^{2+}, which is then taken up and lost in a cyclic manner. This ability is probably dependent on the formation of a lipid soluble electroneutral complex of the form $(PG)_2Ca$; it unfortunately requires rather high prostaglandin (PG) concentrations and leads *in vitro* to a Ca^{2+}-dependent uncoupling effect. However, it should be borne in mind that other processes in the cell competing for cytoplasmic calcium might well render any such energy-dissipating cycle short-lived.

MITOCHONDRIAL Ca^{2+} TRANSPORT AND CELL PHYSIOLOGY

As mentioned in the preceding section, it is rather logical to suggest that a process as remarkable as the transport of Ca^{2+} by mitochondria should have a prominent role in cell physiology, particularly when one considers the impressive list of processes that are influenced by Ca^{2+} in the cell. These include membrane-linked functions like the coupling between excitation and contraction, and between excitation and secretion at nerve endings, cell contact, exogenous secretion, the Na^+–K^+ transport ATPases, the action potential, the release of cyclic–AMP, and of several hormones. Among the contractile and motile systems, one could mention the contraction and relaxation of the myofibrils, the movement of cilia, flagella, and microtubules, and the formation of pseudopodia. The activation and deactivation of several key enzymes are also important targets for Ca^{2+}. It is thus clear that any structure capable of modulating reversibly the levels of Ca^{2+} in the cell may have profound effects on these processes. In considering mitochondria, important values in this respect are their total capacity for rapid and reversible Ca^{2+} transport, the affinity of the transport system for Ca^{2+}, and the velocity of the influx and efflux processes. The importance of the total capacity is obvious, since some of the processes mentioned above (e.g. contraction and relaxation of muscle) involve the movement of large (and known) amounts of Ca^{2+}. Given the very impressive figures (see above) of their Ca^{2+} storage capacity, even in the absence of permeant complexing anions, little doubt exists on the theoretical ability of mitochondria to satisfy the demands of the various processes mentioned in term of total capacity. The affinity for Ca^{2+} is also of paramount importance, since most Ca^{2+}-dependent cell reactions are regulated by the cation in the μM concentration range; as discussed in the preceding section, the problem of the affinity of mitochondria for Ca^{2+} can now be considered as conclusively settled on a K_m value between 1 and 12 μM (Carafoli & Azzi, 1972; Spencer & Bygrave, 1973; Reynafarje & Lehninger, 1974; E. Sigel, M. Crompton & E. Carafoli, in preparation). What remains to be decided, however, is whether this theoretically very high affinity is fully exploited *in vivo*, or whether the ability of mitochondria to take up Ca^{2+} is depressed by some physiological modulator. As discussed above, this could perhaps be the case for Mg^{2+} in heart cells (E. Sigel, M. Crompton & E. Carafoli, in preparation). The velocity of Ca^{2+} transport by mitochondria is the least explored among the important parameters of the process, and the measurements of the uptake rate in heart mitochondria reported above (Sigel et al., in preparation) are to our knowledge the only direct result so far obtained on the problem. Even less is known on the rate of the release of Ca^{2+} from mitochondria, although Carafoli (1975b) has obtained data indicating that the rate is proportional to the amount of Ca^{2+} present in mitochondria. It is thus clear that the main properties of mitochondrial Ca^{2+} transport are theoretically adequate to place mitochondria in a central role in the regulation of a variety of Ca^{2+}-dependent reactions. It is necessary to stress, however, that their actual involvement has not yet been conclusively established for any reaction (a difficulty which is not only mitochondrial, however, but common to all other cell organelles capable of transporting Ca^{2+}). At the present stage, therefore, extrapolations to the situation in vivo must be made with caution. One could nevertheless discuss in this connection the problem of the relaxation of heart myofibrils, mainly because it is becoming increasingly evident that sarcoplasmic reticulum, at variance with skeletal muscle, is not sufficiently active in heart (see Carafoli, 1975c, for a discussion). About 25 nmole Ca^{2+} are mobilised from troponin (the Ca^{2+} receptor in the myofibril) per g heart tissue per beat (400 msec per beat, i.e. 200 msec for the relaxation process, in rodent hearts). Since 1 g heart contains about 90 mg mitochondrial protein (Scarpa & Graziotti, 1973) 1 mg mitochondrial protein would have to bind between 0.2 and 0.3 nmole of Ca^{2+} during relaxation, that is, between 1.0 and 1.5 nmole of $Ca^{2+} sec^{-1}$. This rate compares favourably with the direct measurements of initial velocities of uptake at 10 μM external Ca^{2+} (reported above), and is probably also within the required limits at lower Ca^{2+} concentrations. In addition, one has to remember that isolated mitochondria are probably not as active as mitochondria in situ and, further, that the free Ca^{2+} concentration in heart sarcoplasm is not known with certainty. It is pertinent to add that even the best measurements of the initial rate of Ca^{2+} uptake by heart sarcoplasmic reticulum (Vinogradov & Scarpa, 1973) fall well short of the requirements for the relaxation process (Table 2).

One area where mitochondrial Ca^{2+} could play a significant role is the mitochondrion itself. At least 2 mitochondrial key enzymes involved in the oxidation of substrates are known to be influenced by Ca^{2+}. One is α-glycerophosphate dehydrogenase, which is located externally on the inner membrane, and which is allosterically regulated by Ca^{2+} (Hansford & Chappell, 1967; Donnellan, Barker, Wood & Beechey, 1970) in the μM range (Ca^{2+} activates by lowering the K_m of the enzyme for its substrate).

Table 2. *Ca²⁺ movements during relaxation in rat heart cells*

Ca^{2+} removed from troponin	25 nmole g^{-1} tissue
Relaxation time	200 msec
Mitochondrial content of heart	80–100 mg protein g^{-1} tissue
Initial rate of mitochondrial Ca^{2+} uptake	32–36 nmole Ca^{2+} g^{-1} tissue
(at 10 μM external Ca^{2+}, at 38 °C)	200 msec^{-1}

The other is pyruvate dehydrogenase-phosphate phosphatase, which activates the dehydrogenase by dephosphorylating it (Denton, Randle & Martin, 1972). This enzyme is located in the mitochondrial matrix, presumably in close functional link to the inner side of the inner membrane, and is also activated by Ca^{2+} in the μM range. It is thus permissible to suggest that Ca^{2+} could influence respiration by acting on either one of these two dehydrogenases. Of course, the problem of how the concentrations of free Ca^{2+} in the vicinity of these dehydrogenases is modulated may well be identical to the problem of the control of Ca^{2+} transport in mitochondria *per se*. At this stage, it is perhaps useful to keep in mind that fluctuations in the level of a number of metabolites (e.g. ATP and ADP) which bind Ca^{2+} with different affinities, are a potential means of finely regulating the concentrations of free Ca^{2+}. Such possibilities are difficult to evaluate at the moment, however, since they require a rather detailed knowledge of the ionic composition of the different intracellular compartments. Another interesting possibility for regulation by Ca^{2+} is the transfer of NADH into the apolar regions of the mitochondrial inner membrane. Vinogradov, Scarpa & Chance (1972) have shown that Ca^{2+} is capable of specifically complexing and of transferring NADH from the aqueous (intramitochondrial) phase to the hydrophobic environment of the inner membrane. It is thus permissible to speculate that variations in the intramitochondrial Ca^{2+} levels may induce fluctuations in the level of NADH at specific intramembrane sites. This in turn may affect the activity of several membrane-bound, NAD-dependent dehydrogenases, and thus influence respiration by a mechanism which would be fundamentally different from that involving α-glycerophosphate dehydrogenase and pyruvate dehydrogenase phosphate phosphatase. Of interest is also the observation that the translocation of several respiratory substrates, particularly α-oxoglutarate, into mitochondria is stimulated by low concentrations of Ca^{2+} (Rasmussen & Bikle, 1975). Lastly, it may be appropriate to mention that Azzi, Sorgato & Montecucco (1975) have recently obtained evidence for a role of Ca^{2+} in the rebinding of cytochrome *c* to deficient inner membrane preparations. The system employed by Azzi *et al.* (1975) requires rather high concentrations of Ca^{2+}, and it is thus difficult to draw conclusions concerning the possible physiological significance of the observation. In general, however, a role of Ca^{2+}, and other divalent cations like Mg^{2+}, in the binding of hydrophylic proteins to mitochondrial and other biological membranes is a possibility that is supported by evidence coming from several lines of investigation (Gitler & Montal, 1972; Montal & Korenbrot, 1973), and which is particularly interesting for the erythrocyte protein spectrin (D. Branton, personal communication). It is relevant to mention at this point that several authors have suggested that Ca^{2+} (and other divalent cations) may play a prominent structural role in biological membranes, influencing their permeability, and the extractability of proteins (Burger, Fujii & Hanahan, 1968; Duggan & Martonosi, 1970; Reynolds & Trayer, 1971; Gilbert, 1972; Reynolds, 1972). Williams (1975) has argued that the preferential binding of different cations (Ca^{2+}, Mg^{2+}) to opposite faces of biological membranes may determine the orientation of membrane molecules, and thus make the outside and inside face structurally dissimilar. A detailed discussion of the problem has been presented by Manery (1966), who has argued that Ca^{2+}, by virtue of its ability to form stable chelates with a variety of (membrane) components, would dehydrate the membranes, thus affecting their architecture and permeability properties. Of special interest in this connection are some observations by Rorive, Nielson & Kleinzeller (1972) showing that Ca^{2+}, *while being transported* across the plasma membrane, alters its permeability to water. It may also be recalled here that actomyosin-like filaments, capable of undergoing dramatic conformational changes under the influence of Ca^{2+}, have been demonstrated in several membranes.

MITOCHONDRIA AS LONG-TERM INTRACELLULAR Ca^{2+} BUFFERS

A role of mitochondrial Ca^{2+} in rapid metabolic regulation, both in the mitochondria, and in the cell cytosol, is an interesting possibility, which however remains to be established. Mitochondria, however, could play a role also in the long-term buffering of the intracellular Ca^{2+} levels, so that they will remain within the limits compatible with the normal functioning of the biochemical machinery of the cell. That mitochondria may act as Ca^{2+} sinks, indeed as the most important Ca^{2+} sinks in many cells, is now a generally accepted fact. Evidence for this conclusion comes from different sources, particularly from experiments on isolated organs exposed to excess Ca^{2+} concentration (Peachey, 1964; Legato, Spiro & Langer, 1968, see below). This is particularly evident in cells where a heavy 'traffic' of Ca^{2+} takes place normally in the cytosol, such as the various cells of bone and the chondrocytes (Gonzales & Karnovsky, 1961; Martin & Matthews, 1969) but it is true also for 'normal' cells occasionally exposed to excess Ca^{2+} concentrations. An elegant demonstration of the role of mitochondria as intracellular Ca^{2+} buffers has been provided very recently by Rose & Loewenstein (1975 a, b) who have used isolated *Chironomus* salivary gland cells, injecting into them the Ca^{2+} indicator protein aequorin, and have monitored the luminescence of the indicator inside the cell with a television camera coupled to an image-intensifier. When the cells were subsequently injected with pulses of Ca^{2+}-buffers, they could observe luminescence of aequorin only in the immediate vicinity of the tip of the injection pipette. Evidently, the injected Ca^{2+} was sequestered in some intracellular store. However, if the cells were injected with cyanide or with ruthenium red (which is known to suppress the uptake of Ca^{2+} by mito-chondria but has no effect, at the concentrations normally employed, on any other intracellular system) prior to Ca^{2+}, then the luminescence of aequorin was diffuse and persistent. Rose & Loewenstein have concluded that energised Ca^{2+} uptake by mitochondria is the dominant factor that constrains the free movement of the injected Ca^{2+} in the cytosol.

It is important to emphasise that the Ca^{2+}-buffering ability of mito-chondria does not exclude the existence of other intracellular Ca^{2+} sinks. The case of sarcoplasmic reticulum in muscles is obvious, but also of considerable interest is the recent observation of Ca^{2+} transporting ability in endoplasmic reticulum of epithelial cells (Moore, Chen, Knapp & Landon, 1975). Recent experiments by Baker (this volume) on squid axons suggest the existence in the axoplasm of a component (possibly not an organelle) which binds Ca^{2+} with low capacity, but very high affinity.

It is pertinent to mention at this point that the sequestration of Ca^{2+} that has entered the cell in the mitochondrial store need not occur in all cells or in all instances. According to recent evidence obtained by Terepka, Coleman, Armbrecht & Gunter (this volume) the transcellular transport of Ca^{2+} in kidney occurs by migration of Ca^{2+}-containing vesicles, formed from the plasma membrane, to the opposite pole of the cell.

The older observations involving mitochondria in the buffering of cell Ca^{2+} were usually based on the electron microscopic demonstration of electron-opaque masses, analogues to those observed in isolated mito-chondria after the accumulation of calcium and phosphate (Greenawalt *et al.*, 1964) within the profiles of mitochondria *in situ*. This is the case for osteoclasts in healing bone fractures (Gonzales & Karnovsky, 1961), for chondrocytes in the calcifying cartilage (Martin & Matthews, 1969), for the cells of the calciferous gland of the earthworm (Crang *et al.*, 1968), for the cells of the egg-shell gland (Hohman & Schraer, 1966), for the muscle and epithelial cells of toad urinary bladders exposed to high Ca^{2+} concentrations (Peachey, 1964), and for the cells of dog hearts perfused with high Ca^{2+} concentrations (Legato et al., 1968). One could also mention the accumulation of Ca^{2+} in the mitochondria of kidney cells of animals exposed to excess doses of vitamin D (Scarpelli, 1965) or parathyroid hormone (Caulfield & Schrag, 1964). Quite clearly, then, when mitochondria in situ are exposed to excess Ca^{2+} concentrations, they may accumulate large amounts of it, and store it in the matrix, presumably as an insoluble phosphate salt. It is perhaps convenient to stress again that the sequestration of Ca^{2+} as an insoluble precipitate in the mitochondrial matrix, although not an irreversible process, certainly cannot be reversed rapidly, and is therefore not interesting from the standpoint of rapid metabolic regulation. One other consideration can be made at this point: the accumulation of large amounts of Ca^{2+} and phosphate by mitochondria in situ is apparently compatible with the normal functioning of the organelle, and certainly does not induce gross alterations to its ultrastructure. This consideration is of some importance, since it has been generally (but incorrectly) assumed that the uptake of Ca^{2+} and phosphate, at least by isolated mitochondria, necessarily leads to severe and irreversible damage to the organelle.

In addition to the conditions mentioned above, in-situ loading of mitochondria with large amounts of Ca^{2+} has been observed during experimental and spontaneous injury to various tissues (see Carafoli, 1974, for an extended discussion of the topic). Mitochondrial calcification in necrotic cells is a common observation, but of considerably more interest is the observation of intramitochondrial Ca^{2+} deposits during the initial phases of injury in cells which appear to be otherwise intact. That an increased cellular uptake of Ca^{2+} is an early manifestation of cell injury has indeed been postulated in the past (Judah, Ahmed & McLean, 1964), although the significance of this early intramitochondrial calcification, and its relevance to the subsequent possible death of the cell, are not clear at the moment. It is reasonable to suggest, however, that mitochondrial calcification does not result from primary alterations of the mitochondrial ion-pumping machinery caused by the injuring condition, but is rather part of a defence mechanism by which cells try to control an increased, and dangerous, influx of Ca^{2+} from the extracellular spaces. Of particular interest in this context is a recent report by Horwith et al., (1975) on intramitochondrial calcifications in osteoblasts of some patients with hereditary hyperphosphatasaemia, a condition caused by an extraordinarily rapid turnover of bone due to abnormally increased resorption and formation. The administration of calcitonin to these patients caused a regression of the symptoms, and a dramatic decrease in the number of intramitochondrial electron-opaque masses in the osteoblasts.

CONCLUSIONS

As mentioned above, one question which is still open in dealing with the process of mitochondrial Ca^{2+} transport – perhaps the most important from the standpoint of cell physiology – is how the process is regulated. It would indeed be difficult to understand why mitochondria do not always calcify in the cells, given the continuous influx of Ca^{2+} from the extracellular spaces, if there were no means to control their interaction with Ca^{2+}. The commonly applied argument that the process could be controlled simply by keeping the free Ca^{2+} concentration in the cytosol below the K_m level of the mitochondrial transport system, by actively pumping out the Ca^{2+} that has entered the cell by passive diffusion, is clearly not valid. Mitochondria would take up Ca^{2+} very slowly, but they would eventually still calcify, if no natural means to induce release of the Ca^{2+}, or to block its uptake, were available. Whether the natural regulation of the interaction between mitochondria and Ca^{2+} takes place on the uptake, or on the release process, or on both, is at the moment a matter of speculation. Of the various release-inducing agents described above, some are potentially of physiological significance, and others may be discovered in the future. Unfortunately, the exciting recent report of a cyclic-AMP-mediated release of mitochondrial Ca^{2+} (Borle, 1974; Matlib & O'Brien, 1974) has proved impossible to reproduce in our as well as in other laboratories (A. Scarpa, personal communication; P. Walter, personal communication), and cannot therefore be considered interesting for the time being. The Mg^{2+}-induced inhibition of the uptake of Ca^{2+} described above certainly deserves further study, and its specificity for heart mitochondria is of particular interest. Other 'natural' inhibitors of the uptake process are at the moment not known, and the search for them will be further complicated by the fact that they must be modulated, under physiological conditions, to permit fluctuations in the uptake of Ca^{2+} (the same difficulty will obviously apply to the possible natural release-inducing agents). However difficult the search, there are nevertheless only two alternatives: either the physiological modulators of the transport process are identified, or the role of mitochondrial Ca^{2+} in cell physiology will remain hypothetical.

Part of the original research described in this article has been carried out with the support of the Swiss National Science Foundation (SNF Grant no. 3.1720.73).

REFERENCES

ADDANKI, S., CAHILL, F. D. & SOTOS, J. F. (1968). Determination of intramito-chondrial pH and intramitochondrial–extramitochondrial pH gradient of isolated heart mitochondria by the use of 5,5-dimethyl-2,4-oxazolidinedione. *J. biol. Chem.*, **243**, 2337–2348.

ÅKERMAN, K. E., SARIS, N. E. L. & JÄRVISALO, J. O. (1974). Mitochondrial 'high-affinity' binding sites for Ca^{2+} – fact or artefact? *Biochem. biophys. Res. Commun.*, **58**, 801–807.

AZZI, A. (1969). Redistribution of the electrical charge of the mitochondrial membrane during energy conservation. *Biochem. biophys. Res. Commun.*, **37**, 254–260.

AZZI, A., GHERARDINI, P. & SANTATO, M. (1971). Fluorochrome interaction with the mitochondrial membrane. *J. biol. Chem.*, **246**, 2035–2042.

AZZI, A., SORGATO, C. M. & MONTECUCCO, C. (1975). Effect of calcium ions on the interaction between cytochrome C and the mitochondrial membrane. In *Calcium Transport in Contraction and Secretion*, ed. E. Carafoli, F. Clementi, W. Drabikowski & A. Margreth, pp. 35–49. North-Holland: Amsterdam and New York.

BIELAWSKI, J. & LEHNINGER, A. L. (1966). Stoichiometric relationship in mito-chondrial accumulation of calcium and phosphate supported by hydrolysis of adenosine triphosphate. *J. biol. Chem.*, **241**, 4316–4322.

BORLE, A. (1974). Cyclic AMP stimulation of calcium efflux from kidney, liver and heart mitochondria. *J. Memb. Biol.*, **16**, 221–236.

BRIERLEY, G., MURER, E., BACHMANN, E. & GREEN, D. E. (1963). Studies on ion transport. II. The accumulation of inorganic phosphate and magnesium ions by heart mitochondria. *J. biol. Chem.*, **238**, 3482–3489.

BRIERLEY, G. & SLAUTTERBACK, D. B. (1964). An electron microscope study of the accumulation of Ca^{2+} and inorganic phosphate by heart mitochondria. *Biochim. biophys. Acta*, **82**, 183–186.

BURGER, S. P., FUJII, T. & HANAHAN, D. J. (1968). Stability of the bovine erythro-cyte membrane: release of enzymes and lipid components. *Biochemistry, N.Y.*, **7**, 3682–3699.

BYGRAVE, F. L., DADAY, A. A. & DOY, F. A. (1975). Evidence for a calcium-ion-transport system in mitochondria isolated from flight muscle of the developing sheep blowfly. *Lucilia cuprina. Biochem. J.*, **146**, 601–608.

BYGRAVE, F. L., REED, K. C. & SPENCER, T. (1971). Sigmoidal kinetics associated with calcium uptake and related ATPase in rat liver mitochondria. In *Energy Transduction in Respiration and Photosynthesis*, ed. E. Quagliariello, S. Papa, C. S. Rossi, pp. 981–988. Aadriatica Editrice: Bari, Italy.

CARAFOLI, E. (1965a). Active accumulation of Sr^{2+} by rat-liver mitochondria. III. Stimulation of respiration by Sr^{2+} and its stoichiometry. *Biochim. biophys. Acta*, **97**, 107–117.

(1965b). Active accumulation of Sr^{2+} by rat-liver mitochondria. II. Competition between Ca^{2+} and Sr^{2+}. *Biochim. biophys. Acta*, **97**, 99–106.

(1973). The transport of calcium by mitochondria. Problems and perspectives. *Biochimie*, **55**, 755–762.

(1974). Mitochondrial uptake of calcium ions and the regulation of cell function. *Biochem. Soc. Symp.*, **39**, 89–109.

(1975a). The interaction of Ca^{2+} with mitochondria with special reference to the structural role of Ca^{2+} in mitochondrial and other membranes. *Molec. cell. Biochem.*, **3**, 133–140.

(1975*b*). The role of mitochondria in the contraction–relaxation cycle and other Ca^{2+}-dependent activities of heart cells. In *Basic Functions of Cations in Myocardial Activity*, ed. A. Fleckenstein & N. S. Dhalla, pp. 151–163. University Park Press: Baltimore, Md.

(1975*c*). Mitochondria, Ca^{2+} transport and the regulation of heart contraction and relaxation. *J. molec. cell. Cardiol.*, 7, 83–89.

CARAFOLI, E. & AZZI, A. (1972). The affinity of mitochondria for Ca^{2+}. *Experientia*, 28, 906–907.

CARAFOLI, E., BALCAVAGE, W. X., LEHNINGER, A. L. & MATTOON, J. R. (1970). Ca^{2+} metabolism in yeast cells and mitochondria. *Biochim. biophys. Acta*, 205, 18–26.

CARAFOLI, E. & CROMPTON, M. (1976). Binding proteins and membrane transport. In *Membrane Bound Enzymes*, ed. A. Martonosi. Plenum Press: New York (in press).

CARAFOLI, E. & CROVETTI, F. (1973). Interaction between prostaglandin E$_1$ and calcium at the level of the mitochondrial membrane. *Archs Biochem. Biophys.*, 154, 40–46.

CARAFOLI, E. & GAZZOTTI, P. (1973). The reaction of Ca^{2+} with the inner and outer membrane of mitochondria. *Experientia*, 29, 408–409.

CARAFOLI, E., HANSFORD, R. G., SACKTOR, B. & LEHNINGER, A. L. (1971). Interaction of Ca^{2+} with blowfly flight muscle mitochondria. *J. biol. Chem.*, 246, 964–972.

CARAFOLI, E. & LEHNINGER, A. L. (1971). A survey of the interaction of calcium ions with mitochondria from different tissues and species. *Biochem. J.*, 122, 681–690.

CARAFOLI, E., ROSSI, C. S. & LEHNINGER, A. L. (1964). Cation and anion balance during active accumulation of Ca^{2+} and Mg^{2+} by isolated mitochondria. *J. biol. Chem.* 239, 3055–3061.

(1965). Uptake of adenine nucleotides by respiring mitochondria during active accumulation of Ca^{2+} and phosphate. *J. biol. Chem.*, 240, 2254–2261.

CARAFOLI, E., TIOZZO, R., LUGLI, G., CROVETTI, F. & KRATZING, C. (1974). The release of Ca^{2+} from heart mitochondria by sodium. *J. molec. cell. Cardiol.*, 6, 361–371.

CARAFOLI, E., WEILAND, S. & LEHNINGER, A. L. (1965). Active accumulation of Sr^{2+} by rat-liver mitochondria. I. General features. *Biochim. biophys. Acta*, 97, 88–98.

CAULFIELD, J. B. & SCHRAG, B. A. (1964). Electron microscope study of renal calcification. *Am. J. Path.*, 44, 365–381.

CHANCE, B. (1965). The energy-linked reaction of calcium with mitochondria. *J. biol. Chem.*, 240, 2729–2748.

CHANCE, B. & MELA, L. (1966*a*). Intramitochondrial pH changes in cation accumulation. *Proc. natn. Acad. Sci. USA*, 55, 1243–1251.

(1966*b*). Hydrogen ion concentration changes in mitochondrial membranes. *J. biol. Chem.*, 241, 4588–4599.

(1966*c*). Calcium and manganese interactions in mitochondrial ion accumulation. *Biochemistry, N.Y.*, 5, 3220–3223.

CHANCE, B. & SCHOERNER, B. (1966). High and low energy state of cytochromes. *J. biol. Chem.*, 241, 4577–4587.

CHAPPELL, J. B., COHN, M. & GREVILLE, G. D. (1963). The accumulation of divalent ions by isolated mitochondria. In *Energy-linked Functions of Mitochondria*, ed. B. Chance, pp. 219–231. Academic Press: New York.

CHAPPELL, J. B. & CROFTS, A. R. (1965). Calcium ion accumulation and volume changes of isolated liver mitochondria. *Biochem. J.*, 95, 378–386.

CHAPPELL, J. B. & GREVILLE, G. P. (1963). Isolated mitochondria and accumulation of divalent metal ions. *Fedn Proc.* **22**, 526.

CHEN, C. & LEHNINGER, A. L. (1973). Ca^{2+} transport activity in mitochondria from some plant tissues. *Archs. Biochem. Biophys.*, **157**, 183–196.

COLONNA, R., MASSARI, S. & AZZONE, G. F. (1972). The problem of cation binding sites in the energized membrane of intact mitochondria. *Eur. J. Biochem.*, **34**, 577–585.

CRANG, R. E., HOLSEN, R. C. & HITT, J. B. (1968). Calcite production in mitochondria of earthworm calciferous gland. *Am. Inst. Biol. Sci. Bull.*, **18**, 299–301.

DAVIDOFF, F. (1974). Effects of guanidine derivatives on mitochondrial function. *J. biol. Chem.*, **249**, 6406–6915.

DENTON, R. M., RANDLE, P. J. & MARTIN, R. B. (1972). Stimulation by calcium ions of pyruvate dehydrogenase phosphate phosphatase. *Biochem. J.*, **128**, 161–163.

DONNELLAN, J. F., BARKER, M. D., WOOD, J. & BEECHEY, R. B. (1970). Specificity and locale of the L-3-glycerophosphate-flavoprotein oxidoreduction of mitochondria isolated from the flight muscle of *Sarcophaga barbata*. *Biochem. J.*, **120**, 467–478.

DRAHOTA, Z., CARAFOLI, E., ROSSI, C. S., GAMBLE, R. L. & LEHNINGER, A. L. (1965). The steady state maintenance of accumulated Ca^{++} in rat liver mitochondria. *J. biol. Chem.*, **240**, 2712–2720.

DRAHOTA, Z., GAZZOTTI, P., CARAFOLI, E. & ROSSI, C. S. (1969). A comparison of the effects of different divalent cations on a number of mitochondrial reactions linked to ion translocation. *Archs. Biochem. Biophys.*, **130**, 267–273.

DUGGAN, P. F. & MARTONOSI, A. (1970). The permeability of sarcoplasmic reticulum membranes. *J. gen. Physiol.*, **56**, 147–167.

GEAR, R. L., ROSSI, C. S., REYNAFARJE, B. & LEHNINGER, A. L. (1967). Acid-base exchanges in mitochondria and suspending medium during respiration-linked accumulation of bivalent cations. *J. biol. Chem.*, **242**, 3403–3413.

GILBERT, I. G. F. (1972). The effect of divalent cations on the ionic permeability of cell membranes in normal and tumour tissue. *Eur. J. Cancer*, **8**, 99–105.

GITLER, C. & MONTAL, M. (1972). Formation of decane-soluble proteolipids: influence of monovalent and divalent cations. *FEBS Lett.*, **28**, 329–332.

GITLER, C., RUBALCAVA, B. & CASWELL, A. (1969). Fluorescence changes of ethydium bromide on binding to erythrocyte and mitochondrial membranes. *Biochim. biophys. Acta*, **193**, 474–481.

GONZALES, F. & KARNOVSKY, M. J. (1961). Electron microscopy of osteoclasts in healing fractures of rat bone. *J. Cell Biol.*, **9**, 299–316.

GRABSKE, R. J. (1968). Barium accumulation by isolated rat-liver mitochondria. *Lawrence Radiation Laboratory, Livermore Rept*, UCRL-50435, pp. 1–10.

GREENAWALT, J. W. & CARAFOLI, E. (1966). Electron microscope studies in the active accumulation of Sr^{2+} by rat liver mitochondria. *J. Cell Biol.*, **29**, 37–61.

GREENAWALT, J. W., ROSSI, C. S. & LEHNINGER, A. L. (1964). Effect of active accumulation of calcium and phosphate on the structure of rat liver mitochondria. *J. Cell Biol.*, **23**, 21–38.

GUNTER, T. E. & PUSKIN, J. S. (1972). Manganous ion as a spin-label in studies of mitochondrial uptake of manganese. *Biophys. J.*, **12**, 625–635.

GUNTER, T. E., PUSKIN, J. S. & RUSSELL, P. R. (1975). Quantitative magnetic resonance studies of manganese uptake by mitochondria. *Biophys. J.*, **15**, 319–333.

HACKENBROCK, C. R. & CAPLAN, A. I. (1969). Ion-induced structural transformations in isolated mitochondria. *J. Cell Biol.*, **42**, 221–234.

HANSFORD, R. G. & CHAPPELL, J. B. (1967). Effect of Ca^{2+} on the oxidation of glycerol phosphate by blowfly flight-muscle mitochondria. *Biochem. biophys. Res. Commun.*, **27**, 686–692.

HOHMAN, W. & SCHRAER, H. (1966). The intracellular distribution of calcium in the mucosa of the avian shell gland. *J. Cell Biol.*, **30**, 317–331.

HORWITH, M., NUNEZ, E., WHALEN, J., KROOK, L., VITERI, F., TORUN, B., MENA, E., MacINTYRE, I., SUH, S. M. & EISENBERG, E. (1975). Synthetic human calcitonin in the treatment of hyperphosphatasemia. *Endocrinology* (in press).

JACOBUS, W. E., TIOZZO, R., LUGLI, G., LEHNINGER, A. L. & CARAFOLI, E. (1975). Aspects of energy-linked Ca^{2+} accumulation by rat heart mitochondria. *J. biol. Chem.*, **250**, 7863–7870.

JASAITIS, A. A., KULIENE, V. V. & SKULACHEV, V. P. (1971). Anilinonaphthalene-sulphonate fluorescence changes induced by nonenzymatic generation of membrane potential in mitochondria and submitochondrial particles. *Biochim. biophys. Acta*, **234**, 177–181.

JUDAH, A. D., AHMED, K. & McLEAN, A. E. M. (1964). Possible role of ion shifts in liver injury. In *Cellular Injury*, Ciba Foundation Symposium, ed. A. V. S. de Reuck & J. Knight, pp. 187–205. Churchill Ltd: London.

KIRTLAND, S. J. & BAUM, H. (1972). Prostaglandin E_1 may act as a 'calcium ionophore'. *Nature, Lond.*, **236**, 47–49.

KRANE, S. M. & GLIMCHER, M. J. (1962). Transphosphorylation from nucleoside di- and triphosphate by apatite crystals. *J. biol. Chem.*, **237**, 2991–2998.

LEGATO, M. J. SPIRO, D. & LANGER, G. A. (1968). Ultrastructural alterations produced in mammalian myocardium by variations in perfusate ionic composition. *J. Cell Biol.*, **37**, 1–12.

LEHNINGER, A. L. (1970). Mitochondria and calcium ion transport. *Biochem. J.*, **119**, 129–138.

(1974). Role of phosphate and other proton-donating anions in respiration-coupled transport of Ca^{2+} by mitochondria. *Proc. natn. Acad. Sci. USA*, **71**, 1520–1524.

LEHNINGER, A. L. & CARAFOLI, E. (1969). Evidence for a membrane-linked Ca^{++} carrier in rat liver and kidney mitochondria. In *Biochemistry of the Phagocytic Process*, ed. J. Schultz, pp. 9–22. North-Holland: New York.

LEHNINGER, A. L., CARAFOLI, E. & ROSSI, C. S. (1967). Energy-linked ion movements in mitochondrial systems. *Adv. Enzymol.*, **29**, 259–320.

MALMSTRÖM, K. & CARAFOLI, E. (1975). Effects of prostaglandins on the inter-action of Ca^{++} with mitochondria. *Archs. Biochem. Biophys.*, **171**, 418–424.

MANERY, J. F. (1966). Effects of Ca ions on membranes. *Fedn Proc.*, **25**, 1804–1810.

MARTIN, J. H. & MATTHEWS, J. L. (1969). Mitochondrial granules in chondrocytes. *Calc. Tiss. Res.* **3**, 184–193.

MATLIB, A. & O'BRIEN, P. J. (1974). Adenosine 3'5'-cyclic monophosphate. Stimulation of calcium efflux from mitochondria. *Biochem. J. Soc. Trans.*, **2**, 997–1000.

MEISSNER, H. & KLINGENBERG, M. (1968). Efflux of adenine nucleotide from rat liver mitochondria. *J. biol. Chem.*, **243**, 3631–3639.

MELA, L. (1968). Interaction of La^{3+} and local anesthetic drugs with mitochondrial Ca^{++} and Mn^{++} uptake. *Archs. Biochem. Biophys.*, **123**, 286–293.

MITCHELL, P. (1966). *Chemiosmotic Coupling in Oxidative and Photosynthetic Phosphorylation*, Glynn Research Ltd: Bodmin, UK.

MITCHELL, P. & MOYLE, J. (1969). Estimation of membrane potential and pH difference across the cristae membrane of rat liver mitochondria. *Eur. J. Biochem.*, **7**, 471–484.

MONTAL, M. & KORENBROT, J. I. (1973). Incorporation of rodopsin proteolipid into bilayer membranes. *Nature, Lond.*, **246**, 219–221.

MOORE, C. L. (1971). Specific inhibition of mitochondrial Ca^{++} transport by ruthenium red. *Biochem. biophys. Res. Commun.*, **42**, 298–305.

MOORE, L., CHEN, T., KNAPP, H. R. & LANDON, E. J. (1975). Energy-dependent calcium sequestration activity in rat liver microsomes. *J. biol. Chem.*, **250**, 4562–4568.

NICHOLLS, D. G. (1974). The influence of respiration and ATP hydrolysis on the proton-electrochemical gradient across the inner membrane of rat liver mitochondria as determined by ion distribution. *Eur. J. Biochem.*, **50**, 305–315.

PADAN, E. & ROTTENBERG, H. (1973). Respiratory control and the proton electrochemical gradient in mitochondria. *Eur. J. Biochem.*, **40**, 431–337.

PANDE, S. V. & BLANCHAER, M. C. (1971). Reversible inhibition of mitochondrial ADP phosphorylation by long chain acyl coenzyme A esters. *J. biol. Chem.*, **246**, 402–411.

PASQUALI-RONCHETTI, I., GREENAWALT, J. W. & CARAFOLI, E. (1969). On the nature of the dense matrix granules of normal mitochondria. *J. Cell Biol.*, **140**, 565–568.

PEACHEY, L. D. (1964). Electron microscopic observations on the accumulation of divalent cations in intramitochondrial granules. *J. Cell Biol.*, **20**, 95–111.

PFAFF, E., HELDT, H. W. & KLINGENBERG, M. (1969). Adenine nucleotide translocation of mitochondria. *Eur. J. Biochem.*, **10**, 484–493.

RASMUSSEN, H. & BIKLE, D. D. (1975). Calcium and non-vesicular secretion in the kidney: calcium and mitochondrial function. In *Calcium Transport in Contraction and Secretion*, ed. E. Carafoli, F. Clementi, W. Drabikowski & A. Margreth, pp. 111–121. North-Holland: Amsterdam and New York.

RASMUSSEN, H., CHANCE, B. & OGATA, E. (1965). A mechanism for the reactions of calcium with mitochondria. *Proc. natn. Acad. Sci. USA*, **53**, 1069–1070.

RASMUSSEN, H. & OGATA, E. (1966). Parathyroid hormone and the reaction of mitochondria to cations. *Biochemistry, N.Y.*, **5**, 733–745.

REED, K. C. & BYGRAVE, F. L. (1975). A kinetic study of mitochondrial calcium transport. *Eur. J. Biochem.*, **55**, 497–504.

REYNAFARJE, B. & LEHNINGER, A. L. (1969). High affinity and low affinity binding of Ca^{2+} by rat liver mitochondria. *J. biol. Chem.*, **244**, 584–593.

—— (1974). Ca^{2+} transport by mitochondria from L-1210 mouse ascites tumor cells. *Proc. natn. Acad. Sci.*, **70** 1744–1748.

REYNOLDS J. A. (1972). Are inorganic cations essential for the stability of biological membranes? *Ann. N.Y. Acad. Sci.*, **195**, 75–85.

REYNOLDS, J. A. & TRAYER, H. (1971). Stability of membrane proteins in aqueous media. *J. biol. Chem.*, **246**, 7337–7342.

RORIVE, G., NIELSON, R. & KLEINZELLER, A. (1972). Effect of pH on the water and electrolyte content of renal cells. *Biochim. biophys. Acta*, **266**, 376–396.

ROSE, B., & LOEWENSTEIN, W. R. (1975a). Permeability of cell junction depends on local cytoplasmic calcium activity. *Nature, Lond.*, **254**, 250–252.

—— (1975b). Calcium ion distribution in cytoplasm visualized by aequorin. Diffusion in the cytosol is restricted due to energised sequestering. *Science, Wash.* (in press).

ROSSI, C. S., AZZI, A. & AZZONE, G. F. (1967). Ion transport in liver mitochondria. *J. biol. Chem.*, **242**, 951–957.

ROSSI, C. S., BIELAWSKI, J. & LEHNINGER, A. L. (1966). Separation of H$^+$ and OH$^-$ in the extramitochondrial and mitochondrial phases during Ca^{++} activated electron transport. *J. biol. Chem.*, **241**, 1919–1921.

ROSSI, C. S. & LEHNINGER, A. L. (1964). Stoichiometry of respiratory stimulation, accumulation of Ca^{++} and phosphate, and oxidative phosphorylation in rat liver mitochondria. *J. biol. Chem.*, **239**, 3971–3980.

ROTTENBERG, H. & SCARPA, A. (1974). Calcium uptake and membrane potential in mitochondria. *Biochemistry, N.Y.*, **13**, 4811–4819.

SCARPA, A. & AZZONE, G. F. (1970). The mechanism of ion translocation in mitochondria. *Eur. J. Biochem.*, **12**, 328–335.

SCARPA, A. & GRAZIOTTI, P. (1973). Mechanisms for intracellular calcium regulation in heart. *J. gen. Physiol.*, **62**, 756–772.

SCARPELLI, D. G. (1965). Experimental nephrocalcinosis. A biochemical and morphologic study. *Lab. Invest.*, **14**, 123–141.

SCHÄFER, G. & BOJANOWSKI, D. (1972). Interaction of biguanides with mitochondrial and synthetic membranes. *Eur. J. Biochem.*, **27**, 364–375.

SELWYN, M. J., DAWSON, A. P. & DUNNETT, S. J. (1970). Calcium transport in mitochondria. *FEBS Lett.*, **10**, 1–5.

SORDAHL, L. A. (1974). Effects of magnesium, ruthenium red and the antibiotic ionophore A23187 on initial rates of calcium uptake and release by heart mitochondria. *Archs. Biochem. Biophys.*, **167**, 104–115.

SORDAHL, L. A. & SCHWARTZ, A. (1967). Effects of dipyridamole on heart muscle mitochondria. *Molec. Pharmacol.*, **3**, 509–521.

SOUTHARD, J. H. & GREEN, D. E. (1974). High affinity binding of Ca^{++} in mitochondria. A reappraisal. *Biochem. biophys. Res. Commun.*, **59**, 30–37.

SPENCER, T. & BYGRAVE, F. L. (1973). The role of mitochondria in modifying the cellular ionic environment. *Bioenergetics*, **4**, 347–362.

STUCKI, J. W. & INEICHEN, E. A. (1974). Energy dissipation by calcium recycling and the efficiency of calcium transport in rat liver mitochondria. *Eur. J. Biochem.*, **48**, 365–375.

THOMAS, R. S. & GREENAWALT, J. W. (1968). Microincineration, electron microscopy and electron diffraction of calcium phosphate-loaded mitochondria. *J. Cell Biol.*, **39**, 55–67.

VASINGTON, F. D., GAZZOTTI, P., TIOZZO, R. & CARAFOLI, E. (1972). The effect of ruthenium red on calcium transport and respiration in rat liver mitochondria. *Biochim. biophys. Acta*, **256**, 43–54.

VASINGTON, F. D. & GREENAWALT, J. W. (1964). Ca^{++} and P_i uptake by nonphosphorylating mitochondrial preparations. *Biochem. biophys. Res. Commun.*, **15**, 133–138.

VINOGRADOV, A. & SCARPA, A. (1973). The initial velocities of calcium uptake by rat liver mitochondria. *J. biol. Chem.*, **248**, 5527.

VINOGRADOV, A., SCARPA, A. & CHANCE, B. (1972). Calcium and pyridine nucleotide interaction in mitochondrial membranes. *Archs. Biochem. Biophys.*, **152**, 646–654.

WEINBACH, E. C. & VON BRAND, T. (1965). The isolation and composition of dense granules from Ca^{++} loaded mitochondria. *Biochem. biophys. Res. Commun.*, **19**, 133–137.

WILLIAMS, R. J. P. (1975). The binding of metal ions to membranes and its consequences. In *Biological Membranes*, ed. D. S. Parsons, pp. 106–121. Clarendon Press: Oxford.

TRANSCELLULAR TRANSPORT
OF CALCIUM

By A. R. TEREPKA, J. R. COLEMAN,
H. J. ARMBRECHT and T. E. GUNTER

Department of Radiation Biology and Biophysics,
School of Medicine and Dentistry, University of Rochester,
Rochester, New York 14642

Most higher animal organisms must maintain the calcium concentration of intra- and extracellular fluids within closely guarded limits for proper metabolic function. No other major plasma electrolyte is under more rigorous control in the body (McLean & Urist, 1968). To maintain this control, a complex system of regulation has been developed, involving absorption of calcium by the small intestine, reabsorption by the kidney, and storage in the skeleton. For this system to function, calcium must be transported across at least three tissues: intestinal epithelium, kidney tubular epithelium, and the cellular linings of bone. The cells in these tissues must presumably transport this calcium without raising their own intracellular calcium ion concentration, since high levels of calcium have adverse effects, such as uncoupling oxidative phosphorylation in mitochondria (Chance, 1965) and inhibiting pyruvate kinase or pyruvate carboxylase activity (Bygrave, 1966; Kimmich & Rasmussen, 1969). Consequently, it is generally accepted that intracellular ionic calcium must be maintained at very low levels, around 10^{-6} M (Borle, 1967), and that most of the calcium present in whole tissues is in the extracellular fluid at concentrations of about 10^{-3} M (Mulryan, Neuman, Neuman & Toribara, 1964).

Models to explain the mechanisms by which cellular calcium is regulated have been proposed by several investigators (Schachter, Kowarski, Finkelstein & Ma, 1966; Borle, 1967, 1973; Wasserman, 1968; Talmage, 1969; Lehninger, 1970; Rasmussen, 1970). The common feature in all of these models (Fig. 1) is the presence of outwardly directed plasma membrane calcium 'pumps'. These pumps maintain intracellular calcium concentration at a low and constant level by actively extruding calcium which has entered passively from the surrounding extracellular fluid. Most models also assign an important role to mitochondria and, in a few cases, to endoplasmic reticulum as well. In situations where the outwardly directed calcium pumps are temporarily 'overloaded', these intracellular organelles

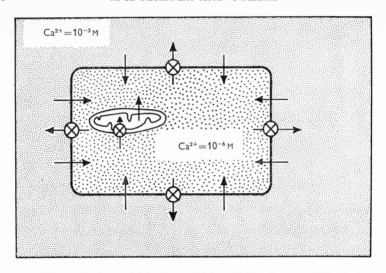

Fig. 1. Basic model for intracellular calcium ion regulation.
For further explanation see text.

serve as 'buffers' by actively sequestering calcium and preventing intracellular calcium from reaching toxic levels.

In order to account for transcellular calcium transport, a slight modification of these calcium homeostatic mechanisms is hypothesized, i.e. the
outwardly directed calcium pumps are predominantly located at the basal
and/or lateral surfaces of the cell. This localization of the calcium pumps
results in a polarized transfer in that any calcium entering passively at the
apex is most likely to be extruded from the basal and/or lateral surfaces of
the cell. Mitochondria and endoplasmic reticulum are still considered to
play a prominent role in this transfer process, serving if necessary to
sequester calcium while in transit. However, our investigations have led
us to conclude that the transcellular transfer of calcium is not a simple
modification of these calcium homeostatic mechanisms.

Studies of two calcium transporting epithelia, embryonic chick chorioallantoic membrane and the small intestine of rat and chick, have strongly
suggested that the transfer of calcium *across* a cell involves processes
distinctly different from intracellular calcium ion regulation. The experimental evidence indicates that calcium in transit through these tissues is
kept separate from the intracellular calcium pools, including the mito-
condria, and that the initial entry of calcium destined for transcellular
transport is not a passive process but a complicated energy-requiring step.
Based on these observations, a tentative model for transcellular calcium
transport will be presented which may reconcile our findings with the
calcium transport models previously proposed.

THE CHICK CHORIOALLANTOIC MEMBRANE

Because of its ready availability, simple morphology and easy adaptability to in-vitro experimentation, the embryonic chick chorioallantoic (CA) membrane has been utilized for most of our studies of transcellular calcium transport. From the thirteenth to the twentieth day of incubation, the CA membrane transfers about 130 mg of calcium from the egg shell into the embryonic circulation to be used in bone formation (Johnston & Comar, 1955; Simkiss, 1967). It lines the interior of the shell after the ninth day and consists of an ectoderm, endoderm and mesoderm (Romanoff, 1961). The ectoderm faces the egg shell and is tightly joined to the acellular inner shell membrane, which is attached to the shell itself. The endoderm is in contact with the allantoic fluid, an excretory product of the developing chick embryo. The mesoderm lies between the ectoderm and endoderm. In addition to absorbing calcium liberated from the egg shell, the CA membrane serves as an extra-embryonic lung and bladder during embryo development. The endodermal cells of the CA membrane actively transport sodium, with chloride and water being reabsorbed in the process (Stewart & Terepka, 1969).

Light and electron microscopical examination of the CA membrane ectodermal cell layer have revealed a close correlation between its full morphological development and the onset of rapid calcium accumulation by the embryo (Stewart & Terepka, 1969; Coleman & Terepka, 1972a). After about 14 days of incubation, the ectodermal surface of the membrane is composed of two distinctive cell types named, because of their location and morphological features, 'capillary-covering' cells and 'villus-cavity' cells. Capillary-covering cells predominate and seem to be the actual calcium-absorbing cells in the tissue, while villus-cavity cells are thought to play a role in dissolving the calcium carbonate of the egg shell (Coleman & Terepka, 1972a). Together they apparently form a tightly joined epithelial layer of cells which does not allow passive movement of calcium to occur. Active transport of calcium is required to transfer the liberated egg shell calcium into the embryonic circulation, and we have studied the physiology and biochemistry of the active transport process extensively in our laboratory (Terepka, Stewart & Merkel, 1969; Moriarty & Terepka, 1969; Terepka, Coleman, Garrison & Spataro, 1971; Garrison & Terepka, 1972a, b).

In a series of experiments with the CA membrane and attached inner shell membrane mounted in an Ussing-type transport chamber, it can be demonstrated that the membrane actively transports calcium from the ectodermal to the endodermal side of the tissue *in vitro* (Terepka *et al.*,

1969). This ability to transport *in vitro* corresponds well with the age of onset of rapid calcium accumulation *in vivo*. The membrane will transport against high chemical gradients with little backflux, against electrical gradients, and under short circuit conditions. The mechanism is saturable and rather specific for calcium. It does transport strontium and manganese but less efficiently. The transport process is energy dependent, requiring oxidative phosphorylation. Metabolic inhibitors, lack of oxygen, and decreased temperature are all strongly inhibitory.

When the oxygen consumption of the membrane was measured either by standard manometric techniques or in a transport chamber fitted with an oxygen electrode (Garrison & Terepka, 1972 a), it was found that the induction of calcium transport was associated with a significant stimulation of oxygen utilization. Divalent cations other than calcium caused stimulation depending on their ability to be transported. Simultaneous measurements of oxygen consumption and calcium transport at many different rates gave a Ca^{2+}/O_2 ratio of 0.5, indicating that this transport system had a high energy requirement compared to other active transport systems.

EVIDENCE FOR COMPARTMENTATION OF CALCIUM

During the course of the investigations described above, it was noted that CA membranes increased their calcium content as much as three-fold under active transport conditions, i.e. when incubated in 1.0 mM Ca^{2+} for 1–3 h. This immediately raised several pertinent questions: (1) What was the spatial distribution of the calcium transferred into the membrane? (2) Was it diffusely distributed throughout, or was it localized to a cell layer, to individual cells, or to only part of a cell? (3) Could the spatial distribution, whether diffuse or localized, tell us something about the membrane's active calcium transfer mechanism. To answer these questions we employed the electron probe X-ray microanalyser, an instrument capable of detecting the in-situ distribution of chemical elements in tissue samples. Our aim was to map sites of calcium accumulation in CA membranes actively transporting this cation. Plates 1 and 2 illustrate some of our results.

For morphological orientation, the lower half of Plate 1 shows a transmission electron micrograph of a region of the CA membrane ectoderm. The capillary bed containing erythrocytes is separated from the noncellular shell membrane by the thin extended cytoplasmic arms of an ectodermal cell. This cell is the so-called capillary-covering cell. For reference, the thickness of the red blood cell in this section is about 1.5 μm. The extended arms of capillary-covering cells are characteristically very thin, 0.1 μm in some places, and are notably free of mitochondria. Most mitochondria tend to cluster around the nucleus in the cell body well away from the cell surface facing the shell membrane and egg shell proper. The cytoplasmic arms do possess a striking morphological feature, i.e. a 'ruffled border' similar to that found in cells engaged in endocytosis, and a 'fuzzy coat' or glycocalyx.

Directly above the electron micrograph in Plate 1 are two sets of images, recorded with the electron probe, of two different CA membrane sections. The upper images (Plate 1*a*, *c*) are sample current images, similar in resolution to light micrographs, which are used for morphological identification. Directly underneath the sample current images is the calcium X-ray distribution in the two tissue sections. The oscilloscope grid lines are 5 μm apart and cover identical areas of each image pair. Calcium-containing areas appear as bright spots in the X-ray images. Random spots in the background represent 'noise' from electronics, bremsstrahlung, and stray environmental radiation. Such calcium localizations were found only in the ectoderm, above the capillary bed, and within or close to the extended arms of the capillary-covering cells.

This localization can be seen more readily in Plate 2 which shows a region of the CA membrane where the cytoplasmic arms that separate the capillaries from the shell membrane are more apparent due to their thickness. The three distinct calcium-containing loci in Plate 2*b* were derived from the tissue section shown in Plate 2*a* which is the specimen current image obtained from the section while the calcium X-ray signals were being recorded. The areas of the membrane containing the calcium loci were re-examined at higher electron probe resolution. The area containing the two loci on the left is shown in Plate 2*c* and *d*, and Plate 2*e* and *f* show the locus on the right. The demarcation of shell membrane from capillary-covering cell cytoplasm is more clearly identifiable with the resolution available in these specimen current images, and it can be seen that the two loci on the left are within the arms of the capillary-covering cell. The localization on the right extends around and above the respiratory capillary with no appreciable extension into the main body of the cell which can be seen below the capillary. Such calcium-rich loci were not found when transport was inhibited by a variety of techniques. From these and many other electron probe studies of CA membranes under transport and non-transport conditions, it was concluded that, during active transcellular transport, calcium was somehow sequestered within a small portion of the cytoplasm of the ectodermal capillary-covering cells.

The intracellular distribution of calcium has been also examined in another calcium-transporting epithelium, the small intestine of rats and chicks, using the electron probe (Warner & Coleman, 1975). These extensive studies revealed that intestines actively transporting calcium also contained calcium-rich loci. These loci were distributed within the absorptive cells of the intestinal epithelium in a pattern remarkably similar to that found in the CA membrane. This can be seen in Plate 3, sample current and X-ray images from two different sections of rat intestine. Plate 3a shows portions of three adjacent villi from an everted gut sac that had been exposed to 2 mM Ca^{2+} for 5 min. Magnification is 19 μm per grid square division. The calcium X-ray image of this section (Plate 3b) shows numerous calcium-rich loci which are located just inside the brush border of the epithelial cells. Their positions can be identified in the specimen current image by the bright spots on each side of the lumen. The tissue section shown in Plate 3c was prepared after 15 min of exposure to calcium. It shows at higher magnification intestinal epithelial cells facing the lumen above and the lamina propria below. The brush border is visible as a bright zone at the apex of the absorptive cells, with the nuclei visible in a row through the approximate cell center. The calcium localizations in this cluster of cells are more deeply situated. Some are in a supra-nuclear position, but most are close to the cell borders and/or within the intercellular space. In the upper left corner of Plate 3c, the tips of the absorptive cells of an adjacent villus are also seen. In these cells calcium is still localized at the luminal border of the cells. The calcium locus seen in the lumen is directly adjacent to a goblet cell and is associated with mucus. As with the CA membranes, these localizations were not found when transport was inhibited by making the animals rachitic, but they reappeared in abundance when rachitic animals were repleted with vitamin D.

Evidence supporting the validity of these electron probe morphological observations has recently been obtained utilizing an entirely different approach, i.e. electron paramagnetic resonance (EPR) spectroscopy (Armbrecht, Gunter, Puskin & Terepka, 1976). Not only is EPR a sensitive spectroscopic tool (as little as 10^{-12} to 10^{-13} mole of a paramagnetic material can be observed) it can provide information concerning the local *milieu* in which the materials reside. For example, using nitroxide radicals as spin labels attached to biologically important molecules, EPR studies have provided information on protein conformation, the active sites of enzymes, and the structure and fluidity of cell membranes (Hamilton & McConnell, 1968; Griffith & Wagner, 1969). A more recent development of the technique has utilized the transition metal manganese to investigate active divalent cation transport (Gunter & Puskin, 1972). Manganous ion, which is paramagnetic, can be studied by EPR spectroscopy; calcium itself cannot. Using manganese and isolated rat liver mitochondria, information has been obtained on the nature of mitochondrial ionic binding sites, the local concentration of the cation, and the viscosity and pH of the medium in which the cation is located (Puskin & Gunter, 1972, 1973).

Analogous to the well-established observation that manganese and calcium uptake by mitochondria involve the same energy-linked process, manganese uptake by the CA membrane also seems to utilize the same energy-dependent system that transports calcium (Armbrecht, Terepka & Gunter, 1976). Uptake of manganese by the tissue was reduced significantly under conditions that reduced calcium uptake, although the reduction was not as large as that seen with calcium. In Ussing-chamber experiments, manganese was taken up preferentially by the ectodermal side of the tissue and this followed the pattern seen with calcium. Kinetic studies of manganese uptake showed that uptake was saturable and had an apparent Michaelis constant of 0.33 mM, only slightly larger than that for calcium (0.28 mM). Also manganese uptake was competitively inhibited by calcium, and calcium uptake by manganese.

Fig. 2a shows the EPR spectrum of a CA membrane exposed on its ectodermal side to manganese for $1\frac{1}{2}$ h in a transport chamber. This first-derivative spectrum can be resolved mathematically into the two spectra shown in Fig. 2b, c which correspond to two different forms of manganese in the tissue. Fig. 2b is very similar to the hyperfine sextet spectrum of manganese in water, i.e. manganese hexahydrate. Quantitatively, the bulk of the manganese (about 95 %) present in the tissue had the spectral form shown in Fig. 2c. The characteristics of this single broad line spectrum have been related to manganese in a 'spin exchange' form. This spin exchange phenomenon only occurs when paramagnetic ions are so closely packed that overlap of electronic orbitals occur. It is a manifestation of magnetic field 'averaging' produced by exchange of electrons between different atomic systems. As a consequence, it may be thought of as a clear-cut indication of a high local concentration of a paramagnetic ion. From the spin exchange line width of manganese in the CA membrane ($\simeq$ 280 oersted) it was calculated that spacing of manganese was about 0.47 ± 0.05 nm. The spin exchange form of manganese could be detected after 5 min of exposure to manganese. Thus, it was concluded that manganese in transit across the CA membrane was kept localized and highly concentrated. Since manganese and calcium appear to be handled by the same membrane transport system, these EPR studies are in accord with the electron probe results indicating intracellular calcium sequestration during transcellular transport.

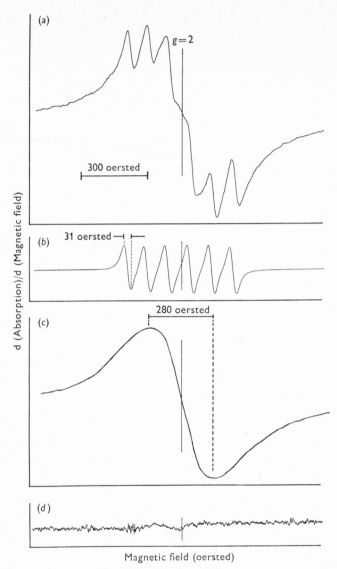

Fig. 2. Electron paramagnetic resonance (EPR) spectra of manganese transported into the CA membrane. (a) EPR spectrum of control CA membrane exposed for 1.5 h to 0.75 mM Mn^{3+} at pH 7.4. (b) Spectrum of manganese hexahydrate in 20 % glycerol–water mixture simulating hexahydrate component of (a). (c) Manganese spin-exchange spectrum obtained when manganese hexahydrate, (b), is subtracted from total spectrum, (a). (d) Background spectrum of standard buffer solution in tissue sample holder taken at 2.5 times the gain of spectrum A. All spectra are first-derivative spectra taken at X-band frequency. The peak to trough distances indicated are the line widths of the absorption lines (from Armbrecht, Gunter, Puskin & Terepka, 1976).

ROLE OF MITOCHONDRIA

It is well established that isolated mitochondria when exposed to calcium are capable of actively transporting and sequestering large amounts of this ion internally. Consequently, the electron probe and EPR observations immediately raised the question of direct mitochondrial involvement in the compartmentation and translocation process. However, the morphological studies with the electron microscope and electron probe initially suggested that this was not the case. The characteristic location of the calcium loci seen in the electron probe calcium X-ray images was in the upper portion or extended arm of the CA membrane capillary-containing cells, areas where mitochondria are conspicuously absent in transmission electron micrographs. In the rat intestine, calcium 'hot spots' were most frequently found at the apical and lateral cell borders of the absorptive cells were mitochondria, again, are generally absent.

Evaluation of published values for mitochondrial calcium uptake in the literature also does not support the suggestion of their involvement in transcellular transport. In fact, as shown in Table 1, if one does some simple calculations utilizing directly determined values of CA membrane dry weight and calcium content during transport, and takes the various mito-chondrial calcium loads given by Borle (1973), Schraer, Elder & Schraer (1973) and Lehninger (1970), one is forced to rely on some implausible assumptions or to accept some impossible conclusions to implicate mito-chondria as the calcium sequestering organelle, at least in the CA mem-brane.

First, consider Borle's kinetically determined values for mitochondrial calcium loads in kidney cells. Under limited loading conditions (first line, Table 1), he gives a value of about 10 nmole Ca mg^{-1} mitochondrial pro-tein. The last column shows the per cent of the total dry weight of the CA membrane that must be mitochondrial protein to accommodate the calcium found in the CA membrane under transport conditions, i.e. when incubated in 1 mM Ca^{2+} for 1–2 h. An impossible value of 750% or 7.5 times the total dry weight of the tissue would be necessary to contain the observed amount of CA membrane calcium. Under Borle's massive loaded conditions (500 nmole Ca^{2+} mg^{-1} mitochondrial protein) the per-centage of the total dry weight of the tissue that must be mitochondrial protein drops to 15% (second line, Table 1). Except for liver, values given for mitochondrial protein in most ordinary tissues is about 1%, but some go as high as 5% of the total protein content of the tissue – not the total dry weight as used here.

The lower part of the Table gives values taken from the work of Schraer

Table 1. *Calculated amounts of mitochondrial protein required to accommodate calcium contained in transporting chorioallantoic membranes*

	Mitochondrial calcium loads (nmole mg⁻¹ protein)	'Mitochondrial Protein' needed for Ca^{2+} in CAM[d] (% total dry weight)
In situ (kinetics)		
Limited load[a]	10	750.0
Massive load[a]	500	15.0
In vitro (isolated)		
Massive load[b]	1500	5.6
Massive load[c]	3000	2.5

[a] Borle (1973).
[b] Schraer *et al.* (1973).
[c] Lehninger (1970).
[d] Chorioallantoic membrane (CAM) Ca^{2+} load $\simeq$ 0.06 μmole cm⁻², dry weight $\simeq$ 0.8 mg cm⁻².

et al. (1973) on isolated mitochondria from avian liver and shell gland and the statement of Lehninger (1970) that under *optimal* conditions isolated mitochondria from liver can accumulate up to 3000 nmole Ca^{2+} mg⁻¹ mitochondrial protein. Using these values for isolated mitochondria, one can obtain a value for CA membrane mitochondrial protein content which is at least within the bounds of the conceivable, i.e. between 2.5–5.0 % of the dry weight of the tissue. However, this would mean that *every* mito-chondrion in the tissue was massively loaded with calcium and presumably completely uncoupled, a rather implausible assumption.

More direct evidence that mitochondria are not involved in calcium sequestration has also been derived from our extensive biochemical studies of actively transporting CA membranes. We have previously reported (Garrison & Terepka, 1972a) that calcium markedly and uniquely stimulates the respiration of isolated CA membranes, and the stimulation of oxygen consumption is directly associated with the active transport of calcium by the membrane (Fig. 3). This stimulation of oxygen uptake is shown by comparing the last bar in the Fig. 3, the respiration of CA membranes in the absence of calcium (about 9 μl O_2 cm⁻² h⁻¹), with the first bar (about 13 μl) which is the oxygen consumption of CA membranes exposed to 1 mM Ca^{2+} under active transport conditions. Now notice the second bar. CA membranes actively transporting calcium and presumably massively loading their mitochondria, when exposed to dinitrophenol, show a striking further increase in oxygen uptake in the presence of the uncoupler. It is difficult for us to envisage calcium-loaded mitochondria as being responsible for the sequestration phenomenon seen during trans-

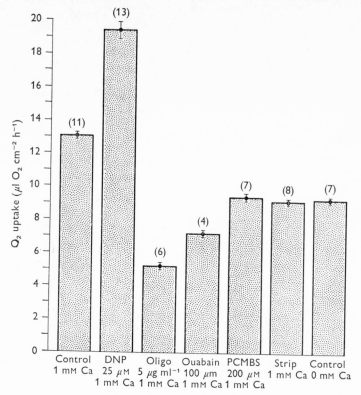

Fig. 3. The effect of various inhibitors of calcium transport on the oxygen uptake of the CA membrane in 1 mM Ca^{2+}. The concentration of inhibitor is shown below each bar. The last bar on the right is the 0 mM Ca^{2+} control value. Experiments performed by manometry. DNP, dinitrophenol; Oligo, oligomycin; PCMBS, p-chloromecuribenzene sulphonate (from Garrison & Terepka, 1972a).

cellular transport when we can uncouple transporting CA membranes with dinitrophenol. Such CA membranes should have had mitochondria already fully uncoupled by the calcium-loading process.

Fig. 3 also demonstrates several other important points relative to possible mitochondrial involvement in calcium translocation. First, it should be noted that oligomycin, which does not directly inhibit mitochondrial calcium uptake has a profound effect on CA membrane oxygen consumption and concomitantly (although not shown in this Figure) on active calcium transport (Garrison & Terepka, 1972a, b). Secondly, notice the oxygen consumption of ouabain-treated CA membranes in the presence of calcium. Isolated mitochondria do not require sodium to effect a massive accumulation of calcium. On the other hand, ouabain treatment or complete sodium withdrawal from the medium stop calcium transport and, as

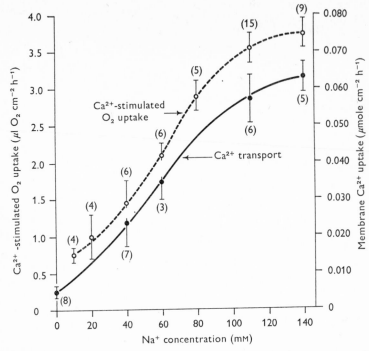

Fig. 4. Calcium-stimulated oxygen uptake (O---O) and calcium transport (●-●) as a function of sodium concentration in the bathing medium. Osmolality of the medium was maintained by replacing sodium chloride with choline chloride (from Garrison & Terepka, 1972b).

shown here, eliminate calcium stimulated oxygen consumption. Actually the rate of ectodermal calcium transport is strictly dependent on the presence of sodium and complex kinetics are observed as a function of sodium concentration as shown in Fig. 4. The data in the Figure indicate a direct and apparently constant relationship between the rate of calcium transport and the rate of calcium-stimulated oxygen uptake. This requirement for sodium, which has also been observed in intestine (Martin & DeLuca, 1969; Adams & Norman, 1970), seems rather inconsistent with the direct involvement of mitochondria in calcium translocation. Parenthetically, it should be mentioned that the CA membrane does not require phosphate in the bathing medium to effect the active transport of calcium transcellularly.

What happens when ectodermal cell mitochondria are exposed directly to calcium? This can be done by mechanically stripping the inner shell membrane from the CA membrane, a procedure that selectively removes the plasmalemma of the capillary-covering cells of the ectoderm (Plate 4).

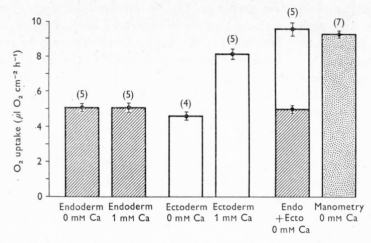

Fig. 5. Oxygen uptake of the endoderm (hatched bars) and the ectoderm (open bars) in the presence and absence of 1 mM Ca^{2+}. Experiments performed in the oxygenstat apparatus of Garrison & Ford (1970). The two bars at the right compare the sum of the oxygen uptake of the ectoderm and the endoderm in 0 mM Ca^{2+} (hatched and open bar) with that measured for the whole membrane in 0 mM Ca^{2+} by manometry (dotted bar) (from Garrison & Terepka, 1972a).

The bulk of the cell remains intact and this includes the nucleus and the mitochondria. Such preparations completely lose their calcium transport ability and, as shown in Fig. 3 (second bar from the right), calcium-stimulated oxygen consumption is completely eliminated.

The oxygen consumption data presented in Figs. 3 and 4 were obtained by standard manometric techniques. Using a specially designed apparatus (Garrison & Ford, 1970), the CA membrane could be mounted in an Ussing-type chamber and the effect of calcium on either the ectoderm or the endoderm could be monitored with an oxygen electrode. As shown in Fig. 5, the stimulation of oxygen consumption when the CA membrane is exposed to 1 mM Ca^{2+} is specifically related to the ectodermal cell layer of the membrane. The two hatched bars on the left in the Figure show the effect of 1 mM Ca^{2+} on the endoderm, and the next two open bars show the effect of calcium on the ectoderm. The endoderm of the CA membrane contains as many, if not more mitochondria (Coleman & Terepka, 1972a), yet there is no stimulation of respiration when 1.0 mM Ca^{2+} is added to the medium. Presumably, these endodermal cells are continually controlling *their* intracellular calcium levels but apparently in some way entirely different from the calcium transport mechanisms present within the ectoderm.

The final point to be made regarding mitochondrial involvement during transcellular calcium transport stems from the previously mentioned EPR studies done during active manganese transport by the CA membrane (Armbrecht, Gunter, Puskin & Terepka, 1976). As with calcium, it can be calculated that if the amount of manganese seen within the tissue in the spin exchange form were *inside* mitochondria, the mitochondria would have to be massively loaded. However, from the work of Puskin & Gunter (1972, 1973) the typical spin exchange line width for massively loaded mito-chondria is 240 oersted, while the narrowest line width seen in the CA membrane was 282 oersted. The observed difference in the spin exchange line widths is significantly different and suggests a different form of close packing of the manganese atoms in the two cases, i.e. in isolated manganese-loaded mitochondria and in intact CA membranes actively transporting manganese.

ROLE OF THE EXTERNAL PLASMA MEMBRANE

The intracellular calcium regulation model shown in Fig. 1 tacitly assumes passive entry of calcium into the cell, an assumption predicated on the 1000-fold downhill calcium concentration difference between extra- and intracellular fluids. If such models apply to transcellular calcium transport, then the initial entry of calcium into the tissue should also be passive. However, even our early studies of this process in the CA membrane indicated that this was not the case. For example, included in the data shown in Table 2 is the effect of *p*-chloromercuribenzene sulphonate (PCMBS) on oxygen consumption and calcium uptake of CA membranes exposed to 1 mM Ca. PCMBS is an organic mercurial that is known to react specifically with sulphydryl groups and does not readily penetrate cell membranes. CA membranes were pretreated with PCMBS for 15 min, washed briefly, and their oxygen and calcium uptake was measured simul-taneously in a medium free of PCMBS. It can be seen that PCMBS decreased membrane calcium uptake dramatically and lowered oxygen consumption to the level observed when the membrane was in a calcium-free medium (first line, Table 2) and, therefore, had no external calcium to transport. Using the oxygenstat apparatus of Garrison & Ford (1970) it could be established that this effect of PCMBS was specific to the calcium transporting ectodermal cell layer of the membrane (Garrison & Terepka, 1972*a*). Endodermal cell oxygen uptake was not affected by PCMBS pre-treatment or calcium removal. The effects of shell membrane stripping, also shown in Table 2, are virtually identical to the effects of PCMBS. Since the common site of action for the two treatments is the external plasma membrane of the ectodermal cells, this suggested that a vital sulphydryl-

Table 2. *Effect of inhibitors on calcium and oxygen uptake by chorioallantoic membranes*

Experiment	Oxygen uptake μl O_2 cm^{-2} h^{-1}	Calcium uptake μmole cm^{-2} h^{-1}	No. of observations
Calcium-free medium	9.3 ± 0.2	—	(7)
Control (1 mM Ca^{2+})	12.8 ± 0.3	0.063 ± 0.004	(5)
PCMBS (0.2 mM)	9.5 ± 0.2	0.005 ± 0.001	(5)
Stripping	9.2 ± 0.2	0.009 ± 0.001	(3)
Ouabain (0.1 mM)	7.2 ± 0.3	0.004 ± 0.001	(4)
Choline$^+$ (140 mM)	7.1 ± 0.2	0.005 ± 0.002	(8)

containing component of the active transcellular calcium transport mechanism in the tissue is located on this surface (Terepka *et al.*, 1971).

Some recent experiments involving ouabain treatment and sodium replacement tend to support this view. Active accumulation of sugars and amino acids by intestine is strictly dependent on external sodium and convincing models based on kinetic studies have been proposed to explain the bioenergetics of the interdependence (Crane, 1965; Kimmich, 1973). As shown in Fig. 4 and Table 2, calcium transport and calcium-stimulated oxygen uptake by the CA membrane also require the presence of sodium in the external medium. Ouabain, in the presence of 140 mM Na$^+$ completely inhibited both processes. Furthermore, it was found that ouabain was most effective when placed on the ectodermal side of the tissue. This would hardly be expected if calcium entry were a strictly passive process.

To investigate initial entry in more detail, calcium uptake by the CA membrane was measured as a function of time under a variety of experimental conditions (H. J. Armbrecht, P. C. Wade & A. R. Terepka, unpublished). Fig. 6 illustrates some of the results. In 0.2 mM Ca^{2+} under control conditions (top line) there was a rapid initial binding within 10 sec followed by a slower linear uptake over the next two minutes. The slope of this line was taken as the initial velocity of calcium entry into the tissue, V_i. With sodium removed or DNP present, initial binding was unaffected, but V_i was reduced to about half that of the controls. V_i was also reduced to the same extent by oligomycin, PCMBS and ouabain indicating that much of the initial uptake of calcium is energy dependent and linked to the presence of normal sodium gradients. When V_i was measured at various external calcium concentrations, it was found that in the physiological concentration range 0.05 to 1 mM the data were consistent with simple Michaelis–Menten kinetics and gave a K_m of 0.24 mM. In previous studies of the kinetics of calcium transport across the whole tissue over several

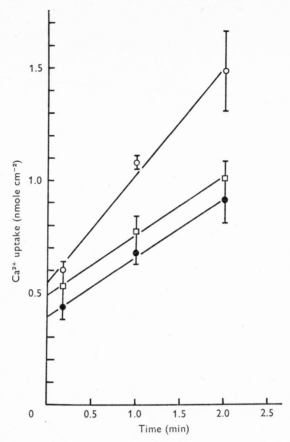

Fig. 6. Effect of dinitrophenol (DNP, ●) and sodium replacement with choline[+] (□) on the rate of calcium uptake by the CA membrane; ○, control. First data points are at 10 sec. The slope of these lines was taken as the initial velocity of calcium entry into the tissue, V_i (H. J. Armbrecht, P. C. Wade & A. R. Terepka, unpublished).

hours (Garrison & Terepka, 1972*a*), a Michaelis constant of 0.28 mM was obtained. The similarity of these K_m values derived in the two experiments suggests that initial entry of calcium into the tissue is a major barrier in the transcellular transport process.

Fig. 7 shows a study in which initial calcium uptake was measured as external sodium concentration was raised from 0 to 140 mM. When the velocity of calcium uptake, expressed as a per cent of the total sodium-dependent uptake, was plotted as a function of sodium concentration, the data followed a sigmoid curve. As shown previously (Fig. 4), when calcium transport and its associated calcium stimulated oxygen uptake was measured over a 1–2 h interval, strict sodium dependence was also

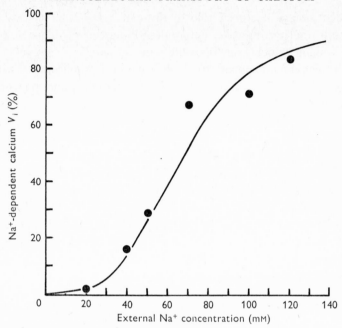

Fig. 7. Effect of external sodium concentration on the initial velocity of calcium entry (V_i) into the CA membrane. V_i is plotted as a percent of the total sodium-dependent calcium uptake (H. J. Armbrecht, P. C. Wade & A. R. Terepka, unpublished).

observed, and the shape of the curve and half-maximal stimulation of these parameters was almost identical, i.e. about 60–70 mM Na⁺. The similarity in results between the initial uptake studies and the studies involving overall calcium transport and oxygen consumption are consistent with the view that transfer of calcium into the interior of the transporting cells is a critical energy dependent process in the CA membrane. When the data in Fig. 7 are expressed in a Hill-type plot to determine the degree of 'cooperativity' between the two ions, the results indicate that about 3 sodium ions are needed for the entry of one calcium. At the present time we cannot say whether the metabolic energy required for transcellular calcium transport is utilized directly for calcium uptake or utilized indirectly to maintain normal extra- to intracellular sodium gradients which are then coupled to transport.

TRANSCELLULAR TRANSPORT MODEL

To reconcile our experimental findings suggesting calcium sequestration, lack of direct mitochondrial involvement, and active initial entry of calcium into the cell, we have formulated a model for transcellular calcium transport which assigns a crucial role to the external plasma membrane of the calcium transporting cells. Before describing this model some experiments involving cytochalasin B should be mentioned.

Intracellular microfilament structures have been implicated as vital for normal cell multiplication (cytokinesis), cell mobility and phagocytosis (Wessells *et al.*, 1971; Carter, 1972). These cell activities have been shown to be selectively inhibited by cytochalasin B, a specific mold metabolic product. In addition, it has been reported (Mizel & Wilson, 1972) that cytochalasin B inhibits the transport of several hexoses in isolated mammalian cells without affecting (in HeLa cells, at least) calcium influx or efflux, i.e. those processes primarily involved in intracellular calcium ion regulation (Fig. 1).

In our electron microscope and electron probe studies of the ectoderm of the CA membrane (Coleman & Terepka, 1972*a*, *b*, *c*) it was noted that the upper portions and extended arms of the capillary-covering cells, the most frequent site of calcium localizations, did not contain mitochondria. They did, however, contain numerous vesicles, bundles of fine fibres which tended to lie oriented in an apical to basal direction, and small granules tentatively identified as glycogen. Another morphological feature was that, coincident with the embryonic age of onset of active calcium transport, the capillary-covering cells became firmly attached to the non-cellular shell membrane and developed a 'ruffled' appearance common to cells carrying out endocytosis. Since cytochalasin B has effects on both endocytosis and microfilament function, we studied the influence of the agent on CA membrane calcium transport. It was found (Fig. 8) that cytochalasin B, in the dose range of 2–20 μg ml^{-1}, was capable of reducing calcium uptake to levels almost as low as those seen when the membranes were exposed to a metabolic inhibitor, dinitrophenol. Dimethyl sulphoxide, the vehicle used to dissolve cytochalasin B, had no effect when used alone. In similar experiments, colchicine did not inhibit CA membrane calcium transport.

The model we are proposing for active transcellular calcium transport starts with the premise that intracellular (cytosol) calcium regulation is of paramount importance to *all* cells, including those in specialized calcium transporting epithelia. Our working hypothesis is that the 'leak in–pump out' transport mechanism that has evolved to maintained strict cytosol calcium control is ubiquitous and inherent to normal cellular function. Transcellular calcium transport, on the other hand, is considered as a specialized process developed only by certain cells in those tissues charged with bulk transfer of calcium.

It seems to us that a cell called upon to transfer calcium in bulk would always be in danger of 'swamping' the active calcium extrusion mechanism

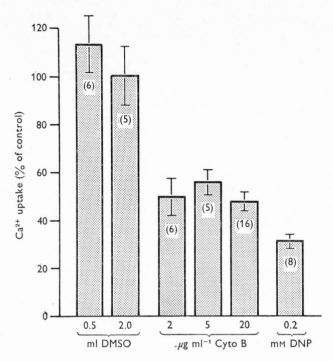

Fig. 8. Effect of cytochalasin B (Cyto B) and dinitrophenol (DNP) on calcium uptake by the CA membrane. Dimethyl sulphoxide (DMSO) was the vehicle used to dissolve the cytochalasin B (A. R. Terepka, P. C. Wade & J. R. Coleman, unpublished).

which, by definition, must have a finite capacity. In non-specialized cells the leak–pump mechanism may be so balanced that mitochondrial buffering is normally not required. While transient small accumulations of calcium by mitochondria may be tolerable, persistent large accumulations are certainly toxic and to be avoided if at all possible. In order to cope with this problem, we are suggesting that an endocytotic mechanism was evolved by those cells continuously faced with the necessity to transfer large amounts of calcium transcellularly. Fig. 9 summarizes the salient features of the proposed model.

The specialized calcium transporting cell shown is tightly bound to its neighbours to prevent indiscriminant leakiness through the epithelial cell layer. Intracellular calcium regulation is maintained by the ubiquitous outwardly directed membrane calcium pumps. An invagination ('ruffling') of the external plasma membrane is shown at location I in the Figure, and step II illustrates the subsequent formation of an endocytotic vesicle. These two steps, which we consider critical to transcellular calcium transport,

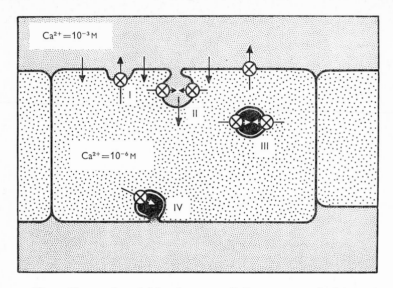

Fig. 9. Proposed model for the transcellular transport of calcium.
For further explanation see text.

are much more complex than shown. Cation binding selectivity, energy
input, sodium dependency and microfilament function are some of the
known factors that come into play. In any case, what the Figure primarily
attempts to show is that by interiorizing plasma membrane in the form of
numerous small vesicles (step III), not only can the number of 'pump'
sites be increased enormously, but their pumping activities would all be
directed toward the interior of the vesicle (Wasserman & Kallfelz, 1970)
and away from both the closely regulated cytosol calcium compartment
and the mitochondria.

What controls the movement of calcium-enriched vesicles through the
cytoplasm is not known. They must somehow be directed to the functionally correct side of the cell, e.g. the surface facing the lamina propria
in the intestine or facing the capillary endothelium in the chick CA
membrane. But once this occurs, net transfer of calcium from the donor
to the acceptor compartments could simply involve refusion of the calcium-
loaded vesicle on the acceptor side (step IV). The overall effect of the
proposed mechanism is bulk calcium movement across a cell, protection
of mitochondria from exposure to high concentrations of calcium, and the
avoidance of wide and potentially toxic fluctuations in cytosol ionic calcium
levels.

It should be noted that the model, as presented, does not require a change
in current concepts regarding the characteristics of calcium transporting

PLATE I

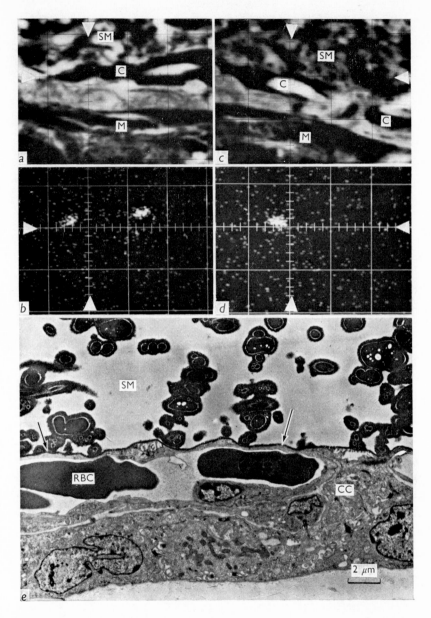

Sample current and corresponding calcium K_α X-ray images of two CA membrane preparations (*a*, *b*; *c*, *d*). The calcium localizations are found between the shell membrane (SM) and lumen of nearby respiratory capillaries (C). M, Mesoderm; △, identical grid marks. A low power electron micrograph (*e*) demonstrates the cell relationships in this region. The usual thin cytoplasmic arms (arrows) of a capillary-covering (CC) cell can be resolved running the entire length of the section and occupying the region between the shell membrane and respiratory capillary. Capillaries are readily identified by their distinctive erythrocytes (RBC). Araldite sections; 10 min integrations (from Coleman & Terepka, 1972*c*).

PLATE 2

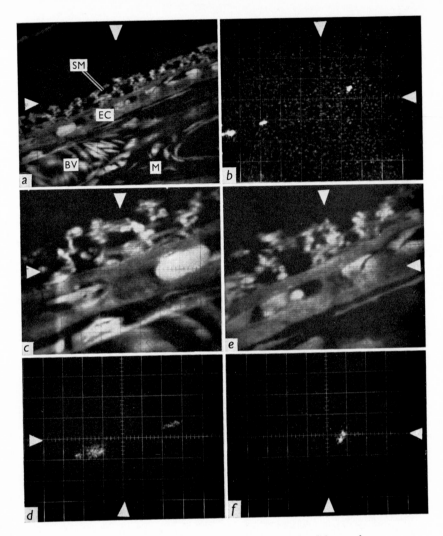

Sample current (*a*) and calcium K_α X-ray images (*b*) of a CA membrane preparation containing unusually wide capillary-covering cell arms. The area containing the two calcium loci on the left in (*b*) are shown at higher resolution (3μm/grid square division) in (*c*, *d*,), and the one on the right is seen in (*e*, *f*). The precise location of the calcium loci can be determined by referring to identical grid marks (△) in each image pair. These loci are clearly situated in the region between the shell membrane and respiratory capillaries, the same region occupied by the capillary-covering arms of the capillary-covering cells (*e*). Araldite sections. Integration times: *b*, 20 min; *d*, *f*, 15 min. EC, Ectoderm; BV, large mesoderm blood vessel; M, mesoderm; SM, shell membrane (from Coleman & Terepka, 1972*c*).

PLATE 3

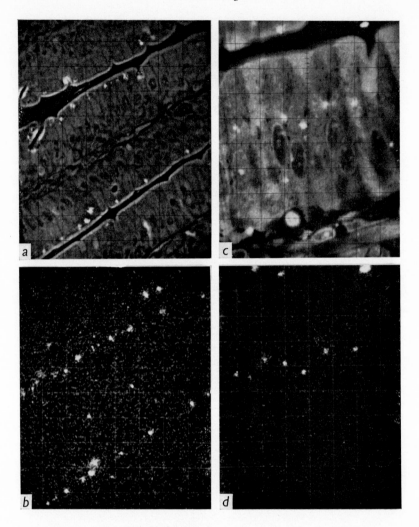

Sample current and calcium K_α X-ray images of sections of rat duodenum (a) Sample current image of portions of three villi in tissue which had been exposed to 2 mM Ca *in vitro* for 5 min before being fixed. The corresponding X-ray image showing the distribution of calcium is seen in (b); 19 μm/grid square division. (c) Sample current image of a portion of two villi from intestine which had been exposed to calcium as above, but for 15 min before being fixed. The corresponding X-ray image showing the distribution of calcium is seen in (d); 7 μm/grid square division (from Warner & Coleman, 1975).

PLATE 4

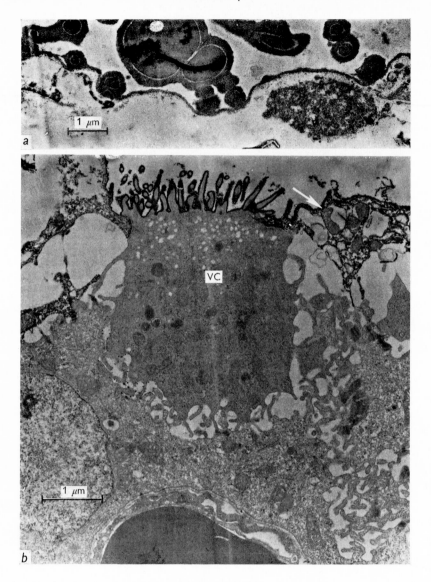

Conventional electron micrographs of mechanically 'stripped' CA membrane. The separated inner shell membrane is seen in (a); the cell membrane has retained its integrity, and small bits of cytoplasmic debris from the damaged ectoderm are seen associated with it. The ectoderm of a stripped CA membrane that was treated with ruthenium red after stripping is seen in (b). An intact villus cavity cell (VC) is seen in the centre of the image, with its villi distinctly outlined by ruthenium red. To either side of the VC cell are remnants of capillary-covering cells. These are broken open, as can be seen by the fact that ruthenium red has stained the cytoplasm remnants. Intact mitochondria can be seen in the broken capillary-covering cells (arrow).

plasma membrane pumps. Although this type of compartmentation model for calcium transport is suggested by the experimental evidence, many of the molecular processes involved in the hypothesis need further investigation and confirmation. In particular we need a better understanding of the way in which calcium is 'packaged' at the cell surface and the form in which it is held while in transit through the cell.

Vesicle formation and sequestration of transported materials by absorptive cells have been proposed by others in the past as part of the mechanism for transcellular transport, but almost exclusively for large organic molecules (Cardell, Badenhausen & Porter, 1967; Weisberg, Rhodin & Glass, 1968; Rodewald, 1971; Walker, Cornell, Davenport & Isselbacher, 1972; Allison & Davies, 1974; Bronk & Leese, 1974). The possibility that a a similar mechanism may play a role in transcellular calcium transport leads to the speculation that such transport processes are of more general biological significance than currently appreciated. Distinct advantages accrue for both the transporting cells and the organism if material required for normal metabolic function is kept sequestered during absorption. Absorbed material would not be diverted to the metabolic needs of the individual cell at the expense of the entire organism, and the transporting cells would not be required to expend excessive energy for protection against potentially toxic intracellular concentrations of materials they are called upon to absorb. If it is established that a small ion such as calcium is transported transcellularly by a process that involves endocytosis and vesicle formation, a radical alteration in current transport concepts may well become a necessity.

This paper is based on work supported, in part, by US PHS Training Grant no. 5To1-DE00175 and in part, by the US Energy Research and Development Administration at the University of Rochester Biomedical and Environmental Research Project and has been assigned Report no. UR-3490-843.

REFERENCES

ADAMS, T. H. & NORMAN, A. W. (1970). Studies on the mechanism of action of calciferol. I. Basic parameters of vitamin D-mediated calcium transport. *J. biol. Chem.*, **245**, 4421–4441.

ALLISON, A. C. & DAVIES, P. (1974). Mechanism of endocytosis and exocytosis. *Symp. Soc. exp. Biol.*, **28**, 419–446.

ARMBRECHT, H. J., GUNTER, T. E., PUSKIN, J. S. & TEREPKA, A. R. (1976). An electron paramagnetic resonance study of Mn^{++} uptake by the chick chorioallantoic membrane. *Biochim. biophys. Acta*, **426**, 557–569.

ARMBRECHT, H. J., TEREPKA, A. R. & GUNTER, T. E. (1976). Energy-dependent Mn^{++} and Ca^{++} uptake by the embryonic chick chorioallantoic membrane. *Biochim. biophys. Acta*, **426**, 547–556.

BORLE, A. B. (1967). Membrane transfer of calcium. *Clin. Orthop. related Res.*, **52**, 267–291.

—— (1973). Calcium metabolism at the cellular level. *Fedn Proc.*, **32**, 1944–1950.

BRONK, J. R. & LEESE, H. J. (1974). Accumulation of amino acids and glucose by the mammalian small intestine. *Symp. Soc. exp. Biol.*, **28**, 283–304.

BYGRAVE, F. L. (1966). The effect of calcium ions on the glycolytic activity of Ehrlich Ascites-tumour cells. *Biochem. J.*, **101**, 480–487.

CARDELL, R. R., JR, BADENHAUSEN, S. & PORTER, K. R. (1967). Intestinal triglyceride absorption in the rat. *J. Cell Biol.*, **34**, 123–152.

CARTER, S. B. (1972). The cytochalasins as research tools in cytology. *Endeavour*, **31**, 77–82.

CHANCE, B. (1965). The energy-linked reaction of calcium with mitochondria. *J. biol. Chem.*, **240**, 2729–2748.

COLEMAN, J. R. & TEREPKA, A. R. (1972a). Fine structural changes associated with the onset of calcium, sodium and water transport by the chick chorioallantoic membrane. *J. Memb. Biol.*, **7**, 111–127.

—— (1972b). Electron probe analysis of the calcium distribution in cells of the embryonic chick chorioallantoic membrane. I. A critical evaluation of techniques. *J. Histochem. Cytochem.*, **20**, 401–413.

—— (1972c). Electron probe analysis of the calcium distribution in cells of the embryonic chick chorioallantoic membrane. II. Demonstration of intracellular location during active transcellular transport. *J. Histochem. Cytochem.*, **20**, 414–424.

CRANE, R. K. (1965) Na^+ dependent transport in the intestine and other animal tissues. *Fedn Proc.*, **24**, 1000–1006.

GARRISON, J. C. & FORD, G. D. (1970) An improved method for measurement of oxygen consumption at constant oxygen tension. *J. appl. Physiol.*, **28**, 685–688.

GARRISON, J. C. & TEREPKA, A. R. (1972a). Calcium-stimulated respiration and active calcium transport in the isolated chick chorioallantoic membrane. *J. Memb. Biol.*, **7**, 128–145.

—— (1972b). The interrelationships between sodium ion, calcium transport and oxygen utilization in the isolated chick chorioallantoic membrane. *J. Memb. Biol.*, **7**, 146–163.

GRIFFITH, O. H. & WAGNER, A. S. (1969). Nitroxide free radicals: spin labels for probing biomolecular structure. *Acc. chem. Res.*, **2**, 17–24.

GUNTER, T. E. & PUSKIN, J. S. (1972). Manganous ion as a spin label in studies of mitochondrial uptake of manganese. *Biophys. J.*, **12**, 625–635.

HAMILTON, C. & McCONNELL, H. (1968). Spin labels. In *Structural Chemistry and Molecular Biology*. ed. A. Rich & N. Davidson, pp. 115–129. W. H. Freeman: San Francisco.

JOHNSTON, P. M. & COMAR, C. L. (1955). Distribution and contribution of calcium from the albumin, yolk and shell to the developing chick embryo. *Am. J. Physiol.*, **183**, 365–370.

KIMMICH, G. A. (1973). Coupling between Na^+ and sugar transport in small intestine. *Biochim. biophys. Acta*, **300**, 31–78.

KIMMICH, G. A. & RASMUSSEN, H. (1969). Regulation of pyruvate carboxylase activity by calcium in intact rat liver mitochondria. *J. biol. Chem.*, **244**, 190–199.

LEHNINGER, A. L. (1970). Mitochondria and calcium ion transport. *Biochem. J.*, **119**, 129–138.

McLean, F. C. & Urist, M. R. (1968). *Bone: An Introduction to the Physiology of Skeletal Tissue*, 3rd edn, pp. 147–172. University of Chicago Press: Chicago.

Martin, D. L. & DeLuca, H. F. (1969). Calcium transport and the role of vitamin D. *Archs. Biochem. Biophys.*, **134**, 139–148.

Mizel, S. B. & Wilson, L. (1972). Inhibition of the transport of several hexoses in mammalian cells by cytochalasin B. *J. biol. Chem.*, **247**, 4102–4105.

Moriarty, C. M. & Terepka, A. R. (1969). Calcium transport by the isolated chick chorio-allantoic membrane. *Archs. Biochem. Biophys.*, **135**, 160–165.

Mulryan, B. J., Neuman, M. W., Neuman, W. F. & Toribara, T. Y. (1964). Equilibration between tissue calcium and injected radiocalcium in the rat. *Am. J. Physiol.*, **207**, 947–952.

Puskin, J. S. & Gunter, T. E. (1972). Evidence for the transport of manganous ion against an activity gradient by mitochondria. *Biochim. biophys. Acta.*, **275**, 302–307.

—— (1973). Ion and pH gradients across the transport membrane of mitochondria following Mn^{++} uptake in the presence of acetate. *Biochem. biophys. Res. Commun.*, **51**, 797–803.

Rasmussen, H. (1970). Cell communication, calcium ion, and cyclic adenosine monophosphate. *Science, Wash.*, **170**, 404–412.

Rodewald, R. (1971). Selective antibody transport in the proximal small intestine of the neonatal rat. *J. Cell Biol.*, **45**, 635–640.

Romanoff, A. L. (1961). *The Avian Embryo*, pp. 1081–1140. Macmillan: New York.

Schachter, D., Kowarski, S., Finkelstein, J. D. & Ma, R. W. (1966). Tissue concentration differences during active transport of calcium by intestine. *Am. J. Physiol.*, **211**, 1131–1136.

Schraer, R., Elder, J. A. & Schraer, H. (1973). Aspects of mitochondrial function in calcium movement and calcification. *Fedn Proc.*, **32**, 1938–1943.

Simkiss, K. (1967). *Calcium in Reproductive Physiology*, pp. 198–213. Chapman & Hall: London.

Stewart, M. E. & Terepka, A. R. (1969). Transport functions of the chick chorioallantoic membrane. I. Normal histology and evidence for active electrolyte transport from the allantoic fluid, *in vivo*. *Expl Cell Res.*, **58**, 93–106.

Talmage, R. V. (1969). Calcium homeostasis–calcium transport–parathyroid action. The effects of parathyroid hormone on the movement of calcium between bone and fluid. *Clin. Orthop.*, **67**, 210–224.

Terepka, A. R., Coleman, J. R., Garrison, J. C. & Spataro, R. F. (1971). Active transcellular transport of calcium by embryonic chick chorioalantoic membrane. In *Cellular Mechanisms for Calcium Transfer and Homeostasis*, ed. G. Nichols & R. H. Wasserman, pp. 371–389. Academic Press: New York.

Terepka, A. R., Stewart, M. E. & Merkel, N. (1969). Transport functions of the chick chorioallantoic membrane. II. Active calcium transport, *in vitro*. *Expl Cell Res.*, **58**, 107–117.

Walker, W. A., Cornell, R., Davenport, L. M. & Isselbacher, K. J. (1972). Macromolecular absorption. Mechanism of horseradish peroxidase uptake and transport in adult and neonatal rat intestine. *J. Cell Biol.*, **54**, 195–205.

Warner, R. W. & Coleman, J. R. (1975). Electron probe analysis of calcium transport by small intestine. *J. Cell Biol.*, **64**, 54–74.

Wasserman, R. H. (1968). Calcium transport by the intestine: a model and comment on vitamin D action. *Calc. Tiss. Res.*, **2**, 301–313.

Wasserman, R. H. & Kallfelz, F. A. (1970). Transport of calcium across biological membranes. In *Biological Calcification: Cellular and Molecular Aspects*, ed. H. Schraer, pp. 313–345. Appleton-Century-Crofts: New York.

WEISBERG, H., RHODIN, J. & GLASS, G. B. J. (1968). Intestinal vitamin B_{12} absorption in the dog. III. Demonstration of the intracellular pathway of absorption by light and electron microscopic autoradiography. *Lab. Invest.*, **19**, 516–525.

WESSELLS, N. K., SPOONER, B. S., ASH, J. F., BRADLEY, M. O., LUDUENA, E. L., TAYLOR, E. L., WRENN, J. T. & YAMADA, K. M. (1971). Microfilaments in cellular and developmental processes. *Science, Wash.*, **171**, 135–143.

A CYBERNETIC VIEW OF
CELL CALCIUM METABOLISM*

By A. B. BORLE and J. H. ANDERSON

Departments of Physiology and Pharmacology, University of
Pittsburgh, School of Medicine, Pittsburgh, Pa 15261, USA

Any attempt to describe the control and the regulation of intracellular calcium must take into account several properties of the cell which are generally well recognized.

First, the concentration of ionized calcium in the cytosol of most cells is three to four orders of magnitude lower than the calcium activity of the extracellular fluids. Therefore, one must postulate the existence of one or several mechanisms able to establish and to maintain steep concentration gradients.

Second, cytoplasmic calcium is a trigger for many cellular events: for excitation–contraction coupling, stimulus-secretion coupling, microtubular function, cell mitosis and cell motility. Calcium is also an inhibitor of several essential enzymes, for instance Na–K ATPase, and adenyl cyclase. Consequently, the cytoplasmic free calcium concentration must be able to rise and fall but only within the limits compatible with other cellular activities.

Third, the cytoplasmic free calcium concentration is well protected from fluctuations in calcium activity of the extracellular fluids. Most metabolic functions are little affected by marked hypocalcaemia. Furthermore, experiments performed *in vitro* with buffers containing 2.5 mM Ca do not show obvious alterations in cellular metabolism, although 2.5 mM is twice the normal concentration of extracellular free calcium. Therefore, it appears that intracellular metabolism is not significantly affected when the extracellular calcium concentration varies within certain limits.

There is of course a whole spectrum of cells, from erythrocytes to muscle. One cannot pretend to describe with a single model the calcium homeostasis of red blood cells and of skeletal muscle cells. It is obvious that an erythrocyte must rely entirely on its plasma membrane to establish and maintain a calcium activity gradient between the intra- and extracellular compartment. Apparently the red blood cell is able to do so with an efficient calcium pump, involving a calcium-dependent ATPase, and a very low

* Supported by a grant from the National Institutes of Health, United States Public Health Service (no. AM 07867).

[141]

membrane permeability to calcium. At the other end of the spectrum, a skeletal muscle cell is able to contract and to relax in an extracellular medium completely free of calcium. Obviously this cell is able to control and regulate its free calcium concentration by mechanisms which are exclusively intracellular. The calcium homeostasis of all other cells must lie somewhere between that of the red blood cell and that of the skeletal muscle cell. To various degrees, all have subcellular structures, mitochondria and perhaps endoplasmic reticulum, capable of buffering their cytosolic calcium, but, in addition, their plasma membrane may play a role in controlling and regulating their cytoplasmic free calcium.

EXPERIMENTAL DATA

All the data that we have collected for the last seven years, in isolated cells and cell systems (Borle, 1969a, b, 1970, 1971a, 1972a, b, 1973, 1974a, b, 1975a) indicate that cytoplasmic calcium is primarily controlled by an intracellular compartment, presumably mitochondrial, and to a much lesser extent by the cell plasma membrane. Our results also suggest that the regulation of the levels of cytoplasmic calcium by phosphate, cyclic AMP, parathyroid hormone (PTH), calcitonin and vitamin D also occurs at the mitochondrial level. Mitochondria appear to be affected first and it is in the mitochondrial pool that the changes are most significant (Borle, 1975c).

Before describing our model of cellular calcium homeostasis I would like to briefly review the most relevant findings of our experiments. First, in all the isolated cell systems studied so far, the calcium fluxes measured by kinetic analysis are the same as those found in nerve, in muscle and in transcellular systems, i.e. between 0.03 and 0.06 pmole cm^{-2} sec^{-1} (Table 1) (Borle, 1969b, 1970, 1971a, 1974a, 1975a). Second, changes in extracellular calcium have a small influence on the intracellular calcium pool (Fig. 1). Intracellular calcium responds to a step increase in extracellular calcium concentration with a steady state gain of 0.65 (Borle, 1970). Third, a rise in cytoplasmic calcium activity increases cellular calcium influx (Fig. 2). This could be due to a calcium–calcium exchange or to a direct effect of intracellular free calcium on the sodium, potassium and calcium conductance of the plasma membrane (Winegrad & Shanes, 1962; Beeler & Reuter, 1970; Lew, 1970; Romero & Whittam, 1971; Blum & Hoffman, 1972; Krnjnevic & Lisiewicz, 1972). A similar effect of intracellular free calcium on calcium influx has also been observed in the squid axon by Blaustein & Russell (1975).

Several ions, nucleotides, and hormones influence the cellular metabolism of calcium and appear to be regulators of cell calcium homeostasis

Table 1. *Unidirectional fluxes of calcium*

Transport system	ECF [Ca²⁺] mM	Calcium influx pmole cm⁻² sec⁻¹	References
Cellular systems			
Squid axon	10.70	0.076	Hodgkin & Keynes, 1957
Sartorius muscle	1.00	0.094*	Bianchi & Shanes, 1959
Atrium	1.25	0.016	Winegrad & Shanes, 1962
	2.50	0.029	
	3.75	0.048	
HeLa cells	1.25	0.030	Borle, 1969*b*
Monkey kidney cells	1.25	0.050	Borle, 1970
Dog kidney cells	1.25	0.059	Borle, 1971*a*
Human intestinal cell	1.25	0.042	Borle, 1971*a*
Chick intestinal cell	1.25	0.057	Borle, 1974*a*
Thymocytes	1.25	0.020	Borle, 1971*a*
Ehrlich ascites cells	0.5	0.014	Levinson & Blumenson, 1970
Transcellular systems			
Intestine	1.25	0.048†	Cramer, 1963
Intestine	2.00	0.025†	Dumont, Curran & Solomon, 1960
Intestine		0.047	Walling & Rothman, 1969
Intestine LCD		0.083	Walling & Rothman, 1970
Frog skin	1.21	0.046	Watlington, Burke & Estep, 1968
Toad bladder	2.7	0.011	Walzer, 1970
Toad bladder	0.25	0.080	Borle, unpublished

* This value would be 0.016 if corrected for the surface area of the T-system.
† Recalculated from the original data.
ECF = extracellular fluids.

(Fig. 3). For instance we have shown by kinetic analysis of ⁴⁵Ca uptake curves that PTH and cyclic AMP increase intracellular calcium and calcium transport across the plasma membrane (Borle, 1970, 1975*b*). On the other hand calcitonin markedly depresses the membrane transport of calcium although it increases the total pool of cell calcium (Borle, 1975*a*). Finally, in intestinal cells vitamin D deficiency decreases both the cell calcium pool and calcium transport (Borle, 1974*a*).

We have further shown by isotopic desaturation experiments that raising the extracellular phosphate concentration depresses calcium efflux (Fig. 4) while reducing extracellular phosphate stimulates calcium efflux (Borle, 1972*a*). Like many other investigators, we have also shown that cyclic

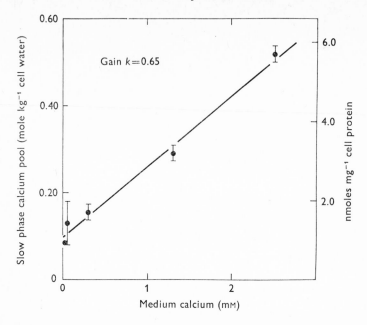

Fig. 1. Effect of the extracellular calcium concentration on the intracellular exchangeable calcium pool (slow phase pool) obtained by kinetic analysis of ^{45}Ca uptake curves in isolated monkey kidney cells (LLC-MK2) grown in culture (after Borle, 1970).

AMP stimulates calcium efflux from the cell (Borle, 1972*b*) while calcitonin inhibits calcium efflux (Borle, 1969*a*; Harrell, Binderman & Rodan, 1973).

Finally, compartmental analyses reveal that a subcellular pool, which we have identified to be the mitochondria, is also affected by the same ions and hormones (Fig. 5), whereas high phosphate, PTH, cyclic AMP and calcitonin increase the size of this pool four- to ten-fold, low phosphate and vitamin D deficiency decrease it to one-third of the control values.

HYPOTHESIS

To explain these results, we have proposed that mitochondria may be the main controllers and the main regulators of cell calcium homeostasis. In our hypothesis, calcium transport and exchange across the mitochondrial membrane determine cytoplasmic calcium activity. Calcium fluxes into and out of the cell, on the other hand, are secondarily regulated by the cytoplasmic calcium levels.

The steady state control of the concentration of free calcium in the cytosol can be described as follows: calcium influx into mitochondria,

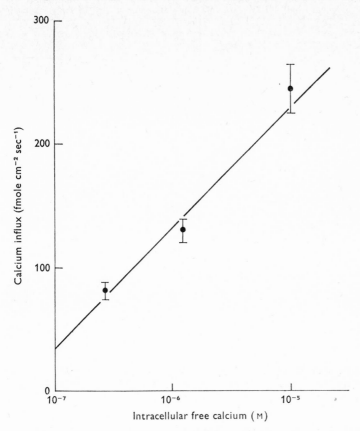

Fig. 2. Effect of the intracellular free calcium concentration on calcium influx measured by kinetic analysis of ^{45}Ca uptake curves. Intracellular calcium was fixed by Ca-EGTA buffers introduced into kidney cells by hypoosomotic treatment (Borle, in preparation).

J_{cm}, is equal to the product of the rate constant of influx, k_{cm}, and the cytoplasmic calcium activity, Ca_c^{2+}:

$$J_{cm} = k_{cm} \cdot Ca_c^{2+}. \qquad (1)$$

Calcium efflux from mitochondria is equal to the product of the rate constant of efflux, k_{mc}, and the mitochondrial calcium activity, Ca_m^{2+}:

$$J_{mc} = k_{mc} \cdot Ca_m^{2+}. \qquad (2)$$

At steady state, influx equals efflux so that

$$k_{cm} \cdot Ca_c^{2+} = k_{mc} \cdot Ca_m^{2+}. \qquad (3)$$

In the presence of precipitates of calcium phosphate in the mitochondrial matrix, Ca_m^{2+} will be a function of a constant of solubility product, K_{sp}, and of the phosphate concentration, P_1:

$$Ca_m^{2+} = f(K_{sp}/P_1). \qquad (4)$$

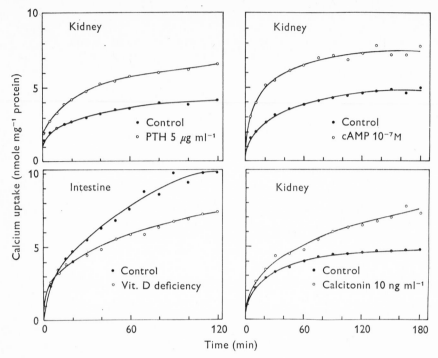

Fig. 3. Effects of PTH, cyclic AMP (cAMP), calcitonin and vitamin D deficiency on calcium uptake in isolated monkey kidney cells (LLC-MK2) and isolated rat intestinal cells (from Borle, 1970, 1974a, 1975a, c).

So that one can write that at steady state

$$\mathrm{Ca_c^{2+}} = \frac{k_{\mathrm{mc}}}{k_{\mathrm{cm}}}\mathrm{Ca_m^{2+}} = \frac{k_{\mathrm{mc}}}{k_{\mathrm{cm}}}\mathrm{f}\left(\frac{K_{\mathrm{sp}}}{\mathrm{P_i}}\right). \qquad (5)$$

We have interpreted the results which we obtained with high and low phosphate concentration, with cyclic AMP, PTH and calcitonin in the following manner:

$$\mathrm{Ca_c^{2+}} \;=\; \frac{k_{\mathrm{mc}}}{k_{\mathrm{cm}}} \;\; \mathrm{Ca_m^{2+}} \;=\; \frac{k_{\mathrm{mc}}}{k_{\mathrm{cm}}}\,\mathrm{f}\left(\frac{K_{\mathrm{sp}}}{\mathrm{P_i}}\right)$$

PTH or cyclicAMP $\xrightarrow{\oplus}$ ... High $\mathrm{P_i}$ $\searrow \ominus$ ... Low $\mathrm{P_i}$ $\swarrow \oplus$

Calcitonin $\xrightarrow{\oplus}$

A high phosphate will increase the calcium phosphate precipitation and lower the free calcium of the mitochondrial matrix, $\mathrm{Ca_m^{2+}}$; consequently

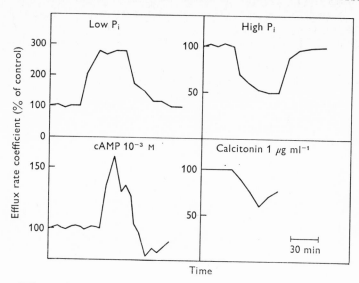

Fig. 4. Effects of low and high extracellular phosphate concentrations, cyclic AMP (cAMP) and calcitonin on the efflux rate coefficient of calcium efflux measured by isotopic desaturation (from Borle, 1969a, 1972a, b).

the cytoplasmic calcium will go down. A low phosphate will produce the opposite effects. Cyclic AMP and PTH increase the free cytoplasmic calcium by increasing the rate of calcium efflux from mitochondria, k_{mc} (Borle, 1974b; Matlib & O'Brien, 1974). Calcitonin depresses the cytoplasmic calcium activity by stimulating the rate of calcium uptake into mitochondria, k_{cm} (Borle, 1975a).

MODEL OF CELLULAR CALCIUM HOMEOSTASIS

Our experimental data and the postulated sites of action of ions, cyclic nucleotides and hormones that we have just outlined form the basis of our computer model of cellular calcium homeostasis (Fig. 6).

The model represents one single cell, 19 μm in diameter, in an infinitely large extracellular fluid compartment. The cell has a volume of 3600 μm^3 and a surface area of 1000 μm^2. It contains 1000 mitochondria with an internal membrane surface area of 30000 μm^2 (Mitchell, 1966). The mitochondria occupy 10 % of the cell volume. Since electron micrographs show that many mitochondria contain granules of amorphous calcium phosphate, we have postulated that the calcium activity of their matrix is in equilibrium with a mineral phase. For the constant of solubility product of the calcium phosphate granules we have selected the K_{sp} of secondary calcium phosphate, 3.3×10^{-6} M^2. Because a mineral phase has no concentration and

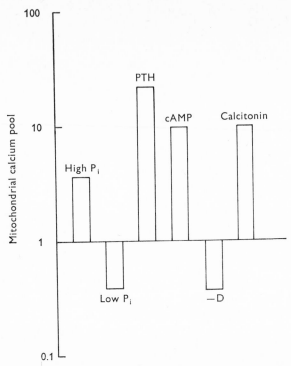

Fig. 5. Effects of high and low extracellular phosphate concentrations, PTH, cyclic AMP (cAMP), vitamin D deficiency ($-$D) and calcitonin on the mitochondrial exchangeable calcium pool measured by kinetic analysis of isotopic desaturation curves (after Borle, 1972*a,b*, 1974*a*, 1975*a*).

since its volume has no effect on the K_{sp} we have given this compartment an arbitrary value of 1 both for mass and concentration. The free calcium concentration of the extracellular fluids is set at 10^{-3} M, that of the cytoplasm at 10^{-6} M. The mitochondrial matrix free calcium is set at 10^{-3} M, phosphate at 3.3×10^{-3} M and both ions are in equilibrium with the K_{sp} of the mineral phase.

The fluxes across the plasma membrane, V_1 and V_2 are taken directly from our experimental measurements. The fluxes across the mitochondrial membrane, V_3 and V_4, are taken from the kinetic measurements of Vinogradov & Scarpa (1973) and of Reed & Bygrave (1975). Exchange between the mitochondrial free calcium and the mineral phase, V_5 and V_6, represents a physico-chemical process and has much higher rate constants. Four functions influence the various calcium fluxes (Table 2): f_1 represents the effect of cytoplasmic free calcium on calcium influx with a half-maximal activation K_a set arbitrarily at 10^{-6} M. The function f_2 represents the calcium pump extruding calcium out of the cell which has been given

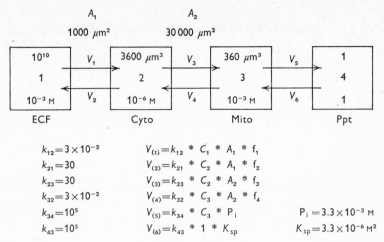

Fig. 6. Computer model of cellular calcium distribution and exchange representing one cell, 19 μm in diameter, in an infinitely large extracellular fluid compartment. A_1 = surface area of the plasma membrane; A_2 = surface area of 1000 mitochondria. Cell volume = 3600 μm^3; mitochondrial volume = 360 μm^3. Compartments 1 = extracellular fluid with a free calcium concentration (C) of 10^{-3} M; 2 = cytoplasmic compartment with free calcium of 10^{-6} M; 3 = mitochondrial matrix compartment with a free calcium of 10^{-3} M; 4 = mitochondrial calcium phosphate precipitate with a mass and a concentration set arbitrarily at 1. All ks are rate constants expressed in min^{-1}. All Vs are fluxes between compartments expressed in pmole cm^{-2} sec^{-1}. P_i = intramitochondrial inorganic phosphate concentration. K_{sp} = constant of solubility product of secondary calcium phosphate. Functions f_1 to f_4 modulate the fluxes V_1 to V_4 and are described in Table 2. Cyto = cytoplasmic free calcium; ECF = extracellular fluid free calcium; Mito = concentration of free calcium in the mitochondrial matrix; Ppt = mass of the calcium phosphate precipitate present in the mitochondria.

a K_m of 10^{-5} M. Finally f_3 and f_4 represent the effects of calcitonin and of cyclic AMP on the mitochondrial calcium exchange: f_3 stimulates the rate of calcium influx k_{cm} into mitochondria (calcitonin), while f_4 stimulates the rate of calcium efflux k_{mc} out of mitochondria (cyclic AMP).

SIMULATION TECHNIQUE

The dynamic and steady state behaviour of this model in response to changes in pool sizes and hormonal signals were simulated using a general simulation program written in Fortran IV. Using subscript notation, the equations of Fig. 6 were inserted into the program together with the initial conditions of the model, and any timed changes in pools or hormones which were to be imposed. The program employs the point-slope method of integration (Garfinkel, Ching, Adelman & Clark, 1966) using a variable step size adjusted to prevent any pool from changing by more than one

Table 2. *Functions modulating cellular calcium fluxes*

$$f_1 = \frac{2C_2}{(10^{-6} + C_2)} \qquad f_3 = 1 + \frac{CT}{(1 + CT)}$$

$$f_2 = \frac{1.1 \times 10^{-5}}{(10^{-5} + C_2)} \qquad f_4 = 1 + \frac{cAMP}{(10^{-6} + cAMP)}$$

C_2 = Cytoplasmic free calcium concentration.
CT = Calcitonin concentration.
cAMP = cyclic AMP concentration.

percent between updates of the model. Typically, we simulated the behaviour of the model of 1000 min using up to 106 integration steps. The behaviour of the model was exceptionally stable despite the presence of a small pool with a large turnover rate (Ca_{cyto}^{2+}), the associated positive feedback loop, the deliberately imposed perturbations, and the inevitable round-off errors introduced by single precision computer arithmetic. The program was executed on a PDP-10 computer and the results displayed and recorded using a Computec 400 terminal and Textronic copier. The computer facilities were made available by the Division of Research Resources, National Institutes of Health, in conjunction with the development of the PROPHET computer system for chemical–biological information handling. The results shown in this paper represent steady state relationships derived from the appropriate portions of the computer-generated dynamics.

RESULTS OF THE SIMULATION OF THE MODEL
CELLULAR HOMEOSTASIS

The results are presented as the per cent change in the calcium activity of each compartment plotted against time. At steady state, the calcium activities of each compartment remain absolutely stable. The capacity of the system to resist imposed perturbations is illustrated in Fig. 7. A sudden 50 % increase in cytoplasmic free calcium is immediately buffered by the mitochondria and the normal calcium activity is restored within 10 msec. A drop in cytoplasmic calcium is also completely and rapidly corrected. The influence of fluctuations in extracellular phosphate concentrations is shown in Fig. 8. A rise in phosphate concentration immediately increases the precipitation of calcium phosphate in the mitochondria. However, the free calcium of the mitochondrial matrix falls as well as the cytoplasmic calcium activity. On the other hand, a fall in phosphate concentration reduces the mineral phase but increases the calcium activity of both cyto-

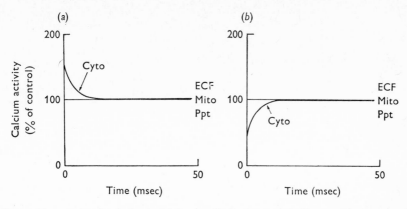

Fig. 7. Computer simulation of (*a*) a 50 % increase and (*b*) a 50 % decrease in the cytoplasmic calcium activity imposed at time zero. For abbreviations see Fig. 6.

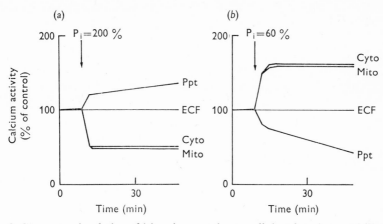

Fig. 8. Computer simulation of (*a*) an increased extracellular phosphate and (*b*) a decreased extracellular phosphate (see note below) concentration on the calcium activity of the different cell compartments. See Fig. 6 for the definitions of abbreviations.

plasm and mitochondria. These results are in complete agreement with the data obtained in isolated cells (Borle, 1972*a*).

Fig. 9 illustrates the effects of cyclic AMP. While cyclic AMP stimulates calcium efflux out of mitochondria, the cytoplasmic free calcium rises and remains elevated. The mitochondrial free calcium, however, is hardly affected, since it is fixed by the solubility product of the calcium phosphate precipitate. The mineral phase is decreasing slowly. Fig. 9*b* shows the effects of a rise in extracellular phosphate* during cyclic AMP stimulation.

* An increase in extracellular phosphate is immediately followed by a rise in intracellular phosphate (Uchikawa & Borle, in preparation).

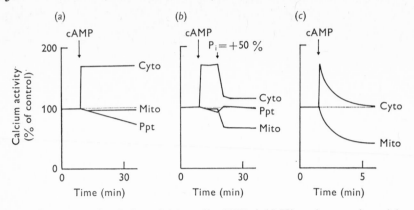

Fig. 9. Computer simulation of (a) cyclic AMP (cAMP) action on the calcium activity on the different cell compartments, (b) a 50 % increase in extracellular phosphate concentration during cyclic AMP stimulation and (c) cyclic AMP action in absence of calcium phosphate precipitate in the mitochondrial compartment. See Fig. 6. for the definitions of abbreviations.

Because it depresses the mitochondrial free calcium, phosphate counter-acts the effects of cyclic AMP and lowers cytoplasmic calcium activity. Fig. 9c shows the effects of cyclic AMP in absence of calcium phosphate precipitate. Without a mineral phase to maintain a fixed mitochondrial free calcium, cyclic AMP triggers a transient rise in cytoplasmic calcium; but at steady state the cytoplasmic free calcium is back to control levels and mitochondrial calcium is depressed, compensating for the rise in the rate constant of calcium efflux from the mitochondria induced by cyclic AMP. Except for the loss of mineral during cyclic AMP stimulation, these results accurately duplicate our experimental data (Borle, 1972b, 1974b, 1975d). A perfect fit between our computer simulation and our experi-mental results can be obtained if the K_m of calcium efflux out of the cell is reduced below K_a, the half-maximal activation of calcium influx by cytoplasmic calcium. In this case cyclic AMP produces an increase in the mineral phase in agreement with our experimental findings.

Fig. 10 presents the effects of calcitonin which stimulates calcium uptake into mitochondria (Borle, 1975a). The model shows that, despite an increase in calcium phosphate precipitation, the mitochondrial free calcium remains at control levels. However, the cytoplasmic calcium is depressed. Fig. 10b illustrates the effect of increasing extracellular phosphate* during calcitonin stimulation. Phosphate enhances the action of calcitonin on the mineral phase and on cytoplasmic calcium by reducing the free calcium of the mitochondrial matrix. On the other hand a 40 % reduction in phos-

* See note on p. 151.

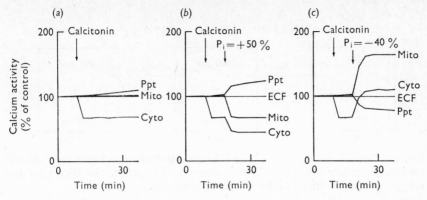

Fig. 10. Computer simulation of (*a*) calcitonin action on the calcium activity of the different cell compartments, (*b*) a 50 % increase in phosphate concentration during calcitonin stimulation and (*c*) of a 40 % decrease in phosphate concentration during calcitonin stimulation. See Fig 6 for the definitions of abbreviations.

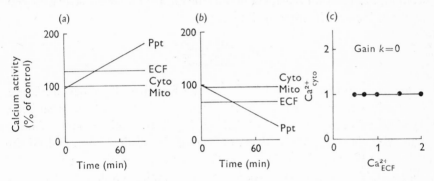

Fig. 11. Computer simulation of (*a*) a 30 % increase in extracellular calcium concentration and (*b*) a 30 % decrease in ECF calcium on the different cell calcium compartments. See Fig. 6 for the definitions of abbreviations. (*c*) Influence of the ECF calcium concentration (or of calcium influx) on the cytoplasmic calcium activity and gain of the model system described in Fig. 6.

phate, completely erases the effects of calcitonin on cytoplasmic calcium and on the mineral phase because it increases the mitochondrial free calcium. (Fig. 10*c*). Again, these results agree with our published data obtained in isolated cells (Borle, 1975*a*).

There is one characteristic of the model which may conflict with the findings of several investigators: the two mitochondrial compartments buffer the cytoplasmic free calcium so tightly that it is completely unaffected by changes in the extracellular calcium concentration or by an increase in calcium influx. Fig. 11 shows that increasing or decreasing the extracellular calcium has absolutely no effect on the cytoplasmic calcium.

6

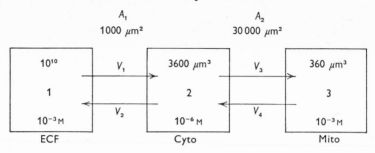

Fig. 12. Computer model of cellular calcium distribution and exchange without calcium phosphate precipitate in the mitochondrial matrix. See Fig. 6 for the values and the definitions of the different parameters.

It only increases or decreases the calcium precipitate pool. Although this is not inconsistent with some of our results (Borle, 1971b, 1972a) the zero gain of the model (Fig. 11c) does not agree with the gain of 0.65 calculated from Fig. 1. The gain is defined here as the ratio of the steady state response to a step increase in stimulus intensity.

$$k = \frac{\Delta Ca_{cyto}^{2+}}{\Delta Ca_{ECF}^{2+}}.$$

This property of the model is due to the fact that all mitochondria are assumed to contain calcium phosphate granules. We recognize that such an assumption may not be justified by the electron micrographs of most cells with the exception perhaps of bone and intestinal cells. Indeed, in kidney or liver cells, calcium granules may appear only in a few mito-chondria. The existence of two populations of mitochondria one with and the other without calcium granules is supported by the existence of heavy and light mitochondria which can be obtained by ultracentrifugation. Consequently we decided to compare the influence of extracellular calcium in a cell model containing mitochondria without calcium phosphate granules (Figs. 12 and 13). In this case a small change in extracellular calcium or in calcium influx produces a dramatic effect on the cytoplasmic free calcium with a gain of 2 to 3. It is obvious that such an amplified response would be incompatible with a normal cellular function and it is in complete disagreement with the data obtained in isolated cells.

However, if our cellular model is modified as in Fig. 14 to include both mitochondria containing calcium phosphate precipitates (mito H = heavy mitochondria) and mitochondria free of mineral (mito L = light mito-chondria), the steady state change in cytoplasmic free calcium in response to a rise or fall in extracellular calcium depends on the ratio of both types of mitochondria (Fig. 15). According to Fig. 15, the gain of 0.65 that we

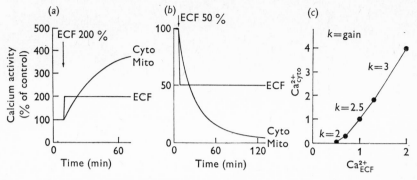

Fig. 13. Computer simulation of (a) a two-fold increase in extracellular fluid (ECF) free calcium concentration and (b) a 50 % reduction in ECF calcium, on the cytoplasmic (Cyto) and mitochondrial (Mito) free calcium concentration. (c) Influence of the ECF calcium concentration (Ca_{ECF}^{2+}) (or of calcium influx) on the cytoplasmic calcium activity (Ca_{cyto}^{2+}) and gain of the model system described in Fig. 12.

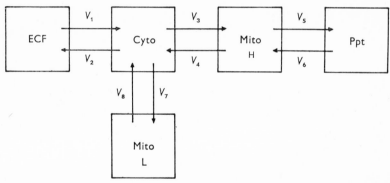

Fig. 14. Computer model of the calcium distribution and exchange in a cell comprising two types of mitochondria: mito H or heavy mitochondria containing calcium phosphate precipitates (Ppt) and mito L or light mitochondria devoid of precipitate. See Fig. 6 for the values and definitions of the various parameters of the model.

observed experimentally can be obtained if 5 % of the mitochondria contain precipitates of calcium phosphate.

Except for a greater influence of the extracellular calcium on the cytoplasmic calcium activity, the behaviour of the model and its response to cyclic AMP, calcitonin and phosphate are not significantly affected even if only 5 % of the mitochondria contain calcium phosphate granules (Fig. 16). Cyclic AMP still increases the cytoplasmic free calcium. Although the calcium activity of the mitochondria containing the precipitate remains fixed by the K_{sp} of the mineral phase, the free calcium of mitochondria devoid of calcium granules is depressed by cyclic AMP (Fig. 16a).

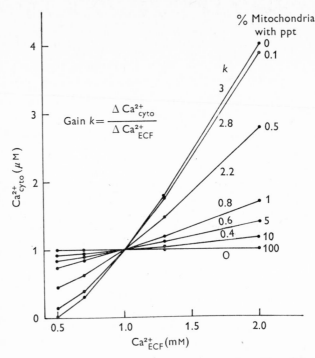

Fig. 15. Influence of extracellular calcium concentration (Ca_{ECF}^{2+}) on the cyto-plasmic calcium activity (Ca_{cyto}^{2+}) of the computer model of the cell described in Fig. 14 related to the ratio of mitochondria with and without calcium phosphate precipitates (expressed as per cent mitochondria with ppt).

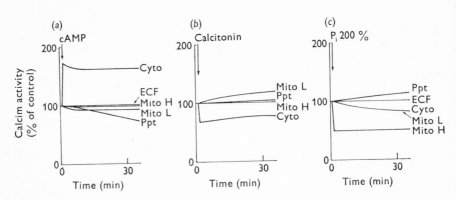

Fig. 16. Computer simulation of the effects of (a) cyclic AMP (cAMP), (b) calcitonin and (c) a two-fold increase in phosphate concentration on the calcium activity of the various cell compartments of the model described in Fig. 14. In this instance, the system has a steady state gain of 0.6 with only 5 % of the mitochondria containing calcium phosphate precipitates (Mito H) and 95 % of the mitochondria free of precipitate (Mito L).

Calcitonin still depresses the cytoplasmic calcium and increases the mineral phase. But in addition, the calcium of the light mitochondria is elevated by calcitonin (Fig. 16b). Finally, an increased concentration of extracellular phosphate* enhances the precipitation of calcium phosphate and lowers the calcium activity in the cytoplasm and in the matrix of both types of mitochondria (Fig. 16c). All these responses again agree with our results obtained in isolated cells and isolated mitochondria.

It is interesting, although perhaps fortuitous, that the patterns of response to cyclic AMP in the three different versions of the model (i.e. without granules, with 1 % and with 100 % of mineral-loaded mitochondria) have been previously observed in isolated mitochondria (Borle, 1974b, 1975d).

CONCLUSIONS

In 1967, Bygrave proposed that cytoplasmic calcium may be an important regulator of the cell metabolic activity (Bygrave, 1967). He later showed that mitochondrial calcium transport could be the major regulator of the cytoplasmic free calcium concentration (Meli & Bygrave, 1972; Spencer & Bygrave, 1973). In 1970, Rasmussen extended the scheme and suggested that intracellular free calcium as well as cyclic AMP might act as universal messengers for hormone release, peptide hormone action, transmitter release as well as for metabolic regulation (Rasmussen, 1970). Our own data, obtained by kinetic analysis of ^{45}Ca movements in isolated cell systems, are in complete agreement with these ideas.

Cellular calcium is distributed among several compartments, of various sizes, of different calcium activity and the magnitude of the calcium fluxes between them differ markedly. It is difficult therefore to integrate the information available without the help of a computer model capable of handling 20 to 30 variables. The model of cellular calcium homeostasis which we have just presented is an attempt to integrate our results and the data of other investigators.

Except for minor details the model fits most of the results of our experiments. We recognize that a good agreement between the model and our data is not a proof of the validity of our original hypothesis. It certainly does not insure that the model is a true representation of cellular calcium homeostasis. However, it shows that our assumptions concerning the metabolism of cell calcium and the mode of action of some hormones are theoretically possible and kinetically sound.

We thought that the model would bring us definite information about the relative importance of the mitochondria versus the plasma membrane

* See note on p. 151.

transport of calcium in controlling and regulating cytoplasmic calcium activity. The model reveals itself to be highly flexible and, depending on several conditions, it gives us a whole spectrum of responses, from a cell markedly affected by calcium transport across the plasma membrane to a cell totally controlled by mitochondria. Among the many variables which affect the model's behaviour are the number of mitochondria, their ability to accumulate calcium, the presence or absence of calcium granules in the mitochondrial matrix and the proportion of mitochondria free of calcium phosphate precipitate.

It is obvious that we need more precise information about several parameters to be introduced in the model: (1) the exact K_m of mitochondrial calcium transport; (2) the K_m of calcium transport across the plasma membrane; (3) the Hill coefficient of these calcium transport processes if their velocity is a sigmoidal function of cytoplasmic calcium; and (4) whether the activation of calcium influx by cytoplasmic calcium is linear, or whether it obeys Michaelis–Menten kinetics.

To us, the real value of this computer model does not reside in its ability to duplicate our experimental results, but in its capacity to teach us the importance of some parameters whose role was underestimated, to suggest new or better experiments and to lead us into fresh avenues of thinking.

REFERENCES

BEELER, G. W. JR & REUTER, H. (1970). The relation between membrane potential, membrane currents and activation of contraction in ventricular myocardial fibres. *J. Physiol.*, **207**, 211–229.

BIANCHI, C. P. & SHANES, A. M. (1959). Calcium influx in skeletal muscle at rest, during activity and during potassium contractures. *J. gen. Physiol.*, **42**, 803–815.

BLAUSTEIN, M. P. & RUSSELL, J. M. (1975). Sodium–calcium exchange and calcium–calcium exchange in internally dialyzed squid giant axons. *J. Memb. biol.*, **22**, 285–312.

BLUM, R. M. & HOFFMAN, J. F. (1972). Ca-induced K transport in human red cells: localization of the Ca-sensitive site to the inside of the membrane. *Biochem. biophys. Res. Commun.*, **46**, 1146–1152.

BORLE, A. B. (1969a). Effects of thyrocalcitonin on calcium transport in kidney cells. *Endocrinology*, **85**, 194–199.

(1969b). Kinetic analyses of calcium movements in HeLa cell cultures. I. Calcium influx. *J. gen. Physiol.*, **53**, 43–56.

(1970). Kinetic analysis of calcium movements in cell cultures. III. Effects of calcium and of parathyroid hormone in kidney cells. *J. gen. Physiol.*, **55**, 163–186.

(1971a). Calcium transport in kidney cells and its regulations. In *Cellular Mechanisms for Calcium Transfer and Homeostasis*, ed. G. Nichols & R. H. Wasserman, pp. 151–174. Academic Press: New York.

(1971b). Effets du phosphate sur les mouvements du calcium en culture cellulaire. In *Phosphate et Métabolisme Phosphocalcique*, ed. D. J. Hioco, pp. 29–45. Expansion Scientifique Francaise: Paris.

(1972*a*). Kinetic analyses of calcium movements in cell cultures. V. Intracellular calcium distribution in kidney cells. *J. Memb. Biol.*, **10**, 45–66.

(1972*b*). Parathyroid hormone and cell calcium. In *Calcium Parathyroid Hormone and the Calcitonins*, ed. R. V. Talmage & P. L. Munson, pp. 484–491. Excerpta Medica: Amsterdam.

(1973). Calcium metabolism at the cellular level. *Fedn Proc.*, **32**, 1944–1950.

(1974*a*). Kinetic studies of calcium movements in intestinal cells: Effects of vitamin D deficiency and treatment. *J. Memb. Biol.*, **16**, 207–220.

(1974*b*). Cyclic AMP stimulation of calcium efflux from kidney, liver and heart mitochondria. *J. Memb. Biol.*, **16**, 221–236.

(1975*a*). Regulation of cellular calcium metabolism and calcium transport by calcitonin. *J. Memb. Biol.*, **21**, 125–146.

(1975*b*). Methods of assessing hormone effects on calcium fluxes *in vitro*. In *Methods in Enzymology*, vol. 39, Hormone action, part D, Isolated cells, tissues and organ systems, ed. J. G. Hardman & B. W. O'Malley, pp. 513–573. Academic Press: New York.

(1975*c*). Regulation of the mitochondrial control of cellular calcium homeostasis and calcium transport by phosphate, parathyroid hormone, calcitonin, vitamin D and cyclic AMP. In *Calcium-Regulating Hormones*, ed. R. V. Talmage, M. Owen & J. A. Parsons, pp. 217–228. Excerpta Medica: Amsterdam.

(1975*d*). Modulation of mitochondrial control of cytoplasmic calcium activity. In *Calcium transport in Contraction and Secretion*. ed. E. Carafoli, F. Clementi W. Drabikowski & A. Magreth, pp. 77–86. North-Holland: Amsterdam & New York.

BYGRAVE, F. L. (1967). The ionic environment and metabolic control. *Nature, Lond.*, **214**, 667–671.

CRAMER, C. F. (1963). Quantitative studies on the absorption and excretion of calcium from Thirty-Vella loops in the dog. In *The Transfer of Calcium and Strontium Across Biological Membranes*, ed. R. G. Wasserman, pp. 75–84. Academic Press: New York.

DUMONT, P. A., CURRAN, P. F. & SOLOMON, A. K. (1960). Calcium and strontium in rat small intestine. Their fluxes and their effect on Na flux. *J. gen. Physiol.*, **43**, 1119–1136.

GARFINKEL, D., CHING, S. W., ADELMAN, M. & CLARK, P. (1966). Techniques and problems in the construction of computer models of biochemical systems including real enzymes. *Ann. N.Y. Acad. Sci.*, **128**, 1054–1068.

HARRELL, A., BINDERMAN, I. & RODAN, G. A. (1973). The effect of calcium concentration on calcium uptake by bone cells treated with thyrocalcitonin (TCT) hormone. *Endocrinology*, **92**, 550–555.

HODGKIN, A. L. & KEYNES, R. D. (1957). Movements of labelled calcium in giant axons. *J. Physiol.*, **138**, 253–281.

KRNJEVIC, K. & LISIEWICZ, A. (1972). Injections of calcium ions into spinal motoneurones. *J. Physiol.*, **225**, 363–390.

LEVINSON, C. & BLUMENSON, L. E. (1970). Calcium transport and distribution in Ehrlich mouse ascites tumor cells. *J. Cell Physiol.*, **75**, 231–240.

LEW, V. L. (1970). Effect of intracellular calcium on the potassium permeability of human red cells. *J. Physiol.*, **206**, 35–36.

MATLIB, A. & O'BRIEN, P. J. (1974). Adenosine 3':5'-cyclic monophosphate stimulation of calcium efflux from mitochondria. *Biochem. Soc. Trans.*, **2**, 997–1000.

MELI, J. & BYGRAVE, F. L. (1972). The role of mitochondria in modifying calcium-sensitive cytoplasmic metabolic activities. *Biochem. J.* **128**, 415–420.

MITCHELL, P. (1966). Metabolic flow in the mitochondrial multiphase system: an appraisal of the chemi-osmotic theory of oxidative phosphorylation. In *Regulation of Metabolic Processes in Mitochondria*, ed. J. M. Tager, S. Papa, E. Quagliariello & E. C. Slater, pp. 65–84. Elsevier: Amsterdam.

RASMUSSEN, H. (1970). Cell communication, calcium ion, and cyclic adenosine monophosphate. *Science, Wash.*, **170**, 404–412.

REED, K. C. & BYGRAVE, F. L. (1975). A kinetic study of mitochondrial calcium transport. *Eur. J. Biochem.*, **55**, 497–504.

ROMERO, P. J. & WHITTAM R. (1971). The control by internal calcium of membrane permeability to sodium and potassium. *J. Physiol.*, **214**, 481–507.

SPENCER, T. & BYGRAVE, F. L. (1973). The role of mitochondria in modifying the cellular ionic environment: studies of the kinetic accumulation of calcium by rat liver motochondria. *Bioenergetics*, **4**, 347–362.

VINOGRADOV, A. & SCARPA, A. (1973). The initial velocities of calcium uptake by rat liver mitochondria. *J. biol. Chem.*, **248**, 5527–5531.

WALLING, M. W. & ROTHMAN, S. S. (1969). Phosphate-independent, carrier-mediated active transport of calcium by rat intestine. *Am. J. Physiol.*, **217**, 1144–1148.

(1970). Apparent increase in carrier affinity for intestinal calcium transport following dietary calcium restriction. *J. biol. Chem.*, **245**, 5007–5011.

WATLINGTON, C. O., BURKE, P. K. & ESTEP, H. L. (1968). Calcium flux in isolated frog skin; the effect of parathyroid substance. *Proc. Soc. exp. biol. Med.*, **128**, 853–856.

WALZER, M. (1970). Calcium transport in the toad bladder: permeability to calcium ions. *Am. J. Physiol.*, **218**, 582–589.

WINEGRAD, S. & SHANES, A. M. (1962). Calcium flux and contractility in guinea-pig atria. *J. gen. Physiol.*, **54**, 371–394.

INTRACELLULAR CALCIUM AND THE CONTROL OF MEMBRANE PERMEABILITY

By R. W. MEECH

ARC Unit of Invertebrate Chemistry and Physiology,
Department of Zoology, University of Cambridge,
Downing Street, Cambridge CB2 3EJ

One of the earliest demonstrations of the role of intracellular calcium in the control of membrane permeability came from work on red blood cells. Erythrocytes poisoned with fluoride ions normally lose potassium ions if there is calcium in the bathing medium (Davson, 1941; Gardos, 1958). However, this potassium loss does not occur if 'ghosts' are prepared in the presence of EDTA (Lepke & Passow, 1968). Since the ghosts were thought to be impermeable to calcium ions, Lepke & Passow concluded that intracellular EDTA could prevent the action of calcium on the outside of the membrane. However, a proposal that calcium was acting at the inside surface of the membrane (Whittam, 1968) was subsequently supported by the finding that ATP-depleted red cells are highly calcium permeable (Romero & Whittam, 1971; Lew, 1971a).

The permeability of certain excitable cells is also dependent on the intracellular calcium concentration. Injection of calcium chloride into *Aplysia* neurones causes an immediate increase in potassium conductance (Meech & Strumwasser, 1970). Similar effects have been reported in cat spinal motoneurones (Krnjević & Lisiewicz, 1972) and cardiac Purkinjé fibres (Isenberg, 1975b). However, as much as 10 mM intracellular calcium has little or no effect on the potassium conductance of perfused squid axons (Begenisich & Lynch, 1974). Intracellular calcium may also increase the chloride permeability of insect salivary glands (Berridge, Lindley & Prince, 1975) and reduce the sensitivity of invertebrate photoreceptors to light (Lisman & Brown, 1972).

There is evidence for intracellular calcium-induced permeability changes in over a dozen different tissues. This review summarises evidence for nine of them. My main conclusion is that membranes with calcium activation coupled to potassium activation can produce a variety of responses to stimulation ranging from graded depolarization to long-lasting action potentials. First, however, I have included a section which summarises some of the methods available to test the effects of intracellular calcium on cell membranes.

METHODS

It may be possible to raise the level of intracellular calcium by blocking the normally active calcium 'sink' or by increasing the calcium permeability of the outer membrane. Unfortunately metabolic poisons such as 2,4-dinitrophenol may have non-specific effects on membrane permeability and many of the ionophores which have been used to increase the permeability of cell membranes to calcium, are not at all specific. Consequently, these indirect techniques have the disadvantage that the intracellular calcium must be *demonstrably* increased.

A number of more direct techniques have been developed which can be applied to individual tissues but these too suffer from certain disadvantages.

Perfusion techniques

It is possible to replace the contents of giant axons with artificial media (Baker, Hodgkin & Shaw, 1962). However, perfused axons do not survive well unless the perfusion fluid contains a high fluoride concentration (Tasaki, Singer & Takenaka, 1965). Since calcium fluoride is only slightly soluble, other less satisfactory anions must be used to study the effects of internal calcium. The main problem with studying perfused muscle fibres is that they contract when the intracellular calcium concentration is higher than about 10^{-6} M (Portzehl, Caldwell & Rüegg, 1964).

Erythrocyte ghosts

The preparation of erythrocyte ghosts has been described by Bodeman & Passow (1972) and Simons (1975). Briefly, ghosts are formed by lysing red cells in hypotonic solution to which salt is subsequently added. The re-sealed cells contain the added salt. The main problem is that in most experiments the potential across the cell membrane is unknown so that the driving forces on the ions cannot be known for certain. Another difficulty is that experiments are carried out on populations of cells. Thus the response of this population to calcium may represent the response of the 'average' cells which are all modified to the same extent, or the average response of cells, some of which are modified and some of which are not.

Injection techniques

Injection is probably the most widely used method for raising the intracellular calcium concentration. The major hazard is 'micropipette blockage'. Apparently calcium chloride-filled micropipettes readily become blocked when inside cells, possibly because calcium causes precipitation of the cell contents at the tip of the pipette. The blocking process is slower in pipettes filled with a mixture of both potassium chloride and calcium chloride. Lisman & Brown (1972) report that during iontophoretic injection experiments calcium–EGTA-filled pipettes remained reliable more often than those filled with calcium chloride alone. However, the transport number of these electrodes for calcium is low (Fein & Lisman, 1975) and so only a small proportion of the current is carried by calcium.

Krnjević & Lisiewicz (1972) give a valuable description of the technical difficulties inherent in experiments with multibarrelled pipettes. In their experiments, calcium was injected by passing a known current between one barrel filled with 200 mM calcium chloride and another filled with 3 M potassium chloride. The third barrel of the triple barrelled pipette (tip diameter 2 μm) was used to record the membrane potential. However, coupling artifacts tended to make the tip of the micropipette positive and since the effect of the injection did not greatly outlast the injection it was often hard to estimate the true membrane potential. It was also difficult to inject calcium without depolarising the membrane.

An important factor in injection experiments is the delivery rate of the calcium. It is possible to inject calcium into the centre of a cell at such a slow rate that the level of calcium at the membrane is not affected (Rose & Loewenstein, 1975). Pressure injection has the advantage that relatively large quantities of calcium can be injected in a very short period. However, the technique is at present limited to cells of 100 μm or more since pipettes with tip diameters of about 1.5 μm are used.

EFFECT OF INTRACELLULAR CALCIUM ON THE PERMEABILITY OF DIFFERENT TISSUES

Squid axons

Chandler, Hodgkin & Meves (1965) have suggested that squid axons have negative fixed charges at the inside surface of the membrane. This means that the potential gradient across the membrane is not equivalent to that measured in bulk solution. If the negative charge density on the membrane is high, the Gouy–Chapman theory predicts that even low concentrations of divalent cations will reduce the surface potential by a screening process (McLaughlin, Szabo & Eisenman, 1971). Consequently the activation curves for sodium and potassium should be shifted along the voltage axis. However, Begenisich & Lynch (1974) were able to find little change in the membrane characteristics even if the intracellular calcium was increased to 10 mM. This report is a little surprising in view of the observation by Tasaki, Watanabe & Lerman (1967) that concentrations of calcium as low as 3 mM caused a reversible conduction block.

Red blood cells

Potassium transport across red cell membranes has been divided into two components (Shaw, 1955; Glynn, 1956): (a) a passive component proportional to the external potassium concentration, (b) an active component which becomes saturated at high external potassium concentrations and which depends on the metabolic state of the cell.

Metabolic poisons such as fluoride and iodoacetate induce a large increase in potassium permeability (Wilbrandt, 1937; Davson, 1941) which is dependent on the presence of external calcium (Gárdos, 1958) and blocked by intracellular EDTA (Lepke & Passow, 1968) or by EGTA (Blum & Hoffman, 1972). It only occurs in ATP-depleted cells. The identification of the component of potassium transport which is altered under these conditions is the subject of considerable discussion. There are two major alternative proposals:

(i) Whittam (1968) argued that metabolic inhibitors should block the outward calcium pump (previously demonstrated by Schatzmann, 1966) and thereby increase the intracellular ionic calcium. He suggested that this raised intracellular calcium caused an increase in the passive permeability of the membrane to potassium. Thus both components of potassium transport should be dependent on the metabolic state of the cell.

(ii) Blum & Hoffman (1970) have proposed that intracellular calcium may modify the active component of potassium transport by increasing the rate at which the sodium pump exchanges potassium.

The situation has been clarified by the finding that ouabain has no effect on the calcium-induced rate of potassium loss when the average concentration of ATP in the red cells is less than 10^{-6} M (Lew, 1971b). Effects of ouabain at higher ATP levels could be interpreted as the result of blocking the ATPase activity of the membrane sodium pump. Since the rate at which cells take up calcium from the medium is related to the rate of depletion of ATP, ouabain-treated cells are likely to take up calcium at a slower rate.

Kregenow & Hoffman (1972) have presented a model in which a neutral carrier becomes charged on combination with potassium and in which the formation of carrier complex on both sides of the membrane is rate limiting. This model has the characteristics of passive transport even though the transport mechanism is carrier mediated. Since formation of the carrier complex could be calcium and ATP dependent this model remains a viable alternative to Whittam's hypothesis. Both suggestions indicate that the distinction between active and passive potassium transport is in some respects unsatisfactory.

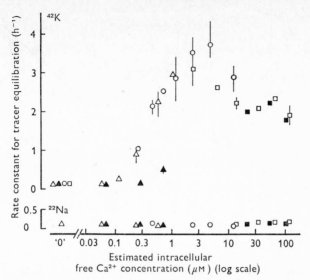

Fig. 1. The effect of intracellular calcium ions on the permeability of red cell membranes to potassium and sodium ions. Ghosts were prepared after lysing red cells in a calcium buffered solution containing ^{42}K (and/or ^{22}Na) *Abscissa*: intracellular free calcium concentration calculated on the assumption that the contents of the ghosts had the same composition as the lysing solution. *Ordinate*: membrane permeability measured as the rate of tracer efflux into a calcium-free medium at 37 °C and pH 7.1. Each point is the average result from up to six separate experiments, and the vertical lines show the extreme range of the results. Open symbols (△, ○, □) indicate that no magnesium was added. Closed symbols (▲, ■) indicate that 2–2.5 mM magnesium was present. Different calcium buffers were used. △, ▲, EGTA; ○, HEDTA (*N*-hydroxyethyl ethylenediamine-triacetic acid); □, ■, citrate. Monovalent cations were at equlibrium during flux measurements, potassium was 90–100 mM (from Simons, 1975).

Whichever component of potassium transport is affected by calcium its sensitivity is clearly very high. An increase in the calcium concentration of only 1 μmole l^{-1} of cells is followed by a considerable increase in the potassium efflux (Lew, 1970). The sensitivity of resealed ghosts to intracellular calcium has been studied in more detail by Simons (1975) who has prepared ghosts in the presence of calcium buffers. As his figure shows (see Fig. 1) the increased potassium permeability is approximately proportional to the intracellular calcium in the range 0.1–3 μM. Simons finds no change in the sodium permeability. This contrasts with the finding of Romero & Whittam (1971) that the membrane of stored red cells may become permeable to both sodium and potassium ions in the presence of intracellular calcium. This may be because red cells stored for four to five weeks are more leaky to calcium than fresh cells and consequently the intracellular calcium can reach high values (Lew, 1974).

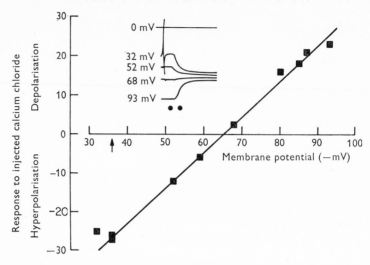

Fig. 2. *Inset*: the response of the membrane potential of an *Aplysia* neurone (R15) to the injection of a small quantity of 3 M calcium chloride. The period of injection is indicated by the two solid circles which are 5 sec apart. The membrane potential was set to four different levels by passing a constant hyperpolarising current and the results obtained were superimposed. *Graph*: the size of each potential response as a function of membrane potential (together with other values obtained from the same cell but not illustrated in the inset). The reversal potential is −65 mV. The arrow indicates the resting potential of the cell (−36 mV) (from Meech, 1972).

T. J. B. Simons (personal communication) finds that both strontium and barium will substitute for calcium although strontium acts at 10× higher concentrations than calcium, and barium acts only slightly at 100× the calcium concentration. Magnesium tends to depress the effect of calcium (Simons, 1975) although it is apparently effective in the presence of fluoride ions (Riordan & Passow, 1973).

Molluscan neurones

The effect of intracellular calcium on membrane resistance has been demonstrated by the pressure injection of calcium salts into *Aplysia* (Meech & Strumwasser, 1970; Meech, 1972) and *Helix* neurones (Meech, 1974a). The response to calcium injection is an immediate reduction in membrane resistance. At normal resting potentials this is associated with a hyperpolarization of the membrane but at more negative potentials the response reverses and injection induces depolarisation (see inset of Fig. 2). Fig. 2 shows that the reversal potential of the response is close to −65 mV. Of the ions tested (sodium, potassium and chloride), only changes in the external potassium concentration have any significant effect on the reversal

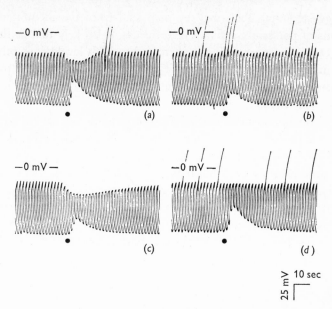

Fig. 3. The response of an *Aplysia* neurone to injected calcium in the presence of 50 mM TEA in the bathing medium. (*a*) Control; (*b*) in the presence of TEA; (*c*) control after 20 min in artificial sea water; (*d*) in the presence of a second application of TEA. Hyperpolarising current pulses (5×10^{-8} A and 1.0 sec duration) were passed through the calcium chloride filled injection micropipette to prevent blockage while it was in the cell and to estimate the membrane resistance. The filled circle indicates the time and duration of the injection. No significance should be placed on the different sizes of the resistance change since it is not possible to be sure that the same amount of calcium was injected each time (from Meech, 1972).

potential. Experiments on cell R15 in the abdominal ganglion of *Aplysia* showed that a 10-fold change of potassium from 10–100 mM produced a 58 mV shift in the reversal potential. The reversal potential may not, however, be a good estimate of the potassium equilibrium potential because it seems likely that repeated injection of calcium may lead to a depletion of the intracellular potassium. Furthermore, in *Aplysia* the reversal potential is not greatly affected by changes in external potassium in the range 1–10 mM.

Fig. 3 shows the effect of tetraethylammonium ions (TEA) on the response to calcium injection. Note that TEA, which blocks potassium activation in many tissues, prevents hyperpolarisation of the membrane by calcium but does not prevent the resistance change. It is possible that the potassium component of the response is blocked by TEA, revealing an effect on other ion conductances. Alternatively, although TEA ions prevent the outward movement of potassium ions, Armstrong & Binstock (1965) have suggested that TEA may be displaced from the membrane under conditions of potassium influx. Nevertheless, it is likely that as the intracellular calcium concentration is increased to high levels, ions other than potassium may become permeable. This would account for the transitory depolarising responses which are seen following massive calcium injections (Meech, 1972, 1974a).

In experiments on identified *Helix aspersa* neurones, it was possible to calculate the amount of injected calcium chloride by estimating the change in the chloride equilibrium potential. This could be done by finding the acetylcholine reversal potential before and after injection because acetylcholine causes a selective increase in chloride permeability in certain cells. In summary, when sufficient calcium chloride was injected to increase the intracellular calcium concentration by 5×10^{-4}–1.4×10^{-3} M the membrane resistance was reduced by 8–32 %. Resistance changes of this order could also be induced by injecting solutions of calcium–EGTA buffer, having about 10^{-6} M ionised calcium. Clearly the calcium chloride is about 1000 times less effective than the calcium–EGTA buffer and there is obviously an active calcium 'sink' in these cells. From a study of injected neurones under the electron microscope, it appears that much of the injected calcium is taken up into the endoplasmic reticulum (G. Nicaise & R. W. Meech, unpublished) although there is also evidence for a sodium/calcium exchange pump (Meech, 1974a).

The calcium influx associated with each action potential in *Aplysia* neurones is best demonstrated by experiments on aequorin-injected cells (Stinnakre & Tauc, 1973). A train of action potentials produces a prolonged increase in the intracellular calcium ion concentration. Such a stimulus also generates a hyperpolarisation of the nerve membrane which is associated with an increase in potassium conductance (Brodwick & Junge, 1972; Moreton, 1972). It is possible to abolish the post-tetanic hyperpolarization in cell R15 by EGTA injection (see Fig. 4). During another series of experiments small quantities of calcium–EGTA buffer were injected. Provided that the volume injected was suitably small, the cell membrane resistance was little affected by the small increase in intracellular calcium. Nevertheless, a prolonged post-tetanic hyperpolarisation was consistently recorded. An hour after the injection the post-tetanic hyperpolarisation was found to be totally abolished, probably because the cell had sequestered the injected calcium and the cell now contained free EGTA. The post-tetanic hyperpolarisation could be restored by further injection of calcium–EGTA buffer. Clearly intracellular calcium plays an important role in the control of membrane permeability in actively firing neurones (Meech, 1974b).

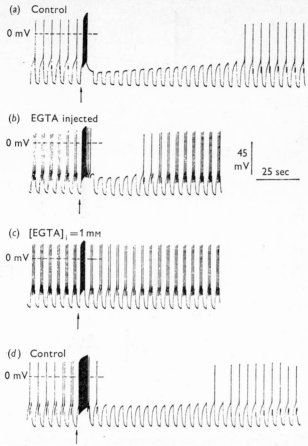

Fig. 4. Effect of intracellular EGTA on the post-tetanic hyperpolarisation of *Aplysia* neurone R15. (*a*) Control: membrane potential recorded with 3 M KCl filled micropipette. The cell was also penetrated with an injection micropipette containing 0.67 M EGTA, 1.3 M KCl and 67 mM histidine hydrochloride buffered to pH 7.26. This was also used to pass hyperpolarising current pulses of 2×10^{-8} A and 2.5 sec duration. The depolarising stimulus (applied at the arrow) was also 2×10^{-8} A and lasted 2.5 sec. The external calcium concentration was 50 % normal. (*b*), (*c*). After injection of EGTA solution. $[EGTA]_i$, final internal concentration of EGTA, was approximately 1 mM since the total volume injected was approximately 3×10^{-10} l. Other experimental conditions as for (*a*). (*d*) The external solution was normal artificial sea water and the depolarising stimulus lasted 5 sec. The dotted line on each record indicates 0 mV. (Note the action potentials are attenuated by the pen recorder (from Meech 1974*b*).)

Many neurones have a repetitive pacemaker-like activity. In cell R15 of *Aplysia* this consists of 'bursts' of action potentials followed by periods of hyperpolarisation (see Fig. 5). This activity occurs even when the cell is ligatured (Alving, 1968). During the hyperpolarising phase the perme-

(a) Control

(b) [EGTA]$_i$=1 mM

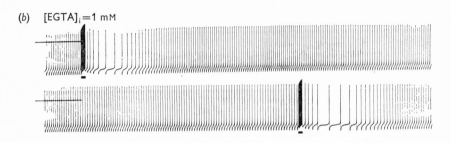

(c) 2×[Ca^{2+}]$_o$

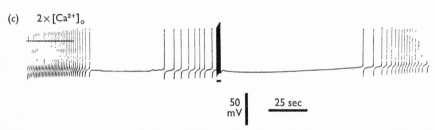

50 | 25 sec
mV

Fig. 5. Effect of intracellular EGTA on the 'bursting' pacemaker activity of *Aplysia* neurone R15. (*a*) Control: membrane potential recorded with a 3 M KCl filled micropipette. Horizontal bar below each record indicates the period of a 2×10^{-8} A depolarising current injected into the cell with a micropipette containing 0.67 M EGTA, 1.3 M KCl and 67 mM histidine hydrochloride buffered to pH 7.26. The external calcium concentration was 50 % normal. (*b*) After EGTA injection. [EGTA]$_i$ as for Fig. 4. (*c*) 'Bursting' pacemaker activity was restored in normal artificial sea water. (Note the action potentials are attenuated by the pen recorder.)

ability of the membrane to potassium is increased (Junge & Stephens, 1973). EGTA injection abolishes the 'bursting' pattern of firing of cell R15 and instead the cell fires continuously. As Fig. 5 shows, 'bursting' returns when the external calcium concentration is increased.

One of the advantages of using *Helix aspersa* is that it is possible to voltage clamp identifiable neurones. The currents recorded under voltage clamp in response to a depolarising pulse consist of a transitory inward current carried by both sodium and calcium ions (Standen, 1975*a*, *b*) and a prolonged outward current which is carried by potassium ions (see Fig. 6; Meech & Standen, 1975). These potassium currents are considerably reduced by the presence in the bathing medium of cobalt ions or the organic calcium antagonist D600 (Meech & Standen, 1974). As shown in

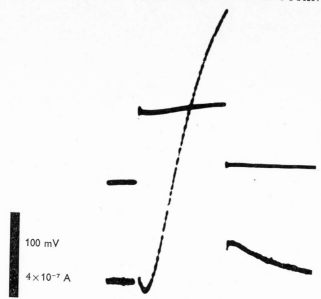

100 mV

4×10^{-7} A

Fig. 6. Membrane currents recorded under voltage clamp from an identified *Helix aspersa* neurone in response to a 20 msec depolarising pulse. Upper trace, membrane voltage; lower trace, membrane current compensated for the series resistance. (The reason for the shape of the voltage trace.) Holding potential, -42 mV; cell diameter, 205 μm, temperature, 18.5 °C (from Meech & Standen, 1975).

Fig. 7 they are made up of two components. There is a normally rectifying component which can be seen in calcium-free saline and a bell-shaped component which depends on the presence of calcium in the bathing medium. This latter component has been called calcium-mediated potassium activation. It results from an influx of calcium through a channel which resembles the 'late calcium channel' described by Baker, Hodgkin & Ridgway (1971) in squid axons.

The point at which there is no calcium-dependent potassium current is termed the 'null point' (-130 mV in Fig. 7). It is displaced 29 mV by a 10-fold change in the external calcium concentration which suggests that the 'null point' is close to the calcium equilibrium potential. There can be no calcium-dependent potassium current at the calcium equilibrium potential because there is no calcium influx. Calculations from 'null points' measured in four cells indicate that the intracellular calcium concentration is in the range of $3-8 \times 10^{-8}$ M in normal saline (10 mM calcium).

During a train of action potentials both components play a significant role in repolarising the membrane after each successive spike. Early in the train the voltage dependent component is more important, (Meech, 1974c), but this appears to be inactivated by prolonged depolarisation so that

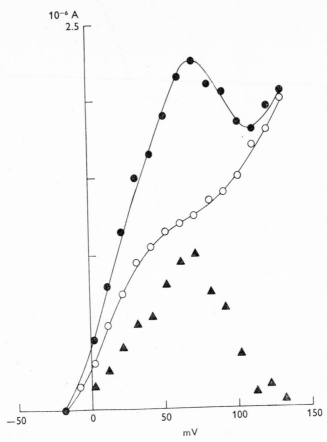

Fig. 7. Effect of calcium-free saline on the relationship between outward membrane current and membrane potential in *Helix* neurones. *Ordinate*: membrane current measured 80 msec after the beginning of the command pulse. *Abscissa*: membrane potential. ●, Normal saline; ○, after three minutes in calcium-free saline; ▲, calcium-dependent component obtained by subtraction of the currents recorded in calcium-free saline from those in normal saline. Holding potential, -48 mV; cell diameter, 210 μm; temperature, 21.5 °C (from Meech & Brown, 1976).

the calcium-dependent component becomes significant for later spikes. This is shown in EGTA-injected neurones. Activation of the calcium-mediated component is delayed by intracellular EGTA and so depolarisation, instead of generating a train of impulses, produces a number of brief action potentials followed by a long-lasting plateau which is dependent on the presence of calcium in the external medium (see Fig. 8.) Repolarization following the plateau presumably occurs when the EGTA buffer is saturated, because the duration of the plateau depends on the amount of

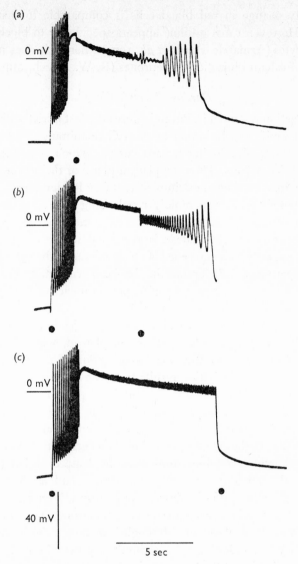

Fig. 8. Prolonged action potentials recorded from *Aplysia* neurone R15 after the injection of EGTA. The filled circles show the duration of the 4×10^{-8} A depolarising current. (Note the action potentials are attenuated by about 10 mV by the pen recorder (from Meech, 1974c).)

EGTA injected (Meech, 1974c). Thus the different properties of the two components of potassium activation are important for the modulation of nervous activity.

It would be attractive to suppose that the calcium-induced potassium permeability change in red blood cells is comparable to that found in neurones. However 1 mM quinine appears specifically to block the effect in erythrocytes (Armando-Hardy *et al.*, 1975) whereas it does not prevent the effect of calcium injection in neurones (R. W. Meech, unpublished).

Cardiac Purkinje fibres

During an action potential in Purkinje fibres the fast initial sodium current generates a rapid depolarisation of the cell membrane. This in its turn activates a slowly developing inward current which is at least partially carried by calcium ions. Thus the plateau phase of the action potential is associated with a prolonged calcium current (see Reuter, 1973). The action potential therefore has two rapidly repolarising steps. In each case the mechanism of repolarisation is a matter of some discussion. The initial step, repolarisation to the plateau is associated with a transient outward current dependent on the presence of chloride ions in the external solution (Dudel, Peper, Rüdel & Trautwein, 1966). Termination of the plateau may be the result of a slowly developing potassium activation or because of inactivation of the calcium current or both.

Isenberg (1975*a*, *b*) reports that iontophoretic injection of calcium hyperpolarises the membrane and shortens the plateau, whereas injection of EGTA produces depolarisation and prolongation of the action potential. He suggests that repolarisation results from the activation of a potassium current by intracellular calcium. A short-lived potassium current can be demonstrated under voltage clamp following calcium injection. The reversal potential of this current is -100 mV and it changes by 60 mV for a 10-fold change in the external potassium concentration (Isenberg, 1975*b*).

Additional evidence comes from work by Kass & Tsien (1975) who found that the calcium antagonists manganese and lanthanum had an inhibitory effect on *both* the inward and outward currents. However, a reduction in the external calcium concentration from 7.2–1.8 mM appeared to *increase* the outward current although it *decreased* the inward calcium current. They conclude that the 'calcium entry hypothesis' is unlikely to account for activation of the late potassium current. The evidence is not entirely convincing, however, because the outward current was estimated from the 'tail' current recorded at the holding potential following different depolarising pulses. The 'tail' currents were increased in low calcium saline. In *Helix* neurones a similar increase results from a change in the 'tail' current reversal potential, not from an increase in potassium activation (see Fig. 10 of Meech & Standen, 1975). Support for the 'calcium entry hypothesis' comes from the work of P. A. McNaughton (personal communication) who finds that calcium-free saline significantly reduces the late outward current. A relationship between the intracellular calcium concentration and potassium permeability has also been suggested for ventricular muscle (McGuigan & Bassingthwaighte, 1974).

Vertebrate central neurones

Calcium-mediated potassium activation has been demonstrated in vertebrate neurones by Krnjević and his co-workers. DNP-treated cat cortical neurones have high membrane potentials and low membrane resistance (Godfraind, Krnjević & Pumain, 1970). The hypothesis proposed by Godfraind et al., that DNP acts by augmenting the intracellular free calcium concentration, is supported by experiments in which calcium was iontophoretically injected into spinal motoneurones of anaesthetised cats (Krnjević & Lisiewicz, 1972). Table 1 shows the changes in the cell input resistance produced by calcium injection and summarises the results from 41 neurones. Clearly the fall in resistance is proportional to the amount of calcium releasing current. In one cell, when conditions were particularly stable Krnjević & Lisiewicz found a reversal potential of 60–65 mV. However in 38 injections into 28 neurones there was no clear effect in 21 % of the cells and a depolarisation in 26 %. Over half (53 %) of the cells became hyperpolarised and the mean hyperpolarisation was − 5 mV (S.E. 0.70 mV). Krnjević & Lisiewicz report that the effect was maximal within 10–30 sec of starting the injection and disappeared within 60 sec. However, they calculate that their injections should each increase the intracellular calcium in the order of 10 mM and suggest that much of the injected calcium was taken up into mitochondria. By comparing the effects of potassium chloride and calcium chloride injection on the membrane resistance and the inhibitory postsynaptic potential reversal level, Krnjević & Lisiewicz conclude that calcium injection must alter the potassium permeability rather than the chloride permeability.

Vertebrate spinal neurones have calcium-dependent action potentials and may generate long-lasting hyperpolarisations after trains of action potentials (see Ransom, Barker & Nelson, 1975). In cat spinal motoneurones, the hyperpolarisation results from a prolonged increase in the potassium conductance (Ito & Oshima, 1962) and it seems likely that a calcium-sensitive potassium system becomes activated (cf. Aplysia neurones).

Cat spinal motoneurones appear to be divided into two main groups: (a) phasic motoneurones, which respond to a sustained muscle stretch with only one or two action potentials and which innervate fast muscle fibres, (b) tonic motoneurones which respond with a more prolonged but slower discharge and which innervate slow muscle fibres (Granit, Henatsch

Table 1. *Changes in cell input resistance produced by intracellular injections of Ca^{2+} in 41 neurones (from Krnjević & Lisiewicz, 1972)*

No. of observations	Ca^{2+}-releasing current (nA)	Mean fall in resistance (%)	S.E. of mean
6	5–8	11.2	4.83
23	10–12	13.6	2.54
32	18–30	36.1	4.07
9	35–50	41.6	6.79

& Steg, 1956; Eccles, Eccles & Lundberg, 1958). Many of the electrical properties are continuously graded, but in general most tonic motoneurones have a more prolonged after-hyperpolarisation than phasic motoneurones, although the membrane potentials and the equilibrium potential of the after-hyperpolarisation are the same in each case (Kuno, 1959). The duration of the after potential in molluscan neurones depends on the intracellular calcium concentration (Meech, 1974*b*) and it is related to the ability of the cell to pump calcium from the cytoplasm (Meech, 1974*a*). Thus the difference between the two groups of motoneurones may be metabolic rather than 'electrophysiological' (cf. Burke, 1967). In fact they are best distinguished by their response to axotomy (Kuno, Miyata & Muñoz-Martinez, 1974*a*). After axotomy the duration of the after-hyperpolarisation is significantly decreased in the tonic motoneurones, whereas that of the fast motoneurone remains unchanged or is increased slightly. The tonic motoneurones regain their original properties when the cut nerve is allowed to regenerate into a muscle (Kuno *et al.*, 1974*b*). The degree of restoration is related to the degree of motor-reinnervation of the muscle and occurred whether the nerve is united to the original slow muscle or to a fast muscle. It seems possible that the calcium metabolism of the entire tonic system is different from that of the phasic system.

Skeletal muscle

Post-tetanic after-potentials associated with an increase in membrane potassium conductance have been described in both mammalian and frog skeletal muscles (Rüdel, Senges & Ehe, 1973; Adrian, Chandler & Hodgkin, 1970). Metabolic poisons combined with repetitive stimulation produce a considerable increase in potassium conductance (Grabowski, Lobsiger & Lüttgau, 1972). Grabowski *et al.* have suggested that a calcium-mediated conductance change is involved.

The currents recorded from frog muscle fibres under voltage clamp resemble those of *Helix* neurones in that the outward potassium current (delayed rectification) is made up of two components (Adrian *et al.*, 1970). It is the slower component which is responsible for the after-potential. This component is negligible at times less than 50 msec after the beginning of small depolarising pulses. The threshold of delayed rectification, measured at 100 or 200 msec, is approximately the same as the threshold of contraction (Costantin, 1968; Kao & Stanfield, 1968). This is true in different external calcium concentrations or in solutions of different external anions. For example the threshold for contraction in normal saline was about -50 mV. The threshold for delayed rectification was -52 mV. When chloride ions were partially replaced by thiocyanate ions the average threshold of contraction was -67.6 mV and that of delayed rectification was -69 mV. Each threshold shifted by about 17 mV (see Kao & Stanfield, 1968). This suggests that the initiation of contraction and the initiation of at least one component of delayed rectification may have a common stimulus which may be the release of calcium from the sarcoplasmic reticulum (Meech, 1972).

A detailed study of the relationship between delayed rectification and contraction at times of 60 msec and less showed that pulses of different duration, with an amplitude just sufficient to give a contraction, did not produce the same amount of delayed rectification (Adrian *et al.*, 1969). The strength–duration curves for contraction and delayed rectification had a significantly different shape. However, a study of 'total' delayed rectification may give a false impression because the time to peak of the 'slow' component appears to be relatively short at more positive potentials. A large proportion of the outward current measured by Costantin (1968) and Kao & Stanfield (1968) probably consisted of the 'slow' component of delayed rectification and it may be this component which is calcium mediated.

Crustacean muscle

Crustacean muscles have a variety of electrical and mechanical responses to motor axon stimulation. This is true not only of the muscles of different species but even within a single muscle since the electrical responses recorded from individual fibres may vary greatly from fibre to fibre (Hoyle & Wiersma, 1958). To some extent the different responses can be accounted for in terms of differences in the properties of the membranes of individual muscle fibres (Fatt & Katz, 1953). Atwood, Hoyle & Smyth (1965) have classified fibres in the walking legs of the crab *Cancer magister* as producing all-or-nothing spikes, graded responses or passive responses. They report that 'only highly aggressive, between-moult male crabs had a high proportion of spiking fibres'. It is possible that 'hormonal and other long-term changes determine the membrane properties of the fibres and cause conversion from one kind to another. The diversity of structure and other properties in one muscle may be related to the need to achieve a wide range of tension control peripherally with only a few available motor axons' (Atwood et al., 1965).

Action potentials can be generated in gradedly responding fibres by treating them with TEA (Fatt & Katz, 1953) or by perfusing them with EGTA (Hagiwara, Chichibu & Naka, 1964). The action potentials occur only in the presence of calcium ions (Fatt & Katz, 1953; Hagiwara et al., 1964) or other bivalent ions such as strontium or barium (Fatt & Ginsborg, 1958; Hagiwara, Fukuda & Eaton, 1974).

Recently, Mounier & Vassort (1975a, b) have described the currents generated by single Carcinus muscle fibres under voltage clamp. The fibres used were isolated from the extensor muscle from the meropodite of walking legs of the crab. They were found to give small, graded, depolarising responses when stimulated electrically. However, the graded responses of some fibres repolarised faster than others and were followed by a pronounced undershoot. By controlling the membrane potential with the double sucrose-gap technique and measuring the currents generated, Mounier & Vassort (1975a, b) have found that these fibres have a fast outward current in addition to the early calcium current and delayed potassium current seen in all fibres. This fast outward current is blocked by TEA but is not seen in calcium-free solution. Mounier & Vassort (1975b) have suggested that it is a potassium current activated by an increase in the calcium concentration at the inside surface of the cell membrane and propose that it can account for the low amplitude and variability of the electrical activity of crab muscle fibres.

In barnacle muscle fibres under voltage clamp, a train of large oscillatory membrane currents are generated during depolarisation of the membrane (Hagiwara & Naka, 1964; Keynes, Rojas, Taylor & Vergara, 1973). These currents can be distinguished from the simple single component identified by Mounier & Vassort (1975b). Nevertheless they disappear when the current of either the calcium or the potassium channel is substantially reduced (Keynes et al., 1973). Keynes et al. have suggested that the oscillations result from the complex anatomy of the muscle fibres. There may be a distributed series resistance and therefore a potential gradient between the tubule and the surface membrane. Furthermore, the sarcolemma may have both calcium and potassium channels whereas the membrane of the tubules may have only calcium channels.

The work by Hagiwara and his co-workers using EGTA-perfused barnacle muscle fibres has shown that calcium spikes can be stimulated when the intracellular concentration of ionic calcium is 8×10^{-8} M, but not when it is 5×10^{-7} M even if the external calcium is adjusted to keep the calcium gradient constant (Hagiwara & Nakajima, 1966). Clearly the role of the chelating agent in the perfusion medium is not simply to increase the calcium equilibrium potential. Another feature of these fibres is that the overshoot of the spike is very sensitive to the concentration of external calcium at low concentrations of calcium, but between 200 and 500 mM external calcium the change in overshoot is much less than that predicted by the Nernst equation. This relative insensitivity can be accounted for by an increase in the contribution of potassium ions to the overshoot. In high external calcium the overshoot changes by 28 mV for a 10-fold change in external potassium – considerably more than in low calcium solutions. Clearly a calcium-mediated potassium activation system could account for these findings as well as the oscillatory currents seen under voltage clamp and the early current identified by Mounier & Vassort (1975b).

Skate electroreceptors

The striking navigational abilities of a specimen of *Gymnarchus miloticus* and its sensitivity to changes in the electric field set up by its electric organ led Lissmann (1958) to suggest that certain lateral line receptors, the ampullae of Lorenzini, were specialised for the detection of electric fields. Ampullae from various species of skate, *Raja*, were subsequently found to be sensitive to voltage gradients of less than 0.1 μV cm^{-1} (Murray, 1962).

In *Raja*, the ampullae of Lorenzini are grouped together in capsules located in the head of the fish. Each ampulla has a canal which goes to the skin and is open to the outside. The canals cover most of the surface of the head. The sensory epithelium has a layer of receptor and supporting cells with occluding tight junctions between them. These junctions prevent the current flowing across the epithelium from leaking through the inter-cellular clefts and they divide the membrane of the receptor cells into lumenal and serosal faces (see Waltman, 1966 and Fig. 9).

The lumenal face of the receptor cells can often generate all-or-none 'receptor spikes' but more frequently excitatory stimuli evoke graded damped oscillations (Obara & Bennett, 1972). The receptor spikes are abolished by perfusing the lumen with cobalt ions (Clusin, Spray & Bennett, 1975).

Afferent fibres of the eighth cranial nerve form typical receptor synapses at the basal face of the receptor cell (Waltman, 1966). The stimulus required to generate impulses in the afferent nerve is usually below the level required to evoke receptor spikes or detectable oscillations (Obara & Bennett, 1972). Stimuli which would tend to hyperpolarise the lumenal face of the receptor and depolarise the serosal face were found to have an inhibitory effect on nerve activity. To account for these effects, Obara &

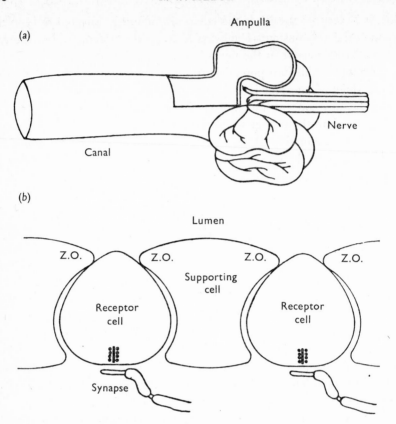

Fig. 9. (a) Skate ampullary electroreceptor and canal (after Waltman, 1966). (b) Details of two receptor cells in the ampullary epithelium. Occluding junctions (Z.O.: zonulae occludentes) occur between receptor cells and adjacent supporting cells. The basal membranes form characteristic ribbon synapses with the afferent nerve fibres (Waltman, 1966) (from Clusin et al., 1975).

Bennett suggested that the lumenal face of each receptor cell was independently spontaneously active. Hyperpolarisation of this face therefore leads to a decreased release of excitatory transmitter at the serosal face.

When the epithelium is voltage clamped, depolarisation of the lumenal face evokes a tetrodotoxin-insensitive inward current followed by an outward current (Clusin et al., 1974, 1975). Both the inward current and delayed rectification were abolished by perfusing the lumen with a calcium-free solution containing EGTA. The late outward current was also suppressed when the epithelium was clamped to the reversal potential of the inward current (see Fig. 10a). Clusin et al. conclude that the inward current is carried by calcium ions and that the outward currents, which are responsible for the repolarising phase of the action potential (since the

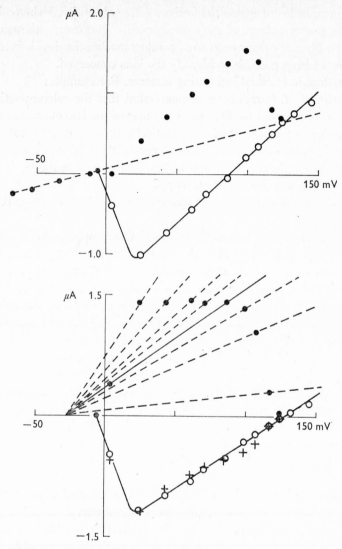

Fig. 10. (a) Current–voltage relationship of skate electroreceptor epithelium during perfusion. *Abscissa*: transepithelial potential. *Ordinate*: epithelial current. Current flowing inward across the lumenal membranes of the receptor cells and outward across the basal membranes is defined as inward current and shown downward. Voltage displacements that depolarise the lumenal membranes (lumen-negative) are defined as positive. The reversal potential for the early current (○), obtained by extrapolation of the leakage current (dashed line), is also the potential at which the late outward current (●) is suppressed (redrawn from Clusin *et al.* 1975). (b) Current–voltage relationship replotted from the data in (a). ○, Early current corrected for leakage current; ●, total late current. This is the difference between the early and the late currents because it appears that the early current is not inactivated by prolonged depolarisation. +, apparent calcium-activated conductance calculated as described in the text. The maximum apparent conductance is plotted so that it corresponds to the maximum early (calcium) current.

inward currents do not appear to inactivate), are only initiated when calcium ions enter the cytoplasm of the excitable cells. However, although it is clear that a distinct calcium activated conductance mechanism is involved, it has not yet been possible to identify the ions concerned.

This system has several interesting features. For example:

(a) Clusin et al. (1975) have demonstrated that the calcium-activated conductance undergoes facilitation. The onset of the late outward current occurs earlier in the second of two identical pulses 2 sec apart unless the conditioning pulse exceeds the reversal potential of the calcium current. This facilitated activation of the late outward current is ascribed to residual cytoplasmic calcium (Clusin et al., 1974).

(b) The calcium-activated conductance is insensitive to voltage (Clusin et al., 1975).

(c) The degree of calcium-activated conductance appears to be a simple function of the amount of inward current. The reasoning is as follows. Fig. 10b shows the calcium current separated from the leak current. It also shows the *total* current activated by calcium influx. (Note that the calcium current is not inactivated.) It is reasonable to assume that the calcium-activated conductance is constant for a given inward calcium current. We can therefore find a value for the effective equilibrium potential of the outward current if we assume that the channel does not rectify to any great extent (see Fig. 10b). The conductance for each outward current point can now be calculated. This conductance has been normalised and plotted on the same axis as the calcium current. As the Figure shows the outward conductance is closely related to the inward calcium current over a wide range of potential. The function of the calcium-activated conductance in this preparation may be similar to that proposed for other synapses (Meech & Standen, 1975). In the resting state the receptor is either on or close to the negative resistance region of the calcium current–voltage relationship, but any tendency to increase calcium activation and therefore depolarisation of the lumenal surface is opposed by the increase in conductance of the outward current system. It is therefore possible to get a steeply graded calcium influx. Consequently small changes in the current flowing across the lumenal membrane will produce large changes in calcium influx. This may well account for the sensitivity of the receptor.

Photoreceptors

The current–voltage curve of a dark-adapted *Limulus* ventral photoreceptor has a negative resistance region when the membrane is depolarised beyond 0 mV (Smith, Stell & Brown, 1968; Lisman & Brown, 1971). In Fig. 11, experimental points taken from the work of Lisman & Brown (1971) have

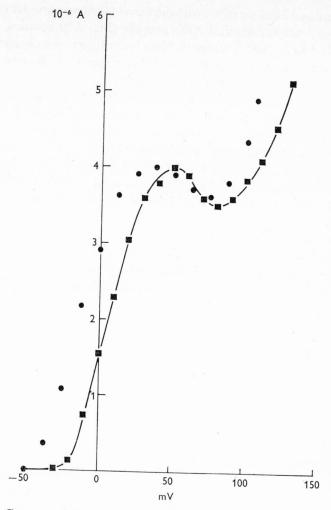

Fig. 11. Current–voltage relationship of a dark-adapted photoreceptor from *Limulus* ventral eye and of an identified neurone from *Helix aspersa*. *Abscissa*: membrane voltage; *ordinate*: membrane current. ■, Data from an identified *Helix aspersa* neurone (from Meech & Standen, 1975, fig. 4*b*). Current measured at 80 msec after the beginning of the command pulse. ●, Data from *Limulus* photoreceptor (replotted from Lisman & Brown, 1971, fig. 1*a*). Voltage clamp experiment in which the membrane voltage was changed slowly (10 mV sec^{-1}) with a triangular wave.

(Note: the points have been plotted so that the peak outward current at $+40$ mV (5.5×10^{-8} A) for the *Limulus* data corresponds to the peak of the *Helix* current–voltage relationship at $+50$ mV (Meech & Brown, 1976).)

been normalised for comparison with the current–voltage curve for delayed rectification from an identified *Helix* neurone (Meech & Standen, 1975). As described above the 'N'-shape in *Helix* is sensitive to the concentration of calcium in the external saline and this also appears to be true in the photoreceptor preparation (Smith *et al.*, 1968).

An interesting feature of these currents in *Limulus* is that the 'N'-shape is depressed for 5–10 min following a light stimulus (Lisman & Brown, 1971). It recovers with a time course comparable to the recovery of light sensitivity during dark adaptation. Lisman & Brown (1972) have proposed that the decrease in sensitivity to light following a light stimulus is a result of increased intracellular calcium, since prolonged iontophoretic injection of calcium also depresses the sensitivity of the photoreceptor. However, if the peak in the dark-adapted current–voltage relationship *is* the result of calcium-mediated potassium activation, it is hard to see why it should be *depressed* when the intracellular calcium is high. An alternative hypothesis, that a light stimulus leads to an increase in the intracellular H^+ concentration, can account for many of the features of light adaptation (Meech & Brown, 1976). It may explain the low sensitivity of the photoreceptor during prolonged light stimuli at a time when the intracellular calcium appears to be at a very low level as judged by the aequorin experiments of Brown & Blinks (1974).

Calliphora salivary gland

Fluid secretion by the salivary gland of the adult blowfly (*Calliphora erythrocephala*) is stimulated by low levels of 5-hydroxytryptamine (5-HT) (Berridge & Patel, 1968). Stimulated epithelial cells secrete a solution of 155 mM potassium chloride into the lumen of the gland (Oschman & Berridge, 1970). In the unstimulated state there is a potential of about +16 mV across the epithelium, but in the presence of 5-HT this potential goes to near 0 mV. The potential change is associated with a considerable decrease in membrane resistance which is dependent on the presence of both calcium and chloride ions. Most of the resistance change takes place at the lumenal surface of the epithelium. In calcium depleted cells or in chloride-free solution, there is little or no resistance change and the potential across the epithelium becomes even more positive. This response is thought to represent the activity of an electrogenic potassium pump (Berridge *et al.*, 1975).

The role of calcium in the generation of this conductance change is not entirely clear because there is no direct evidence that there is a significant increase in ionised intracellular calcium. Although there is an increase in both the influx and efflux of calcium in the presence of 5-HT, it is possible that 5-HT simply increases the exchange of calcium between different compartments of the cytoplasm. It may be that the calcium ions are required as a co-factor for the action of 5-HT. However, in the presence of calcium ions the ionophore A23187 also produces a decrease in membrane resistance comparable to the effect of 5-HT (Prince, Rasmussen & Berridge, 1973). At present there is a possibility that the ionophore induces a generalised increase in calcium permeability over both surfaces of the epithelium. This would account for the fact that the change develops over a period of 10 min. However, if the permeability change proves to be as chloride-dependent as the 5-HT response there will be a persuasive argument that intracellular calcium can modify membrane conductance in these cells.

Other tissues

In this review I have omitted references to many other tissues which also show signs of having calcium-induced potassium activation. For example, certain insect neurones are normally inexcitable, but generate calcium spikes in the presence of TEA or injected chelating agents (Pitman, 1975); calcium-dependent potassium currents have been identified in guinea-pig smooth muscle (Vassort, 1975); scallop photoreceptors generate a light-evoked hyperpolarisation which may be calcium mediated (Gorman & McReynolds, 1974).

CONCLUSIONS

A striking feature of the wide range of tissues considered in this account is the extensive distribution of 'calcium action potentials'. A dozen years ago it seemed possible that, because of their high charge, calcium ions would simply stick to the outside surface of the cell and be unable to carry current across the membrane. Perhaps one reason why calcium action potentials have sometimes appeared to be so mysterious is because calcium activation is often closely linked with potassium activation so that the calcium spike can only be seen in the absence of the potassium component. Excitable membranes which process information in some way require the added flexibility that a calcium activation/potassium activation system provides. Prolonged calcium action potentials are generated by a system dominated by calcium activation while one dominated by potassium activation is non-regenerative. The nature of the dominating component may depend on the activity of the calcium pump and therefore on the level of metabolic activity of the cell or fibre. Calcium-mediated potassium activation contributes to pacemaker activity and adaptation phenomena. When regenerative calcium activation is balanced by potassium activation, steeply graded responses can be generated for example in the skate electroreceptor/giant synapse preparation.

It seems that in most systems intracellular calcium induces a specific increase in the membrane potassium permeability. However, it is possible that the large effects of calcium injection on potassium permeability in *Helix* and *Aplysia* neurones may mask an effect on the sodium inactivation system.

The evidence for calcium-induced chloride activation is not strong at present because anions appear to have a considerable influence on calcium-dependent responses in muscle. It has been suggested that anions may alter the electric field across muscle membranes by adsorption at the outer surface (Hodgkin & Horowicz, 1960). Thus simple exchange of one anion for another may have an effect on the permeability of many other ions.

The best evidence that calcium acts directly on the cell membrane comes from work with red blood cell ghosts. Additional evidence comes from injection experiments using *Helix* neurones. The time between calcium injection and membrane response is approximately that expected for calcium to diffuse from the centre to the surface of a sphere of cytoplasm (R. W. Meech, unpublished).

At present only preliminary evidence is available concerning the receptor molecule. The site has a very high affinity for calcium and it resembles the calcium binding proteins extracted from muscle. Like troponin, a critical factor in the selectivity of the receptor is the ionic radius of the binding ion. Thus, when injected into *Helix* neurones, barium, lead, strontium, mercury, calcium, cadmium and possibly manganese are effective in increasing potassium conductance, whereas magnesium, nickel, cobalt, copper, zinc, iron and lanthanum are not. There is reason to believe that strontium and barium are much less effective than calcium since action potentials in strontium or barium are greatly prolonged as if the repolarising potassium system is incompletely activated. Of course, any of these ions may simply displace calcium from intracellular sites but it is nevertheless interesting that the properties of the receptor appear to match those of the troponin calcium binding site (Fuchs, 1971).

SUMMARY

1. Intracellular calcium alters the membrane permeability of invertebrate as well as vertebrate tissues and of inexcitable as well as excitable cells. It appears to be effective in muscles, neurones and receptors.

2. In most tissues intracellular calcium affects the potassium permeability of the membrane although there is a suggestion that if the intra-

cellular calcium concentration is high the sodium permeability of the red blood cells is also affected.

3. Calcium-mediated potassium activation can be induced by calcium influx, as in *Helix* neurones, but possibly also by intracellular release in muscle systems.

4. A calcium activation/potassium activation system is very flexible. Thus a system dominated by calcium activation gives prolonged action potentials whereas a system dominated by potassium activation is non-regenerative. The system may oscillate and therefore produce pacemaker activity or it may be balanced and give steeply graded calcium responses. The nature of the dominating component may depend on the activity of the calcium 'sink' and therefore on the level of metabolic activity of the cell or fibre.

REFERENCES

ADRIAN, R. H., CHANDLER, W. K. & HODGKIN, A. L. (1969). The kinetics of mechanical activation in frog muscle. *J. Physiol., Lond.*, **204**, 207–230.
(1970). Slow changes in potassium permeability in skeletal muscle. *J. Physiol., Lond.*, **208**, 645–668.
ALVING, B. O. (1968). Spontaneous activity in isolated somata of *Aplysia* pacemaker neurons. *J. gen. Physiol.*, **51**, 29–45.
ARMANDO-HARDY, M., ELLORY, J. C., FERREIRA, H. G., FLEMINGER, S. & LEW, V. L. (1975). Inhibition of the calcium-induced increase in the potassium permeability of human red blood cells by quinine. *J. Physiol., Lond.*, **250**, 32–33P.
ARMSTRONG, C. M. & BINSTOCK, L. (1965). Anomalous rectification in the squid giant axon injected with tetraethylammonium chloride. *J. gen. Physiol.*, **48**, 859–872.
ATWOOD, H. L., HOYLE, G. & SMYTH, T. (1965). Mechanical and electrical responses of single innervated crab-muscle fibres. *J. Physiol., Lond.*, **180**, 449–482.
BAKER, P. F., HODGKIN, A. L. & RIDGWAY, E. B. (1971). Depolarization and calcium entry in squid giant axons. *J. Physiol., Lond.*, **218**, 709–755.
BAKER, P. F., HODGKIN, A. L. & SHAW, T. I. (1962). Replacement of the axoplasm of giant nerve fibres with artificial solutions. *J. Physiol., Lond.*, **164**, 330–354.
BEGENISICH, T. & LYNCH, C. (1974). Effects of internal divalent cations on voltage-clamped squid axons. *J. gen. Physiol.*, **63**, 675–689.
BERRIDGE, M. J., LINDLEY, B. D. & PRINCE, W. T. (1975). Membrane permeability changes during stimulation of isolated salivary glands of *Calliphora* by 5-hydroxytryptamine. *J. Physiol., Lond.*, **244**, 549–567.
BERRIDGE, M. J. & PATEL, N. G. (1968). Insect salivary glands: stimulation of fluid secretion by 5-hydroxytryptamine and adenosine-3',5'-monophosphate. *Science, Wash.*, **162**, 462–463.
BLUM, R. M. & HOFFMAN, J. F. (1970). Carrier mediation of Ca induced K transport and its inhibition in red blood cells. *Fedn Proc.*, **29**, 663.
(1972). Ca-induced K transport in human red cells: localization of the Ca-sensitive site to the inside of the membrane. *Biochem. biophys. Res. Commun.*, **46**, 1146–1152.
BODEMAN, H. & PASSOW, H. (1972). Factors controlling the resealing of the

membrane of human erythrocyte ghosts after hypotonic hemolysis. *J. Memb Biol.*, **8**, 1–26.

BRODWICK, M. S. & JUNGE, D. (1972). Post-stimulus hyperpolarisation and slow potassium conductance increase in *Aplysia* giant neurone. *J. Physiol., Lond.*, **223**, 549–570.

BROWN, J. E. & BLINKS, J. R. (1974). Changes in intracellular free calcium concentration during illumination of invertebrate photoreceptors. Detection with aequorin. *J. gen. Physiol.*, **64**, 643–665.

BURKE, R. E. (1967). Motor unit types of cat triceps surae muscle. *J. Physiol., Lond.*, **193**, 141–160.

CHANDLER, W. K., HODGKIN, A. L. & MEVES, H. (1965). The effect of changing the internal solution on sodium inactivation and related phenomena in giant axons. *J. Physiol., Lond.*, **180**, 821–836.

CLUSIN, W., SPRAY, D. C. & BENNETT, M. V. L. (1974). Activation of a voltage-insensitive conductance by inward calcium current. *Biol. Bull.*, **147**, 472.

—— (1975). Activation of a voltage-insensitive conductance by inward calcium current. *Nature, Lond.*, **256**, 425–427.

COSTANTIN, L. L. (1968). The effect of calcium on contraction and conductance thresholds in frog skeletal muscle. *J. Physiol., Lond.*, **195**, 119–132.

DAVSON, H. (1941). The effect of some metabolic poisons on the permeability of the rabbit erythrocyte to potassium. *J. cell comp. Physiol.*, **18**, 173–185.

DUDEL, J., PEPER, K., RÜDEL, R. & TRAUTWEIN, W. (1966). Excitatory membrane current in heart muscle (Purkinjé fibers). *Pflüg. Arch. ges. Physiol.*, **292**, 255–273.

ECCLES, J. C., ECCLES, R. M. & LUNDBERG, A. (1958). The action potentials of the alpha motoneurones supplying fast and slow muscles. *J. Physiol., Lond.*, **142**, 275–291.

FATT, P. & GINSBORG, B. L. (1958). The ionic requirements for the production of action potentials in crustacean muscle fibres. *J. Physiol., Lond.*, **142**, 516–543.

FATT, P. & KATZ, B. (1953). The electrical properties of crustacean muscle fibres. *J. Physiol., Lond.*, **120**, 171–204.

FEIN, A. & LISMAN, J. (1975). Localized desensitization of *Limulus* photoreceptors produced by light or intracellular calcium ion injection. *Science, Wash.*, **187**, 1094–1096.

FUCHS, F. (1971). Ion exchange properties of the calcium receptor site of troponin. *Biochim. biophys. Acta*, **245**, 221–229.

GÁRDOS, G. (1958). The function of calcium in the potassium permeability of human erythrocytes. *Biochim. biophys. Acta*, **30**, 653–654.

GLYNN, I. M. (1956). Sodium and potassium movements in human red cells. *J. Physiol., Lond.*, **134**, 278–310.

GODFRAIND, J. M., KRNJEVIĆ, K. & PUMAIN, R. (1970). Unexpected features of the action of dinitrophenol on cortical neurones. *Nature, Lond.*, **228**, 562–564.

GORMAN, A. L. F. & McREYNOLDS, J. S. (1974). Control of membrane K+ permeability in a hyperpolarizing photoreceptor: similar effects of light and metabolic inhibitors. *Science, Wash.*, **185**, 620–621.

GRABOWSKI, W., LOBSIGER, E. A. & LÜTTGAU, H. CH. (1972). The effect of repetitive stimulation at low frequencies upon the electrical and mechanical activity of single muscle fibres. *Pflüg. Arch. ges. Physiol.*, **334**, 222–239.

GRANIT, R., HENATSCH, H.-D. & STEG, G. (1956). Tonic and phasic ventral horn cells differentiated by post-tetanic potentiation in cat extensors. *Acta physiol. scand.*, **37**, 114–126.

HAGIWARA, S., CHICHIBU, S. & NAKA, K. (1964). The effects of various ions on resting and spike potentials of barnacle muscle fibers. *J. gen. Physiol.*, 48, 163–179.

HAGIWARA, S., FUKUDA, J. & EATON, D. C. (1974). Membrane currents carried by Ca, Sr, and Ba in barnacle muscle fiber during voltage clamp. *J. gen. Physiol.*, 63, 564–578.

HAGIWARA, S. & NAKA, K. (1964). The initiation of spike potential in barnacle muscle fibers under low intracellular Ca^{++}. *J. gen. Physiol.*, 48, 141–162.

HAGIWARA, S. & NAKAJIMA, S. (1966). Effects of the intracellular Ca ion concentration upon the excitability of the muscle fiber membrane of a barnacle. *J. gen. Physiol.*, 49, 807–818.

HODGKIN, A. L. & HOROWICZ, P. (1960). The effect of nitrate and other anions on the mechanical response of single muscle fibres. *J. Physiol., Lond.*, 153, 404–412.

HOYLE, G. & WIERSMA, C. A. G. (1958). Excitation at neuromuscular junctions in Crustacea. *J. Physiol., Lond.*, 143, 403–425.

ISENBERG, G. (1975a). Is potassium conductance of cardiac Purkinje fibres controlled by $[Ca^{2+}]_i$? *Nature, Lond.*, 253, 273–274.

(1975b). Is potassium conductance controlled by $[Ca^{2+}]_i$? *5th Int. Biophys. Congr.* p. 140.

ITO, M. & OSHIMA, T. (1962). Temporal summation of after-hyperpolarisation following a motoneurone spike. *Nature, Lond.*, 195, 910–911.

JUNGE, D. & STEPHENS, C. L. (1973). Cyclic variation of potassium conductance in a burst-generating neurone in *Aplysia*. *J. Physiol., Lond.*, 235, 155–181.

KAO, C. Y. & STANFIELD, P. R. (1968). Actions of some anions on electrical properties and mechanical threshold of frog twitch muscle. *J. Physiol., Lond.*, 198, 291–309.

KASS, R. S. & TSIEN, R. W. (1975). Multiple effects of calcium antagonists on plateau currents in cardiac Purkinje fibers. *J. gen. Physiol.*, 66, 169–192.

KEYNES, R. D., ROJAS, E., TAYLOR, R. E. & VERGARA, J. (1973). Calcium and potassium systems of a giant barnacle muscle fibre under membrane potential control. *J. Physiol., Lond.*, 229, 409–455.

KREGENOW, F. M. & HOFFMAN, J. F. (1972). Some kinetic and metabolic characteristics of calcium-induced potassium transport in human red cells. *J. gen. Physiol.*, 60, 406–429.

KRNJEVIĆ, K. & LISIEWICZ, A. (1972). Injections of calcium ions into spinal motoneurones. *J. Physiol., Lond.*, 225, 363–390.

KUNO, M. (1959). Excitability following antidromic activation in spinal motoneurones supplying red muscles. *J. Physiol., Lond.*, 149, 374–393.

KUNO, M., MIYATA, Y. & MUÑOZ-MARTINEZ, E. J. (1974a). Differential reaction of fast and slow α-motoneurones to axotomy. *J. Physiol., Lond.*, 240, 725–739.

(1974b). Properties of fast and slow alpha motoneurones following motor reinnervation. *J. Physiol., Lond.*, 242, 273–288.

LEPKE, S. & PASSOW, H. (1968). Effects of fluoride on potassium and sodium permeability of the erythrocyte membrane. *J. gen. Physiol.*, 51, 365–372s.

LEW, V. L. (1970). Effect of intracellular calcium on the potassium permeability of human red cells. *J. Physiol., Lond.*, 206, 35–36P.

(1971a). On the ATP dependence of the Ca^{2+}-induced increase in K^+ permeability observed in human red cells. *Biochim. biophys. Acta*, 233, 827–830.

(1971b). Effect of ouabain on the Ca^{2+}-dependent increase in K^+ permeability in depleted guinea-pig red cells. *Biochim. biophys. Acta*, 249, 236–239.

(1974). On the mechanism of the Ca-induced increase in K permeability observed in human red cell membranes. In *Comparative Biochemistry and Physiology of Transport*, ed. L. Bolis, K. Bloch, S. E. Luria & F. Lynen, pp. 310–316. North-Holland: Amsterdam.

LISMAN, J. E. & BROWN, J.E. (1971). Two light-induced processes in the photo-receptor cells of *Limulus* ventral eye. *J. gen. Physiol.*, 58, 544–561.

(1972). The effects of intracellular iontophoretic injection of calcium and sodium ions on the light response of *Limulus* ventral photoreceptors. *J. gen. Physiol.*, 59, 701–719.

LISSMANN, H. W. (1958). On the function and evolution of electric organs in fish. *J. exp. Biol.*, 35, 156–191.

LUX, H. D. & ECKERT, R. (1974). Inferred slow inward current in snail neurones. *Nature, Lond.*, 250, 574–576.

McGUIGAN, J. A. S. & BASSINGTHWAIGHTE, J. B. (1974). Possible relationship between [Ca_i] and P_k in ventricular muscle. *Experientia*, 30, 680.

McLAUGHLIN, S. G. A., SZABO, G. & EISENMAN, G. (1971). Divalent ions and the surface potential of charged phospholipid membranes. *J. gen. Physiol.*, 58, 667–687.

MEECH, R. W. (1972). Intracellular calcium injection causes increased potassium conductance in *Aplysia* nerve cells. *Comp. Biochem. Physiol.*, 42A, 493–499.

(1974a). The sensitivity of *Helix aspersa* neurones to injected calcium ions. *J. Physiol., Lond.*, 237, 259–277.

(1974b). Calcium influx induces a post-tetanic hyperpolarisation in *Aplysia* neurones. *Comp. Biochem. Physiol.*, 48A, 387–395.

(1974c). Prolonged action potentials in *Aplysia* neurones injected with EGTA. *Comp. Biochem. Physiol.*, 48A, 397–402.

MEECH, R. W. & BROWN, H. M. (1976). Invertebrate photoreceptors: a survey of recent experiments on photoreceptors from *Balanus* and *Limulus*. In *Perspectives in Experimental Biology*, Vol. 1, ed. P. Spencer Davies, pp. 331–351. Pergamon Press: Oxford.

MEECH, R. W. & STANDEN, N. B. (1974). Calcium-mediated potassium activation in *Helix* neurones. *J. Physiol., Lond.*, 237, 43–44P.

(1975). Potassium activation in *Helix aspersa* neurones under voltage clamp: a component mediated by calcium influx. *J. Physiol., Lond.*, 249, 211–239.

MEECH, R. W. & STRUMWASSER, F. (1970). Intracellular calcium injection activates potassium conductance in *Aplysia* nerve cells. *Fedn Proc.*, 29, 834.

MORETON, R. B. (1972). Electrophysiology and ionic movements in the central nervous system of the snail, *Helix aspersa*. *J. exp. Biol.*, 57, 513–541.

MOUNIER, Y. & VASSORT, G. (1975a). Initial and delayed membrane currents in crab muscle fibre under voltage-clamp conditions. *J. Physiol., Lond.*, 251, 589–608.

(1975b). Evidence for a transient potassium membrane current dependent on calcium influx in crab muscle fibre. *J. Physiol., Lond.*, 251, 609–625.

MURRAY, R. W. (1962). The response of the ampullae of Lorenzini of elasmobranchs to electrical stimulation. *J. exp. Biol.*, 39, 119–128.

OBARA, S. & BENNETT, M. V. L. (1972). Mode of operation of ampullae of Lorenzini of the skate, *Raja*. *J. gen. Physiol.*, 60, 534–557.

OSCHMAN, J. L. & BERRIDGE, M. J. (1970). Structural and functional aspects of salivary fluid secretion in *Calliphora*. *Tissue and Cell*, 2, 281–310.

PITMAN, R. M. (1975). Calcium-dependent action potentials in the cell body of an insect motoneurone. *J. Physiol.*, 251, 62–63P.

PORTZEHL, H., CALDWELL, P. C. & RÜEGG, J. C. (1964). The dependence of contraction and relaxation of muscle fibres from the crab *Maia squinado* on the internal concentration of free calcium ions. *Biochim. biophys. Acta*, 79, 581–591.

PRINCE, W. T., RASMUSSEN, H. & BERRIDGE, M. J. (1973). The role of calcium in fly salivary gland secretion analysed with the ionophore A-23187. *Biochim. biophys. Acta*, **329**, 98–107.

RANSOM, B. R., BARKER, J. L. & NELSON, P. G. (1975). Two mechanisms for poststimulus hyperpolarisations in cultured mammalian neurones. *Nature, Lond.*, **256**, 424–425.

REUTER, H. (1973). Divalent cations as charge carriers in excitable membranes. *Progr. Biophys.*, **26**, 1–43.

RIORDAN, J. R. & PASSOW, H. (1973). The effects of calcium and lead on the potassium permeability of human erythrocytes and erythrocyte ghosts. In *Comparative Physiology* ed. L. Bolis, K. Schmidt-Nielsen & S. H. P. Maddrell, pp. 543–581. North-Holland: Amsterdam.

ROMERO, P. J. & WHITTAM, R. (1971). The control by internal calcium of membrane permeability to sodium and potassium. *J. Physiol., Lond.*, **214**, 481–507.

ROSE, B. & LOEWENSTEIN, W. R. (1975). Permeability of cell junction depends on local cytoplasmic calcium activity. *Nature, Lond.*, **254**, 250–252.

RÜDEL, R., SENGES, J. & EHE, L. (1973). A post-tetanic decrease of membrane resistance in mammalian skeletal muscle fibres and its antimyotonic effects. *Pflüg. Arch. ges. Physiol.*, **341**, 121–130.

SCHATZMANN, H. J. (1966). ATP-dependent Ca^{++}-extrusion from human red cells. *Experientia*, **22**, 364–365.

SHAW, T. I. (1955). Potassium movements in washed erythrocytes. *J. Physiol., Lond.*, **129**, 464–475.

SIMONS, T. J. B. (1975). Resealed ghosts used to study the effect of intracellular calcium ions on the potassium permeability of human red cell membranes. *J. Physiol., Lond.*, **246**, 52–54P.

SMITH, T. G., STELL, W. K. & BROWN, J. E. (1968). Conductance changes associated with receptor potentials in *Limulus* photoreceptors. *Science, Wash.*, **162**, 454–456.

STANDEN, N. B. (1975a). Calcium and sodium ions as charge carriers in the action potential of an identified snail neurone. *J. Physiol., Lond.*, **249**, 241–252.

(1975b). Voltage-clamp studies of the calcium inward current in an identified snail neurone: comparison with the sodium inward current. *J. Physiol., Lond.*, **249**, 253–268.

STINNAKRE, J. & TAUC, L. (1973). Calcium influx in active *Aplysia* neurones detected by injected aequorin. *Nature, New Biol.*, **242**, 113–115.

TASAKI, I., SINGER, I. & TAKENAKA, T. (1965). Effects of internal and external ionic environment on excitability of squid giant axon. *J. gen. Physiol.*, **48**, 1095–1123.

TASAKI, I., WATANABE, A. & LERMAN, L. (1967). Role of divalent cations in excitation of squid giant axons. *Am. J. Physiol.*, **213**, 1465–1474.

VASSORT, G. (1975). Voltage-clamp analysis of transmembrane ionic currents in guinea-pig myometrium: evidence for an initial potassium activation triggered by calcium influx. *J. Physiol., Lond.*, **252**, 713–734.

WALTMAN, B. (1966). Electrical properties and fine structure of the ampullary canals of Lorenzini. *Acta physiol. scand.*, **66**, suppl. 264, 1–60.

WHITTAM, R. (1968). Control of permeability to potassium in red blood cells. *Nature, Lond.*, **219**, 610.

WILBRANDT, W. (1937). A relation between the permeability of the red cell and its metabolism. *Trans. Faraday Soc.*, **33**, 956–959.

THE ROLE OF CALCIUM IN
SECRETORY PROCESSES:
MODEL STUDIES IN MAST CELLS

By J. C. FOREMAN, L. G. GARLAND*
AND J. L. MONGAR

Department of Pharmacology, University College London,
London WC1E 6BT, and *Pharmacology Laboratory,
Wellcome Research Laboratories, Langley Court,
Beckenham BR3 3BS, Greater London

The number of secretory processes which have been shown to be dependent on calcium ions has grown rapidly in the last two decades. The intense interest in calcium and secretion followed the observation that acetylcholine release from motor nerve terminals requires extracellular calcium (del Castillo & Stark, 1952) – an observation which had been made, but without the realisation of its significance, at the turn of the century (Locke, 1894). In 1968, Douglas reviewed the functional importance of calcium in secretion and suggested the concept of stimulus–secretion coupling based on the model of excitation–contraction coupling in muscle (Sandow, 1952). His hypothesis stated that the role of calcium in secretion was to link the event which triggered the cell, such as electrical depolarisation, with the secretory response which might be the exocytosis of granular material stored in the cytoplasm.

Since Douglas proposed the stimulus–secretion coupling hypothesis for the role of calcium, evidence has rapidly accumulated, and we now recognise the coupling action of calcium in the secretion of fluid, from salivary gland for example, as well as in the secretion of preformed granular stores of secretory material. It is also becoming apparent that the calcium required for the secretory process does not come from the extracellular compartment in every case, and that some secretory tissues, platelets for example, utilise an intracellular calcium pool for their secretory activity. In the fly salivary gland, there is evidence that calcium for secretory activity is available from both intra- and extracellular compartments in the same way as calcium is available for smooth muscle activity (Prince, Rasmussen & Berridge, 1973). For comprehensive reviews of the role of calcium in secretory processes the reader is referred to the writings of Rubin (1970, 1974) and Berridge (1975).

The purpose of this chapter is to describe the characteristics of one particular secretory system, the mast cell, and to outline some of the evidence relating to the control of secretion. Since the realisation of a role for calcium in secretion, further questions have arisen. For example: What is the site of action of calcium? What controls its action? There is growing evidence that the coupling of stimulus with secretion is initiated by a rise in the free calcium ion concentration in the cytoplasm and thus the various factors influencing cytoplasmic calcium are possible sources of control in secretory processes. Cell membrane permeability to calcium, mitochondrial calcium pumps and membrane calcium pumps may all have a part in the regulation of cell calcium.

Superimposed on the general concept of calcium as a link between stimulus and actual release of secretory material, is the role of cyclic nucleotides, in particular cyclic 3',5'-adenosine monophosphate (cyclic AMP), but there is increasing interest in the functional significance of the guanosine compound, cyclic GMP. Certainly, cyclic AMP influences a number of secretory systems though no general picture emerges here, and in some systems cyclic AMP stimulates secretory activity, for example in fly salivary gland (Prince, Berridge & Rasmussen, 1972), whilst in other systems, of which the mast cell is an example, compounds which stimulate cyclic AMP production inhibit secretion (Schild, 1936; Lichtenstein & Margolis, 1968; Assem & Schild, 1969). There are also calcium-dependent secretory processes in which cyclic AMP does not appear to be involved, and the adrenal medulla is one example. Berridge (1975) has produced a comprehensive review of the role of cyclic nucleotides in calcium-dependent secretory systems and outlines three basic types of secretory mechanism (Table 1). One thing seems clear: it is not possible at this time to fit cyclic AMP and calcium into a general scheme for the mechanism of secretion because of the clear-cut tissue differences. Rasmussen (1970) has produced a model for secretory systems in which calcium is required, and cyclic AMP activates secretion as a 'second messenger'. This proposes that the stimulus activates adenyl cyclase to generate cyclic AMP and that the cyclic AMP activates a protein kinase which phosphorylates an element in the cytoskeleton. The phosphorylation converts the cytoskeleton from a calcium-insensitive to a calcium-sensitive state, and the influx of calcium into the cytoplasm, also initiated by the stimulus, allows a calcium–cytoskeleton interaction which expels secretory material.

In contrast to this model, we wish to discuss the roles of calcium and cyclic AMP in the process of histamine secretion from mast cells, a system in which calcium is required and cyclic AMP inhibits secretion. We shall generate a model for this second type of secretory activity where cyclic

Table 1. *Classification of secretory processes according to the roles of calcium and cyclic AMP (see Berridge, 1976)*

Type 1. Calcium-dependent	Cyclic AMP has no role
2. Calcium-dependent	Cyclic AMP also activates release as 'second messenger'
3. Calcium-dependent	Cyclic AMP inhibits secretion

NOTE: the calcium required for the secretion may be derived from either extracellular or intracellular compartments.

AMP appears to act as a controlling influence rather than a 'second messenger'. Mast cells can be studied by following histamine release, which can be assessed by means of a sensitive assay for histamine (Boura, Mongar & Schild, 1954). The experimental work described here has been performed on mast cells isolated from the peritoneal cavities of rats, which have all the advantages of free-floating cells, and in particular, precise and rapid control over the cellular environment. Also, a simple density gradient centrifugation over albumin allows high purity samples of mast cells to be obtained for calcium uptake measurements. However, when histamine secretion alone is measured it is unnecessary to purify the cells since only mast cells contain histamine. The stimulus used to release histamine in these experiments was either the antigen–antibody reaction or dextran (mol. wt: 110000). In the case of the antigen–antibody reaction, the rats were given sensitizing injections of antigen (ovalbumin), 15 to 30 days before the experiment, and this results in the generation and fixation to mast cells of IgE antibody. Challenge of the cells by antigen then causes histamine secretion. Dextran releases histamine from cells without the sensitisation procedure by a direct action on the cell membrane, possibly by interaction with glucose-receptor moieties which could form receptors for dextran. In either case, cross-linking of receptors (antibody or glucose receptors) by the stimulus (antigen or dextran) appears to induce the change in membrane conformation which initiates the secretory mechanism (Ishizaka & Ishizaka, 1969; Siraganian, Hook & Levine, 1975).

CALCIUM AND HISTAMINE SECRETION

In 1958, Mongar & Schild observed that histamine secretion from mast cells in chopped lung stimulated by an antigen–antibody reaction required the presence of extracellular calcium. This observation was subsequently confirmed in isolated rat mast cells (Foreman & Mongar, 1972a) and in other systems involving antigen-induced histamine release from mast cells

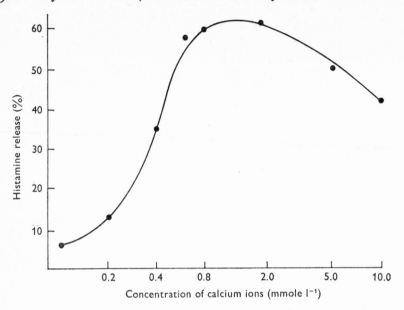

Fig. 1. Concentration–response curve for calcium in the antigen-induced release of histamine from rat mast cells.

or basophil leucocytes in blood (Lichtenstein & Osler, 1964; Greaves & Mongar, 1968, Yamamoto & Greaves; 1973). Fig. 1 shows a typical concentration–response relationship for the activation by calcium of antigen-induced histamine secretion from isolated rat mast cells. The effective concentration range is 0.1 to 1.0 mmole l^{-1}, with a maximum effect occurring at 1–2 mmole l^{-1}. A small component of antigen-induced histamine release is not dependent on extracellular calcium and is resistant to the presence of EDTA. It is not certain whether this represents true calcium-independent release or whether an intracellular calcium pool is being utilised, but this component of release is usually less than 10 % of total cell histamine. Relative to the antigen-induced release of histamine (usually 20–60 % of total) dextran is a weak stimulus and releases only 5–20 % of total cell histamine (see Fig. 2). However, the release induced by dextran and also by antigen can be enhanced by the addition of a low concentration, 10 μg ml^{-1}, of phosphatidyl serine (Fig. 2). The effect of phosphatidyl serine on histamine release from peritoneal mast cells was first reported by Goth, Adams, & Knoohuizen (1971) and was confirmed by Mongar & Svec (1972). The enhancing effect of this particular phospholipid is quite specific and is not shown by any other phospholipids.

Of particular interest is the observation that the potentiating effect of phosphatidyl serine on dextran- and antigen-induced histamine release is

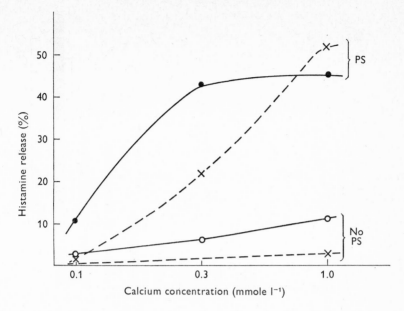

Fig. 2. Concentration–response curves for calcium in the release of histamine induced by antigen, 10 μg ml^{-1} (—), and dextran, 6 mg ml^{-1} (···). The action of phosphatidyl serine (PS), 10 μg ml^{-1}, on these concentration–response curves for calcium. Note that the degree of antigen-induced histamine release in the experiment shown in Fig. 1 is greater than that shown in Fig 2; this is due to a varying degree of antibody fixation to the cells during the sensitisation process and the factors controlling this have not been established.

dependent on the presence of calcium in the extracellular medium (Foreman & Mongar, 1972*b*, 1973; Mongar & Svec, 1972; Garland & Mongar, 1974). It is also well known that the acidic phospholipids bind calcium ions, and this led us to postulate the hypothesis that phosphatidyl serine might be involved in the interaction of calcium with the secretory mechanism in the mast cell. We shall return to this hypothesis in the light of other experimental results. In addition to the physiological significance which phosphatidyl serine might have, it also provides a tool for increasing dextran-induced histamine release to useful proportions.

In view of the dependence of histamine release on calcium ions, it was clearly of importance to gain some insight into the permeability of the mast cell membrane to calcium, and this we did using ^{45}Ca. Association of ^{45}Ca with mast cells was measured directly by incubating the cells with the tracer and then separating cells from tracer by a high speed centrifugation ($\sim$ 12000 *g*) through silicone oil. The pellet was assayed for ^{45}Ca by liquid scintillation spectroscopy after digestion (Foreman, 1973; Foreman, Mongar & Gomperts, 1973). Antigen caused an increase in the amount of ^{45}Ca

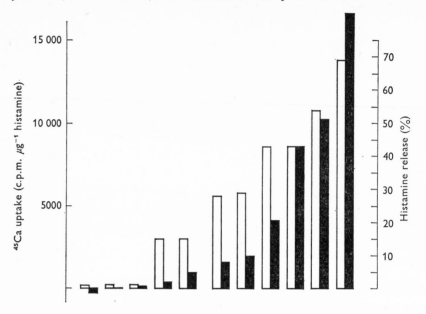

Fig. 3. The relationship between antigen-induced histamine release (□) and the increase in ^{45}Ca associated with mast cells following antigen stimulation (■) in 11 separate experiments. There is a positive correlation between histamine release and ^{45}Ca uptake with $r = 0.90$ ($P < 0.001$).

associated with the cells (Fig. 3) and the magnitude of the uptake of ^{45}Ca was correlated with the magnitude of the secretion of histamine. The increase in ^{45}Ca associated with antigen-challenged cells was not due to increased surface area generated by exocytosis (Foreman, 1973); neither was it due to binding of calcium to granular material because, when histamine release is totally inhibited by antimycin A (an inhibitor of oxidative phosphorylation), calcium uptake is maintained at 65 % of its control level in the absence of inhibitor (Foreman, Hallett & Mongar, 1975). It appears then that this association of ^{45}Ca with antigen-challenged cells represents an increased membrane permeability to calcium following stimulation. The tracer experiments do not distinguish calcium exchange between intra- and extracellular compartments from actual net uptake of calcium but other experiments with a calcium ionophore lead us to believe that it is the entry of calcium into the mast cell resulting from increased membrane permeability which triggers secretion. It is interesting to note at this point that phosphatidyl serine, which potentiates histamine release, also increases the change in membrane permeability to calcium which results from antigen challenge, thus supporting the hypothesis that this phospholipid influences histamine release by its effect on the calcium mast cell interaction.

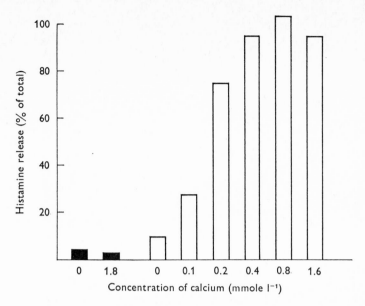

Fig. 4. Activation of histamine release from rat mast cells by calcium in the presence of the calcium ionophore A23187, 0.6 μmole l^{-1}. $\square$, Histamine release in the presence of A23187; $\blacksquare$, histamine release in the absence of A23187.

Some of the most powerful evidence to suggest that entry of calcium into mast cells triggers histamine secretion has come from experiments with the calcium ionophore A23187. This molecule is a specific transporting substance for calcium, which dissolves in cell membranes or other organic phases and forms hydrophobic complexes with calcium, enabling the calcium to enter and traverse the organic phase, down a concentration gradient (Caswell & Pressman, 1972; Reed & Lardy, 1972). A23187 will dissolve in the organic phase of mast cells but this alone has no histamine-releasing action. However, when calcium is added to cells containing ionophore in their membrane, the mast cells take up calcium from the extracellular medium and release histamine (Fig. 4). The calcium uptake induced by the ionophore has been demonstrated using murexide as an extracellular calcium indicator and also by following ^{45}Ca movement (Foreman *et al.*, 1973; M. B. Hallett, unpublished).

The evidence indicates, therefore, that one of the primary events in the initiation of histamine secretion from mast cells is a stimulus-induced increase in the cell membrane permeability to calcium, resulting in an influx of calcium into the cytoplasm down a concentration gradient. One of the questions which follow from this model is whether control over calcium entry determines the magnitude of histamine release.

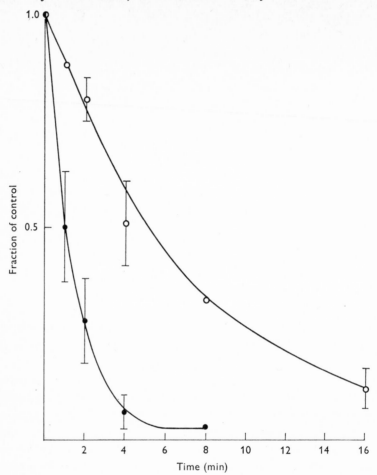

Fig. 5. Time course of the decay in the response to calcium of rat mast cells stimulated by the antigen-antibody reaction. ●—●, without phosphatidyl serine (PS); ○—○, phosphatidyl serine 10 μg ml⁻¹. The response is expressed as a fraction of the control histamine release obtained when antigen, 10 μg ml⁻¹, was added to cells in complete Tyrode solution containing calcium, 1 mmole l⁻¹. The control histamine releases were between 18 and 54 % of total in the absence of phosphatidyl serine, and 37–64 % in the presence of phosphatidyl serine.

DECAY OF THE ACTIVATED STATE

It has already been pointed out that if cells are challenged with antigen in the absence of calcium less than 10 % of the histamine is released. If calcium is added after antigen-challenge in calcium-free conditions, the response that is obtained depends on the interval between antigen-challenge and the addition of calcium. Fig. 5 shows that the response to calcium under these conditions decays quite rapidly with time so that, when calcium is added 4 min after antigen, the response has almost completely decayed. The decay of the response to calcium in antigen-activated cells is approximately exponential and has a half-time of about 1 min. The decay cannot be attributed to destruction of antigen since a second antigen-challenge after calcium addition does not restore the decayed response. Neither have the cells totally lost their ability to secrete because, after complete loss of response to antigen, it is possible to obtain a full response to the calcium ionophore A23187 (Foreman & Garland, 1974). In other words, decay of the response to calcium in antigen-treated cells does not reduce the response to ionophore. Phosphatidyl serine has a marked slowing effect on the decay in the response to calcium, decreasing the rate of decay by about four-fold (Fig. 5), in addition to its effect of increasing antigen-induced membrane permeability to calcium.

Turning now to dextran-stimulated cells, it is possible to show that these also exhibit a decay phenomenon which is similar to that described for antigen-challenged cells. It must be re-emphasised that, in the case of dextran, the phospholipid is required to achieve a measurable response, and so the experimental protocol is different from that employed when antigen is used as the stimulus. Fig. 6 shows the effect of adding phosphatidyl serine at various times after stimulation of the cells with dextran and calcium. Dextran with calcium produces no response in the absence of phosphatidyl serine but when the phospholipid is added at various times after challenge with dextran and calcium, the response to the phospholipid declines with a time-course similar to that for the decay in antigen-challenged cells. Again, cells exhibiting decay after dextran stimulation are fully responsive to the calcium ionophore (Garland, 1975).

The fact that decay following stimulation by antigen or dextran does not alter the release induced by the calcium ionophore suggests that the decay phenomenon involves a step at or before the change in membrane permeability to calcium caused by the antigen or dextran stimuli. Clearly, if decay occurred at a stage after calcium entry into the cell, the ionophore-induced release should be affected by prior decay to antigen or dextran. The experiments shown in Fig. 7 support the hypothesis that the decay phenomenon is, in fact, a shutting-off of membrane permeability to calcium following dextran or antigen stimulation. When cells in calcium-free medium are treated with ionophore, no release occurs, and adding back calcium at various times after the addition of the ionophore, gives a response which does not decay with time, in complete contrast to the experiments with antigen and dextran. In other words, if a calcium transporting molecule is inserted into the mast cell membrane, no decay in response with time is observed, and this implies that the decay seen after

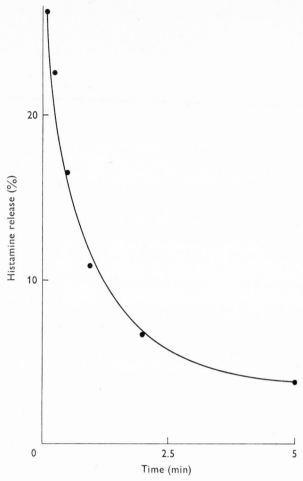

Fig. 6. Histamine release induced by adding phosphatidyl serine, 10 μg ml^{-1}, to rat mast cells at various times after the addition of dextran, 6 mg ml^{-1}, and calcium, 1.8 mmole l^{-1}.

antigen and dextran stimulation is a result of a change in the calcium channels opened by these stimuli.

The decay in response following antigen-stimulation in rat mast cells has been confirmed by Baxter & Adamik (1975) and also by Diamant and his colleagues (Diamant, Grosman, Stahl Skov & Thomle, 1974). Siraganian & Osler (1970) and Lichtenstein & Osler (1964) have observed this phenomenon in leucocytes.

Furthermore, a similar decay phenomenon has been observed in neurohypophysis (Nordmann, 1975) and adrenal medulla (Borowitz, Leslie & Baugh, 1975) where the secretory response declines despite continuing

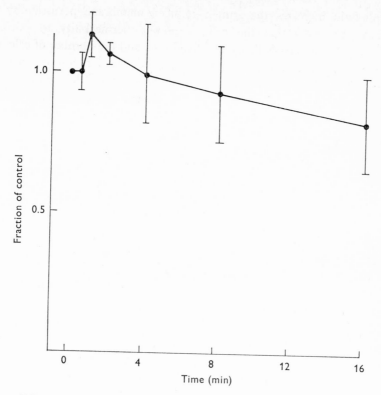

Fig. 7. Effect of adding calcium, 1 mmole l⁻¹, to cells suspended in calcium-free Tyrode solution, at various times after the addition of A23187, 0.6 μmole l⁻¹. The release is expressed as a fraction of the control release obtained when A23187 was added to complete Tyrode solution (calcium, 1 mmole l⁻¹). Control histamine releases in the four experiments ranged from 44 to 77 %.

stimulation, and this may be related to decreasing membrane permeability to calcium in these systems.

An answer to the question whether or not control over calcium entry into the mast cell limits histamine secretion is beginning to form. The experiments in the previous section support the view that antigen and dextran stimuli initiate release by raising cell membrane permeability to calcium and allowing calcium to enter the cell. In this section, evidence is emerging which suggests that almost as soon as release has been initiated in this way another process begins which shuts down the membrane permeability to calcium again and it is possible that this limits calcium entry and regulates the degree of histamine secretion.

The slowing of decay by phosphatidyl serine in the case of antigen-induced histamine release is consistent with the observation that this phospholipid increases the antigen-induced membrane permeability to calcium. Assuming that the cell membrane permeability to calcium declines after antigen-challenge at a fixed rate, and the number of calcium channels open at any time determines the degree of secretion, then, if phosphatidyl serine acts by increasing calcium transport through the open channels, the degree of secretion will be greater in the presence of the phospholipid than in its absence at any given time and the rate of fall of the secretory response will thus be reduced.

DO ANTIGEN AND DEXTRAN ACTIVATE COMMON PATHWAYS?

Since both antigen and dextran induce histamine release which exhibits a decay phenomenon, it was of interest to investigate (1) whether or not the two stimuli interacted with one another, (2) whether or not decay after one stimulus altered the response to the other stimulus. Figs. 8 and 9 show experiments related to the first of these questions. In the experiment shown in Fig. 8, similar responses were obtained to the two stimuli. Another sample of the same pool of cells was stimulated with antigen and 10 min later with dextran and phosphatidyl serine. The response obtained with this sequential addition of the two stimuli to the same cells is much less than the sum of the stimuli added to separate cells (cf. the double cross-hatched rectangle with the dashed rectangle of Fig. 8). In other words the response of the cells to dextran and phosphatidyl serine has been modified by the prior application of the antigen stimulus. (It should be pointed out that the phosphatidyl serine added with the dextran does not influence the antigen response when added 10 min after antigen (Stechschulte & Austen, 1974).) Fig. 9 shows a similar type of experiment, except that in this case the first stimulus was dextran and phosphatidyl serine, followed after 10 min by antigen and phosphatidyl serine. (Under these conditions the phosphatidyl serine added with the dextran does influence the antigen response, hence the control with antigen and phosphatidyl serine.) Again, a certain amount of interaction between the stimuli has occurred, though it is not as great as when dextran with phosphatidyl serine was given after antigen. Thus, even when the two stimuli produce equal histamine release, dextran produces less of a change in the ability of the cells to respond than does antigen.

The experiments depicted in Figs. 10 and 11 also relate to stimulus interaction but involve the decay process which follows stimulation. In the experiment shown in Fig. 10, the decay in response to calcium after antigen

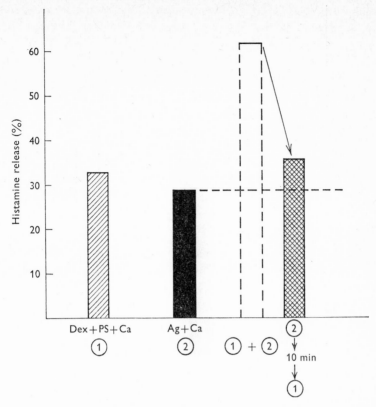

Fig. 8. Histamine release from rat mast cells obtained with dextran (Dex), 6 mg ml^{-1} and phosphatidyl serine (PS), 10 μg ml^{-1} (▨), and with antigen (Ag), 10 μg ml^{-1} (■); both stimuli were given in the presence of calcium, 1 mmole l^{-1}. The effect of adding the dextran stimulus 10 min after the antigen stimulus, on the same cell sample (▧), is compared with the algebraic sum of the release obtained by using the two stimuli separately (⬚).

stimulation has been demonstrated (filled columns), then cells whose response was allowed to decay after antigen stimulation were treated with dextran, and the response compared with the response of non-pretreated cells towards dextran. It can be seen that cells whose response to antigen stimulation had been allowed to decay show a reduced capacity to respond or are 'desensitised' to dextran stimulation. In fact, after decay of the response to antigen, the response to dextran with phosphatidyl serine is reduced by 67% of the response to dextran with phosphatidyl serine in cells not pretreated with antigen. Fig. 11 shows the comparable experiment with the two stimuli reversed. The decay of the response after dextran stimulation has been demonstrated (hatched rectangles) and then cells whose response to dextran has been allowed to decay are given an antigen

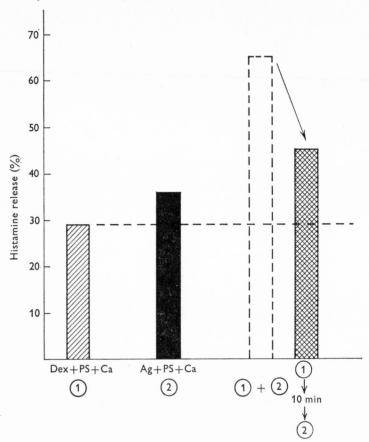

Fig. 9. Histamine release from rat mast cells obtained with dextran (Dex), 6 mg ml^{-1}, and phosphatidyl serine (PS), 10 μg ml^{-1} (▨), and with antigen, 10 μg ml^{-1}, and phosphatidyl serine, 10 μg ml^{-1} (■): both stimuli were given in the presence of calcium, 1 mmole l^{-1}. The effect of adding the antigen stimulus 10 min after the dextran stimulus, on the same cell sample (▧), is compared with the algebraic sum of the release obtained by using the two stimuli separately (□).

stimulus. The response to this antigen-stimulation is compared with the response to antigen of cells which had not been pretreated with dextran. Again it can be seen that prior decay of the response to one stimulus influenced the response to the second stimulus, but in this case the fall in response was only 20%. This is related to the observation noted above that dextran produces less of a refractory state than does antigen, and this will be discussed later in the chapter in terms of a model.

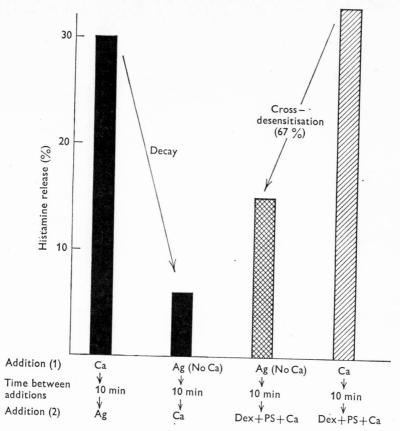

Fig. 10. The decay in the response of the mast cells when calcium, 1 mmole l⁻¹ was added 10 min after the cells were stimulated by antigen (Ag) in the absence of calcium, ■ (see also Fig. 5). The response obtained with dextran (Dex), 6 mg ml⁻¹, and phosphatidyl serine (PS), 10 μg ml⁻¹ (▨), in cells whose response after antigen-stimulation had been allowed to decay; this is compared with the response of cells to the dextran stimulus when no antigen pretreatment was given (▨).

CYCLIC AMP AND ITS RELATIONSHIP TO THE DECAY OF THE RESPONSE

During histamine release induced by an antigen–antibody reaction, changes in intracellular levels of cyclic AMP can be recorded, and such changes are depicted in Fig. 12, which is taken from the work of Kaliner & Austen (1974). Following antigen stimulation there is a fall in the intracellular cyclic AMP level which precedes the release of histamine. The level of cyclic AMP then returns to the pre-stimulus level within 5 min. We wish to point out that the return of the cyclic AMP level towards basal level after stimulation has a time course similar to the time course of the decay

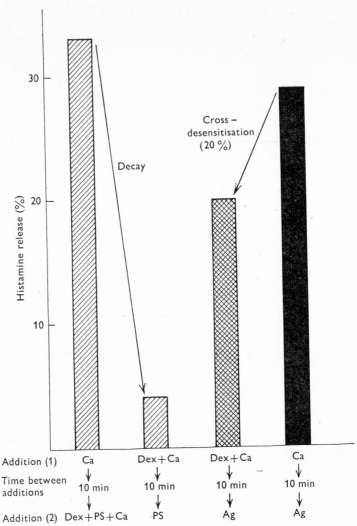

Fig. 11. The decay in the response of mast cells when phosphatidyl serine (PS), 10 μg ml⁻¹, was added 10 min after the cells were stimulated by dextran, 6 mg ml⁻¹, in the presence of calcium, 1 mmole l⁻¹, ▨ (see also Fig. 6). The response obtained with antigen (Ag), 10 μg ml⁻¹, in cells whose response to dextran stimulation had been allowed to decay (▨); this is compared with the response of cells to the antigen stimulus when no dextran pretreatment was given (■).

phenomenon described above and we shall examine further the hypothesis that cyclic AMP might mediate this decay process.

It is well known that cyclic AMP inhibits antigen- and dextran-induced histamine release, as do agents which raise intracellular levels of cyclic AMP (Lichtenstein & Margolis, 1968; Koopman, Orange & Austen, 1970:

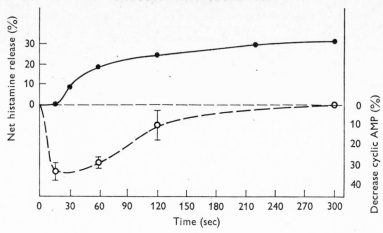

Fig. 12. Comparison of the time courses of histamine release (●—●) and changes in intracellular cyclic AMP level (○—○) in rat mast cells stimulated by an antigen–antibody reaction. The stimulus was given at zero time (from Kaliner & Austen, 1974).

Assem & Schild, 1971; Baxter, 1972; Bourne, Lichtenstein & Melmon, 1972; Johnson, Moran & Mayer, 1974). However, no explanation of the site of action of cyclic AMP in inhibiting histamine secretion was available until the ionophore was used as a tool for probing the action of cyclic AMP. In the experiments shown in Fig. 13, it can be seen that dibutyryl cyclic AMP inhibits both antigen- and dextran-induced histamine release in the concentration range 0.01 to 10 mmole dibutyryl cyclic AMP per litre. At these concentrations, dibutyryl cyclic AMP was without effect on the histamine release induced by the calcium ionophore A23187 (Foreman, Mongar, Gomperts & Garland, 1975; Garland & Mongar, 1976). The evidence outlined above suggests that antigen and dextran release histamine by opening calcium channels in the mast cell membrane, allowing calcium to enter the cell, and this is also how the ionophore acts except that it is itself a calcium carrier and so by-passes the physiological membrane channels operated by the other stimuli. The fact that the ionophore-mediated histamine release is not inhibited by dibutyryl cyclic AMP puts the site of action of this cyclic nucleotide, in antigen- or dextran-induced release, at or before the entry of calcium into the cell since, if it inhibited a stage after calcium entry, the release mediated by the ionophore would also be reduced. The experiments in Fig. 13 thus indicate that cyclic AMP inhibits antigen- and dextran-induced histamine release by interfering with the calcium entry into the mast cell which these stimuli induce. Some direct evidence for this hypothesis has been gained by assessing the mast cell membrane permeability to calcium using ^{45}Ca as described above.

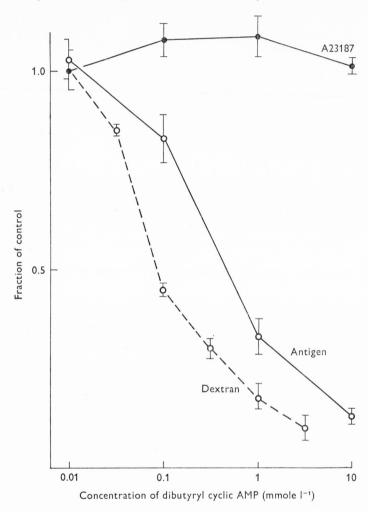

Fig. 13. Concentration–response relationship for the action of dibutyryl cyclic AMP on histamine release induced by antigen, 10 μg ml^{-1} (O—O); dextran, 6 mg ml^{-1}, with phosphatidyl serine, 10 μg ml^{-1} (O---O), and A23187, 5 μmole l^{-1} (●—●). Response is expressed as a fraction of control histamine release from cells suspended in a medium containing calcium, 1 mmole l^{-1} and no dibutyryl cyclic AMP. Control releases were 22 % of total for antigen, 47 % for dextran and 77 % for A23187. Cells with or without dibutyryl cyclic AMP were pre-incubated for 30 min at 37 °C before the addition of the releasing agent.

We have already stated that it is possible to dissociate calcium transport across the cell membrane and histamine release by using antimycin A, which prevents histamine release without blocking calcium transport (Foreman, Hallett & Mongar, 1975). Fig. 14 shows that dibutyryl cyclic AMP inhibits antigen-induced histamine release and prevents the calcium

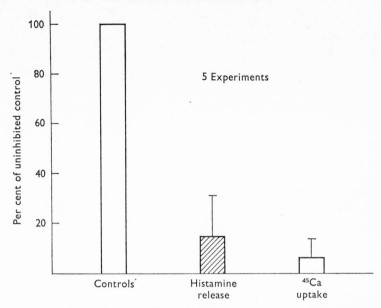

Fig. 14. The effect of dibutyryl cyclic AMP, 10 mmole l⁻¹, on histamine release induced by antigen and an antigen-induced ^{45}Ca-uptake in the same cell population. The responses are expressed as fractions of the histamine releases and ^{45}Ca-uptake measured in the absence of dibutyryl cyclic AMP which were: histamine release 11–54 %; ^{45}Ca-uptake 0.07–0.27 nmole Ca per 10⁶ cells.

movement across the cell membrane, as one would expect if the interpretation of the experiment in Fig. 13 was the correct one, namely that cyclic AMP exerts its inhibitory effect at or before the level of calcium entry into the cell.

How is the action of cyclic AMP related to decay? In the section dealing with decay we discussed the evidence suggesting that the decay phenomenon was the result of a shutting down of the membrane permeability to calcium after an initial increase in the calcium permeability induced by stimulation. The experiments described in this section would support the view that cyclic AMP inhibits histamine release by inhibiting calcium transport across the mast cell membrane. Since the time courses of the decay process and of rising intracellular cyclic AMP levels after stimulation appear to be parallel, it is temping to suggest that cyclic AMP mediates the decay phenomenon by reducing membrane permeability to calcium. It is interesting to recall that histamine release following stimulation of the mast cell either with antigen or a pharmacological histamine-releasing agent, compound 48/80, is accompanied by an initial fall in intracellular cyclic AMP levels (Gillespie, 1973; Kaliner & Austen, 1974; Sullivan,

Parker, Eisen & Parker, 1975), followed by a return to basal level. This would be consistent with the hypothesis that cyclic AMP regulates calcium entry into the cell, and in order for calcium entry to occur and initiate secretion, there must be an initial fall in intracellular cyclic AMP. The degree of secretion is then modulated by regeneration of cyclic AMP and a reduction of calcium entry into the cell.

A MODEL FOR ONE OF THE MECHANISMS
OF SECRETION

It was pointed out in the introduction that the variety of actions of cyclic nucleotides and their interactions with calcium in secretory systems makes it unlikely that a general scheme encompassing all secretory mechanisms will be discovered. We outline in this section a model for the type of secretory activity in which calcium is required and cyclic AMP has an inhibitory role, in contrast to the model described by Rasmussen (1970) for systems in which cyclic AMP and calcium both activate secretion. Our model is based on the experiments which have been performed in mast cells and basophil leucocytes (see also Foreman & Mongar, 1975). Fig. 15 is a diagrammatic representation of a mast cell sharing two stimuli to histamine secretion, antigen and dextran, both of which require calcium in the extracellular medium before they can induce secretory activity. Both stimuli are depicted as acting on a common membrane unit, which we will call the operator unit. The evidence for describing a common site of action comes from the stimulus interaction and cross-desensitisation experiments. The model proposes that the stimulus, either antigen or dextran, activates the operator unit which itself initiates simultaneously two sequences of events, one of which is opening of calcium channels in the cell membrane and the other is the action on the adenylcyclase–phosphodiesterase system which results in a transient fall in the intracellular cyclic AMP level. Thus the two criteria for histamine secretion to occur, entry of calcium into the cell and reduction of intracellular cyclic AMP, are fulfilled. Cellular homeostasis and the modulation of the degree of histamine secretion is then brought about, at least in the first instance, by the reaccumulation of intracellular cyclic AMP, which acts to close the calcium channels and so to prevent further calcium entry.

The ionophore is shown as a calcium conductance in the membrane parallel to the physiological channels and outside the operator unit and the cyclic nucleotide system. Once calcium has entered the cell it seems that a source of ATP is also necessary before the histamine-containing granules can be extruded. This conclusion is based on experiments with

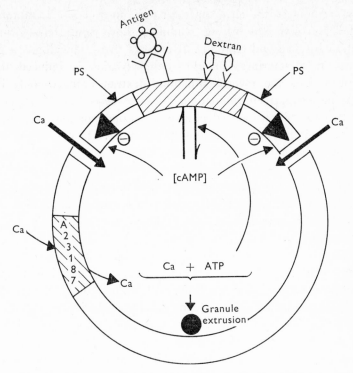

Fig. 15. A model for the mechanism of histamine release from mast cells. The stimuli are shown acting on a common membrane site (hatched area) which, when activated, is considered to mediate the opening of membrane calcium-channels and a transient fall in the intracellular cyclic AMP (cAMP) level. Cyclic AMP reduces calcium influx and thus the degree of release is controlled by the return of the cyclic AMP levels towards the resting value, after the transient fall induced by the stimulus. Phosphatidyl serine (PS) is shown acting on the opening of the calcium-channels. The calcium ionophore bypasses the stimulus-operated, physiological calcium-channels. The influx of calcium into the cell triggers the extrusion of histamine-containing granules, a process which also utilises a source of ATP within the cell.

metabolic inhibitors and measurement of cellular ATP levels (Mongar & Schild, 1957; Diamant & Uvnäs, 1961; Foreman *et al.*, 1973; Diamant, Norn, Felding, Olsen, Ziebell & Nissen, 1974; Johansen & Chakravarty, 1974; Peterson, 1974; Garland & Mongar, 1976).

The exact nature of the calcium channel and the operator unit is unknown but the experiments with phosphatidyl serine strongly support the view that this phospholipid acts by increasing calcium transport across the mast cell membrane, and it is interesting to speculate that this phospholipid forms part of the physiological calcium-transporting system which is operated in the mast cell membrane by stimuli such as antigen or dextran.

We have employed the term 'operator unit' to explain a membrane unit which, when activated by one stimulus, shows some refractoriness to activation by a second stimulus. It appears that after antigen a greater amount of refractoriness or cross-desensitisation is induced towards dextran than is the reverse situation; dextran produces only limited refractoriness to antigen stimulation. The cross-desensitisation experiments imply that the two stimuli affect a common operator unit since decay to one stimulus influences the response to the other. The operator unit could conceivably be a membrane-bound phosphodiesterase or a phosphodiesterase closely associated with the membrane, and this possibility has been suggested before (Gillespie, 1973; Sullivan & Parker, 1973). Alternatively, the operator unit could be adenyl cyclase or a combination of both enzymes. Only further work will establish just how cyclic nucleotide metabolism is controlled in the mast cell.

Although we have depicted calcium channels and the operator unit as separate entities, there is at least circumstantial evidence to initiate speculation about whether these two functional entities might not occupy the same membrane site. There are reports that phosphatidyl serine, which we believe to be involved with calcium transport in mast cells, influences cyclic nucleotide metabolism in other systems (Bublitz, 1973), and it is conceivable that calcium transporting units and cyclic nucleotide-metabolising units of the membrane are more closely linked than we have shown them to be.

SUMMARY

1. There appear to be distinct mechanisms for secretory activity depending on the nature of the interaction of cyclic nucleotide and calcium in a particular tissue. We have discussed a system in which calcium is required for secretion and cyclic AMP inhibits secretion.

2. Rat peritoneal mast cells release histamine when stimulated by an antigen–antibody reaction or by dextran, provided calcium is present. Phosphatidyl serine augments the response of the mast cells to these stimuli by increasing the activity of calcium.

3. Movement of ^{45}Ca and experiments with the calcium ionophore, A23187, suggest that the entry of calcium into the cell is a primary event in the secretory mechanism.

4. After stimulation with antigen or dextran, the cells do not maintain responsiveness, but the activated state decays almost completely in a period of 4 to 5 min.

5. The decay of the response following stimulation is related to a reduc-

tion in membrane permeability to calcium and may be the means of controlling the degree of secretion.

6. Cyclic AMP inhibits calcium transport across the mast cell membrane through the physiological calcium channels. It is possible that changes in cyclic AMP levels within the cell control the degree of secretion by limiting calcium entry. The decay process after stimulation may be mediated by changes in intracellular cyclic AMP levels.

7. Interaction experiments with antigen and dextran stimuli suggest that these stimuli both act on a common site in the initiation of histamine release, and this site becomes refractory to one of the stimuli after the other one has been allowed to act.

8. A model is discussed which incorporates the activation of secretion by the entry of calcium into the cell and the control of this by the level of cyclic AMP within the cell.

REFERENCES

Assem, E. S. K. & Schild, H. O. (1969). Inhibition by sympathomimetic amines of histamine release by antigen in passively sensitized human lung. *Nature, Lond.*, **224**, 1028–1029.

(1971). Antagonism by β-adrenoceptor blocking agents of the antianaphylactic effect of isoprenaline. *Br. J. Pharmac*, **42**, 620–630.

Baxter, J. H. (1972). Histamine release from rat mast cells by dextran: effects of adrenergic agents, theophylline and other drugs. *Proc. Soc. exp. Biol. Med.*, **141**, 576–581.

Baxter, J. H. & Adamik, R. (1975). Control of histamine release: Effects of various conditions on rate of release and rate of cell desensitization. *J. Immun.*, **114**, 1034–1041.

Berridge, M. J. (1975). Interaction of cyclic nucleotides and calcium in the control of cellular activity. *Adv. cyclic Nucleotide Res.*, **6**, 1–98.

(1976). The role of cyclic nucleotides and calcium in the control of secretion. *Proceedings of the sixth International Congress of Pharmacology.* IUPHAR: Helsinki, **1**, 213–221.

Borowitz, J. L., Leslie, S. W. & Baugh, L. (1975). Adrenal catecholamine release: possible mechanisms of termination. In *Calcium Transport in Contraction and Secretion*, ed. E. Carafoli, F. Clementi, W. Drabikowski & A. Margreth, pp. 227–234. North-Holland: Amsterdam & New York.

Boura, A. L., Mongar, J. L. & Schild, H. O. (1954). Improved automatic apparatus for pharmacological assays on isolated preparations. *Br. J. Pharmac.*, **9**, 24–30.

Bourne, H. R., Lichtenstein, L. M. & Melmon, K. L. (1972). Pharmacologic control of allergic histamine release in vitro: evidence for an inhibitory role of 3′5′-adenosine monophosphate in human leucocytes. *J. Immun.*, **108**, 695–705.

Bublitz, C. (1973). Effects of lipids on cyclic nucleotide phosphodiesterases. *Biochem. biophys. Res. Commun.*, **52**, 173–180.

Caswell, A. H. & Pressman, B. C. (1972). Kinetics of transport of divalent cations across sarcoplasmic reticulum vesicles induced by ionophores. *Biochem. biophys. Res. Commun.*, **49**, 292–298.

DEL CASTILLO, J. & STARK, L. (1952). The effect of calcium on the motor endplate potentials. *J. Physiol., Lond.*, **116**, 507–515.

DIAMANT, B., GROSMAN, N., STAHL SKOV, P. & THOMLE, S. (1974). Effect of divalent cations and metabolic energy on the anaphylactic histamine release from rat peritoneal mast cells. *Int. Archs. Allergy appl. Immun.*, **47**, 412–424.

DIAMANT, B., NORN, S., FELDING, P., OLSEN, N., ZIEBELL, A. & NISSEN, J. (1974). ATP level and CO_2 production of mast cells in anaphylaxis. *Int. Archs. allergy appl. Immun.*, **47**, 894–908.

DIAMANT, B. & UVNÄS, B. (1961). Evidence for energy requiring processes in histamine release and mast cell degranulation in rat tissue induced by compound 48/80. *Acta physiol., scand.*, **53**, 315–329.

DOUGLAS, W. W. (1968). Stimulus–secretion coupling: the concept and clues from chromaffin and other cells. *Br. J. Pharmac.*, **34**, 451–474.

FOREMAN, J. C. (1973). A pharmacological approach to the study of the role of calcium in the secretion of histamine from mast cells. Ph.D. thesis, University of London.

FOREMAN, J. C. & GARLAND, L. G. (1974). Desensitization in the process of histamine secretion induced by antigen and dextran. *J. Physiol., Lond.*, **239**, 381–391.

FOREMAN, J. C., HALLETT, M. B., & MONGAR J. L. (1975). 45-calcium uptake in rat peritoneal mast cells. *Br. J. Pharmac.*, **55**, 283–284P.

FOREMAN, J. C. & MONGAR, J. L. (1972a). The role of the alkaline earth ions in anaphylactic histamine secretion. *J. Physiol., Lond.*, **224**, 753–769.

(1972b). Effect of calcium on dextran-induced histamine release from isolated mast cells. *Br. J. Pharmac.*, **46**, 767–769.

(1973). The interaction of calcium and strontium with phosphatidyl serine in the anaphylactic secretion of histamine. *J. Physiol., Lond.*, **230**, 493–507.

(1975). Calcium and the control of histamine secretion from mast cells. In *Calcium Transport in Contraction and Secretion*, ed. E. Carafoli, F. Clementi, W. Drabikowski & A. Margreth, pp. 175–184. North-Holland: Amsterdam & New York.

FOREMAN, J. C., MONGAR, J. L. & GOMPERTS, B. D. (1973). Calcium ionophores and movement of calcium ions following the physiological stimulus to a secretory process. *Nature, Lond.*, **245**, 249–251.

FOREMAN, J. C., MONGAR, J. L., GOMPERTS, B. D. & GARLAND, L. G. (1975). A possible role for cyclic AMP in the regulation of histamine secretion and the action of cromoglycate. *Biochem. Pharmac.*, **24**, 538–540.

GARLAND, L. G. (1975). An investigation into the action of drugs that suppress histamine release from mast cells. Ph.D. thesis, University of London.

GARLAND, L. G. & MONGAR, J. L. (1974). Inhibition by cromoglycate of histamine release from rat peritoneal mast cells induced by mixtures of dextran, phosphatidyl serine and calcium ions. *Br. J. Pharmac.*, **50**, 137–143.

(1976). Differential histamine release by dextran and the ionophore A23187: the actions of inhibitors. *Int. Archs. Allergy appl. Immun.*, **50**, 27–42.

GILLESPIE, E. (1973). Compound 48/80 decreases adenosine $3'5'$-monophosphate formation in rat peritoneal mast cells. *Experientia*, **29**, 447–448.

GOTH, A., ADAMS, H. R. & KNOOHUIZEN, M. (1971). Phosphatidyl serine: selective enhancer of histamine release. *Science, Wash.*, **173**, 1034–1035.

GREAVES, M. W. & MONGAR, J. L. (1968). The mechanism of anaphylactic histamine release from rabbit leucocytes. *Immunology*, **15**, 743–749.

ISHIZAKA, K. & ISHIZAKA, T. (1969). Immune mechanism of reversed type reaginic hypersensitivity. *J. Immun.*, **103**, 588–595.

JOHANSEN, T. & CHAKRAVARTY, N. (1974). Adenosine triphosphate content of mast cells in relation to histamine release induced by anaphylactic reaction. *Int. Archs. Allergy appl. Immun.*, **49**, 208.

JOHNSON, A. R., MORAN, N. C. & MAYER, S. E. (1974). Cyclic AMP content and histamine release in mast cells. *J. Immun.*, **112**, 511–519.

KALINER, M. & AUSTEN, K. F. (1974). Cyclic AMP, ATP and reversed anaphylactic histamine release from rat mast cells. *J. Immun.*, **112**, 664–674.

KOOPMAN, W. J., ORANGE, R. P. & AUSTEN, K. F. (1970). Modulation of the IgE mediated release of SRS-A by agents influencing the level of cyclic 3'5'-AMP. *J. Immun.*, **105**, 1096–1102.

LICHTENSTEIN, L. M. & MARGOLIS, S. (1968). Histamine release in vitro: inhibition by catecholamines and methylxanthines. *Science, Wash.*, **161**, 902–903.

LICHTENSTEIN, L. M. & OSLER, A. G. (1964). Studies of the mechanism of hypersensitivity phenomena. IX. Histamine release from human leucocytes by Ragweed pollen antigen. *J. exp. Med.*, **120**, 507–530.

LOCKE, F. S. (1894). Notiz über des Einfluss physiologischer Kochsalzlösung auf die elektrische Erregbarkeit von Muskel und Nerv. *Zentralbl. Physiol.*, **8**, 166–167.

MONGAR, J. L. & SCHILD, H. O. (1957). Inhibition of the anaphylactic reaction. *J. Physiol., Lond.*, **135**, 301–319.

(1958). The effect of calcium and pH on the anaphylactic reaction. *J. Physiol., Lond.*, **140**, 272–284.

MONGAR, J. L. & SVEC, P. (1972). The effect of phospholipids on anaphylactic histamine release. *Br. J. Pharmac.*, **46**, 741–752.

NORDMANN, J. J. (1975). Hormone release and Ca-entry inactivation in the rat neurohypophysis. In *Calcium Transport in Contraction and Secretion*, ed. E. Carafoli, F. Clementi, W. Drabikowski & A. Margreth, pp. 281–286. North-Holland: Amsterdam & New York.

PETERSON, C. (1974). Role of energy metabolism in histamine release. A study in isolated rat mast cells. *Acta Physiol. scand.*, suppl., **413**.

PRINCE, W. T., BERRIDGE, M. J. & RASMUSSEN, H. (1972). Role of calcium and adenosine 3'5' cyclic monophosphate in controlling fly salivary gland secretion. *Proc. natn Acad. Sci USA*, **69**, 553–557.

PRINCE, W. T., RASMUSSEN, H. & BERRIDGE, M. J. (1973). The role of calcium in fly salivary gland secretion analysed with the ionophore A23187. *Biochim. biophys. Acta*, **329**, 98–107.

RASMUSSEN, H. (1970). Cell communication, calcium ion, and cyclic adenosine monophosphate. *Science, Wash.*, **170**, 404–412.

REED, P. W. & LARDY, H. A. (1972). Antibiotic A23187 as a probe for the study of calcium and magnesium function in biological systems. In *The Role of Membranes in Metabolic Regulation*, ed. M. A. Mehlman & R. W. Hanson, pp. 111–131. Academic Press: New York.

RUBIN, R. P. (1970). The role of calcium in the release of neurotransmitter substances and hormones. *Pharmac. Rev.*, **22**, 389–428.

(1974). *Calcium and the Secretory Process*. Plenum Press: New York.

SANDOW, A. (1952). Excitation–contraction coupling in muscular response. *Yale J. biol. Med.*, **25**, 176–201.

SCHILD, H. O. (1936). Histamine release and anaphylactic shock in isolated lungs of guinea pigs. *Q. Jl exp. Physiol.*, **26**, 166–179.

SIRAGANIAN, R. P., HOOK, W. A. & LEVINE, B. B. (1975). Specific *in vitro* histamine release from basophils by bivalent haptens: evidence for activation by simple bridging of membrane bound antibody. *Immunochemistry*, **12**, 149–157.

218   J. C. FOREMAN, L. G. GARLAND AND J. L. MONGAR

SIRAGANIAN, R. P. & OSLER, A. G. (1970). Antigenic release of histamine from rabbit leucocytes. *J. Immun.*, **104**, 1340–1347.
STECHSCHULTE, D. J. & AUSTEN, K. F. (1974). Phosphatidyl serine enhancement of antigen-induced mediator release from rat mast cells. *J. Immun.*, **112**, 970–978.
SULLIVAN, T. J. & PARKER, C. W. (1973). Cyclic AMP phosphodiesterase activation by the histamine releasing agent compound 48/80. *Biochem. biophys. Res. Commun.*, **55**, 1334–1339.
SULLIVAN, T. J., PARKER, K. L., EISEN, S. A. & PARKER, C. W. (1975). Modulation of cyclic AMP in purified rat mast cells. II. Studies on the relationship between intracellular cyclic AMP concentrations and histamine release. *J. Immun.*, **114**, 1480–1485.
YAMAMOTO, S. & GREAVES, M. W. (1973). The role of calcium in human cutaneous anaphylaxis. *Int. Archs. Allergy appl. Immun.*, **44**, 797–803.

CALCIUM, CYCLIC NUCLEOTIDES
AND CELL DIVISION

By M. J. BERRIDGE

Unit of Invertebrate Chemistry and Physiology, Department of Zoology,
University of Cambridge, Downing Street, Cambridge CB2 3EJ

Calcium is an essential link in both excitation–contraction and stimulus–secretion coupling. External signals arriving at the cell are converted into an increase in the intracellular level of calcium which then triggers the various mechanical and secretory events. The cyclic nucleotides (cyclic AMP and cyclic GMP) are examples of another group of intracellular second messengers that regulate cell function, particularly various metabolic events (Robison, Butcher & Sutherland, 1971). There is growing interest in the possibility that these various internal signals function as regulators of cell division. However, the presence of different kinds of internal signals has caused considerable confusion. Cyclic AMP, cyclic GMP and calcium have all been implicated in the control of growth and it is difficult to determine the relative importance of these different candidates. One of the candidates can probably be removed from contention because recent studies on lymphoma cells suggest that cyclic AMP is not an 'essential regulator of growth' (Bourne, Coffino & Tomkins, 1975; Coffino, Gray & Tomkins, 1975). The opposite effects of cyclic AMP on cell division depending on the cell type under investigation also argues against an essential function for cyclic AMP. We are left with calcium and cyclic GMP. On the basis of recent experiments in several different cell types (reported in detail later), calcium is the most likely candidate as a primary regulator of cell division. A unifying hypothesis will be developed which integrates current information on both the cyclic nucleotides and calcium. The main feature of this hypothesis is that cell division is triggered by calcium – the apparent role of cyclic nucleotides in cell growth can be accounted for on the basis of their ability to interact with calcium. An important feature of the hypothesis is the existence of a complex web of feedback reactions operating between these various second messengers (Berridge, 1975a, b). Before considering how cell division is regulated, therefore, it is necessary to describe some examples of how cyclic nucleotides and calcium interact with each other.

THE ABILITY OF CALCIUM TO MODULATE THE LEVEL OF CYCLIC NUCLEOTIDES

There are a number of ways that calcium can influence the intracellular levels of both cyclic AMP and cyclic GMP. The first indication of such an interaction was obtained from broken cell preparations where calcium was found to inhibit adenyl cyclase (Streeto, 1969; Taunton, Roth & Pastan, 1969; Drummond & Duncan, 1970). This inhibitory effect may also operate in the intact cell. In the rat parotid gland, the normal increase in cyclic AMP concentration induced by β-adrenergic agents is severely reduced if the glands are simultaneously treated with an α-adrenergic agent (Butcher, 1975). Such α-adrenergic agents are thought to act by increasing the influx of calcium (Selinger, Eimerl & Schramm, 1974). Since the divalent ionophore A23187 causes a similar depression of the cyclic AMP level, there is further reason for implicating calcium as an inhibitor of adenyl cyclase (Butcher, 1975). Steer & Levitzki (1975) have uncovered a similar inhibitory effect of calcium on the adenyl cyclase of turkey erythrocyte ghosts. The physiological significance of this inhibitory effect remains to be determined. At much lower concentrations of calcium adenyl cyclase is stimulated rather than inhibited. This activation of adenyl cyclase by calcium requires the participation of a calcium-dependent receptor (CDR) protein (Brostrom, Huang, Breckenridge & Wolff, 1975). An intriguing feature of this protein is that it appears to be identical to the modulator protein which activates phosphodiesterase (Brostrom *et al.*, 1975; Teo & Wang, 1973; Kakiuchi *et al.*, 1975). Therefore, when calcium combines with this receptor protein the complex can simultaneously activate both adenyl cyclase and phosphodiesterase. The picture is further complicated by the observation that the modulator protein seems to exert a larger inhibitory effect against the phosphodiesterase which degrades cyclic GMP (Kakiuchi *et al.*, 1975).

In addition to activating adenyl cyclase, there is increasing evidence that calcium may stimulate guanyl cyclase. Agents that elevate the level of cyclic GMP in intact cells have absolutely no effect on either the soluble or particulate forms of the isolated enzyme (Goldberg, O'Dea & Haddox, 1973). In the absence of any direct effect, it was proposed that such agents may activate guanyl cyclase indirectly by altering the intracellular concentration of ions, in particular calcium (Schultz *et al.*, 1973). In the presence of both GTP and manganese, calcium can stimulate guanyl cyclase in a cell-free system (Chrisman, Garbers, Parks & Hardman, 1975; Garbers, Dyer & Hardman, 1975). This activation of guanyl cyclase by calcium may occur in the intact cells as well because the ability of a range of stimulants to increase cyclic GMP levels is totally dependent on the presence of external calcium (Ferrendelli, Kinscherf & Chang, 1973; Schultz et al., 1973; Schultz & Hardman, 1975). The ionophore A23187, which increases calcium entry into cells, can markedly stimulate the synthesis of cyclic GMP in dog thyroid slices (Van Sande, Decoster & Dumont, 1975) and the rat parotid gland (Butcher, 1975).

THE EFFECT OF CYCLIC NUCLEOTIDES ON INTRACELLULAR CALCIUM HOMEOSTASIS

The ability of cyclic AMP to influence calcium homeostasis is a very significant feedback effect common to many different cell types (Berridge, 1975a). Some good examples of this feedback effect have been uncovered in mammalian heart during the action of catecholamines. Each heart beat is driven by the cardiac action potential. There is controversy concerning the origin of the phasic calcium signal which mediates each contraction cycle. Some calcium enters from outside during the slow phase of the action potential and is augmented by a release of calcium from an intracellular reservoir which is most likely to be the sarcoplasmic reticulum. Relaxation is achieved by rapidly clearing the cytoplasm of this calcium signal by means of pumps located both on the surface and internal membranes. Adrenaline can modulate cardiac contractility by inducing subtle alterations in this basic ebb-and-flow of calcium which occurs during each heart beat. These effects of adrenaline on calcium homeostasis are mediated by cyclic AMP. One important action of adrenaline is to prolong the entry of calcium which is responsible for the slow inward current and this effect can be mimicked by dibutyryl cyclic AMP (Reuter, 1974). This stimulation of calcium entry may account for the increased uptake of ^{45}Ca recorded during the action of adrenaline or dibutyryl cyclic AMP (Meinertz, Nawrath & Scholz, 1973). In addition to this effect on the sarcolemma, cyclic AMP also modulates calcium movement across the sarcoplasmic reticulum by stimulating the Ca^{2+}-ATPase which sequesters calcium during relaxation (Kirchberger, Tada, Repke & Katz, 1972; Tada, Kirchberger, Repke & Katz, 1974; Tada, Kirchberger & Katz, 1975). These authors have shown that cyclic AMP stimulates a protein kinase which phosphorylates a small protein (phospholamban) which then activates the Ca-ATPase. Cyclic AMP not only enhances calcium entry, which probably contributes to an increase in tension during contraction, but it also stimulates calcium transport into the sarcoplasmic reticulum which speeds up relaxation.

In contrast to its ability to promote calcium uptake into the heart, cyclic

AMP seems to inhibit the entry of calcium into lymphocytes and mast cells. The plant lectin concanavalin A (Con A) can increase the uptake of calcium into lymphocytes (Freedman, Raff & Gomperts, 1975). If the intracellular level of cyclic AMP is elevated by treating lymphocytes with cholera toxin, theophylline, or simply by adding dibutyryl cyclic AMP, this Con A-induced uptake of calcium is greatly reduced. The addition of dibutyryl cyclic AMP to mast cells causes a similar inhibition of the normal antigen-induced uptake of calcium (Foreman, Garland & Mongar, this volume).

Cyclic AMP has also been implicated as a regulator of calcium movement across mitochondria (Borle, 1974). When treated with low concentrations of cyclic AMP (3×10^{-6} M) isolated mitochondria of liver, kidney and heart release calcium into the surrounding medium. A similar effect of cyclic AMP has been reported in subcellular fractions prepared from rat islet cells (Howell & Montague, 1975). The ability of cyclic AMP to release calcium stored within mitochondria could account for the increase in calcium efflux which occurs during stimulation of rat liver (Friedmann & Park, 1968; Friedmann, 1972), toad bladder (Thorn & Schwartz, 1965), rat β-cells (Brisson & Malaisse, 1973), *Calliphora* salivary gland (Prince, Berridge & Rasmussen, 1972) and mammalian salivary glands (Nielsen & Petersen, 1972).

There clearly is a complex web of feedback interactions operating between the cyclic nucleotides and calcium. The full extent of these interactions remains to be uncovered but it is already apparent that they are of central importance in determining how cell activity is regulated (Berridge, 1975 a). In particular, cyclic AMP seems to play a significant role in modulating the intracellular level of calcium. In some systems cyclic AMP seems to exert an inhibitory effect as occurs in lymphocytes, mast cells and in heart during relaxation. Conversely, in other systems such as mammalian and insect salivary glands, the rat liver and β-cells, cyclic AMP acts together with calcium and it may even augment the calcium signal. This ability of cyclic AMP to exert opposite effects on calcium homeostasis in different cell types has provided a framework for understanding its involvement in the control of cell division (Berridge, 1975 a, b).

CONTROL OF CELL DIVISION

At some stage in their life, all cells must decide whether or not to divide. There is increasing evidence that the second messengers previously associated with various hormonal responses in differentiated cells may also function as internal regulators of cell division. A possible role for cyclic

AMP in the control of cell division first emerged from studies on fibro-blasts where it seemed to exert an inhibitory effect (Abell & Monahan, 1973). Unfortunately, there are other cell types where rapid cell growth is associated with very high levels of cyclic AMP (Thomas, Murad, Looney & Morris, 1973). As mentioned earlier, these contradictory effects would seem to rule out cyclic AMP as an essential regulator of cell growth. A similar view has been advanced by Coffino et al. (1975) on the basis of their studies on an S49 lymphoma cell line; mutant cells which lack the cyclic AMP-dependent protein kinase were capable of growing normally. These mutant cells were also insensitive to dibutyryl cyclic AMP which was capable of arresting the growth of normal cells. Therefore, although cyclic AMP is clearly not required for cells to progress through the cell cycle, this nucleotide may have an important function as a negative regula-tor of growth. It will be argued later that this function of cyclic AMP might depend on its ability to modulate the intracellular level of calcium.

Cyclic GMP has also been implicated as a regulator of cell growth (Hadden, Hadden, Haddox & Goldberg, 1972). An increase in the intra-cellular level of cyclic GMP is a consistent feature associated with the initiation of cell division in lymphocytes (Hadden et al., 1972), fibroblasts (Seifert & Rudland, 1974) and mammalian salivary glands (Durham, Baserga & Butcher, 1974). However, if we wish to invoke cyclic GMP as an internal regulator, it is necessary to describe how the various division stimuli activate guanyl cyclase. Previously, I summarized the evidence suggesting that guanyl cyclase may be sensitive to the intracellular level of calcium. In the light of this feedback interaction operating between calcium and guanyl cyclase, I think it is important to examine the possi-bility that an increase in the intracellular level of calcium, at a specific stage in the cell cycle, is a universal stimulus for cell growth. Such a model is attractive because it can accommodate most previous observations con-necting the cyclic nucleotides with the control of division. The high levels of cyclic GMP, which are associated with division stimuli in many cells, would be a direct consequence of the calcium signal. The opposite effects of cyclic AMP on cell growth can be accounted for in terms of its ability to modulate the intracellular level of calcium in a positive or negative manner depending on the cell type as described earlier. In this model, therefore, cyclic AMP plays an indirect role by regulating the intracellular level of calcium which is considered to be the primary intracellular regulator of growth.

One of the first suggestions that calcium was an important regulator of growth was put forward by Mazia (1937) who considered that calcium was responsible for triggering cell division after fertilization. Support for this view has come from studies with the divalent ionophore A23187 which can duplicate all the events normally initiated by fertilization (Chambers, Pressman & Rose, 1974; Steinhardt & Epel, 1974). It has been claimed that there is an increased uptake of calcium immediately after fertilization (Clothier & Timourian, 1972) which may account for the proposed increase in free calcium concentration (Nakamura & Yasumasu, 1974).

A similar uptake of calcium may mediate the mitogenic action of plant lectins, such as Con A and phytohaemagglutinin (PHA), on lymphocytes. The current idea is that these mitogens act on the plasma membrane to open up specific calcium gates and the resulting influx of calcium is responsible for initiating DNA synthesis and cell division (Freedman et al., 1975). The mitogenic effect of PHA is totally dependent upon external calcium (Alford, 1970; Whitney & Sutherland, 1972a) and there is a marked uptake of calcium during its action (Allwood, Asherson, Davey & Goodford; 1971; Whitney & Sutherland, 1972b; Freedman et al., 1975). A role for calcium is also apparent from the fact that the divalent ionophore A23187 can also stimulate DNA synthesis in lymphocytes (Maino, Green & Crumpton, 1974). Lectin-induced transformation of lymphocytes can be inhibited if the intracellular level of cyclic AMP is increased either by direct addition of the nucleotide or by treatment with PGE_1 (prostaglandin E_1), cholera toxin or theophylline (DeRubertis, Zenser, Adler & Hudson, 1974; Weinstein, Chambers, Bourne & Melmon, 1974). This inhibitory effect of cyclic AMP might be mediated through an effect on the calcium signal because Freedman et al. (1975) have found that high cyclic AMP levels can reduce the Con A-induced uptake of calcium.

The action of high calcium concentrations resulting from these mitogens may activate guanyl cyclase (see earlier, p. 220) to account for the large but phasic rise in cyclic GMP level which occurs after the addition of lectins (Hadden et al., 1972). This rise in cyclic GMP level may be an important link between the rise in calcium and the initiation of early events of the division programme. Johnson & Hadden (1975) consider that cyclic GMP may induce lymphocyte proliferation by phosphorylating nuclear acidic proteins.

Hyperplasia of rat salivary glands following hyperstimulation is another interesting example where cell division is induced by a well-defined stimulus. Injecting rats with a large dose of isoproterenol or hyperstimulation of the superior cervical sympathetic nerve leads to an increase in proliferation (Selye, Veilleux & Cantin, 1961; Baserga, 1970; Muir, Pollock & Turner, 1973). These salivary glands have both α- and β-adrenergic receptors and there is evidence that both cyclic AMP and calcium signals develop during adrenergic stimulation. An apparent relationship between adrenergic stimulation, cyclic AMP and cell division has led to the suggestion that cyclic AMP is the mitogenic agent (Guidotti, Weiss & Costa, 1972). However, by studying a range of adrenergic agents Durham *et al.* (1974) found that certain agents could induce DNA synthesis with no increase in the level of cyclic AMP. A temporal analysis of the change in cyclic AMP level produced by prolonged supramaximal nervous stimulation revealed that the early rise in cyclic AMP level did not persist for the minimum period of stimulation required to trigger cell division (Muir & Templeton, 1975). If cyclic AMP is not responsible for switching on division it is conceivable that calcium is the signal.

There are at least two ways whereby adrenergic agents could elevate the intracellular level of calcium. First, by interacting with the α-receptor they may increase the calcium permeability of the plasma membrane thus promoting an influx of external calcium (Selinger *et al.*, 1974). Secondly, the increase in the level of cyclic AMP may stimulate the release of calcium from internal reservoirs (Nielsen & Petersen, 1972). Such a release mechanism may account for the decline in the calcium content of mitochondria and microsomes prepared from glands which have been treated with adrenaline (Dormer & Ashcroft, 1974). The release of internal calcium together with an influx of external calcium may provide the stimulus to divide when salivary glands are hyperstimulated.

Studies on the growth of cells in tissue culture provide many opportunities for studying the control of cell division. When first seeded, most cells grow exponentially until they fill up the available space whereupon cell division stops and the cells enter a stationary phase. Such stationary cells can be induced to divide again by various stimuli such as fresh serum or brief digestion of the cell surface with enzymes. This induction of cell division with specific stimuli has some interesting analogies with what happens in lymphocytes. There is some indication that calcium may be the primary signal. Simply raising the calcium concentration of the bathing medium will switch stationary cell cultures back into the cell cycle (Balk, Whitfield, Youdale & Braun, 1973; Boynton, Whitfield, Isaacs & Morton, 1974; Dulbecco & Elkington, 1975). These experiments suggest that fresh serum or enzyme digestion may induce cell division through an effect on the permeability of the plasma membrane to calcium which then floods into the cell to provide a mitogenic signal. Such an increase in the permeability of the membrane to calcium is certainly consistent with the large membrane depolarization observed in embryonic rat cells within minutes of adding serum (Hülser & Frank, 1971). The immediate rise in cyclic GMP recorded when fresh serum is added to stationary fibroblasts (Seifert &

Rudland, 1974), is also consistent with a sudden rise in the intracellular level of calcium. It is more difficult to account for the reciprocal fall in cyclic AMP level when these cells are triggered to divide. The sudden rise in calcium may exert a negative feedback effect on adenyl cyclase as described earlier. Alternatively, the same membrane changes which lead to an influx of calcium may also inhibit adenyl cyclase. Indeed the fall in the intracellular level of cyclic AMP may be a necessary prerequisite for the rise in calcium. If the level of cyclic AMP is kept high (by adding exogenous cyclic AMP, or by treating cells with agents such as PGE_1 or cholera toxin), there is no cell division (Froehlich & Rachmeler, 1972; Bombik & Burger, 1973; Rudland, Seeley & Seifert, 1974). Furthermore, when PGE_1 raises the intracellular level of cyclic AMP there is a corresponding fall in the level of cyclic GMP (Rudland et al., 1974). This reciprocal change between cyclic AMP and cyclic GMP is readily explicable if these two nucleotides are linked to calcium by feedback loops as discussed earlier. Cyclic AMP may exert a negative feedback effect on the level of calcium which, in turn, regulates the level of cyclic GMP through a positive feedback effect on guanyl cyclase.

There is less information about the role of second messengers during normal growth and in the contact-inhibition of growth which occurs at confluency. Measurement of cyclic AMP levels during the cell cycle have shown that there is a consistent fall at the time of mitosis in a variety of cells (Burger, Bombik, Breckenridge & Sheppard, 1972; Millis, Forrest & Pious, 1972; Otten, Johnson & Pastan, 1972; Sheppard & Prescott, 1972; Yasumasu, Fujiwara & Ishida, 1973). During this fall in cyclic AMP level there may be a corresponding rise in the level of calcium. Baker & Warner (1972) have shown that calcium is essential for cleavage in Xenopus eggs. The mechanical events associated with mitosis, such as chromosome movement and cell cleavage, are probably mediated by actin filaments (Gawadi, 1971; Schroeder, 1973; Arnold, 1975; Sanger, 1975). Arnold (1975) has some evidence that calcium stimulates the contraction of a band of microfilaments responsible for cleavage. This high level of calcium during mitosis may then be carried through into early G_1 to provide the signal for another division. In order to account for the cessation of growth at confluency, we must speculate that there is a rapid reduction of the calcium level following the last mitosis so that the cells pass through the critical period of decision early in G_1 with a low level of calcium which is not mitogenic.

CONCLUSION

This very speculative model attempts to integrate current information on the apparent involvement of both the cyclic nucleotides and calcium in the control of cell division. The model is built around the feedback interactions which operate between these different second messengers. The primary signal responsible for switching on cell division is postulated to be an increase in the intracellular level of calcium. The ability of cyclic AMP to influence growth in many cells can be accounted for on the basis of its positive or negative feedback effects on calcium. A sudden increase in the level of calcium could also account for the parallel rise in cyclic GMP which occurs consistently when cells are induced to divide. We are still a long way from understanding how second messengers such as calcium and cyclic GMP are responsible for switching on the complicated sequence of events which constitutes the cell cycle.

REFERENCES

ABELL, C. W. & MONAHAN, T. M. (1972). The role of adenosine 3',5'-cyclic monophosphate in the regulation of mammalian cell division. *J. Cell Biol.*, **59**, 549–558.

ALFORD, R. H. (1970). Metal cation requirements for phytohemagglutinin-induced transformation of human peripheral blood lymphocytes. *J. Immun.*, **104**, 698–703.

ALLWOOD, G., ASHERSON, G. L., DAVEY, M. J. & GOODFORD, P. J. (1971). The early uptake of radioactive calcium by human lymphocytes treated with phyto-hemagglutinin. *Immunology*, **21**, 509.

ARNOLD, J. M. (1975). An effect of calcium in cytokinesis as demonstrated with ionophore A23187. *Cytobiologie*, **11**, 1–9.

BAKER, P. F. & WARNER, A. E. (1972). Intracellular calcium and cell cleavage in early embryos of *Xenopus laevis*. *J. Cell Biol.*, **53**, 579–581.

BALK, S. D., WHITFIELD, J. F., YOUDALE, T. & BRAUN, A. C. (1973). Roles of calcium, serum, plasma, and folic acid in the control of proliferation of normal and Rous sarcoma virus-infected chicken fibroblasts. *Proc. natn. Acad. Sci. USA*, **70**, 675–679.

BASERGA, R. (1970). Induction of DNA synthesis by a purified compound. *Fedn Proc.*, **29**, 1443–1446.

BERRIDGE, M. J. (1975a). The interaction of cyclic nucleotides and calcium in the control of cellular activity. *Adv. cyclic Nucleotide Res.*, **6**, 1–98.

(1975b). Control of cell division: a unifying hypothesis. *J. cyclic Nucleotide Res.*, **1**, 305–320.

BOMBIK, B. M. & BURGER, M. M. (1973). c-AMP and the cell cycle: inhibition of growth stimulation. *Expl Cell Res.*, **80**, 88–94.

BORLE, A. B. (1974). Cyclic AMP stimulation of calcium efflux from kidney, liver and heart mitochondria. *J. Memb. Biol.*, **16**, 221–236.

BOURNE, H. R., COFFINO, P. & TOMKINS, G. M. (1975). Somatic genetic analysis of cyclic AMP action: characterization of unresponsive mutants. *J. Cell Physiol.*, **85**, 611–619.

BOYNTON, A. L., WHITFIELD, J. F., ISAACS, R. J. & MORTON, H. J. (1974). Control of 3T3 cell proliferation by calcium. *In Vitro*, **10**, 12–17.

BRISSON, G. R. & MALAISSE, W. J. (1973). The stimulus secretion coupling of glucose-induced insulin release. XI. Effects of theophylline and epinephrine on ^{45}Ca efflux from perifused islets. *Metabolism*, **22**, 455–465.

BROSTROM, C. O., HUANG, Y.-C., BRECKENRIDGE, B. McL. & WOLFF, D. J. (1975). Identification of a calcium-binding protein as a calcium-dependent regulator of brain adenylate cyclase. *Proc. natn. Acad. Sci. USA*, **72**, 64–68.

BURGER, M. M., BOMBIK, B. M., BRECKENRIDGE, B. McL. & SHEPPARD, J. R. (1972). Growth control and cyclic alterations of cyclic AMP in the cell cycle. *Nature New Biol.*, **239**, 161–163.

BUTCHER, F. R. (1975). The role of calcium and cyclic nucleotides in α-amylase release from slices of rat parotid: studies with the divalent cation ionophore A 23187. *Metabolism*, **24**, 409–418.

CHAMBERS, E. L., PRESSMAN, B. C. & ROSE, B. (1974). The activation of sea urchin eggs by the divalent ionophore A 23187 and X-537 A. *Biochem. biophys. Res. Commun.*, **60**, 126–132.

CHRISMAN, T. D., GARBERS, D. L., PARKS, M. A. & HARDMAN, J. G. (1975). Characterization of particulate and soluble guanylate cyclase from rat lung. *J. biol. Chem.*, **250**, 374–381.

CLOTHIER, G. & TIMOURIAN, H. (1972). Calcium uptake and release by dividing sea urchin eggs. *Expl Cell Res.*, **75**, 105–110.

COFFINO, P., GRAY, J. W. & TOMKINS, G. M. (1975). Cyclic AMP, a non-essential regulator of the cell cycle. *Proc. natn. Acad. Sci. USA*, **72**, 878–882.

DeRUBERTIS, F. R., ZENSER, T. V., ADLER, W. H. & HUDSON, T. (1974). Role of cyclic adenosine 3′,5′-monophosphate in lymphocyte mitogenesis. *J. Immun.*, **113**, 151–161.

DORMER, R. L. & ASHCROFT, S. J. H. (1974). Studies on the role of calcium ions in the stimulation by adrenaline of amylase release from rat parotid. *Biochem. J.*, **144**, 543–550.

DRUMMOND, G. I. & DUNCAN, L. (1970). Adenyl cyclase in cardiac tissue. *J. biol. Chem.*, **245**, 976–983.

DULBECCO, R. & ELKINGTON, J. (1975). Induction of growth in resting fibroblastic cell cultures by Ca^{++}. *Proc. natn. Acad. Sci. USA*, **72**, 1584–1588.

DURHAM, J. P., BASERGA, R. & BUTCHER, F. R. (1974). The effect of isoproterenol and its analogs upon adenosine 3′,5′-monophosphate and guanosine 3′,5′-monophosphate levels in mouse parotid gland *in vivo*. *Biochim. biophys. Acta*, **372**, 196–217.

FERRENDELLI, J. A., KINSCHERF, D. A. & CHANG, M. M. (1973). Regulation of levels of guanosine cyclic 3′,5′-monophosphate in the central nervous system: effects of depolarizing agents. *Molec. Pharm.*, **9**, 445–454.

FREEDMAN, M. H., RAFF, M. C. & GOMPERTS, B. (1975). Induction of increased calcium uptake in mouse T lymphocytes by concanavalin A and its modulation by cyclic nucleotides. *Nature, Lond.*, **255**, 378–382.

FRIEDMANN, N. (1972). Effects of glucagon and cyclic AMP on ion fluxes in the perfused liver. *Biochim. biophys. Acta*, **274**, 214–225.

FRIEDMANN, N. & PARK, C. R. (1968). Early effects of 3′,5′-adenosine monophosphate on the fluxes of calcium and potassium in the perfused liver of normal and adrenalectomized rats. *Proc. natn. Acad. Sci. USA*, **61**, 504–508.

FROEHLICH, J. E. & RACHMELER, M. (1972). Effect of adenosine 3′,5′-cyclic monophosphate on cell proliferation. *J. Cell Biol.*, **55**, 19–31.

GARBERS, D. L., DYER, E. L. & HARDMAN, J. G. (1975). Effects of cations on guanylate cyclase of sea urchin sperm. *J. biol. Chem.*, **250**, 382–387.

GAWADI, N. (1971). Actin in the mitotic spindle. *Nature, Lond.*, **234**, 410.

GOLDBERG, N. D., O'DEA, R. F. & HADDOX, M. K. (1973). Cyclic GMP. *Adv. Cyclic Nucleotide Res.*, **3**, 155–223.

GUIDOTTI, A., WEISS, B. & COSTA, E. (1972). Adenosine 3',5'-monophosphate concentrations and isoproterenol induced synthesis of deoxyribonucleic acid in mouse parotid gland. *Molec. Pharm.*, **8**, 521–530.

HADDEN, J. W., HADDEN, E. M., HADDOX, M. K. & GOLDBERG, N. D. (1972). Guanosine 3',5'-cyclic monophosphate: a possible intracellular mediator of mitogenic influence in lymphocytes. *Proc. natn. Acad. Sci. USA*, **69**, 3024–3027.

HOWELL, S. L. & MONTAGUE, W. (1975). Regulation by nucleotides of ^{45}calcium uptake in homogenates of rat islets of Langerhans. *FEBS Lett*, **52**, 48–52.

HÜLSER, D. F. & FRANK, W. (1971). Stimulierung von Kulturen embryonaler Rattenzellen durch eine Proteinfraktion aus fötalem Kälberserum. *Z. Naturf.*, **26b**, 1045–1048.

JOHNSON, E. M. & HADDEN, J. W. (1975). Phosphorylation of lymphocyte nuclear acidic proteins: regulation by cyclic nucleotides. *Science, Wash.*, **187**, 1198–1200.

KAKIUCHI, S., YAMAZAKI, R., TESHIMA, Y., UENISHI, K. & MIYAMOTO, E. (1975). Ca^{2+}/Mg^{2+}-dependent cyclic nucleotide phosphodiesterase and its activator protein. *Adv. Cyclic Nucleotide Res.*, **5**, 163–178.

KIRCHBERGER, M. A., TADA, M., REPKE, D. I. & KATZ, A. M. (1972). Cyclic adenosine 3',5'-monophosphate-dependent protein kinase stimulation of calcium uptake by canine cardiac microsomes. *J. molec. cell. Cardiol.*, **4**, 673–680.

MAINO, V. C., GREEN, N. M. & CRUMPTON, M. J. (1974). The role of calcium ions in initiating transformation of lymphocytes. *Nature, Lond.*, **251**, 324–327.

MAZIA, D. (1937). The release of calcium in *Arbacia* eggs on fertilization. *J. cell. comp. Physiol.*, **10**, 291–304.

MEINERTZ, T., NAWRATH, H. & SCHOLZ, H. (1973). Dibutyryl cyclic AMP and adrenaline increase contractile force and ^{45}Ca uptake in mammalian cardiac muscle. *Nauyn-Schmiedebergs Arch. exp. Path. Pharmak.*, **277**, 107–112.

MILLIS, A. J. T., FORREST, G. & PIOUS, D. A. (1972). Cyclic AMP in cultured human lymphoid cells: relationship to mitosis. *Biochem. biophys. Res. Commun.*, **49**, 1645–1649.

MUIR, T. C., POLLOCK, D. & TURNER, C. J. (1973). The effects of electrical stimulation of the sympathetic nerves on the size and mitotic index of rat salivary glands. *J. Physiol., Lond.*, **232**, 43P.

MUIR, T. C. & TEMPLETON, D. (1975). The role of adenosine 3',5'-monophosphate (cyclic AMP) in the catecholamine-induced growth of rat salivary glands. *J. Physiol., Lond.*, **246**, 85P.

NAKAMURA, M. & YASUMASU, I. (1974). Mechanism for increase in intracellular concentration of free calcium in fertilized sea urchin egg. *J. gen. Physiol.*, **63**, 374–388.

NIELSEN, S. P. & PETERSEN, O. H. (1972). Transport of calcium in the perfused submandibular gland of the cat. *J. Physiol., Lond.*, **223**, 685–697.

OTTEN, J., JOHNSON, G. S. & PASTAN, I. (1972). Regulation of cell growth by cyclic adenosine 3',5'-monophosphate. *J. biol. Chem.*, **247**, 7082–7087.

PRINCE, W. T., BERRIDGE, M. J. & RASMUSSEN, H. (1972). Role of calcium and adenosine-3',5'-cyclic monophosphate in controlling fly salivary gland secretion. *Proc. natn. Acad. Sci. USA*, **69**, 553–557.

REUTER, H. (1974). Exchange of calcium ions in the mammalian myocardium. *Circulation Res.*, **34**, 599–605.

ROBISON, G. A., BUTCHER, R. W. & SUTHERLAND, E. W. (1971). *Cyclic AMP*. Academic Press: New York & London.

RUDLAND, P. S., SEELEY, M. & SEIFERT, W. (1974). Cyclic GMP and cyclic AMP levels in normal and transformed fibroblasts. *Nature, Lond.*, **251**, 417–419.

SANGER, J. W. (1975). Changing patterns of actin localization during cell division. *Proc. natn. Acad. Sci. USA*, **72**, 1913–1916.

SCHROEDER, T. E. (1973). Actin in dividing cells: contractile ring filaments bind heavy meromyosin. *Proc. natn. Acad. Sci. USA*, **70**, 1688–1692.

SCHULTZ, G. & HARDMAN, J. G. (1975). Regulation of cyclic GMP levels in the ductus deferens of the rat. *Adv. Cyclic Nucleotide Res.*, **5**, 339–351.

SCHULTZ, G., HARDMAN, J. G., SCHULTZ, K., BAIRD, C. E. & SUTHERLAND, E. W. (1973). The importance of calcium ions for the regulation of guanosine 3′,5′-cyclic monophosphate levels. *Proc. natn. Acad. Sci. USA*, **70**, 3889–3893.

SEIFERT, W. E. & RUDLAND, P. S. (1974). Possible involvement of cyclic AMP in growth control of cultured mouse cells. *Nature, Lond.*, **248**, 138–140.

SELINGER, Z., EIMERL, S. & SCHRAMM, M. (1974). A calcium ionophore simulating the action of epinephrine on the α-adrenergic receptor. *Proc. natn. Acad. Sci. USA*, **71**, 128–131.

SELYE, H., VEILLEUX, R. & CANTIN, M. (1961). Excessive stimulation of salivary gland growth by isoproterenol. *Science, Wash.*, **133**, 44–45.

SHEPPARD, J. R. & PRESCOTT, R. R. (1972). Cyclic AMP levels in synchronized mammalian cells. *Expl Cell Res.*, **75**, 293–296.

STEER, M. L. & LEVITZKI, A. (1975). The control of adenylate cyclase by calcium in turkey erythrocyte ghosts. *J. biol. Chem.*, **250**, 2080–2084.

STEINHARDT, R. A. & EPEL, D. (1974). Activation of sea-urchin eggs by a calcium ionophore. *Proc. natn. Acad. Sci. USA*, **71**, 1915–1919.

STREETO, J. M. (1969). Renal cortical adenyl cyclase: effect of parathyroid hormone and calcium. *Metabolism*, **18**, 968–973.

TADA, M., KIRCHBERGER, M. A. & KATZ, A. M. (1975). Phosphorylation of a 22,000-dalton component of the cardiac sarcoplasmic reticulum by adenosine 3′,5′-monophosphate-dependent protein kinase. *J. biol. Chem.*, **250**, 2640–2647.

TADA, M., KIRCHBERGER, M. A., REPKE, D. I. & KATZ, A. M. (1974). The stimulation of calcium transport in cardiac sarcoplasmic reticulum by adenosine 3′,5′-monophosphate-dependent protein kinase. *J. biol. Chem.*, **249**, 6174–6180.

TAUNTON, O. D., ROTH, J. & PASTAN, I. (1969). Studies on the adreno-corticotropic hormone-activated adenyl cyclase of a functional adrenal tumour. *J. biol. Chem.*, **244**, 247–253.

TEO, T. S. & WANG, J. H. (1973). Mechanism of activation of a cyclic adenosine 3′,5′-monophosphate phosphodiesterase from bovine heart by calcium ions. *J. biol. Chem.*, **248**, 5950–5955.

THOMAS, E. W., MURAD, F., LOONEY, W. B. & MORRIS, H. P. (1973). Adenosine 3′,5′-monophosphate and guanosine 3′,5′-monophosphate: concentrations in Morris hepatomas of different growth rates. *Biochim. biophys. Acta*, **297**, 564–567.

THORN, N. A. & SCHWARTZ, I. L. (1965). Effect of antidiuretic hormone on washout curves of radiocalcium from isolated toad bladder tissue. *Gen. comp. Endocr.*, **5**, 710.

VAN SANDE, J., DECOSTER, C. & DUMONT, J. E. (1975). Control and role of cyclic 3′,5′-guanosine monophosphate in the thyroid. *Biochem. biophys. Res. Commun.*, **62**, 168–175.

WEINSTEIN, Y., CHAMBERS, D. A., BOURNE, H. R. & MELMON, K. L. (1974). Cyclic GMP stimulates lymphocyte nucleic acid synthesis. *Nature, Lond.*, **251**, 352–357.

WHITNEY, R. B. & SUTHERLAND, R. M. (1972a). Requirement for calcium ions in lymphocyte transformation stimulated by phytohemagglutinin. *J. Cell Physiol.*, **80**, 329–338.

(1972b). Enhanced uptake of calcium by transforming lymphocytes. *Cell. Immun.*, **5**, 137–147.

YASUMASU, I., FUJIWARA, A. & ISHIDA, K. (1973). Periodic change in the content of adenosine 3′,5′-cyclic monophosphate with close relation to the cycle of cleavage in the sea urchin egg. *Biochem. biophys. Res. Commun.*, **54**, 628–632.

CALCIUM IN THE BIOELECTRIC AND MOTOR FUNCTIONS OF *PARAMECIUM*

By R. ECKERT, Y. NAITOH* and H. MACHEMER†

Department of Biology, University of California at
Los Angeles, Los Angeles, California 90024, USA

Paramecium offers special advantages for biophysical, genetic, and molecular biological approaches to problems of membrane organization and function and the regulation of cell motility by the surface membrane. This ciliate has a combination of features which facilitate a multidisciplinary approach. As a unicellular microorganism which has bioelectric properties resembling muscle and nerve, it manifests its membrane responses while swimming free in culture through its locomotor responses to various chemical stimuli, and it is susceptible to both genetic and electrophysiological manipulation. In a series of recent studies we have found that Ca^{2+} is of central importance in several roles in the bioelectric and locomotor functions of *Paramecium*.

Since the ciliates are evolutionarily distant from excitable metazoan tissues such as nerve and epithelium, etc., it is especially interesting that Ca^{2+} appears in these 'lower' forms as a highly adapted regulatory agent performing functions akin to those it performs in metazoan cells.

OVERVIEW OF THE SENSORY-MOTOR SYSTEM OF *PARAMECIUM*

The physiology of sensory-motor behaviour in this ciliate has been reviewed elsewhere (Eckert, 1972; Eckert & Naitoh, 1972; Naitoh, 1974); the most familiar stereotyped behaviour of *Paramecium* is the 'avoiding reaction' first described by Jennings (1906) and illustrated in Fig. 1 (top). The major roles of Ca^{2+} in this sensory-motor sequence (Fig. 1, bottom) can be summarized as follows. When the ciliate collides with an object the deformation of the membrane at the cell anterior activates receptor channels which carry an inward current presumed to be carried by Ca^{2+}. This local receptor current propagates electrotonically (i.e. by simple cable spread), depolarizing by some millivolts the entire cell membrane. Other ion

---

* Present address: University of Tsukuba, Department of Biological Sciences, Sakura-mura, Ibaraki 300-31, Japan.
† Present address: Ruhr-Universität Bochum, Abt. für Biologie, D 463 Bochum, German Federal Republic.

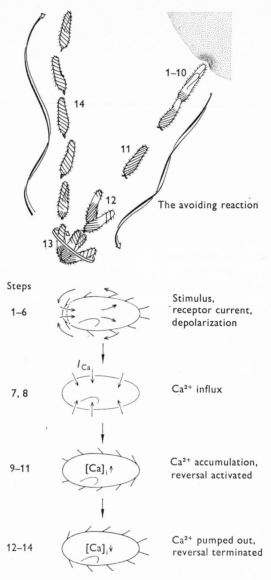

Fig. 1. The avoiding reaction. *Top*: *Paramecium* collides with object, ret reats rotates at random, resumes forward swimming. (after Grell, 1973). *Bottom*:, sequence of steps underlying this behaviour. Step 1, stretch of anterior membrane upon collision with obstacle; step 2, local increase in membrane conductance; and step 3, inward current through rest of membrane (arrows show current flow). Step 6, depolarization of cell membrane (receptor potential) produces step 7, increase in Ca^{2+} conductance. Step 8, inward Ca^{2+} current; step 9, rise in intracellular Ca^{2+} concentration; step 10, cilia reverse beat; and step 11, cell swims backward. Step 12, Ca^{2+} pumped out; step 13, intracellular concentration of Ca^{2+} drops, cilia resume normal orientation; and step 14, cell swims forward. (From Eckert, 1972. Copyright © 1972 from the American Association for the Advancement of Science.)

channels, specific for Ca^{2+} and sensitive to membrane potential changes, are located primarily in those portions of membrane covering the cilia. These respond to the depolarization produced by the receptor current originating at the cell anterior by becoming specifically permeable to Ca^{2+}. Since free Ca^{2+} is several orders of magnitude more concentrated outside the cell than inside, a net influx of Ca^{2+} occurs which causes a further depolarization of the membrane. The influx of Ca^{2+} has four major consequences. First, it produces a spike-like potential termed the 'calcium response'; second, the rise in internal Ca^{2+} concentration, $[Ca]_i$, activates an ATP-dependent process which causes the direction of the power stroke of the cilia to shift toward the cell anterior. This 'ciliary reversal' makes the ciliate retreat momentarily from the obstacle with which it collided. The ciliate resumes forward locomotion after a brief and variable bowling-pin rotation at the end of its retreat. The third action of Ca^{2+} which enters the cell through the surface membrane is to cause the cilia to beat with increased frequency, an effect which may be closely associated with the shift in orientation of the power stroke. Finally, the rise in $[Ca^{2+}]_i$ appears to produce or contribute to a temporary rise in potassium conductance (g_K), as a result of which the membrane undergoes a rapid repolarization to the resting level, terminating the calcium response.

The relation between ciliary activity and electric phenomena is most readily investigated by artificially altering the membrane potential with an intracellular current-passing electrode and monitoring ciliary activity by cinematographic or photometric means (Machemer, 1974*a*, *b*; Machemer & Eckert, 1973, 1975). In the absence of stimulation the cilia beat at 5 to 15 Hz with the power stroke directed at an angle of 120° to 150° (i.e. *c*. 4 or 5 o'clock) viewing the cell with the anterior end at 0° (i.e. 12 o'clock). When the membrane potential is displaced the cilia exhibit the following responses:

(1) *Stimuli which depolarize the cell membrane evoke a shift in orientation of the power stroke from the 'resting' angle of 4 to 5 o'clock toward about 12 o'clock (i.e. ciliary reversal) and an increase in frequency of beating of up to two to four times the resting rate.* In the free-swimming ciliate, depolarization arises from encounters with water of higher salt concentrations or from mechanical stimulation, by collision, of the mechano-receptor membrane of the cell anterior (Eckert, Naitoh & Friedman, 1972).

(2) *Hyperpolarization evokes a shift in orientation toward 6 o'clock with a concomitant rise in frequency of beating.* The rise in frequency is equivalent to that which occurs during depolarization, but the direction of beating is not reversed. Hyperpolarization occurs whenever the posterior end of the cell is mechanically stimulated. Ion substitution experiments have

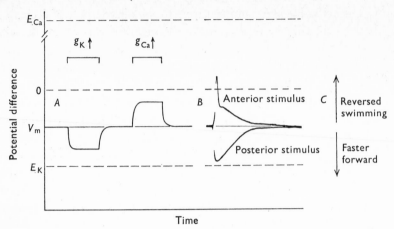

Fig. 2. Regulation of locomotor activity by conductance changes of surface membrane. *A*, resting potential lies between Ca^{2+} and K^+ equilibrium potentials and shifts toward either one with changes in conductance according to equation (2). *B*, mechanical stimulation of cell anterior causes g_{Ca} to rise, while stimulation of posterior causes g_K to rise. *C*, corresponding locomotor responses. (From Eckert, 1972. Copyright © 1972 from the American Association for the Advancement of Science.)

shown that mechanical stimulation of the rear produces a local outward K^+ current (Naitoh & Eckert, 1973). The hyperpolarization leads to accelerated forward swimming. The bioelectric aspects of sensory-motor function in *Paramecium* are summarized in Fig. 2, which emphasizes the point that changes in membrane permeability primarily to Ca^{2+} and K^+ provide the modulation of membrane potential which in turn regulates the motor function of the cilia.

ROLE OF CALCIUM IN THE ELECTRIC PROPERTIES OF THE SURFACE MEMBRANE

Ca^{2+} has several known interactions with the cell membrane of *Paramecium* which affect its electric properties. The quantitative relations of these interactions are not yet well understood. Moreover, since these various effects of Ca^{2+} occur simultaneously, the relations between Ca^{2+} and membrane properties such as resting potential, zero-current resistance, current–voltage relations, regenerative response (Naitoh & Eckert, 1968*a*; Naitoh, Eckert & Friedman, 1972) are complex. We will first consider some interactions of Ca^{2+} with the membrane of *Paramecium*.

Calcium must be actively removed from the cell

The Ca^{2+}-sensitivity of cilia of reactivated, extracted paramecia (Naitoh & Kaneko, 1972) suggest that the intraciliary concentration of Ca^{2+} in living, forward-swimming paramecia must be below 10^{-6} M. Assuming a resting concentration of 10^{-7} M, the concentration gradient in a medium of 10^{-3} M Ca^{2+}, is 10^4. Without active transport through the cell surface or via the contractile vacuole, $[Ca]_i$ would eventually approach the extracellular calcium concentration, $[Ca]_o$, since it is unlikely that all the Ca^{2+} which leaks in can be permanently sequestered within the cell. It is not known if active transport of Ca^{2+} involves net charge transfer (i.e. contributes to the resting potential). Since cells survive for many hours in Na^+-free medium, immediate Ca^{2+} balance appears not to depend entirely on a Ca^{2+}–Na^+ exchange similar to that which has been demonstrated in squid axon (Baker, Blaustein, Hodgkin & Steinhardt, 1967). However, increased $[Na]_o$ does produce a drop in membrane conductance (Naitoh & Eckert, 1968a), consistent with a reduced g_K. The resting K^+ conductance is, in part, a function of $[Ca]_i$ in some cells (Meech & Strumwasser, 1970); and there is evidence (see below) that this is also the case in *Paramecium*. Thus, since increased $[Na]_o$ may lead to lowered $[Ca]_i$, part of the Ca^{2+} flux may be coupled to the Na^+ gradient.

The extracellular:intracellular concentration gradient produces an EMF

The first intracellular recordings from animal cells were reported in the *Journal of Experimental Biology* by Kamada (1934) who had used saline-filled capillary glass microelectrodes to measure the resting potential of *Paramecium* in Sir James Gray's laboratory. He noted that the resting potential varied as a function of extracellular cation concentration and was relatively independent of the anion. Kamada found that cations of similar valence had somewhat similar effects on the resting potential, the monovalent cations Na^+ and K^+ producing a greater depolarization for a five-fold rise in external concentration than the divalent cations Ca^{2+} and Mg^{2+}. The differences were of the order predicted by the Nernst relation:

$$E_x = \frac{0.058}{z} \text{V} \log [x]_o/[x]_i \qquad (1)$$

in which E_x is the membrane potential at which ion x is in electrochemical equilibrium and z is the valence of ion x, at a temperature of 20 °C. With K^+ and Ca^{2+} both present in extracellular concentrations of 10^{-3} M each, and as free ions in the intracellular milieu at 2×10^{-2} and 10^{-7} M respectively (as appears to be the case; Naitoh & Eckert, 1973; Naitoh & Kaneko, 1972) E_K is calculated to be -77 mV and $E_{Ca} = +116$ mV. As noted by Kamada (1934), the resting freshwater ciliate membrane exhibits poor cation selectivity. Thus, the resting potential (V_{rest}) of *Paramecium* changes about $+30$ mV for a 10-fold increase in $[K]_o$ and about $+15$ mV for a 10-fold increase in $[Ca]_o$ (Fig. 3, bottom) compared with theoretical Nernst slopes of $+58$ and $+29$ mV respectively (Naitoh & Eckert, 1968a). This suggests that g_{Ca} and g_K in the resting membrane are of the same order of magnitude and account for a large part of the total resting conductance in KCl–$CaCl_2$ media. In this case, $g_K \approx g_{Ca}$ and $g_K + g_{Ca} \approx G$, the aggregate conductance; i.e. $g_K \approx g_{Ca} \approx G/2$. However, according to Hodgkin & Horowicz (1959) the resting potential is related to conductances as

$$V_{rest} = [g_K/G] E_K + [g_{Ca}/G] E_{Ca} \tag{2}$$
$$= 1/2(-77 \text{ mV}) + 1/2(+116 \text{ mV})$$
$$= +19.5 \text{ mV}.$$

Instead, a *Paramecium* bathed in a 1 mM K + 1 mM Ca solution has a negative resting potential of at least -30 mV. The negative potential could arise from one or more of the following: (1) The resting membrane may in fact be much more permeable to K^+ than to Ca^{2+}, the slope (Fig. 3) obtained in the Ca^{2+} concentration series arising not from an appreciable resting Ca^{2+} conductance, but from a drop in g_K with increased $[Ca]_o$. Evidence that this occurs is presented below. (2) Possible electrogenic pumping of Ca^{2+} and/or other ions. (3) Undoubtedly ions such as Na^+ and Mg^{2+}, with concentration gradients such as to produce negative equilibrium potentials, contribute their diffusion potentials to the resting membrane potential.

The conductance to calcium ions is a function of membrane potential

A step depolarization by voltage clamp produces a transient inward Ca^{2+} current (Naitoh & Eckert, 1974). This weakly regenerative current, I_{Ca}, which results from a transient rise in Ca^{2+} conductance (g_{Ca}) of the membrane, is responsible for the 'calcium response' of the unclamped membrane as seen in Fig. 4 (Naitoh et al., 1972). All-or-none calcium spikes or graded responses occur in a number of metazoan cells (Hagiwara, 1973).

An ionic membrane current carried by a single species x can be formally described by the ohmic relation

$$I_x = g_x(V_m - E_x) \tag{3}$$

in which g_x is the membrane conductance for X; V_m, the membrane potential; and E_x, the equilibrium potential of x. For *Paramecium* the

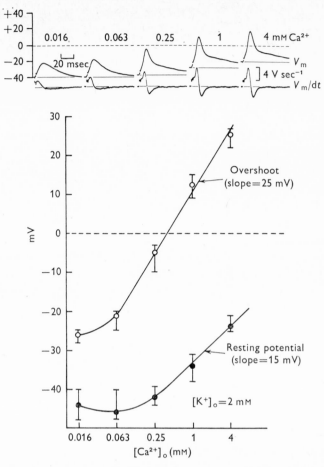

Fig. 3. Effect of $[Ca]_o$ on resting potential and calcium responses. Characteristic voltage records at top correspond to Ca^{2+} concentrations along abscissa (from Naitoh *et al.*, 1972).

term $V_m - E_{Ca}$ is rather large under most conditions, since the intracellular concentration of free Ca^{2+} appears, as in other known cases, to be very low (i.e. $< 10^{-6}$ M). Thus, I_{Ca} should be a steep function of g_{Ca}, and be less sensitive to changes in $[Ca]_o$.

Unlike a full action potential, the amplitude of the response is graded with stimulus intensity, very much as in many arthropod muscle fibres. Failure for the regenerative depolarization to become all-or-none apparently stems from the quantitative relations between g_{Ca} and the leakage conductance (largely g_K) of the membrane. These relations can be altered by addition of Ba^{2+} to the medium, which renders the response all-or-none. The reasons for this will be considered below (p. 243).

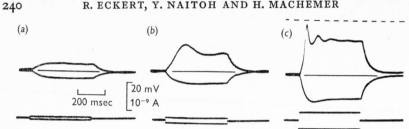

Fig. 4. Graded responses to 500 msec injected current pulses. Upper trace, membrane potential; lower trace, injected current. Dashed line gives zero reference. Purely passive potential changes are seen in (a). Graded calcium responses are seen in (b) and (c). Cell bathed in 4 mM KCl, 1 mM CaCl₂ and 1 mM Tris-HCl. Temperature, 19 °C (from Naitoh et al., 1972).

The amplitude and overshoot of the calcium response at saturation stimulation varies with ionic conditions. As seen in Fig. 3, both increase as $[Ca]_o$ increases, the peak potential changing with a slope of about $+25$ mV per 10-fold rise in $[Ca]_o$. However, the analysis is complicated by the concomitant shift in V_{rest} which occurs with changes in extracellular cation concentration in *Paramecium*. Of special interest is the constancy of the amplitude and shape of the response under different ionic conditions but constant $[K]_o:[Ca]_o^{\frac{1}{2}}$ ratio (Fig. 5). The shape and duration of the calcium response remains constant with $[K]:[Ca]^{\frac{1}{2}}$ over a 63-fold range of $[Ca]_o$ (e.g. 16 mM K + 1 mM Ca versus 2 mM K + 0.016 mM Ca). Recent insight into the role of $[Ca]_i$ in membrane K⁺ conductance (see below) lends new significance to the duration of the Ca²⁺ response, which now appears to be related to the amount of Ca²⁺ entering during the response. From the constancy of response duration with constant ratio of $[K]:[Ca]^{\frac{1}{2}}$ it appears that $[Ca]_o$ influences not merely the term $V_m - E_{Ca}$ (see equation (3)), but also the calcium conductance, g_{Ca}. This may occur through an interaction between Ca²⁺ and ionic binding sites on the membrane which influence the behaviour of voltage-sensitive calcium channels. The relations of the Donnan ratio $[x^+]_o:[Ca^{2+}]_o^{\frac{1}{2}}$ to the electric properties of the membrane may arise from competitive binding to surface sites of Ca²⁺, K⁺, and other cations. Ion exchange behaviour has been demonstrated in *Paramecium* in a study of washout kinetics by Naitoh & Yasumasu (1967). In addition to the fast transient Ca²⁺ current, there appears to be a slowly relaxing weak and prolonged Ca²⁺ current associated with steady depolarization. The slower calcium current can be inferred from the behaviour of cilia during prolonged potential changes, since maintained depolarization by voltage clamp produces reversed beating with a relaxation time of many seconds or minutes (Machemer & Eckert, 1975). Long-lasting, voltage-sensitive calcium conductances have recently

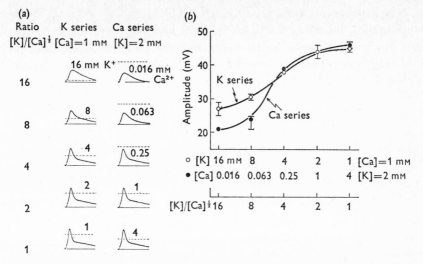

Fig. 5. Calcium response under different combinations of $[K]_o$ and $[Ca]_o$. (*a*) $[Ca]_o$ and $[K]_o$ each varied while the other ion held constant so as to compare responses at similar $[K]:[Ca]^{\frac{1}{2}}$ ratios but different Ca^{2+} concentrations. Note similarity of two columns except for zero references (dashed lines). (*b*) Base-to-peak amplitudes plotted (from Naitoh *et al.*, 1972).

been discovered in cardiac tissue (Trautwein, 1973) and molluscan neurones (Eckert & Lux, 1975*a*).

Spontaneous firing can be induced in *Paramecium* by addition of Ba^{2+} to the bath (Kinosita, Murakami & Yasuda, 1965). The Ba^{2+}-induced spontaneity seen in *Paramecium* may result from a decrease in leakage conductance (largely g_K) seen in the presence of Ba^{2+} (see below) together with an enhanced inward Ca^{2+}, Ba^{2+} current through a slowly inactivating $Ca^{2+}-Ba^{2+}$ conductance. With the leakage conductance depressed, a weak inward current with slow inactivation kinetics could, in principle, produce a pacemaker depolarization which reaches the firing level of the all-or-none $Ca^{2+}-Ba^{2+}$ action potential. A net late inward current which persists for many seconds during depolarization appears to play a role in the pacemaker waves which underlie spontaneous 'bursting' in certain snail neurones (Eckert & Lux, 1975*b*).

A persistent, voltage-sensitive Ca^{2+} conductance may also contribute to the inward-going ('anomalous') rectification seen in *Paramecium* (Naitoh & Eckert, 1968*a*) as it does in snail neurones (Eckert & Lux, 1975*b*). Thus the steady inward current could contribute to a high slope resistance at small to moderate depolarizations (< 20 mV). Ca^{2+} influx could also contribute to inward rectification at higher potentials by increasing g_K, and hence reducing the slope resistance.

The usual agents variously successful in blocking Ca^{2+} currents in other membranes, including La^{3+}, Mn^{2+}, Co^{2+} and Mg^{2+} are not very useful for this purpose in the unclamped membrane of *Paramecium*, since they are either toxic (La^{3+}) or depolarize the cell at concentrations which might otherwise block the Ca^{2+} conductance.

Membrane calcium currents produce significant changes in $[Ca]_i$

Because Ca^{2+} is maintained at low ($< 10^{-6}$ M) cytoplasmic free concentration by cellular regulatory mechanisms, membrane currents can produce significant percentage changes in the intracellular concentration of Ca^{2+} as compared with Mg^{2+}, Na^+ or K^+, which are present in concentrations several orders of magnitude higher. Nonetheless, the concentration changes which a brief (< 1 sec) Ca^{2+} current could produce throughout the entire intracellular bulk of a *Paramecium* would be trivial, because of the small surface:volume ratio of the total cell compartment. Thus, it can be calculated that the net influx of positive charge in the form of Ca^{2+} sufficient to depolarize the surface membrane by 10 mV (a 10 mV Ca^{2+} response is sufficient to produce a vigorous reversal) would produce an increment in $[Ca]_i$ of 5×10^{-7} M, assuming instantaneous uniform distribution throughout the cell, no short circuiting by an opposite K^+ current, and no binding or sequestering. The results of Naitoh & Kaneko (1972) imply resting cytoplasmic concentrations of this magnitude, and indicate that the cilia require an increment of at least 10^{-6} M for reversed beating. On the other hand, if one considers only the surface and internal volume of the cilia while ignoring the remainder of the cytoplasmic mass, the intraciliary Ca^{2+} concentration should undergo a transient rise of 10^{-5} M if Ca^{2+} alone carries the charge for a 10 mV depolarization (Eckert, 1972).

Unpublished experiments of Ms K. Dunlap in our laboratory indicate that most of the calcium current enters the cell through the membrane areas covering the cilia, because the calcium current varies with the regeneration of cilia on denuded paramecia. Consideration of the electrical equivalents and the dimensions of a cilium (Eckert, 1972, Note 24) suggests that the number of free Ca^{2+} within each cilium rises from about 10 in the resting state (assuming 50 % free space and a concentration of 10^{-7} M) to 1000 or more during a 10 mV calcium response. The latter figure assumes no partial short-circuiting by simultaneous K^+ efflux, and is therefore a low estimate.

The potassium conductance appears to be modulated by $[Ca]_i$

A rise in $[Ca]_i$ is known to cause an increase in g_K in some cell types (see Meech, this volume). Several observations suggest that this also occurs in

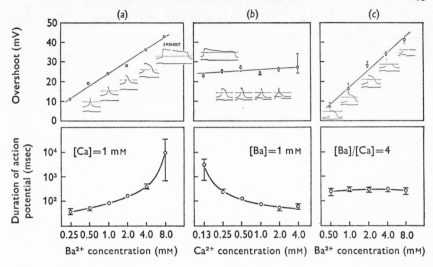

Fig. 6. Calcium–barium effects on plateau and overshoot of all-or-none action potential. Overshoot shows nearly ideal Nernst slope with [Ba]$_0$ (a), and little calcium dependence in presence of Ba^{2+} (b). Duration of plateau plotted below. Note constancy of duration with constant [Ba]:[Ca] ratio (c) (after Naitoh & Eckert, 1968b).

Paramecium. Ba^{2+} can substitute for Ca^{2+} in carrying depolarizing current during the regenerative response of the membrane (Naitoh & Eckert, 1968b). Inspection of Fig. 6 shows, in fact, that Ba^{2+} is more effective in this respect than Ca^{2+}. A second effect of Ba^{2+} is to prolong the action potential by delaying repolarization. As the ratio [Ba]$_0$:[Ca]$_0$ increases the repolarization is progressively delayed (Fig. 6a, b). This may be due to the failure of Ba^{2+} which enters the cell to substitute effectively for Ca^{2+} in activating the Ca^{2+}-sensitive component of g_K. If the ratio [Ba]$_0$:[Ca]$_0$ is kept constant while the concentrations of both cations are raised the overshoot of the response increases as predicted by the Nernst relation (equation (1)) but the duration of the plateau phase remains constant (Fig. 6c). Under those conditions the proportions of Ca^{2+} and Ba^{2+} carrying inward current should remain constant, and so the internal increment in Ca^{2+} and Ba^{2+} concentrations during the response should also remain constant.

During prolonged (i.e. many seconds) depolarization by voltage clamp part of the net membrane current (primarily outward K$^+$ current) relaxes approximately in parallel with the relaxation of ciliary reversal (Machemer & Eckert, 1975). The parallel behaviour of the cilia and I_K may have a common basis, namely a temporal relaxation of g_{Ca} with corresponding changes in I_{Ca}.

The surface membrane shows a steady state decrease in conductance when Na$^+$ is added to the standard extracellular solution of CaCl$_2$, KCl and Tris-HCl (Naitoh & Eckert, 1968a). This is consistent with stimulation of Ca^{2+} efflux by extracellular Na^+ (Baker, this volume), and a consequent reduction of g_K by the reduction in $[Ca]_i$ (Meech, this volume). A reduction of the resting g_K by $[Na]_o$ in this manner might also explain the induction by extracellular Na^+ of spontaneous excitability in *Paramecium* (Kung *et al.*, 1975).

$[Ca]_o$ influences g_{Ca} and other ionic conductances of the membrane

Calcium ions appear to bind to the cell surface in inverse proportion to the ratio $[x]_o^{1/a} : [Ca]_o^{\frac{1}{2}}$ in which a is the valence of competing cation x (Naitoh & Yasumasu, 1967). The amount of Ca^{2+} bound to the membrane appears to influence the electric behaviour and in particular the ionic conductances of the membrane (Naitoh & Eckert, 1968a). The effect of $[Ca]_o$ on the membrane conductance is seen in Fig. 7. In the lower range of $[Ca]_o$ the membrane showed a more or less linear drop in conductance with log $[Ca]_o$. Beyond that the conductance increased steeply. The subsequent increase at high $[Ca]_o$ could be due to the concomitant depolarization brought on by increased $[Ca]_o$ (Naitoh & Eckert, 1968a) and consequent rise in the potential-dependent g_K (Machemer & Eckert, 1975). This appears not to include a rise in g_{Ca} and $[Ca]_i$ since this would induce ciliary reversal, which at high $[Ca]_o$ does not occur (Kamada & Kinosita, 1940; Naitoh, 1968). Failure of depolarization by elevated $[Ca]_o$ to produce ciliary reversal may result from a negative shift in the current–voltage relations of the Ca^{2+} system with increased $[Ca]_o$, similar to that exhibited by the Na^+ system with increased $[Ca]_o$ in the squid axon (Frankenhaeuser & Hodgkin, 1957).

'Stabilization' of the membrane

The integrity of the membrane requires the presence of extracellular Ca^{2+}. *Paramecium* lyses when placed in a 'Ca²⁺-free' (i.e. no Ca^{2+} added to medium) solution containing several millimoles of KCl or other salt. Survival is longer in distilled water, suggesting that Ca^{2+} bound to the membrane is important to membrane integrity and can be displaced by other cations (Naitoh & Yasumasu, 1967). Other divalent cations such as Mg^{2+} do not substitute for Ca^{2+} in this respect. This property of the *Paramecium* membrane restricts experiments with low $[Ca]_o$.

ROLE OF CALCIUM IN COUPLING CILIARY
ORIENTATION TO MEMBRANE ACTIVITY

Several lines of evidence, listed below, favour the view that calcium is the agent that couples the orientation of ciliary beating to changes in the state

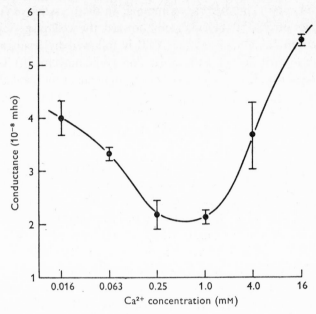

Fig. 7. Extracellular calcium concentration and steady-state, zero-current input conductance of *Paramecium caudatum*. KCl was 2 mM throughout (replotted from Naitoh & Eckert, 1968a).

of the surface membrane. In this view the surface membrane influences the ciliary activity by regulating the influx, and hence the internal concentration of Ca^{2+}. The quantitative relations between Ca^{2+} influx and $[Ca]_i$ varies with the unidirectional Ca^{2+} flux. Since short-term ion fluxes are most readily monitored as currents, it will be useful to recall equation (3).

The free calcium ion concentration to which the ciliary apparatus is exposed determines the orientation of the power stroke

Naitoh & Kaneko (1972, 1973) extracted *P. caudatum* with Triton X-100 to render the surface membrane permeable to small molecules and ions, and then reactivated the extracted 'models' with ATP and Mg^{2+}. The usefulness of this approach stems from the elimination of the cell membrane as a functional barrier regulating the internal milieu, so that the latter simply equilibrates with the experimental solution. With this method it was found that both Mg^{2+} and Ca^{2+} act as co-factors for ATPases involved in ciliary movement. Mg^{2+} is required together with ATP for the cyclic beating movements. Below 10^{-6} M Ca^{2+} the effective stroke is directed more-or-less posteriorly producing forward locomotion. At 10^{-6} M Ca^{2+} or higher the orientation of the power stroke is shifted toward the cell anterior, producing reverse swimming. In the presence of ATP alone (i.e. no Ca^{2+} or Mg^{2+}) the cilia point toward the cell posterior without beating. Addition of Ca^{2+} to the ATP is followed by a single shift of orientation toward the cell anterior (i.e. no cyclic movement.) Thus $[Ca]_i$ appears to control the orientation of ciliary movement by regulating ATP-dependent processes. Cyclic movement (i.e. beating) of the cilia requires the presence of Mg^{2+} together with ATP in the reactivation medium. The molecular bases of the ionically regulated movements and changes in orientation are not known.

If it is assumed that the Ca^{2+}-sensitive ATPase(s) are involved in this Ca^{2+} dependency, which behave with similar kinetics in both living and extracted cells, these results suggest that in living paramecia forward swimming occurs when the internal (i.e. intraciliary) free calcium concentration is below 10^{-6} M, and that the cilia beat in reverse whenever the concentration exceeds this level. The increments in intraciliary Ca^{2+} concentration produced by fluxes associated with reversed beating in the living cell (see p. 242) are in agreement with those concentrations. Since Ca^{2+}-induced reverse swimming in extracted models continues steadily for some minutes, the ciliary activity of living paramecia provides a qualitative assay for changes in the intraciliary free Ca^{2+} concentration.

Reversed beating is closely linked to stimuli which depolarize the membrane

These include outward current injected into the cell with a microelectrode, depolarizing receptor current evoked by mechanical stimulation of the cell anterior (Eckert et al., 1972), depolarization of the cell by an increase in concentration of extracellular cations (Kinosita, 1954; Naitoh & Eckert, 1968a), and spontaneous depolarization (Kinosita, Murakami & Yasuda, 1965). Larger, prolonged depolarizations produce stronger, more prolonged beating in reverse (Kinosita, 1954; Machemer & Eckert, 1975). Depolarization can produce a net influx of Ca^{2+} (pp. 238–242). One exception appears to be depolarization by increased $[Ca^{2+}]_o$.

Reversed beating is associated with an influx of Ca^{2+} during depolarization

Stimulation with brief (< 25 msec) stimuli evoked reversed beating only if the depolarization evoked a Ca^{2+} response (see p. 239) correlated with a net influx of Ca^{2+} (Machemer & Eckert, 1973). Passive (i.e. simple ohmic-capacitive) depolarization was not sufficient for eliciting reversed beating.

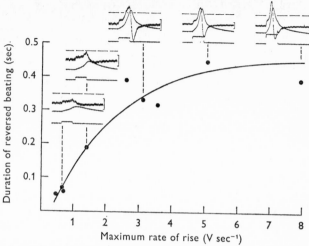

Fig. 8. Duration of reversed beating as a function of the maximum rate of rise of the calcium response. Representative recordings are shown corresponding to points along plot. Upper trace is dV_m/dt, middle trace is V_m and lower trace indicates a 20 msec current pulse. The vertical calibration marks equal 2 V sec^{-1}. Note the change in scale for the fourth and fifth recordings (from Machemer & Eckert, 1973).

The duration of reversed beating in response to a brief stimulus increases with increased amplitude of the Ca^{2+} response (Fig. 8). This is consistent with an increase in the time required to lower [Ca]$_i$ to prestimulus levels after a larger influx of the ion.

Suppression of Ca^{2+} influx suppresses reversed beating

Reversed beating induced by increased intracellular positive potential fails to occur as the membrane potential approaches large positive values (Fig. 9), consistent with the presumptive equilibrium potential for Ca^{2+} (i.e. the term $V_m - E_x$ of equation (3) becomes small). The calcium equilibrium potential (E_{Ca}) in forward-swimming paramecia can be estimated for a given [Ca^{2+}]$_o$ by assuming that [Ca]$_i$ during forward swimming is below 10^{-6} (i.e. 10^{-7} M). This is based on the assumption that the Ca^{2+}-sensitivity of ciliary activity in living cells is no less sensitive than in extracted cells. Thus, with [Ca]$_o$ = 10^{-3} M, E_{Ca} at rest should be about +120 mV. At the minimum [Ca]$_i$ for reversed beating (e.g. 10^{-6} M), E_{Ca} should drop to about +90 mV. Thus, at some potential approaching +90 mV the EMF on Ca^{2+} (i.e. $V_m - E_{Ca}$) will be sufficiently low so that the unidirectional Ca^{2+} influx will be insufficient to overcome the active efflux of Ca^{2+}, with the result that [Ca]$_i$ will remain near its resting value in spite of the rise in g_{Ca}. In *Paramecium* and *Euplotes* complete suppression

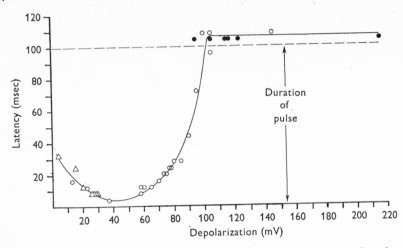

Fig. 9. Latency from onset of pulse to onset of ciliary reversal as a function of positive potential shifts. Stimuli consisted of 100 msec outward current pulses of 4×10^{-10} to 1.4×10^{-7} A. The latency is minimal for steady state potential shifts of between +30 and +60 mV. Reversed beating suppressed until after end of pulse for displacements above +100 mV. Latencies of less than 4 msec could not be resolved at the framing rate of 250 sec⁻¹. Each of the three symbols represents a different specimen (from Machemer & Eckert, 1973).

occurs above about +70 mV (Epstein & Eckert, 1973; Machemer & Eckert, 1973). At the end of the suppression pulse there is a transient period of reversed beating, suggesting that the Ca^{2+} gates which remain activated during the pulse, require some time to close. It is presumed that during this period there is an influx of Ca^{2+} because the term $V_m - E_{Ca}$ is again large. Similar behaviour, i.e. suppression at large positive potentials and a post-pulse response, has been reported for the Ca^{2+}-regulated release of transmitter from presynaptic terminals in squid giant axon (Katz & Miledi, 1966), and for Ca^{2+} influx in *Aplysia* neurone somata detected by injected aequorin (Stinnakre & Tauc, 1973).

Membrane mutants with a defective calcium system fail to show
ciliary reversal in response to membrane depolarization

A behavioural mutant of *P. aurelia* named 'pawn' fails to swim in reverse upon depolarization with KCl or electric current (Kung, 1971). Significantly, it shows little or no electric sign of an increased Ca^{2+} influx during depolarization (Fig. 10), although Triton-extracted pawns show the same Ca^{2+} sensitivity as wild type (Kung & Naitoh, 1973). This supports the conclusion that the locomotor and bioelectric abnormalities of that mutant are both due to a defect in the potential-dependent Ca^{2+} conductance of

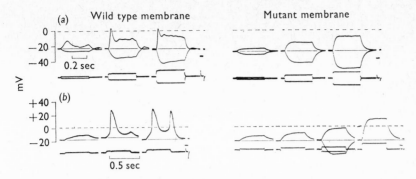

Fig. 10. Electric excitability of a wild type *Paramecium aurelia* (*left*) compared with the inexcitability of the membrane mutant pawn (*right*). (*a*) Bath solution consisted of 1 mM $CaCl_2$, 4 mM KCl and 1 mM Tris-HCl (pH 7.2). The wild type shows graded calcium responses to depolarizing current, while the pawn mutant shows only electrotonic responses. (*b*) Bath solution consisted of 1 mM $CaCl_2$, 4 mM $BaCl_2$, 1 mM Tris-HCl. The wild type exhibits all-or-none Ba^{2+} spikes (Naitoh & Eckert, 1968*b*), while the pawn shows only electrotonic responses. The lower set of traces in each record indicates the stimulus current (after Kung & Eckert, 1972).

the membrane. In pawn, a single gene mutation has eliminated, or greatly reduced, the influx of Ca^{2+} normally evoked by depolarization and, as a result, ciliary reversal is uncoupled from the behaviour of the surface membrane. A number of pawn clones have been generated in various laboratories. A deletion mutant would be useful for comparison with the wild type in order to isolate and identify the molecular constituents associated with the voltage-sensitive Ca^{2+} conductance.

There are additional observations consistent with the view that membrane-gated Ca^{2+} regulates ciliary activity. Reversed beating becomes prolonged at reduced temperatures (Oliphant, 1938; Machemer, 1974*a*, *b*), which is predicted if cooling lowers the rate of Ca^{2+} extrusion or sequestration. Strong hyperpolarization induces an anomolous reversed beat in *Opalina* (Naitoh, 1958), which is predicted if the term $(V_m - E_{Ca})$ in equation (3) were to rise more rapidly with hyperpolarization than a concomitant drop in g_{Ca}. Reversed beating has not been seen with strong hyperpolarization in *Paramecium*, however, suggesting that g_{Ca} is, in fact, depressed during hyperpolarization in that ciliate. Such a drop in Ca^{2+} conductance with hyperpolarization has recently been confirmed in unpublished voltage clamp experiments by Y. Naitoh. Proposed relations between membrane potential, Ca^{2+} and ciliary responses are summarized in Fig. 11.

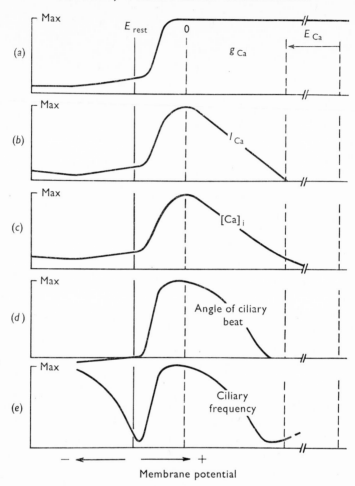

Fig. 11. Inferred relations between membrane responses and ciliary responses to potential changes in *Paramecium*. E_{rest} indicates resting potential; o indicates zero potential difference, and E_{Ca} the Ca^{2+} equilibrium potential which drops as $[Ca]_i$ rises during depolarization. Plots are approximations based in whole or part on actual measurements (*b, d, e*) or inference (*a, c*). (*a*) Conductance for Ca^{2+} as a function of membrane potential. (*b*) Ca^{2+} current through the membrane varies as a function of membrane potential and produces corresponding concentration changes in the cilium (*c*). The Ca^{2+} concentration changes produce the observed changes in angle of power stroke (*d*) and frequency (*e*) as proposed in the text (from Eckert & Machemer, 1975).

ROLE OF CALCIUM IN REGULATING FREQUENCY OF BEATING

The regulation of ciliary frequency is understood less well than the control of ciliary orientation, but it is clear that the membrane is involved. Naitoh (1958) noted that in *Opalina* frequency increased upon both hyper- and depolarization. This has been reinvestigated in *Paramecium* more recently, using ciné analysis (Machemer, 1974*a*; Eckert & Machemer, 1975; Machemer & Eckert, 1975) based on an improved understanding of ciliary movement patterns (Machemer, 1974*b*). As seen in Fig. 11*e*, the change of frequency with membrane potential is not symmetrical about the resting potential. While hyperpolarization produced a monotonic rise in frequency, with the effective stroke directed approximately toward the cell posterior, a more complex sequence occurs with depolarization and increased positive potential. With small increments of depolarization (< 5 mV) there is a progressive drop in beating frequency until often there is complete arrest at a given depolarization. Further depolarization produces progressive increments in frequency of beating with the orientation of the power stroke shifting progressively counterclockwise toward the cell anterior. Both negative and positive potential displacements of 30 mV or more evoke beating frequencies of up to 50 Hz, compared to resting frequencies of 5 to 15 Hz.

In the previous section, evidence was reviewed that the orientation of the ciliary movements is coupled to membrane potential by Ca^{2+} fluxes. Recent evidence (Machemer & Eckert, 1975) outlined below suggests that the frequency of beating may also be coupled to membrane potential through calcium fluxes.

Frequency changes are closely linked to calcium-regulated shifts in direction of beating (Machemer, 1974a, b; Machemer & Eckert, 1975)

At several millivolts depolarization, the cilia are inactivated or greatly slowed. Frequency and orientation change together in a graded manner upon departure from this 'null' potential, positive displacement producing progressive increases in frequency along with a rotation of the beating direction from about 3 o'clock toward 12 o'clock, and negative displacement producing progressive increases in frequency with rotation toward 6 o'clock (cell anterior being 12 o'clock). The close linkage of graded frequency changes with graded orientational changes suggests that both phenomena are regulated by a common agent.

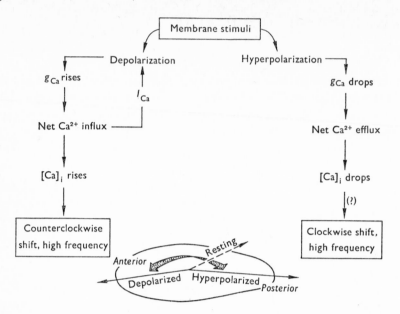

Fig. 12. Overview of membrane-regulated changes in beating orientation and frequency. Anterior is taken as 12 o'clock. In the resting state the power stroke is directed toward about 4 or 5 o'clock with a frequency below 15 Hz (broken arrow). With progressive depolarization, the rise in [Ca]ᵢ produces first a decrease, then an arrest, and finally an increase in frequency (indicated by width of stippled arrow) and a shift in orientation toward 11 or 12 o'clock. With hyperpolarization there is a parallel increase in frequency and clockwise shift of power stroke toward about 6 o'clock. The relation between a drop in [Ca]ᵢ and the ciliary response to hyperpolarization is tentative, as indicated by the question mark.

Approach to the presumptive calcium equilibrium potential suppresses high frequency beating

This is seen both in *Paramecium* (Machemer & Eckert, 1973) and *Euplotes* (Epstein & Eckert, 1973). The question arises as to whether Ca^{2+} acts directly on the ciliary apparatus to influence beating frequency. Naitoh & Kaneko (1973) plotted frequency of beating in the extracted cells against Mg^{2+}, ATP, and Ca^{2+} concentrations. The effects of Mg^{2+} and ATP on beating frequency were most striking, both producing a monotonic increase in the physiological concentration ranges. However, the frequency–Ca^{2+} plot shows a dip between 10^{-7} and 10^{-6} M Ca^{2+}. The plot consists of averaged readings from many cells. Thus, the dip in frequency may in fact be much deeper, having been smoothed out as the result of variations between individual specimens. If this is in fact the case, the actual relations between [Ca^{2+}] and frequency could resemble those between membrane potential and frequency (Fig. 11e). One could then argue that in living cells the frequency minimum which occurs several millivolts positive to the resting potential may correspond to an intraciliary Ca^{2+} concentration equal to that which produces the dip in the frequency–Ca^{2+} curve of extracted paramecia.

Magnesium must also be considered in the ionic control of beating frequency, since beating frequency in extracted models rises as Mg^{2+} concentration is increased from 0.25 to 4 mM (Naitoh & Kaneko, 1973). However, since the potential-related changes in frequency in the living cell occur in the absence of extracellular Mg^{2+}, an influx of this ion cannot be the basis for potential-evoked frequency increases. It is unlikely, moreover, that brief membrane fluxes of Mg^{2+} could significantly alter the intracellular concentration of that ion or the Ca^{2+}–Mg^{2+} balance, as Mg^{2+} undoubtedly occurs with an intracellular activity several orders of magnitude greater than that of Ca^{2+}.

The present view of Ca^{2+} and the regulation of ciliary activity in *Paramecium* is summarized in Fig. 12. Changes in $[Ca]_i$ due to altered membrane Ca^{2+} conductance with more-or-less constant active removal of Ca^{2+} from the cilia leads to changes in the activity of chemo-mechanical transducer sites in the axoneme which are expressed in altered orientation and frequency of ciliary movements. At present, the molecular bases for the coordination of axonemal components to produce changes in frequency and orientation of the ciliary movement remain unknown.

We are indebted to Ms Kathleen Dunlap for permission to refer to unpublished findings. Support for the studies which form the bases of this review were provided by NSF Grant BMS 74-19464 and NIH Grants NS 08364 and NS 05670, and by an equipment grant from the Deutsche Forschungsgemeinshaft.

REFERENCES

BAKER, P., BLAUSTEIN, M., HODGKIN, A. & STEINHARDT, R. (1967). The effect of sodium concentration on calcium movements in giant axons of *Loligo forbesi*. *J. Physiol., Lond.*, **192**, 43*P*.

ECKERT, R. (1972). Bioelectric control of ciliary activity. *Science, Wash.*, **176**, 473–481.

ECKERT, R. & LUX, H. (1975*a*). A non-inactivating inward current recorded during small depolarizing voltage steps in snail pacemaker neurons. *Brain Res.*, **83**, 486–489.

    (1975*b*). A voltage-sensitive persistent calcium conductance in neuronal somata of *Helix. J. Physiol., Lond.*, **254**, 129–151.

ECKERT, R. & MACHEMER, H. (1975). Regulation of ciliary beating frequency by the surface membrane. In *Symposium of the Society for General Physiology*, ed. R. Stephens & S. Inoue, pp. 151–163. Raven Press: New York.

ECKERT, R. & NAITOH, Y. (1972). Bioelectric control of locomotion in the ciliates. *J. Protozool.*, **19**, 237–243.

ECKERT, R., NAITOH, Y. & FRIEDMAN, K. (1972). Sensory mechanisms in *Paramecium*. I. Two components of the electric response to mechanical stimulation of the anterior surface. *J. exp. Biol.*, **56**, 683–694.

EPSTEIN, M. & ECKERT, R. (1973). Membrane control of ciliary activity in the protozoan *Euplotes*. *J. exp. Biol.*, **58**, 437–462.

FRANKENHAEUSER, B. & HODGKIN, A. (1957). The action of calcium on the electrical properties of squid axon. *J. Physiol., Lond.*, **137**, 218–244.

GRELL, K. (1973). *Protozoology*. Springer: Berlin.

HAGIWARA, S. (1973). Ca spike. *Adv. Biophys.*, **4**, 71–102.

HODGKIN, A. & HOROWICZ, P. (1959). The influence of potassium and chloride ions on the membrane potential of single muscle fibres. *J. Physiol., Lond.*, **148**, 127–160.

JENNINGS, H. (1906). *Behavior of the Lower Organisms*. Columbia University Press: New York.

KAMADA, T. (1934). Some observations on potential differences across the ectoplasm membrane of *Paramecium*. *J. exp. Biol.*, **11**, 94–102.

KAMADA, T. & KINOSITA, H. (1940). Calcium–potassium factor in ciliary reversal of *Paramecium*. *Jap. Acad. Proc.*, **16**, 125–130.

KATZ, B. & MILEDI, R. (1966). Input–output relation of a single synapse. *Nature, Lond.*, **212**, 1242–1245.

KINOSITA, H. (1954). Electric potentials and ciliary response in *Opalina*. *J. Fac. Sci. Tokyo Univ.*, IV, **7**, 1–14.

KINOSITA, H., MURAKAMI, A. & YASUDA, M. (1965). Interval between membrane potential change and ciliary reversal in *Paramecium* immersed in Ba–Ca mixture. *J. Fac. Sci. Tokyo Univ.*, IV, **10**, 421–425.

KUNG, C. (1971). Genic mutants with altered system of excitation in *Paramecium aurelia*. I. Phenotypes of the behavioral mutants. *Z. vergl. Physiol.*, **71**, 142–164.

KUNG, C., CHANG, S. Y., SATOW, Y., HOUTEN, J. & HANSMA, H. (1975). Genetic dissection of behavior in *Paramecium*. *Science, Wash.*, **188**, 898–904.

KUNG, C. & ECKERT, R. (1972). Genetic modification of electric properties in an excitable membrane. *Proc. natn. Acad. Sci. USA*, **69**, 93–97.

KUNG, C. & NAITOH, Y. (1973). Calcium-induced ciliary reversal in the extracted models of 'Pawn', a behavioral mutant of *Paramecium*. *Science, Wash.*, **179**, 195–196.

MACHEMER, H. (1974*a*). Frequency and directional responses of cilia to membrane potential changes in *Paramecium*. *J. comp. Physiol.*, **92**, 293–316.

(1974*b*) Ciliary activity and metachronism in protozoa. In *Cilia and Flagella*, ed. M. Sleigh, pp. 199–286. Academic Press: London.

MACHEMER, H. & ECKERT, R. (1973). Electrophysiological control of reversed ciliary beating in *Paramecium*. *J. gen. Physiol.*, **61**, 572–587.

(1975). Ciliary frequency and orientational responses to clamped voltage steps in *Paramecium*. *J. comp. Physiol.*, **104**, 247–260.

MEECH, R. & STRUMWASSER, F. (1970). Intracellular calcium injection activates potassium conductance in *Aplysia* nerve cells. *Fedn Proc.*, **29**, 834.

NAITOH, Y. (1958). Direct current stimulation of *Opalina* with intracellular microelectrode. *Annotnes zool. jap.*, **31**, 59–73.

(1968). Ionic control of the reversal response of cilia in *Paramecium caudatum*. A calcium hypothesis. *J. gen. Physiol.*, **51**, 85–103.

(1974). Bioelectric basis of behavior in protozoa. *Am. Zool.*, **14**, 883–893.

NAITOH, Y. & ECKERT, R. (1968*a*). Electrical properties of *Paramecium caudatum*: Modification by bound and free cations. *Z. vergl. Physiol.*, **61**, 427–452.

(1968b). Electrical properties of *Paramecium caudatum*: All-or-none electrogenesis. *Z. vergl. Physiol.*, **61**, 453–472.

(1973). Sensory mechanisms in *Paramecium*. II. Ionic basis of the hyperpolarizing receptor potential. *J. exp. Biol.*, **59**, 53–65.

(1974). The control of ciliary activity in Protozoa. In *Cilia and Flagella*, ed. M. Sleigh, pp. 305–352. Academic Press: London.

NAITOH, Y., ECKERT, R. & FRIEDMAN, K. (1972). A regenerative calcium response in *Paramecium*. *J. exp. Biol.*, **56**, 667–681.

NAITOH, Y. & KANEKO, H. (1972). ATP–Mg-reactivated triton-extracted models of *Paramecium*: Modification of ciliary movement by calcium ions. *Science, Wash.*, **176**, 523–524.

(1973). Control of ciliary activities by adenosinetriphosphate and divalent cations in triton-extracted models of *Paramecium caudatum*. *J. exp. Biol.*, **58**, 657–676.

NAITOH, Y. & YASUMASU, I. (1967). Binding of Ca ions by *Paramecium caudatum*. *J. gen. Physiol.*, **50**, 1303–1310.

OLIPHANT, J. (1938). The effects of chemicals and temperature on reversal in ciliary action in *Paramecium*. *Physiol. Zööl.*, **11**, 19–30.

STINNAKRE, J. & TAUC, L. (1973). Calcium influx in active *Aplysia* neurones detected by injected aequorin. *Nature, New Biol.*, **242**, 113–115.

TRAUTWEIN, W. (1973). Membrane currents in cardiac muscle fibers. *Physiol. Rev.*, **53**, 793–835.

CELL-TO-CELL COMMUNICATION, GAP JUNCTIONS AND CALCIUM

By N. B. GILULA and M. L. EPSTEIN

The Rockefeller University, New York, New York 10021, USA

All cells must interact with other cells as well as with their environment. These interactions are regulated by specialized functions of the cell surface membrane, generally termed the plasma membrane. The plasma membrane contains a variety of transport systems that enable the cell to interact with its environment so as to regulate selectively its internal ionic and nutritional requirements. In addition, the plasma membrane can generate specialized regions for the specific purpose of interacting with other cells. These direct, physical, cell-to-cell interactions are undoubtedly important for cell-to-cell adhesion, permeability regulation across epithelial and endothelial layers, and cell-to-cell communication. The purpose of this report will be to focus on the basic phenomenon of cell-to-cell communication. In particular, we will present and discuss information concerning (1) the biological phenomenon of cell-to-cell communication; (2) the gap junctional pathway for communication; and (3) the role of calcium in cell communication.

CELL-TO-CELL COMMUNICATION

Cell-to-cell communication is a basic biological phenomenon that is present between most metazoan cells, in both the plant and animal kingdoms. This sociological phenomenon is expressed biologically as an exchange of small ions and metabolites between cells. This exchange is facilitated by the presence of a specialized physical continuity between cells – a communication pathway. As a result of this pathway, cells can exchange materials directly without the involvement of normal release and uptake mechanisms. Experimentally, it is possible to define communication as the cell-to-cell transfer of ions (current), the transfer of metabolites, or both. Thus, communicating cells may be referred to as ionically coupled or metabolically coupled.

Ionic coupling

Ionically coupled cells were first described between invertebrate neurones by Furshpan & Potter (1959). This type of coupling was later found between myocardial cells (Barr, Dewey & Berger, 1965; Dreifuss, Girardier &

Forssmann, 1966), other excitable cells (Bennett, 1966; Payton, Bennett & Pappas, 1969; Llinás, Baker & Sotelo, 1974), as well as between non-excitable cells in epithelia and other tissues (Loewenstein et al., 1965; Loewenstein, 1966; Sheridan, 1966; Sheridan, 1971). To date, ionic coupling has been widely documented in a variety of normal and abnormal tissues (Loewenstein & Penn, 1967; Sheridan, 1970; Bennett, 1973) as well as between many different cell types grown in culture (Furshpan & Potter, 1968; Borek, Higashino & Loewenstein, 1969; Hyde et al., 1969; Johnson & Sheridan, 1971; Azarnia, Michalke & Loewenstein, 1972; Gilula, Reeves & Steinbach, 1972; Hülser & Webb, 1973). In fact, with only a few exceptions (Azarnia et al., 1972; Gilula et al., 1972; Hülser & Webb, 1973), ionic coupling appears to be a general property of cells grown in culture.

Ionic coupling can be demonstrated between cells by utilizing intra-cellular microelectrode techniques. When two adjacent cells are impaled with microelectrodes (capable of passing current and recording voltage changes), coupling is determined by examining the voltage attenuation of a pulse of current that is injected between the two cells. A slight voltage attenuation indicates the presence of a low-resistance pathway between the current-pulsing electrode in one cell and the voltage-recording electrode in the adjacent cell. Conversely, a severe voltage attenuation indicates a high-resistance barrier between the two microelectrodes. The high-resistance barrier is typical of the normal resistance of a non-specialized plasma membrane; however, the low-resistance pathway is a characteristic feature of a specialized membrane between the adjacent cells. The resistance measurements actually imply that there is cytoplasmic continuity between the cells. This type of specialized mechanism is generally referred to as an electrical synapse, an electrotonic synapse, or a low-resistance synapse. Katz (1966) has suggested that as few as 20 cytoplasmic bridges (10 nm in diameter) could be responsible for the low-resistance measurements at these sites.

The current that is transferred between ionically coupled cells actually reflects the transfer of small ions, primarily K^+, Na^+, and Cl^-. On the basis of the physiological observations, the pathway for ionic transfer must have water-like properties, i.e. a hydrophilic character. Based on the hydrated ion size for molecules like K^+, Na^+, and Cl^-, the low-resistance pathway should contain 'channels' that are at least 1–1.5 nm in diameter.

The permeability properties of the ionic coupling channels have been examined by ionophoretic injections of several exogenous materials, primarily fluorescent dyes (Loewenstein, 1966; Furshpan & Potter, 1968; Bennett, 1973). These studies have extended the permeability of the coupling channels to include molecules such as fluorescein (mol. wt 330),

Procion Yellow (mol. wt 625), and Chicago Sky Blue (mol. wt 993), in addition to Na^+, K^+, and Cl^-. In most cases, ionic permeability is associated with dye permeability. However, in certain developing embryos, ions can be transferred while dyes can not (Slack & Palmer, 1969; Bennett, Spira & Pappas, 1972; Tupper & Saunders, 1972). A recent study has indicated that cyclic AMP (mol. wt 330) can also be transferred between cells, presumably via the ionic coupling mechanism (Tsien & Weingart, 1974).

Thus far, no direct evidence has been obtained to indicate that ionic coupling requires a localized enzymatic activity. In addition, the transfer of material, such as dyes, follows a time course of passive diffusion. In most cases, the ionic coupling is bidirectional between cells. Ironically, one of the few exceptions to this exists at the crayfish giant motor synapse, the first electrical synapse that was described (Furshpan & Potter, 1959). In this case, the coupling is primarily unidirectional, and the structure is referred to as a rectifying electrical synapse.

Metabolic coupling

The cell-to-cell transfer of metabolites was first described by Subak-Sharpe, Burk & Pitts (1969). They initially termed this phenomenon as 'metabolic cooperation between cells'. In their initial observations, these authors clearly demonstrated that this type of metabolic exchange required direct physical contact between cells.

The phenomenon of metabolic coupling was discovered in culture by co-cultivating two types of cells that differed with respect to their abilities to metabolize (incorporate) exogenous purines into their nucleic acids. One cell type (normal or wild type) had the metabolic capacity to incorporate the exogenous purine, while the other cell type (mutant) was defective in this capacity. The authors observed that upon co-cultivation the normal cell could transmit the metabolic capacity to the mutant cell after they physically interacted. At present, it is clear that a metabolic product, a nucleotide, is transferred from the normal to the mutant cell during this process (J. Pitts, unpublished observations). Recent studies by John Pitts and others (Rieske, Schubert & Kreutzberg, 1975) indicate that a variety of molecules, such as amino acids, sugars, phosphorylated sugars, nucleotides, and cyclic nucleotides, may be transferred between cells.

This phenomenon is relatively insensitive to a variety of treatments that perturb normal cell functions (Cox, Krauss, Balis & Dancis, 1974), and it is widespread between cells in culture (Cox et al., 1970; Pitts, 1972; Goldfarb, Slack, Subak-Sharpe & Wright, 1974). Furthermore, the exchange of molecules is also on the time scale of passive diffusion. One estimate by John Pitts (unpublished observation) indicates that about 10^5–10^6 molecules sec^{-1} can be transferred between cells.

In two different studies (Azarnia *et al.*, 1972; Gilula *et al.*, 1972), it has been demonstrated that metabolically coupled cells are also ionically coupled. This fact was further strengthened by the observation in those studies that communication-incompetent cells lack both ionic and metabolic coupling. Thus, the present evidence indicates that a communication-competent phenotype possesses the ability to transfer both ions and metabolites. Therefore, the communication pathway is probably the same for both metabolic and ionic coupling, with some qualitative qualifications.

Biological relevance

Cell-to-cell communication clearly plays a significant role in excitable tissues. In the myocardium, communication is essential for the synchronization of myocardial cell beating. Without communication, the myocardium cannot function properly (DeHaan & Sachs, 1972). In the nervous system, electrical interactions between neurones rely on the communication event. In many of these instances, electrical communication provides the neuronal network with a mechanism for responding rapidly to stimuli (Bennett, 1972). The synaptic delay that is present in chemical communication does not exist in ionic communication. In recent studies, there have been indications that cell communications may play a regulatory role in two important differentiation processes: (1) myoblast fusion in muscle differentiation (Rash & Fambrough, 1973); and (2) neuromuscular junction formation (Fischbach, 1972).

In non-excitable tissues the biological role of communication is less clear. Different studies have led to suggestions that communication plays a significant role in developing and adult tissues (for review, Bennett, 1973) and in regulating growth in normal and neoplastic tissues (Loewenstein & Kanno, 1967; Johnson & Sheridan, 1971; Azarnia, Larsen & Loewenstein, 1974). Since the presence of communication can provide a mechanism for the exchange of small molecules between cells, it is extremely attractive to associate communication with a number of different regulatory functions. However, at present, there is no significant information that enables us to understand a specific biological function of communication between non-excitable cells. Warner & Lawrence (1973) have found that communication, as well as the gap junctional distribution (Lawrence & Green, 1975), does not perform an obvious function in the differentiation (pattern formation) of abdominal segments in the insect *Oncopeltus*, while in an earlier study on squid embryo development, Potter, Furshpan & Lennox (1966) found that a loss of communication *is* temporally associated with organogenesis. Therefore, much of the future efforts to understand the biological role of communication may rely on the development of more sensitive probes and the characterization of qualitative differences in communication, such as those previously reported in developing embryos (Slack & Palmer, 1969; Tupper & Saunders, 1972; Bennett *et al.*, 1972).

THE COMMUNICATION PATHWAY – THE GAP JUNCTION

Structural characterization

The gap junction has been implicated as the structural pathway for cell-to-cell communication by several different studies that combined electro-physiological and electron microscopic techniques (Dreifuss *et al.*, 1966; Payton *et al.*, 1969; Hyde *et al.*, 1969; Johnson & Sheridan, 1971; Rose, 1971; Gilula *et al.*, 1972; Hülser & Demsey, 1973; Azarnia *et al.*, 1974). These studies have demonstrated this prevalent role of gap junctions between invertebrate cells (Payton *et al.*, 1969; Rose, 1971), vertebrate cells (Dreifuss *et al.*, 1966), and between a variety of cells in culture (Johnson & Sheridan, 1971; Gilula *et al.*, 1972; Hülser & Demsey, 1973; Azarnia *et al.*, 1974). There are a large number of pleiomorphic forms of gap junctions that have been described (for reviews, McNutt & Weinstein, 1973; Gilula, 1974*a*; Staehelin, 1974). However, they all share basic structural features with the exception of the gap junctions in arthropods. The arthropod gap junctions are strikingly different from those found in other phyla (Flower, 1972; Peracchia, 1973*b*; Johnson, Herman & Preus, 1973; Gilula, 1974*a*), and this structural difference may be responsible for certain functional differences in communication.

The gap junction, in its present form, was first described by Revel & Karnovsky in 1967. Prior to that time, the structure (in various forms) was studied in several different tissues (Dewey & Barr, 1962; Robertson, 1963). One of these studies led to the use of the term 'nexus' to describe the junction between smooth muscle cells (Dewey & Barr, 1962). Currently, the term nexus is used synonymously with the term gap junction. In thin-section electron microscopy, the gap junction appears as a complex of two adjacent plasma membranes separated by a small space or gap (Plates 1 and 2). The gap is about 2–4 nm and the entire width of the junction is 15–19 nm. The gap between the two junctional membranes can be penetrated by electron-dense materials, such as lanthanum and ruthenium red, to reveal the presence of a polygonal lattice of 8–9 nm subunits (Revel & Karnovsky, 1967). The gap junction exists as a plaque-like contact between cells, and the junction size and number can vary considerably between different cells. In thin sections, the arthropod gap junctions share the same characteristic features (Plate 4) (Payton *et al.*, 1969; Hudspeth & Revel, 1971; Rose, 1971; Peracchia, 1973*a*).

With freeze-fracture electron microscopy, the gap junctional membrane is characterized by two internal complementary elements (Plate 2). The inner membrane half contains a polygonal lattice of 8–8.5 nm intramembrane particles, while the outer membrane half contains a complementary arrangement of pits or depressions (Chalcroft & Bullivant, 1970; Goodenough & Revel, 1970; McNutt & Weinstein, 1970; Gilula & Satir, 1971). Most gap junctions contain these fracture face components, with the exception of the arthropod junction. The arthropod gap junction is characterized by a loose arrangement of 10–30 nm particles on the outer membrane half, and a complementary arrangement of pits or depressions on the inner membrane half (Plate 4) (Flower, 1972; Johnson *et al.*, 1973; Peracchia, 1973*b*; Gilula, 1974*a*). In essence, the particles are not normally homogeneous in size, and they are associated with the opposite fracture face.

Negative-stain studies on isolated gap junctions have also revealed a similar polygonal lattice of 8–8.5 nm particles (Plate 3) (Benedetti & Emmelot, 1968; Goodenough & Stoeckenius, 1972; Zampighi & Robertson, 1973). In certain preparations, a small 1.5–2 nm electron-dense dot occupies the central region of the 8–8.5 nm particles. This 1.5–2 nm region has been interpreted in thin sections and freeze-fracturing to represent the site of the hydrophilic 'channel' for cell-to-cell communication (Payton *et al.*, 1969; McNutt & Weinstein, 1970). It has been possible recently to demonstrate that the 1.5–2 nm region corresponds to a continuous channel that traverses the entire junctional width in appropriate negative stain preparations (Gilula, 1974*b*).

Biochemical characterization

Gap junctions have been isolated as enriched subcellular fractions, primarily from rat and mouse liver (Benedetti & Emmelot, 1968; Goodenough & Stoeckenius, 1972; Evans & Gurd, 1972; Gilula, 1974*b*). Some of these fractions are highly enriched; however, the only criterion for purity, at present, is ultrastructural analysis.

Goodenough & Stoeckenius (1972) have devised a procedure based on enzymatic digestion (collagenase and hyaluronidase) and detergent (Sarkosyl NL 97) treatment to obtain a highly enriched fraction of gap junctions from mouse liver. This fraction contains both lipid and protein. The lipid fraction includes both phospholipid and neutral lipid. The protein was initially separated into two major electrophoretic bands: one corresponding to 36000 daltons, and the other to 18000 daltons. Subsequently, Goodenough reported that the 18000 dalton protein can be resolved into two

10 000 dalton proteins by reduction with mercaptoethanol or dithiothreitol (Goodenough, 1974).

In our laboratory, we have found that a purified fraction of gap junctions from rat liver contains two major proteins that can be resolved electrophoretically (with reduction and alkylation) (Gilula, 1974*b*). One of these proteins has a mobility that corresponds to 25 000 daltons, while the other corresponds to 18 000 daltons. Amino acid analysis of these two proteins indicates that they have very different amino acid compositions. The 25 000 dalton protein is relatively hydrophobic, while the 18 000 dalton protein is hydrophilic. We are currently trying to relate the amino acid contents of these two proteins to the hydrophilic communication property of the gap junction.

ROLE OF CALCIUM IN CELL COMMUNICATION

It has been firmly established, primarily with invertebrate tissues, that calcium can play an important regulatory role in cell-to-cell communication. In some instances, magnesium or strontium can mimic the effects of calcium. Since much of the data concerning calcium and communication has been obtained from invertebrate tissues, we have decided to discuss the results obtained on invertebrates and vertebrates separately. Also, it is important to recall that the gap junctions in arthropods are very different from those in mammalian tissues (Gilula, 1974*a*).

Invertebrate studies

Practically all of the information about calcium and cell communication has been generated by the extensive studies of Werner Loewenstein and his colleagues. In 1966, they found that removal of calcium from a classic cell system, the *Chironomus* salivary gland epithelium, caused an uncoupling of the previously coupled cells (Nakas, Higashino & Loewenstein, 1966; Loewenstein, 1966). From these observations, they suggested that the cellular content of calcium was an important factor in intercellular communication; hence a depletion of intracellular calcium leads to a loss of coupling. Uncoupling can also be the result of long-term exposure ($>$ 1 h) to calcium-, magnesium-free medium (Rose & Loewenstein, 1971). In subsequent studies, Loewenstein and his colleagues found that elevated levels of intracellular calcium (estimated to be > 5–8×10^{-5} M) also was associated with a loss of coupling (Loewenstein, Nakas & Socolar, 1967; Politoff, Socolar & Loewenstein, 1969; Oliveira-Castro & Loewenstein, 1971). The calcium levels for these studies were experimentally elevated by (1) ionophoretic injection of calcium; (2) producing a hole in the surface membrane; and (3) using a metabolic inhibitor that affects the pumping mechanism for maintaining a low intracellular calcium concentration. Thus, coupling could be regulated by maintaining a normal level of intracellular calcium, while an elevation or depression of calcium levels causes uncoupling (Loewenstein, 1967). A similar uncoupling effect of varying calcium has been reported for the crayfish septate axon (Asada & Bennett, 1971).

Recently, Rose & Loewenstein (1975) have elegantly demonstrated a specific effect of free intracellular calcium on the coupling (permeability) between *Chironomus* salivary gland cells. They utilized the calcium-specific bioluminescence of the compound aequorin (Shimomura & Johnson, 1967) to demonstrate that uncoupling is associated with an elevation in cytoplasmic calcium. Cytoplasmic calcium levels were elevated by (1) utilizing an energy metabolism inhibitor, cyanide; (2) applying ionophores; and (3) direct intracellular calcium injections. All three treatments caused an increased calcium level that was clearly defined by the aequorin luminescence. Further, from the resolution of this system, they could demonstrate that the uncoupling occurs when the elevated calcium concentration approaches the junctional region. In effect, the calcium-induced uncoupling appears to result from a local effect of calcium on the junctional region.

Mammalian studies

In general, there is little information available on the role of calcium in mammalian coupled cell systems. One study clearly demonstrated that a calcium–magnesium chelator, EDTA, has no effect on coupling between cells in the heart (Dreifuss *et al.*, 1966). Recently, there have been reports indicating that intracellular calcium injections can uncouple cells in canine Purkinjé fibres (De Mello, 1974) and human lymphocytes (Oliveira-Castro & Barcinski, 1974). In the Purkinjé fibres, De Mello has observed that magnesium does not cause uncoupling, while injections of sodium and strontium, in addition to calcium, are effective uncouplers (De Mello, 1974, 1975).

Recent observations

We have been interested in comparing the effects of calcium on insect cells and mammalian cells grown in culture since (1) these cells from two different phyla have phenotypically different junctions; and (2) we have already observed that little, if any, significant communication or ionic coupling exists between co-cultivated insect and mammalian cells (Epstein & Gilula, 1975). For these reasons, we are attempting to ascertain if the

PLATE I

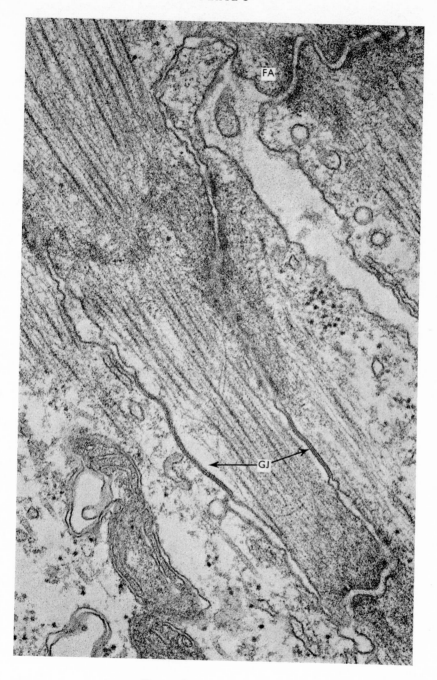

For explanation see p. 272

PLATE 2

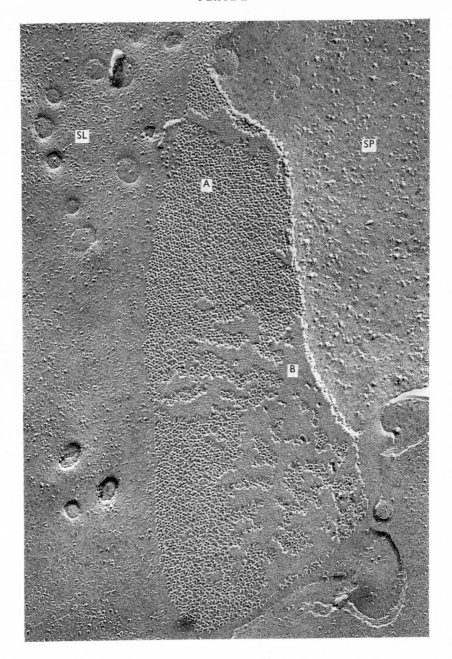

For explanation see p. 272

PLATE 3

For explanation see p. 272

PLATE 4

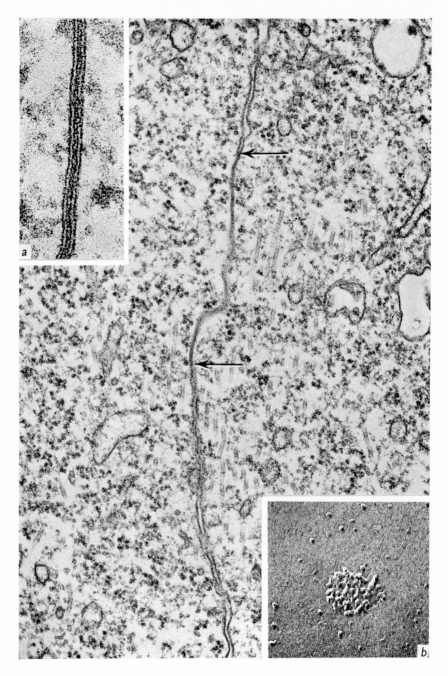

For explanation see p. 272

PLATE 5

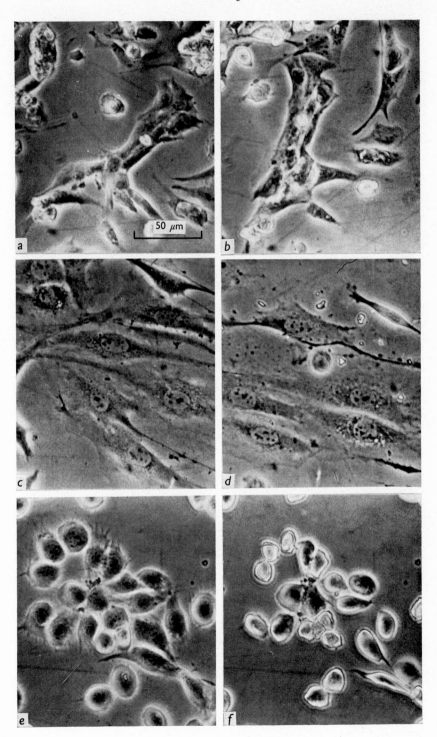

50 μm

a

b

c

d

e

f

For explanation see p. 272

substantiated role of calcium in insect cell coupling is also involved in mammalian cell coupling.

For these studies, we have tried to elevate the cytoplasmic calcium levels by treating three cell culture populations with the calcium ionophore A23187. The three cell populations are (1) TN cells – insect-derived cell line (McIntosh, Maramorosch & Rechtoris, 1973); (2) 3T3 mouse cell line; and (3) MC cells – neonatal mouse ventricular cells (3 days in culture). The ionophore A23187 appears to be relatively selective for calcium and magnesium over other cations (Reed & Lardy, 1972), although a recent report indicates that the compound can form a complex with leucine (Hovi, Williams & Allison, 1975). The A23187 ionophore produces responses in a number of biological systems where calcium is involved, such as fluid secretion in fly salivary glands (Prince, Rasmussen & Berridge, 1973) and histamine release from mast cells (Foreman, Mongar & Gomperts, 1973). Most significantly, the study of Rose & Loewenstein (1975) demonstrates that the ionophore increases intracellular calcium, leading to uncoupling in *Chironomus* salivary gland cells.

The TN insect cell line is rapidly affected by treatment with the iono-phore at concentrations of 10^{-5} M and 2×10^{-6} M (Table 1). Within minutes (10–15), coupling is lost between the cells (Fig. 1) and there is a profound change in cell morphology (Plate 5). Essentially all of the insect cells become round and uncoupled after a 1 h exposure to the ionophore. These observations are definitely compatible with the results of Rose & Loewen-stein (1975) that indicate a rise in cytoplasmic calcium leads to uncoupling.

Ionophore-treated 3T3 fibroblasts respond quite differently. There is practically no effect on coupling of a high concentration (10^{-5} M) for a short period of time (0.5 h). After 1.5 h at the same concentration, a significant reduction in coupling is present (Fig. 1, Table 1). At this time, vacuoles are present in some of the cells and processes begin to retract (Plate 5). Lower concentrations of the ionophore (2×10^{-6} M) have a minimal effect on coupling even after 3 h (Table 1). These results indicate that the mammalian 3T3 cells have a significantly different sensitivity to ionophore-induced uncoupling.

Myocardial cells continue to beat synchronously after 3 h treatment with 10^{-5} M ionophore (Table 1). Hence, ionic coupling between myocardial cells is practically unaffected by this treatment (Fig. 1, Table 1). In addi-tion, the ionophore does not significantly alter the myocardial cell morpho-logy (Plate 5).

In conclusion, the ionophore treatment significantly uncouples the insect cells in culture, and this is presumably due to an elevated cytoplasmic calcium concentration. However, the ionophore has a reduced effect on

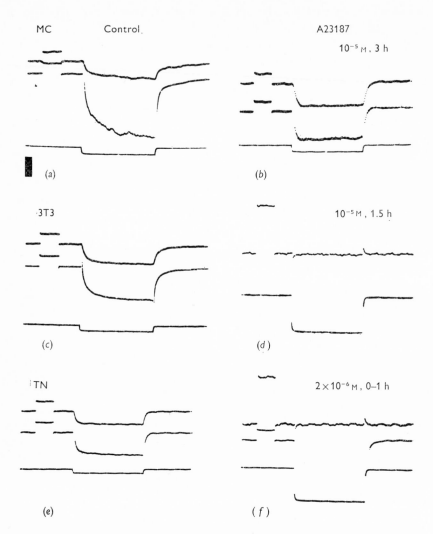

Fig. 1. Electrophysiological records from control and ionophore-treated cultures of myocardial (mouse), 3T3 (mouse), and TN (insect) cells. (*a*) Ionic coupling is observed between non-adjacent myocardial cells. A pulse of current whose intensity is given in the bottom trace is injected by means of a bridge circuit through a micro-electrode into one cell. This current produces a transmembrane voltage deflection in that cell (V_1, middle trace) and in a different cell (V_2, upper trace). The voltage deflection on the upper trace indicates current can flow from one cell to another (ionic coupling). (In *a* and *b* the bridge did not balance so the voltage deflections shown on the middle trace do not represent the actual voltage deflections in that cell.) The calibration pulse at the beginning of the voltage traces represents 5 mV and 20 msec and the vertical line indicates 5×10^{-9} A for all records. Recordings in (*a*) and (*b*) are from 3-day-old cultures taken from mouse neonatal ventricles. (*b*) Ionic coupling between non-adjacent myocardial cells incubated for 3 h in

Table 1. *Effect of ionophore A23187 on ionic coupling*

Cell type	Ionophore concentration (M)	Time (h)	Coupling Before	Coupling After
TN (insect)	1×10^{-5}	0–1	4/4	0/2
	2×10^{-6}	0–1	4/5	1/7
	2×10^{-6}	0–1	8/8	0/6
			12/13	1/13
3T3 (mouse)	1×10^{-5}	0.5	8/8	9/10
	1×10^{-5}	1.5		5/17
	2×10^{-6}	3		14/14
	2×10^{-6}	3		5/10
				19/24
MC (mouse)	1×10^{-5}	2		5/6
	1×10^{-5}	3		7/7
	1×10^{-5}	3		4/5
				16/18

The coupling results are expressed as the number of cell pairs coupled/number of cell pairs examined.

coupling between mouse 3T3 cells, and this may result from either (1) a toxic effect of the ionophore, or (2) a reduced sensitivity to the ionophore-induced calcium permeability. Finally, the ionophore has practically no effect on myocardial cell coupling, and this may indicate (1) that the ionophore is ineffective on myocardial cells, or (2) that an elevated intra-cellular calcium concentration does not result in myocardial cell uncoupling. Further experiments are in progress to attempt to clarify the possible explanations for the differential effects of the ionophore on the insect and mammalian cell populations. These observations certainly increase our suspicions that different junctional phenotypes may have different per-meability control mechanisms.

Fig. 1 (*cont.*)

10^{-5} M A23187 at 37 °C. (*c*) Coupling records from adjacent control 3T3 cells. (*d*) No ionic coupling is observed between adjacent 3T3 cells treated with 10^{-5} M A23187 for 1.5 h at 37 °C as indicated by the lack of a voltage deflection. Note the greater current intensity and voltage amplification in this record. (*e*) Ionic coupling between adjacent control TN cells. (*f*) Lack of coupling between adjacent TN cells treated with 2×10^{-6} M A23187 for 0–1 h at room temperature. The current intensity and voltage amplification for the upper trace have been increased.

268   N. B. GILULA AND M. L. EPSTEIN

The authors gratefully acknowledge the excellent assistance of Asneth Kloesman and Eleana Sphicas. These studies were supported by the Andrew Mellon Fund, the Irma T. Hirschl Trust, and a US Public Health Service Research grant (no. HL 16507). The ionophore was a generous gift of Dr Otto K. Behrens, Eli Lilly Research Laboratories, Indianapolis, Indiana. The insect cells were kindly supplied by Dr A. H. McIntosh of Rutgers University.

REFERENCES

ASADA, Y. & BENNETT, M. V. L. (1971). Experimental alteration of coupling resistance at an electrotonic synapse. *J. Cell Biol.*, **49**, 159–172.
AZARNIA, R., LARSEN, W. J. & LOEWENSTEIN, W. R. (1974). The membrane junctions in communicating and noncommunicating cells, their hybrids, and segregants. *Proc. natn. Acad. Sci. USA*, **71**, 880–884.
AZARNIA, R., MICHALKE, W. & LOEWENSTEIN, W. R. (1972). Intercellular communication and tissue growth. VI. Failure of exchange of endogenous molecules between cancer cells with defective junctions and noncancerous cells. *J. Memb. Biol.*, **10**, 247–258.
BARR, L., DEWEY, M. M. & BERGER, W. (1965). Propagation of action potentials and the structure of the nexus in cardiac muscle. *J. gen. Physiol.*, **48**, 797–823.
BENEDETTI, E. L. & EMMELOT, P. (1968). Hexagonal array of subunits in tight junctions separated from isolated rat liver plasma membranes. *J. Cell Biol.*, **38**, 15–24.
BENNETT, M. V. L. (1966). Physiology of electrotonic junctions. *Ann. N.Y. Acad. Sci.*, **137**, 509–539.
(1972). A comparison of electrically and chemically mediated transmission. In *Structure and Function of Synapses*, ed. G. D. Pappas & D. P. Purpura, pp. 221–256. Raven: New York.
(1973). Function of electrotonic junctions in embryonic and adult tissues. *Fedn Proc.*, **32**, 65–75.
BENNETT, M. V. L., SPIRA, M. E. & PAPPAS, G. D. (1972). Properties of electrotonic junctions between embryonic cells of *Fundulus*. *Develop. Biol.*, **29**, 419–435.
BOREK, C., HIGASHINO, S. & LOEWENSTEIN, W. R. (1969). Intercellular communication and tissue growth. IV. Conductance of membrane junctions of normal and cancerous cells in culture. *J. Memb. Biol.*, **1**, 274–293.
CHALCROFT, J. P. & BULLIVANT, S. (1970). An interpretation of liver cell membrane and junction structure based on observations of freeze-fracture replicas of both sides of the fracture. *J. Cell Biol.*, **47**, 49–60.
COX, R. P., KRAUSS, M., BALIS, M. E. & DANCIS, J. (1970). Evidence for transfer of enzyme product as the basis of metabolic cooperation between tissue culture fibroblasts of Lesch-Nyhan and normal cells. *Proc. natn. Acad. Sci. USA*, **67**, 1573–1579.
(1974). Metabolic cooperation in cell culture: Studies of the mechanisms of cell interaction. *J. Cell Physiol.*, **84**, 237–252.
DEHAAN, R. L. & SACHS, H. G. (1972). Cell coupling in developing systems: The heart-cell paradigm. *Curr. Topics Develop. Biol.*, **7**, 193–228.
DE MELLO, W. (1974). Intracellular Ca injection and cell communication in heart. *Fedn Proc.*, **33**, 445.

(1975). Uncoupling of heart cells produced by intracellular sodium injection. *Experientia*, **31**, 460–461.

DEWEY, M. M. & BARR, L. (1962). Intercellular connection between smooth muscle cells: the nexus. *Science, Wash.*, **137**, 670–672.

DREIFUSS, J. J., GIRARDIER, L. & FORSSMANN, W. G. (1966). Propagation de l'excitation dans le ventricule de rat du moyen de solutions hypertoniques. *Pflügers Arch. ges. Physiol.*, **292**, 13–33.

EPSTEIN, M. L. & GILULA, N. B. (1975). Does cell-to-cell communication exist between co-cultured mammalian and arthropod cells? *J. Cell Biol.*, **67**, 109a.

EVANS, W. H. & GURD, J. W. (1972). Preparation and properties of nexuses and lipid-enriched vesicles from mouse liver plasma membranes. *Biochem. J.*, **128**, 691–700.

FISCHBACH, G. D. (1972). Synapse formation between dissociated nerve and muscle cells in low density cell cultures. *Develop. Biol.*, **28**, 407–429.

FLOWER, N. E. (1972). A new junctional structure in the epithelia of insects of the order Dictyoptera. *J. Cell Sci.*, **10**, 683–691.

FOREMAN, J. C., MONGAR, J. L. & GOMPERTS, B. D. (1973). Calcium ionophores and movement of calcium ions following the physiological stimulus to a secretory process. *Nature, Lond.*, **245**, 249–251.

FURSHPAN, E. J. & POTTER, D. D. (1959). Transmission at giant motor synapses of the crayfish. *J. Physiol., Lond.*, **143**, 289–325.

(1968). Low-resistance junctions between cells in embryos and tissue culture. *Curr. Topics Develop. Biol.*, **3**, 95–127.

GILULA, N. B. (1974a). Junctions between cells. In *Cell Communication*, ed. R. P. Cox, pp. 1–29. John Wiley and Sons, Inc.: New York.

(1974b). Isolation of rat liver gap junctions and characterization of the polypeptides. Abstract. *J. Cell Biol.*, **63**, 111a.

GILULA, N. B., REEVES, O. R. & STEINBACH, A. (1972). Metabolic coupling, ionic coupling, and cell contacts. *Nature, Lond.*, **235**, 262–265.

GILULA, N. B. & SATIR, P. (1971). Septate and gap junctions in molluscan gill epithelium. *J. Cell Biol.*, **51**, 869–872.

GOLDFARB, P. S. G., SLACK, C., SUBAK-SHARPE, J. H. & WRIGHT, F. D. (1974). Metabolic co-operation between cells in tissue culture. *Symp. Soc. exp. Biol.*, **28**, 463–484.

GOODENOUGH, D. A. (1974). Bulk isolation of mouse hepatocyte gap junctions. *J. Cell Biol.*, **61**, 557–563.

GOODENOUGH, D. A. & REVEL, J. P. (1970). A fine structural analysis of intercellular junctions in the mouse liver. *J. Cell Biol.*, **45**, 272–290.

GOODENOUGH, D. A. & STOECKENIUS, W. (1972). The isolation of mouse hepatocyte gap junctions. Preliminary chemical characterization and X-ray diffraction. *J. Cell Biol.*, **54**, 646–656.

HOVI, T., WILLIAMS, S. C. & ALLISON, A. C. (1975). Divalent cation ionophore A23187 forms lipid soluble complexes with leucine and other amino acids. *Nature, Lond.*, **256**, 70–72.

HUDSPETH, A. J. & REVEL, J. P. (1971). Co-existence of gap and septate junctions in an invertebrate epithelium. *J. Cell Biol.*, **50**, 92–101.

HÜLSER, D. & DEMSEY, A. (1973). Gap and low-resistance junctions between cells in culture. *Z. Naturf.*, **28c**, 603–606.

HÜLSER, D. & WEBB, D. J. (1973). Relationship between ionic coupling and morphology of established cells in culture. *Expl Cell Res.*, **80**, 210–222.

HYDE, A., BLONDEL, B., MATTER, A., CHENEVAL, J. P., FILLOUX, B. & GIRARDIER, L. (1969). Homo- and heterocellular junctions in cell structures: an electrophysiological and morphological study. *Prog. Brain Res.*, **31**, 283–311.

JOHNSON, R. G., HERMAN, W. S. & PREUS, D. M. (1973). Homocellular and heterocellular gap junctions in *Limulus*: a thin-section and freeze-fracture study. *J. Ultrastruct. Res.*, **43**, 398–412.

JOHNSON, R. G. & SHERIDAN, J. D. (1971). Junctions between cancer cells in culture: ultrastructure and permeability. *Science, Wash.*, **174**, 717–719.

KATZ, B. (1966). *Nerve, Muscle and Synapse*. McGraw-Hill: New York.

LAWRENCE, P. A. & GREEN, S. M. (1975). The anatomy of a compartment border. The intersegmental boundary in *Oncopeltus*. *J. Cell Biol.*, **65**, 373–382.

LLINÁS, R., BAKER, R. & SOTELO, C. (1974). Electrotonic coupling between neurons in cat inferior olive. *J. Neurophysiol.*, **37**, 560–571.

LOEWENSTEIN, W. R. (1966). Permeability of membrane junctions. *Ann. N.Y. Acad. Sci.*, **137**, 441–472.

(1967). Cell surface membranes in close contact. Role of calcium and magnesium ions. *J. Colloid Interface Sci.*, **25**, 34–46.

LOEWENSTEIN, W. R. & KANNO, Y. (1967). Intercellular communication and tissue growth. I. Cancerous growth. *J. Cell Biol.*, **33**, 225–234.

LOEWENSTEIN, W. R., NAKAS, M. & SOCOLAR, S. (1967). Junctional membrane uncoupling. Permeability transformations at a cell membrane junction. *J. gen. Physiol.*, **50**, 1865–1891.

LOEWENSTEIN, W. R. & PENN, R. D. (1967). Intracellular communication and tissue growth. II. Tissue regeneration. *J. Cell Biol.*, **33**, 235–242.

LOEWENSTEIN, W. R., SOCOLAR, S. J., HIGASHINO, Y., KANNO, Y. & DAVIDSON, N. (1965). Intercellular communication: renal, urinary, bladder, sensory and salivary gland cells. *Science, Wash.*, **149**, 295–298.

McINTOSH, A. H., MARAMOROSCH, K. & RECHTORIS, C. (1973). Adaptation of an insect cell line (*Agallia constricta*) in a mammalian cell culture medium. *In Vitro*, **8**, 375–378.

McNUTT, N. S. & WEINSTEIN, R. S. (1970). The ultrastructure of the nexus. A correlated thin-section and freeze-cleave study. *J. Cell Biol.*, **47**, 666–687.

(1973). Membrane ultrastructure at mammalian intercellular junctions. *Prog. Biophys.*, **26**, 45–101.

NAKAS, M., HIGASHINO, S. & LOEWENSTEIN, W. R. (1966). Uncoupling of an epithelial cell membrane junction by calcium ion removal. *Science, Wash.*, **151**, 89–91.

OLIVEIRA-CASTRO, G. M. & BARCINSKI, M. A. (1974). Calcium-induced uncoupling in communicating human lymphocytes. *Biochim. biophys. Acta*, **352**, 338–343.

OLIVEIRA-CASTRO, G. M. & LOEWENSTEIN, W. R. (1971). Junctional membrane permeability. Effects of divalent cations. *J. Memb. Biol.*, **5**, 51–77.

PAYTON, B. W., BENNETT, M. V. L. & PAPPAS, G. D. (1969). Permeability and structure of junctional membranes at an electronic synapse. *Science, Wash.*, **166**, 1641–1643.

PERACCHIA, C. (1973a). Low-resistance junctions in crayfish. I. Two arrays of globules in junctional membranes. *J. Cell Biol.*, **57**, 54–65.

(1973b). Low resistance junctions in crayfish. II. Structural details and further evidence for intercellular channels by freeze-fracture and negative staining. *J. Cell Biol.*, **57**, 66–76.

PITTS, J. D. (1972). Direct interaction between animal cells. In *Cell Interactions*, Proceedings of the third Lepetit Colloquium, ed. L. G. Silvestri, pp. 277–285. North-Holland: Amsterdam & London.

POLITOFF, A. L., SOCOLAR, S. J. & LOEWENSTEIN, W. R. (1969). Permeability of a cell membrane junction. Dependence on energy metabolism. *J. gen. Physiol.*, **53**, 498–515.

POTTER, D. D., FURSHPAN, E. J. & LENNOX, E. S. (1966). Connections between cells of the developing squid as revealed by electrophysiological methods. *Proc. natn. Acad. Sci. USA*, **55**, 328–336.

PRINCE, W. T., RASMUSSEN, H. & BERRIDGE, M. J. (1973). The role of calcium in fly salivary gland secretion analyzed with the ionophore A-23187. *Biochim. biophys. Acta*, **329**, 98–107.

RASH, J. E. & FAMBROUGH, D. (1973). Ultrastructural and electrophysiological correlates of cell coupling and cytoplasmic fusion during myogenesis *in vitro*. *Develop. Biol.*, **30**, 166–186.

REED, P. W. & LARDY, H. A. (1972). A23187: a divalent cation ionophore. *J. biol. Chem.*, **247**, 6970–6977.

REVEL, J. P. & KARNOVSKY, M. J. (1967). Hexagonal array of subunits in intercellular junctions of the mouse heart and liver. *J. Cell Biol.*, **33**, C7–C12.

RIESKE, E., SCHUBERT, P. & KREUTZBERG, G. W. (1975). Transfer of radioactive material between electrically coupled neurons of the leech central nervous system. *Brain Res.*, **84**, 365–382.

ROBERTSON, J. D. (1963). The occurrence of a subunit pattern in the unit membranes of club endings in Mauthner cell synapses in goldfish brains. *J. Cell Biol.*, **19**, 201–221.

ROSE, B. (1971). Intercellular communication and some structural aspects of membrane junctions in a simple cell system. *J. Memb. Biol.*, **5**, 1–19.

ROSE, B. & LOEWENSTEIN, W. R. (1971). Junctional membrane permeability. Depression by substitution of Li for extracellular Na, and by long-term lack of Ca and Mg; restoration by cell repolarization. *J. Memb. Biol.*, **5**, 20–50.

(1975). Permeability of cell junction depends on local cytoplasmic calcium activity. *Nature, Lond.*, **254**, 250–252.

SHERIDAN, J. D. (1966). Electrophysiological study of special connections between cells in the early chick embryo. *J. Cell Biol.*, **31**, C1–C5.

(1970). Low-resistance junctions between cancer cells in various solid tumors. *J. Cell Biol.*, **45**, 91–99.

(1971). Electrical coupling between fat cells in newt fat body and mouse brown fat. *J. Cell Biol.*, **50**, 795–893.

SHIMOMURA, O. & JOHNSON, F. H. (1967). Properties of the bioluminescent protein aequorin. *Biochemistry, N.Y.*, **8**, 3991–3997.

SLACK, C. & PALMER, J. F. (1969). The permeability of intercellular junctions in the early embryo of *Xenopus laevis*, studied with a fluorescent tracer. *Exptl Cell Res.*, **55**, 416–419.

STAEHELIN, L. A. (1974). Structure and function of intercellular junctions. *Int. Rev. Cytol.*, **39**, 191–283.

SUBAK-SHARPE, J. H., BURK, R. R. & PITTS, J. D. (1969). Metabolic cooperation between biochemically marked mammalian cells in tissue culture. *J. Cell Sci.*, **4**, 353–367.

TSIEN, R. W. & WEINGART, R. (1974). Cyclic AMP: cell-to-cell movement and ionotropic effect in ventricular muscle, studied by a cut-end method. *J. Physiol., Lond.*, **242**, 95–96P.

TUPPER, J. T. & SAUNDERS, J. W., JR (1972). Intercellular permeability in the early *Asterias* embryo. *Develop. Biol.*, **27**, 546–554.

WARNER, A. E. & LAWRENCE, P. A. (1973). Electrical coupling across developmental boundaries in insect epidermis. *Nature, Lond.*, **245**, 47–48.

ZAMPIGHI, G. & ROBERTSON, J. D. (1973). Fine structure of the synaptic discs separated from the goldfish medulla oblongata. *J. Cell Biol.*, **56**, 92–105.

EXPLANATION OF PLATES

PLATE I

Myocardial cells cultured from neonatal mouse ventricular myocardium. In thin-sections, the myocardial cells form an intercellular specialization, the intercalated disc, that is comprised of desmosomal elements (FA, fascia adhaerens) and gap junctions (GJ). The desmosomal elements are utilized for cell-to-cell adhesion and for the insertion of myofilaments. The gap junction or nexus provides a structural pathway for cell-to-cell communication. × 61 800.

PLATE 2

Freeze-fracture replica of mouse ventricular myocardium. The fracture process has exposed the intramembrane features of the sarcolemma (SL) and the gap junction (A and B). The sarcoplasm is also apparent (SP). The sarcolemma membrane is characterized by a random distribution of smooth regions and heterogeneous particles, while the gap junction exists as a plaque of homogeneous 8–8.5 nm particles that are frequently polygonally packed. The inner membrane half (A) contains particles, while the complementary outer membrane half (B) contains pits or depressions. Note that the particle plaque is frequently interrupted by smooth regions. × 100 800.

PLATE 3

Negative-stain and thin-section appearance of gap junctions isolated from rat liver. (*a*) Isolated gap junctions after negative staining with 1 % uranyl formate. The entire isolated membrane structure consists of 8–8.5 nm subunits (particles). A 1.5–2 nm electron-dense region is present in the centre of the subunits. This 1.5–2 nm region corresponds to the proposed location of hydrophilic channels within the intact junctional membrane. (*b*) The structural integrity of gap junctions can be maintained during isolation. In thin-sections, the isolated junction is a complex of two plasma membranes separated by a 2–4 nm gap. (*a*) × 240 000; (*b*) × 260 000.

PLATE 4

Gap junctions between cells of the TN insect cell line. The gap junctions between TN cells are usually small; however, large ones, such as the one that extends between the two arrows in this micrograph, are also present. Inset (*a*): in thin-sections, the junction is comprised of two membranes separated by a 4 nm gap. Thus, the TN cell gap junctions closely resemble those present in non-arthropod tissues (cf. Plate 3). Inset (*b*): the freeze-fractured gap junction between TN cells exists as a plaque with heterogeneous particles (10–30 nm in diameter) associated with the outer membrane half. Complementary depressions are present on the inner membrane half (not shown in this image). The internal membrane organization of the TN cell gap junctions is therefore strikingly different from the non-arthropod structure (cf. Plate 2). × 51 500; inset (*a*) × 205 200; inset (*b*) × 100 000.

PLATE 5

Phase contrast photomicrographs of cultured myocardial (mouse), 3T3 (mouse), and TN (insect) cells before and after treatment with the calcium–magnesium ionophore A23187. (*a*) Aggregate of synchronously beating neonatal mouse myocardial cells. (*b*) Beating myocardial cells incubated for 3 h at 37 °C in the presence of 10^{-5} M A23187. No morphological changes are detectable, and the cells continue to beat synchronously. (*c*) Control 3T3 cells. (*d*) 3T3 cells after 1.5 h at 37 °C in the presence of 10^{-5} M A23187. Vacuoles are present in some cells at this time. (*e*) Control TN cells. (*f*) Same TN cells as in (*e*), after a 15 min incubation in 2×10^{-6} M A23187, at room temperature. This treatment produces distinct morphological changes in cell processes and size of the TN cells. All images are at the same magnification.

THE SPASMONEME AND CALCIUM-DEPENDENT CONTRACTION IN CONNECTION WITH SPECIFIC CALCIUM BINDING PROTEINS

By W. B. AMOS, L. M. ROUTLEDGE,
T. WEIS-FOGH and F. F. YEW

Department of Zoology, University of Cambridge,
Downing Street, Cambridge CB2 3EJ

The chief interest of the spasmoneme, a contractile organelle found in certain ciliated protozoa, lies in the fact that calcium ions appear to drive the contraction rather than merely serve as a signal. The process is fundamentally different from muscular contraction, the beating of flagella and cytoplasmic streaming.

It has long been known that a type of contraction can occur when protein fibres are exposed to high salt concentrations. In the case of collagen, the contraction is due to the destruction of hydrogen bonds and is readily reversed by washing out the salt. The high concentrations necessary for this effect make it unlikely that it is of any physiological significance. However, Levine (1956) and Hoffmann-Berling (1958) discovered that the spasmoneme of *Vorticella* contracted dramatically at micromolar calcium ion concentrations and extended in calcium-free solution, a process which, like the contraction of collagen, could be repeated apparently indefinitely. This and subsequent work (Amos, 1971) has indicated that in these protozoa the energy for extension and contraction is supplied to the contractile apparatus in the form of a change in intracellular calcium ion concentration. According to this theory, the change in concentration is produced by an active mechanism, such as a membrane calcium pump, which consumes metabolic energy. The idea that the calcium ion concentration in itself could serve as an important medium for the storage or transmission of cellular energy is unfamiliar, perhaps because normal intracellular calcium ion levels are considered too low for a significant effect. However, the low concentration is no bar to this mechanism, and quantitative evidence which supports the theory has now been obtained (Routledge, Amos, Gupta, Hall & Weis-Fogh, 1975).

The molecular basis of this effect of calcium is of obvious interest and relevance, but progress on this aspect is hampered by the small size of the organelles in most ciliates and the difficulty of separating the contractile substance from extraneous material. Our rediscovery of a ciliate with a giant organelle (Weis-Fogh & Amos, 1972) has made it possible to examine the effect of calcium on the mechanical and optical properties of the spasmoneme directly, after dissecting the organelle out of the cell. A new type of calcium binding protein, which we have named spasmin (Amos, Routledge & Yew, 1975), has been isolated from the giant organelle. Spasmin is almost certainly part of the contractile mechanism of the spasmoneme, and it may well have some relation to the calcium-regulated proteins described elsewhere in this symposium.

THE SPASMONEME

Although many ciliated protozoa contain contractile fibres, particularly in the orders Gymnostomatida, Peritrichida, Heterotrichida and Entodiniomorphida, the fibres are most amenable to observation in the peritrich family Vorticellidae. The cell body in these organisms is attached to a solid support, such as a submerged leaf, by a stalk. A contractile organelle, named the spasmoneme by Entz (1892) runs longitudinally within the stalk. The spasmoneme is readily visible with the light microscope because of its high mass concentration (20 % w/v) and because it is not obscured by any cytoplasmic structures except the spherical mitochondria which are clustered against it (Plate 1a, b, c). When the spasmoneme contracts the stalk is thrown into a plane curve, a zig-zag or a helix, depending on species. This is because the sheath is never quite symmetrical: it is stiffened on one side by extracellular fibres (Fauré-Fremiet, 1905; Randall & Hopkins, 1962; Amos, 1972). The fine structure and calcium-sensitivity of the spasmoneme appears to be the same in all vorticellid species.

In the common, helically coiling species *Vorticella* and *Carchesium*, the diameter of the spasmoneme is usually 1 μm and 8–10 μm respectively, though it may be more than 1 mm long in both species. These organisms exist as separate cellular entities, though in *Carchesium* they are linked together by extracellular material. Both genera have been used in studies of the calcium-induced contraction, but we have also made extensive use of *Zoothamnium geniculatum* which is especially suitable for several types of experiment because it is large enough for the spasmoneme to be dissected out of the cell by hand (Weis-Fogh & Amos, 1972). This species forms branching colonies (Plate 2) containing several thousand cell bodies or zooids which individually resemble *Vorticella* but have cytoplasmic continuity throughout the colony. Each zooid is connected by a small branch of spasmonemal material to the main spasmoneme trunk, which is

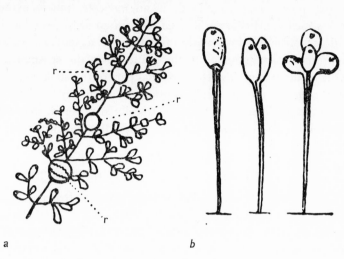

a   b

Fig. 1. Drawings of a large species of *Zoothamnium*. (*a*) A branch of the colony, showing the ordinary zooids and the spherical reproductive individuals (r). (*b*) Division of the reproductive cell, which has settled and secreted a stalk approximately 60 μm in diameter and destined to become the giant main stalk of the colony (from Trembley, 1748).

1 mm long and 30–40 μm in diameter. If a glycerinated *Zoothamnium* colony is cut at two points with fine scissors, the spasmoneme can quite easily be pulled out of the stalk sheath with needles (Plate 2*b*). It is a substantial rubbery rod, which remains intact during stretching and other manipulations.

The origin of the giant spasmoneme can be understood in terms of the development of *Zoothamnium*, which was first observed by Trembley (1748), as discussed by Baker (1952). Certain individuals within the colony swell and become transformed into spherical giant cells, 0.2 mm in diameter (Fig. 1*a*, *b*), which detach themselves and swim away by means of cilia. Each of these cells ultimately settles and develops within a few hours a stalk containing the giant spasmoneme, which seems to be derived from preformed granular material. Trembley's discovery of the division of the giant zooid at this stage into daughter cells (Fig. 1*b*) is among the first observations of cell division. In spite of this distinction, and the fine studies by Engelmann (1875), Wesenberg-Lund (1925) and Furssenko (1929) the giant species of *Zoothamnium* have not attracted the attention of physiologists, though they seem to be highly suitable for experimental work. The species we have used is *Zoothamnium geniculatum* Ayrton, which owes its name to the presence of a knee-joint in the main stalk of the colony (Plate 2*a*). This joint (which the species observed by Trembley and

Engelmann seems to have lacked) is an intricate, totally extracellular structure which is articulated like the limb of an arthropod. An inelastic extracellular structure serves the function of a tendon, transmitting tension from the spasmoneme to the joint. When the spasmoneme contracts, the result is complex: the proximal part of the stalk coils helically, as in *Vorticella* and the joint flexes through more than 90°. The pattern of bending is clearly determined by the extracellular material, since when the spasmoneme is dissected out it shows a simple linear contraction.

VELOCITY AND ENERGY OF CONTRACTION

A microscope produces an illusion of high velocity. It makes quite slow events, such as the passage of the bending wave down the tail of a sperm at 0.8 mm sec^{-1} (Gray, 1955) impossible to follow by eye. Perhaps because of this, the unusually high speed of contraction of the spasmoneme was not appreciated until recently (Ueda, 1954). Jones, Jahn & Fonseca (1970) obtained high speed films of *Vorticella* and found that contraction took only 4 msec whereas extension took several seconds. Their measurements, indicating a very small degree of shortening of the spasmoneme, are probably in error because of the inevitably poor resolution of the high-speed film, since in *Carchesium*, which has an almost identical but larger stalk, the spasmoneme became reduced to 36 % of its initial length (Amos, 1972). The organelle is so large in *Zoothamnium* that it can be measured accurately in high-speed films. It contracts to 45 % of its initial length and the rate of contraction may be as high as 172 lengths per second, as in the specimen shown in Plate 2c. Even with its natural viscous and elastic loads increased by compression between slide and coverslip the *Zoothamnium* spasmoneme shown in Plate 2c contracted more than 15 times faster (in lengths per second) than the fastest known striated muscle unloaded (Close, 1965).

Rahat, Pri-Paz & Parnas (1973) have succeeded in measuring the force developed by the stalk of living specimens of *Carchesium*, and obtained values for the tension in the spasmoneme of 4×10^4 and 8×10^4 N m^{-2}. This agrees reasonably well with an indirect estimate made by calculation from the film data of Jones *et al.* (1970) by the following method (Amos, 1971): when *Vorticella* contracts, the cell body, which may be approximated to a sphere of radius 20 μm, may be pulled 80 μm at an average velocity of 23 mm sec^{-1}. Assuming a viscosity of 10^{-3} N sec m^{-2} (0.01 poise), the viscous drag on the body may be calculated from Stokes' formula ($F = 6\pi a \eta v$) as 0.86×10^{-8} N. If the radius of the spasmoneme is taken as 0.5 μm, the tension per unit cross-sectional area becomes 1.1×10^4 N m^{-2}.

(a minimum value since it takes no account of energy lost in overcoming the resistance of the stiffening fibres or the internal viscosity of the stalk).

These observed and calculated tensions in the spasmoneme are low in comparison to the maximum isometric tensions of muscles from a variety of animals, which are in the range 10^5 to 10^6 N m^{-2} (Prosser & Brown, 1962) but the spasmoneme tension is probably of the same order as the tensions in working muscles. It is the speed of shortening that marks out the spasmoneme as quite different from muscle. Because of this high rate of shortening the spasmoneme achieves a high power output (2.7 kW per kg of wet weight) but this is maintained for only a small fraction of the contraction/relaxation cycle. Repeated contractions can occur in *Vorticella* but they are separated by periods of extension at least 500 times longer than the contraction phase.

EFFECT OF CALCIUM AND OTHER DIVALENT CATIONS

Levine reported in 1956 that calcium and apparently also magnesium and manganese ions could induce contraction in glycerinated *Vorticella*. He found that the contraction could be reversed by treatment with the calcium-chelating agent ethylene diamine tetra-acetic acid (EDTA) and, most significantly, the cycle of contraction and relaxation could be repeated many times by repeating the treatment, without the need to add ATP or any other apparent fuel. Hoffmann-Berling (1958) made similar experiments, but added EDTA to the solution at a slightly lower concentration than the divalent ion under test in order to reduce the concentration of calcium ions which he suspected were present as a contaminant. With this precaution, he found that Mg^{2+} and Be^{2+} were not effective in inducing contraction. Less than 10^{-5} M Ca^{2+} was sufficient to cause contraction. Among the other alkaline earth ions, Sr^{2+} was effective, Ba^{2+} less so.

Hoffmann-Berling discovered that these effects of calcium were not prevented by cyanide or by the mercurial poison salyrgan (mersalic acid). He suggested that the molecular basis of the effect might be that calcium ions were able to neutralise negative charges on long polymer molecules, releasing the molecules from electrostatic repulsive forces and allowing them to fold thermokinetically. He observed more complex effects when ATP and magnesium ions were present in the solution but it was subsequently found (Amos, 1971) that these effects were abolished if the calcium level was adequately controlled by an EGTA buffer (EGTA = ethyleneglycol bis (β-aminoethyl ether)-N,N'-tetra-acetic acid). For the convenience of other workers, the apparent association constants of the calcium and magnesium complexes of EDTA and EGTA are tabulated in the Appendix (p. 298).

In the presence of a Ca–EGTA buffer, the glycerinated *Vorticella* stalks coiled if the free calcium level was above 4×10^{-7} M and remained coiled indefinitely if the calcium level was kept high, implying that the spasmoneme maintained tension indefinitely under these circumstances, since the elastic sheath was present. If the calcium level was lowered to 10^{-8} M the stalks extended. In these glycerinated preparations the stalks took several seconds to contract and extend. It was confirmed that the stalks could be taken through many successive cycles of contraction and extension, as many as 35 being recorded over a period of 2 days at room temperature. Detergents (digitonin, saponin, Tween 80), metabolic inhibitors (KCN, dinitrophenol, fluorodinitrobenzene) and mercurials (mersalic acid, *para*-chloromercuribenzoate) were without effect on the number of cycles that could be obtained, even when present continuously throughout the experiment. Lanthanum and terbium induced contraction but the threshold concentrations were not measured.

Magnesium ions did not alter the threshold level for calcium, even when present in great excess, though a high concentration of magnesium (5×10^{-2} M) blocked contraction irreversibly. It seems unlikely that enough endogenous ATP was present to drive the contractile process in these experiments by some mechanism which is resistant to such a wide range of inhibitors. However, it may seem at first sight that the alternative hypothesis, that the difference in chemical potential of calcium ions which exists between the high and low calcium solutions serves as the source of energy, requires an unreasonably large amount of calcium to be bound by the contracting organelle. The following argument shows that this is not so.

The energy output of the spasmoneme in a single contraction can be calculated from the velocity and tension (Amos, 1971), as previously mentioned. It is 11 joule per kg of wet weight. The glycerinated spasmoneme progresses from almost full extension to contraction as the calcium level is increased over a 100-fold range (see Fig. 7). It is reasonable to assume that the cell could vary its internal calcium ion level a 100-fold, for instance from 10^{-8} to 10^{-6} M, which would involve a chemical potential change of approximately 10^4 joule per mole of calcium ions. This is calculated from the equation

$$\Delta\mu_{Ca} = RT \log_e \frac{[Ca^{2+}]_{upper}}{[Ca^{2+}]_{lower}},$$

where $\Delta\mu_{Ca}$ is the change in chemical potential, R the gas constant and T the absolute temperature. To produce the observed output of energy the spasmoneme must bind at least $11 \div 10^4$ mole of calcium per kg of wet weight, or 0.04 g kg^{-1}: a modest amount. It can be seen that to produce such an effect, a cell need not generate a high concentration of calcium ions. It need only be able to vary the level over a large range, which the calcium pump in the sarcoplasmic reticulum of muscle is known to be capable of doing (Hasselbach, Makinose & Fiehn, 1970).

MICROPROBE MEASUREMENTS OF CALCIUM CONTENT

Recently it has been possible to measure the amount of calcium taken up in contraction by the glycerinated spasmoneme directly by means of an electron microprobe (Routledge *et al.*, 1975). This instrument allows one to measure the mass fraction of calcium in a specimen, that is the number of grams of calcium per kilogram of dry mass (Hall, 1971). The measurements were made on the isolated spasmoneme of *Zoothamnium geniculatum*, which can easily be positioned and located for microprobe analysis (Plate 3a). Before being used for this purpose the *Zoothamnium* colonies were glycerinated for 1–3 months in a 50 % glycerol medium containing 0.1 M KCl, 4×10^{-3} M EDTA, 2×10^{-2} M histidine buffer, pH 7.0. Dissection was performed in the glycerol medium, and resulted in preparations containing chiefly the contractile material with a small amount of the surrounding cytoplasm.

Extended spasmonemes were obtained by rinsing in a low calcium solution containing 5×10^{-2} M KCl, 2×10^{-2} M Tris-HCl buffer, pH 7.0, and a Ca–EGTA buffer adjusted to give a free calcium ion level of 10^{-8} M. To obtain contracted organelles a similar solution was used but with 10^{-6} M Ca^{2+}. In both cases the organelles were dried after blotting with pieces of Millipore filter to remove as much fluid as possible. In the microprobe analyser, the calcium K$_\alpha$ radiation was selected by means of a diffracting spectrometer. A continuum signal, giving a measure of the local mass per unit area, was obtained from a second spectrometer set in a background position near to the potassium line or from a solid-state detector.

As well as calcium bound to the organelle before drying, a dried specimen might contain an accumulation of calcium deposited in it along with other solutes during the evaporation of the surrounding fluid. In order to distinguish between true binding and an accretion of this type, the total calcium concentrations in the solutions with 10^{-6} and 10^{-8} M free calcium were made equal. Accretion should have raised the calcium content of the contracted and extended organelles equally. With a total calcium level of 2×10^{-3} M, the dried contracted organelles nevertheless contained signifi-

Table 1. *Microprobe measurements of calcium in the spasmoneme of Zoo-thamnium. The figures in parentheses are standard errors (from Routledge et al., 1975)*

Total calcium $M \times 10^{-6}$	Calcium in organelle (g kg^{-1} dry mass)		Calcium bound during contraction (g kg^{-1} dry mass)
	Contracted	Extended	
0	—	0.21 (0.04)	—
21	1.39 (0.06)	0.27 (0.07)	1.12 (0.09)
104	1.58 (0.27)	0.46 (0.13)	1.12 (0.30)
208	2.59 (0.09)	0.89 (0.06)	1.70 (0.11)
2075	3.81 (0.22)	2.12 (0.12)	1.69 (0.25)

cantly more calcium than the extended ones: 3.81 g per kg of dry mass as against 2.12 g kg^{-1} (as on the bottom line of Table 1).

The question arises of whether the calcium content of the extended organelles is due entirely to accretion. To test this point, solutions were used with various total calcium concentrations but with the free calcium ion levels in each pair of solutions at 10^{-6} and 10^{-8} M as before. As would be expected if it were due to accretion, the calcium content of the extended organelles was found to decline as the total calcium concentration of the solutions was reduced (see Table 1 and Fig. 2). However, there was evidence for some strong specific binding of calcium to the extended organelle, namely that the decline was not linear and the calcium content was not zero even at very low levels of total calcium. The difference in calcium content between contracted and extended organelles remained approximately constant when the total calcium concentration of the solutions was varied. This difference presumably represents a binding of calcium which occurs during contraction, amounting to 1.7 g of calcium per kg of dry mass. Since the glycerinated spasmoneme contains approximately 20 % dry mass (measured by interference microscopy) the calcium taken up by the hydrated organelle on contraction is 0.34 g kg^{-1}. This is more than eight times the amount needed to supply the work done against viscous forces on the chemical potential theory, which, as mentioned above, requires 0.04 g kg^{-1}.

It may be objected, however, that several crude approximations are made in this comparison. The work done by the spasmoneme is clearly under-estimated, since only the external viscosity is taken into account. Also, the thermodynamic efficiency cannot be 100 %. On the other hand, the calcium uptake measured in the microprobe may be expected to be lower than in life, since the spasmoneme's ability to bind calcium may well become reduced during glycerination.

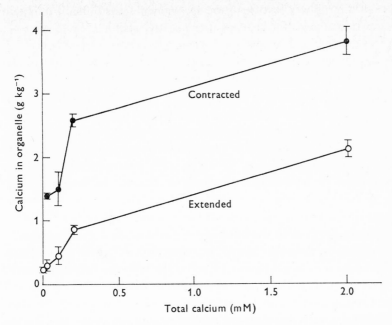

Fig. 2. Measurements of the calcium content of spasmonemes made with the JEOL electron probe microanalyser. The total calcium concentration of the bathing solution, including complexed forms, is plotted on the abscissa. The contracted spasmonemes were dried down from solutions with 10^{-6} M Ca^{2+}. The extended ones were from solutions with 10^{-8} M Ca^{2+}, except for the point on the ordinate, where a 2×10^{-2} M EDTA solution was used without any addition of calcium. Each point represents an average of at least 20 measurements made on five or six different spasmonemes. The limits are standard deviations (from Routledge et al., 1975).

LOCATION AND TRANSPORT OF CALCIUM
IN THE SPASMONEME

The fine structure of the spasmoneme is essentially the same in the three genera of vorticellids that have been studied (Favard & Carasso, 1965; Amos, 1972; Allen, 1973a, b). All the material examined so far with the electron microscope has been contracted, as a result of fixation.

The most likely candidate for the contractile apparatus is the dense mass of filaments which the organelle contains. Each filament has a diameter of 2–3 nm. In spite of the contracted state of the material, these filaments appear chiefly longitudinal in orientation (Plate 1e). There is no membrane separating the filamentous mass from the rest of the cytoplasm, which contains mitochondria and structures which resemble the basal bodies of cilia. In Zoothamnium geniculatum, there are mitochondria embedded in the interior of the spasmoneme as well as in the peripheral layer of cytoplasm.

Because of the rubbery texture of the spasmoneme it is difficult to obtain fragments thin enough for viewing after negative staining. Fragments can sometimes be obtained by soaking isolated glycerinated spasmonemes in distilled water before transferring them to saturated aqueous uranyl acetate as negative stain. These appear to contain filaments of similar size to those seen in sections. They have an indistinct beaded appearance (Plate 4a) with a longitudinal periodicity of about 3.5 nm (Amos, 1975a). Allen (1973b) has found that the corresponding filaments in the peritrich *Opercularia* show a longitudinal periodicity of 12 nm in sectioned material. As yet, the origin of these periodicities is not clear: they may represent subunits, regular thickenings or perhaps cross-bridges. This will be discussed later.

It is of interest in relation to calcium storage that wherever spasmonemal material occurs it has membranous compartments embedded in the filamentous mass or, as in the cell body of *Vorticella* and *Epistylis*, closely pressed against it. Fauré-Fremiet, Favard & Carasso (1962) showed that the membranes of compartments of the latter type are continuous with rough endoplasmic reticular membranes. The spasmonemes in the cell body of *Vorticella*, which are branches of the stalk spasmoneme, have such compartments, each in the form of an extensive flattened sac pressed against the filamentous material on one side. Allen (1973a, b) has discovered intricate bilaterally symmetrical disc-shaped structures on the membrane of this sac, arranged regularly at 1 μm intervals. He has named these *linkage complexes*. They are composed of fibrils of several characteristic types arranged to form a complex structure. One type of fibril appears to form part of the membrane.

The spasmoneme of the stalk also has a system of membranous sacs (Plate 1d), but in this case they are in the form of regular tubules embedded in it (Sotelo & Trujillo-Cenóz, 1959; Favard & Carasso, 1965). They appear not quite longitudinal in sectioned material but some of them have been followed for several micrometres along the length of the organelle. It seems likely that they run helically throughout the entire length of the stalk. But they do not make contact with the plasma membrane or open to the exterior (Amos, 1972). Although the walls of the tubules sometimes present a trilaminar appearance like that of many other cell membranes, this appearance is rare. Allen (1973b) in micrographs of remarkably high resolution, has shown that the reason for this is that fibrous elements identical to those in the linkage complexes are embedded in this membrane, making it a structure of great complexity. The diameter of the tubules is usually 38–70 nm but they can become dilated locally, probably as a result of bad fixation.

The resolution of the microprobe method has not yet allowed us to compare the calcium content of the tubules with that of the intervening filamentous material. However, Carasso & Favard (1966) showed cytochemically that the tubules are capable of accumulating calcium in their lumina. They found crystals in this location after the oxalate treatment, which was developed to explore the distribution of calcium in striated muscle (Costantin, Franzini-Armstrong & Podolsky, 1965). They suggested that the tubules might function like the sarcoplasmic reticulum of muscle in controlling the calcium ion concentration around the filaments, which they considered analogous to myofilaments.

To summarise: the chief components of the spasmoneme appear to be *longitudinal filaments*, which are of a single type, and *membranous tubules* which run longitudinally through the mass of filaments, and probably store calcium.

It takes approximately 1 sec for tension to develop in a glycerinated spasmoneme when the bathing solution is changed as rapidly as possible to a high calcium solution. This could be interpreted as the time taken for calcium to diffuse into the interior of the organelle. Clearly, however, the rate of activation of the spasmoneme in life is not limited by diffusion of calcium from the exterior, since it is as rapid in *Zoothamnium geniculatum* where the spasmoneme may be 40 μm in diameter in a large specimen as in *Vorticella* where the organelle is only 1 μm in diameter. The method given by Hill (1948) can be used to calculate that if calcium ions were released instantaneously at the surface of a spasmoneme 40 μm in diameter it would take 100 msec for their concentration at the centre to reach 1/100 of the final concentration. Yet *Zoothamnium* starts to develop tension 4 msec after an electrical stimulus and is fully contracted within a further 4 msec. Diffusion from outside is obviously inadequate to supply the calcium. Release from the internal tubules seems likely, since the diffusion distance is in that case never greater than 200 nm.

Whatever the internal source of calcium, a rapid conduction of a signal for contraction into the interior of the spasmoneme must occur, and this remains completely mystifying, since no transverse membranous structures which might correspond to the T-system in striated muscle have been described in the spasmoneme: the tubules are longitudinal. The only visible entities which run from the cell surface into the interior of the spasmoneme are fibrils, parts of Allen's linkage complex, which he suggests are involved in conducting the activation signal. He also suggests that the other parts of the linkage complex which appear to be embedded in the tubule membrane are sites of calcium pumping activity (Allen, 1973b). It is interesting that the branches of the spasmoneme which pass into the cell body can be activated independently of the part which lies in the stalk

(Jones *et al.*, 1970) though the branches are identical in fine structure with the stalk spasmoneme and are fully continuous with it (Amos, 1972). This suggests that calcium release can be finely localised in different parts of the cell.

If, as is suggested here, the filamentous mass of the spasmoneme is a mechanochemical engine shortening in response to an increase in calcium ion concentration, it is possible to explain the disproportionately long extension time (of the order of seconds, compared with 2–4 msec for contraction) as the phase when metabolic energy is expended in pumping calcium into the tubule system, while the filamentous mass extends because calcium is removed from it. The rapid release of calcium from the tubules, by means as yet unknown, would cause the filamentous mass to shorten rapidly, in effect releasing chemical potential energy accumulated over a long period by the pump (Amos, 1971).

Jones & Morley (1969) found that the speed of extension increases rapidly with temperature (Q_{10} approximately 2.5 between 10 and 20 °C). The speed was reduced by high external Ca^{2+} or Na^+ concentrations.

Hoffmann-Berling (1958) discovered that ATP, in the absence of a calcium buffer, can produce an extension of the glycerinated spasmoneme even in the presence of millimolar concentrations of calcium. The relaxation is followed by cyclical contractions and extensions. The Mg–ATP complex appears to power these oscillations, since their vigour is greatest with equimolar concentrations of ATP and Mg^{2+} (W. B. Amos, unpublished observations). It was clearly shown by Hoffmann-Berling that these effects could not be due to the chelating action of ATP (i.e. the formation of the Ca–ATP complex) and that they were blocked by ATPase inhibitors. As previously mentioned, these somewhat complex effects disappear if a Ca–EGTA buffer is added to control the calcium level. They vanish if glycerination is prolonged for more than a few days, whereas the calcium-induced contraction survives at least two years of glycerination. The ATP-induced relaxation of glycerinated preparations seems likely to be due to the reactivation of a calcium pump in the tubule membranes.

SPASMONEME PROTEINS

Our work on the protein composition of the spasmoneme has furnished the most direct evidence that this motile system is fundamentally different from others such as muscular contraction or flagellar beating (Amos *et al.*, 1975; Routledge, Amos, Yew & Weis-Fogh, 1976).

Zoothamnium geniculatum has proved the most convenient species for chemical work since the contractile material forms a high proportion of the total dry mass of the colony and a substantial part of it may be isolated by dissection after glycerination. The isolated preparation, consisting chiefly of the spasmoneme, but coated with a thin layer of cytoplasm, shows only a feeble periodic acid–Schiff reaction and probably consists almost entirely of protein. Unlike a myofibril, it is insoluble in KCl solutions but it is evidently not covalently cross-linked since it can be dissolved totally in 1% sodium dodecyl sulphate (SDS) and partially in 8 M urea or 3 M guanidine hydrochloride (GuCl). In order to determine the molecular weights of the proteins present, small numbers of isolated spasmonemes were dissolved in 2% SDS and subjected to electrophoresis in 15% polyacrylamide. Supplies of *Zoothamnium* are unfortunately seasonal and limited, but since the dry mass of the spasmonemal material from a single colony is as much as 0.5 μg, it has been possible to perform electrophoresis on small numbers of colonies. A micro-slab gel apparatus was devised for for this purpose (Amos, 1976). A discontinuous SDS gel system was used, following the formulae of Laemmli & Favre (1973). The resulting pattern (Plate 4*b*) showed a prominent band corresponding to a molecular weight of 20000 which contained 60% of the stainable material. The band was not dissociated into components of lower molecular weight by preheating or adding mercaptoethanol. Most of the remaining protein had molecular weights above 100000. It is particularly significant that when the spasmoneme sample was run in parallel with actin and tubulin in an SDS slab gel, no bands co-migrating with these proteins were found in the spasmoneme pattern (Plate 4*b*). It may be concluded that neither actin nor tubulin, if they are present, forms more than a few per cent of the total spasmoneme protein, that is, the basic elements of myofibrillar and flagellar motility are lacking. Since, during contraction, the spasmoneme equals striated muscle in power output per unit mass (Amos, 1971) it is extremely unlikely that traces of actin or tubulin below the limits of detection are responsible for the contraction. The high power output suggests that our attention should be directed to the major protein components and in particular to the 20000 mol. wt band which is characteristic of the spasmoneme. A recent study (Routledge *et al.*, 1976) has shown that a similar band predominates in the patterns from the spasmonemes of *Vorticella* and *Carchesium*.

Recent work has shown that the 20000 mol. wt band contains a distinct class of calcium binding proteins of a new type, which we have named *spasmin* (Amos *et al.*, 1975). All the major proteins of the spasmoneme were found to dissolve in a 3 M GuCl solution, leaving a remnant from which no protein could be extracted by SDS. Since the proteins remained in solution when freed of salt by dialysis, the spasmin could be subjected to isoelectric focusing. It proved to be quite acid, with an isoelectric point of

4.7–4.8. In this respect it resembles troponin C. When spasmin obtained by GuCl extraction was subjected to electrophoresis in 15 % acrylamide gels without SDS it was resolved into a fast component (A) and a slightly slower component (B). A proved to have a slightly lower molecular weight than B when the two were run in an SDS gel. The relationship between A and B is at present not clear, but since the proportions of A and B remain approximately constant it seems more likely that two types of spasmin are present than that A is a degradation product of B.

A direct test of whether the isolated spasmins bind calcium has not yet been possible because of technical problems arising from the small quantity of material available, but this work is now in progress. An indirect test of this point has been made by measuring the effect of calcium ions on the electrophoretic mobility of spasmin. A series of experiments was performed with polyacrylamide gels containing Ca–EGTA buffers. In a gel buffered at 10^{-8} M Ca^{2+}, the spasmin formed a complex leading band consisting of a prominent peak with a trailing shoulder (see Fig. 3). In another gel with the free calcium ion concentration increased to 10^{-6} M, the leading band was markedly retarded, while the mobility of spasmoneme proteins other than spasmin was unchanged. Also, the leading band became more distinctly divided into two principal components at the higher calcium level (Fig. 3). These effects were not observed with magnesium ions in the same concentration range, though a decrease in mobility occurred with 5×10^{-4} M Mg^{2+}. The decrease in mobility which is induced by calcium with a high degree of specificity must be due to an alteration of charge or conformation. The effect seems likely to have a bearing on the mechanism of contraction, since it occurs in the same range of concentrations of calcium as contraction in glycerinated preparations.

The microprobe measurements allow a comparison of the amount of calcium bound during contraction with the amount of protein present. If it is assumed that all the calcium is bound to the spasmin, which has a molecular weight of 20000, and constitutes between 40 and 60 % of the dry mass, the number of calcium atoms taken up by each spasmin molecule is between 1.4 and 2.1.

An amino acid analysis of microgram samples of spasmin (Amos *et al.*, 1975) has been obtained (by F. F. Yew), using sensitive fluorimetric methods. The spasmin was first eluted from SDS gels and hydrolysed in acid. In one method, the amino acids were separated on an ion exchange column and allowed to react with o-phthaldehyde to produce fluorescent compounds. Proline was determined separately, by converting it to an oxidation product before reaction with o-phthaldehyde.

The results (see Table 2) appear to be similar for the spasmins A and B,

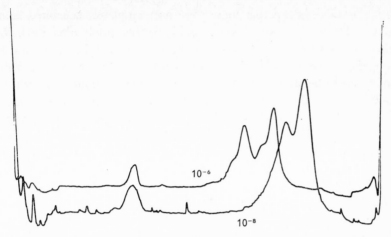

Fig. 3. Reduction of the electrophoretic mobility of spasmin by free calcium ions: superimposed densitometer tracings of stained polyacrylamide gels which were run at 10^{-6} and 10^{-8} M free calcium ion concentrations. The concentrations were maintained by means of a Ca–EGTA buffer. Migration was to the right (anode).

Table 2. *Amino acid composition of components A and B of the 20 000 mol. wt protein band from the spasmoneme of* Zoothamnium (*Amos* et al., *1975*)

The results are expressed as residues per 100 residues. Proline was determined separately from the other amino acids. Each value is the average of 6 determinations. Standard errors are given.

	Fast band (A)	S.E.	Slow band (B)	S.E.
Aspartic acid + asparagine	12.48	1.24	11.30	2.03
Threonine	7.95	0.17	6.20	0.94
Serine	15.05	4.15	12.58	2.25
Glutamic acid + glutamine	8.57	1.09	7.01	0.50
Proline	6.33	1.47	5.83	1.38
Glycine	9.32	2.07	8.90	2.08
Alanine	6.87	0.89	6.99	1.11
Valine	3.98	0.47	4.15	0.74
Cystine	0		0	
Methionine	0		0	
Isoleucine	2.82	0.26	4.08	0.53
Leucine	4.34	0.41	4.92	1.11
Tyrosine	3.89	0.80	3.30	0.76
Phenylalanine	1.95	0.37	2.20	0.74
Tryptophane	1.15	0.23	1.77	0.36
Lysine	6.95	1.10	7.85	1.35
Histidine	3.74	0.71	7.63	1.64
Arginine	4.59	0.71	6.33	2.42

but it should be noted that these were not completely separated before analysis. The amino acid composition is neither predominantly hydrophobic nor hydrophilic. Aspartic acid and glutamic acid, which are not distinguished by the method of analysis from asparagine and glutamine, were found to be abundant, as was serine. There were no significant unknown peaks in the chromatogram. We have found no correspondence between the amino acid composition of spasmin and of proteins from muscle. Unlike the calcium binding parvalbumin from carp muscle and troponin C, spasmin lacks cysteine.

MECHANICAL AND OPTICAL CHANGES
INDUCED BY CALCIUM

The extended spasmoneme is positively birefringent with respect to its length. The birefringence is approximately 4×10^{-3} in the intact spasmoneme of *Carchesium polypinum*. This may be compared with 2.3×10^{-3} for the A band of an intact muscle fibre (Noll & Weber, 1934). Schmidt (1940) observed that in normal and in osmium-fixed *Carchesium* the birefringence falls to zero during contraction. This has been confirmed in similar material and also in caffeine-treated specimens in which extension occurs, but is sufficiently slowed for reliable observations to be made (Amos, 1971). The birefringence in the intact spasmoneme of a living *Zoothamnium geniculatum* colony does not fall to zero during contraction, possibly because the reduction in length during contraction is less in this species than in *Carchesium*. A similar reduction in birefringence occurs in the contractile fibres of the heterotrich ciliate *Stentor* (Kristensen, Nielsen & Rostgaard, 1974). The fall in birefringence strongly suggests a folding rather than a sliding filament mechanism.

An optical and mechanical analysis of the isolated, glycerinated spasmoneme of *Zoothamnium* has been made by Weis-Fogh & Amos (1972). The organelle was suspended between a movable glass plate and a glass platform supported on springy quartz fibres in such a way that it could be viewed in a polarising microscope. The length of the spasmoneme was varied by moving the glass plate and the tension could be measured by the deflection of the platform (Fig. 4).

The spasmoneme proved to be highly extensible. It could be stretched up to four times its resting length with complete elastic recovery, though extension beyond this resulted in a permanent decrease in width. At a high calcium level (10^{-5} M) the organelle behaved mechanically as a rubber. The tensile force plotted against extension fitted a theoretical curve for a long-chain rubber and the birefringence was approximately proportional to

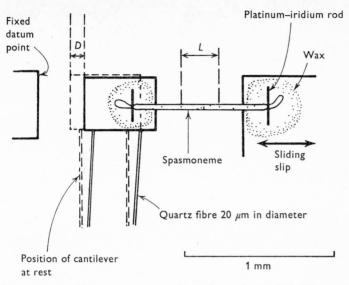

Fig. 4. Apparatus used for measurement of length, tension and birefringence of the isolated glycerinated spasmoneme of *Zoothamnium*. The entire apparatus is assembled on a microscope slide beneath a coverslip. The wax and platinum rods are used to attach the spasmoneme to a glass plate (on the right), the position of which can be adjusted, and to a sprung glass platform (centre). Tension is measured by recording the deflection (D) of the platform (from Weis-Fogh & Amos, 1972).

stress, falling to zero at the unstressed length (see Figs. 5 and 6). This behaviour is usual for a soft rubbery solid. However, when the high calcium solution was replaced by one with a free calcium ion level of 10^{-8} M, the spasmoneme elongated actively, developing a measurable pushing force, and became birefringent even when unstressed. In spite of this profound change in the nature of the material at low calcium levels, it remained extensible, though the length/tension curve moved to the right (Fig. 5). The Figure shows a series of experimental points which represent equilibrium states determined by any two of the three variables; length, tension and calcium ion concentration. By varying the length and calcium level the spasmoneme could be made to follow an anticlockwise loop on the length/tension diagram, that is, to do work, proportional to the area of the loop. Although the positions of the in-vivo resting and contracted lengths on the abscissa of Fig. 5 were not determined, it may be pointed out that the area between the curves defined by the limits $\lambda = 1$ and $\lambda = 2$ represents approximately 20 joule per kg of wet weight of spasmoneme. It is quite consistent with the calculated figure of 11 joule kg^{-1} for the energy expended by the intact spasmoneme in one in-vivo twitch against viscous

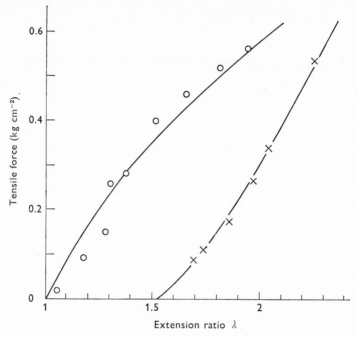

Fig. 5. Relation between length and tension in an isolated spasmoneme from a glycerinated *Zoothamnium* colony, at 10^{-5} M Ca^{2+} (○) and at 10^{-8} M (×). The extension ratio, λ, is the length expressed as a fraction of the unstressed length at 10^{-5} M Ca^{2+}. The preparation was allowed to equilibrate for several minutes before each measurement was made. The line through the low calcium points was drawn by eye, but that through the high calcium points is the theoretical curve for a long-chain rubber (see Weis-Fogh & Amos, 1972).

forces. This gives us confidence that a substantial part of the contractile mechanism survives in the glycerinated material.

There is some evidence that the unusual property of active extension may occur in living peritrichs as well as in the experiments with glycerinated preparations described here. Engelmann (1875) described how, if the stalk sheath of *Zoothamnium* was physically prevented from unfolding after contraction by some obstruction, the spasmoneme nevertheless elongated and became bent into a sinuous form within the confines of the stalk. A process of this type has been observed and photographed in *Carchesium* (Plate 1*b*, *c*). The contractile fibres in the ciliate *Stentor* also become sinuous in the early stages of extension (Bannister & Tatchell, 1968; Huang & Pitelka, 1973; Kristensen *et al.*, 1974), possibly for the same reason.

In spite of the high shortening rate of the spasmoneme, equilibrium measurements on the glycerinated spasmoneme do not indicate that calcium ions act cooperatively in producing shortening. If the slack length

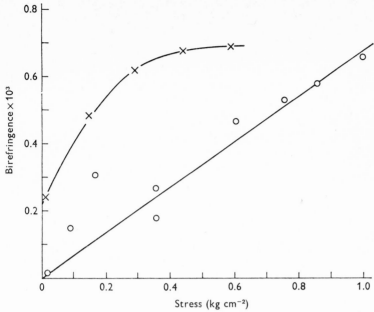

Fig. 6. Variation of birefringence with stress in the same preparation as was used for Fig. 5. In the high calcium solution (○) the birefringence is proportional to stress, but this is not so in the low calcium medium (×) (from Weis-Fogh & Amos, 1972).

of the spasmoneme is plotted against calcium concentration (Fig. 7), the maximum gradient of the curve is consistent with a simple kinetic model in which the binding of one calcium ion produces unit shortening.

MOLECULAR BASIS OF
THE CALCIUM-INDUCED CONTRACTION

Several lines of evidence now lead to the conclusion that the contraction of the spasmoneme is brought about by a direct interaction between calcium and the contractile apparatus. It seems likely that the basic contractile event is a conformational change which the characteristic protein, spasmin, undergoes when it binds calcium ions. The affinity of spasmin for divalent cations fits this role: it binds calcium in the range of concentrations where contraction in the glycerinated preparation occurs and it has a high specificity for calcium relative to magnesium, which neither induces contraction nor competes with calcium. The change in electrophoretic mobility of spasmin may be a consequence of a conformational change which occurs when calcium becomes bound.

The microprobe experiments allow the tentative conclusion that a small

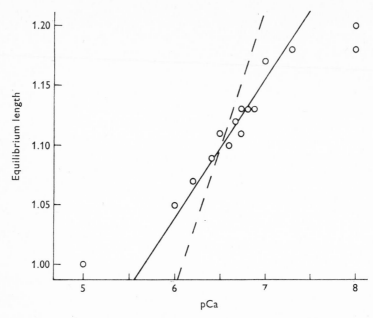

Fig. 7. Variation of unstressed length with calcium ion concentration in the isolated glycerinated spasmoneme of *Zoothamnium* (pCa = $-\log_{10}[\text{Ca}^{2+}]$). The lengths are given relative to the maximally contracted length. The solid line shows the gradient at the centre of the sigmoid curve expected if the binding of one calcium ion produces unit shortening, while the dashed line shows the gradient if two are required for each shortening event (the gradient increases in direct proportion to the number of ions required) (from Amos, 1975*b*).

number (one or two) calcium ions are bound to each molecule. In this respect, and also in being an acidic protein of low molecular weight, spasmin resembles carp muscle parvalbumin (Kretsinger & Nockolds, 1973) and troponin C (Hartshorne & Pyun, 1971). In these proteins (Collins, this volume; Kretsinger, 1975) it is known that the calcium atoms are bound by an octahedral arrangement of coordinate bonds. The binding of calcium to troponin C produces a substantial structural change (Van Eerd & Kawasaki, 1972).

On the grounds of abundance, it seems likely that spasmin is present in the filaments of which the spasmoneme is largely composed. Assuming a protein density of 810 daltons nm^{-3} (Lake & Leonard, 1974) the diameter of a spherical spasmin molecule is 3.6 nm, which compares well with the longitudinal spacing of 3.5 nm observed in negatively stained filaments. This suggests that the filaments may consist of linear aggregates of spasmin molecules. A linear arrangement has the attraction that it would result in the series summation of unit contractile events occurring independently in each molecule. This could explain the high shortening rate of the spasmoneme relative to muscle, where the units of series summation correspond to half-sarcomeres and are therefore far fewer. A linear arrangement would also be consistent with the simple non-cooperative relation between calcium concentration and degree of shortening.

As yet, no structural change in the filaments in the spasmoneme of vorticellids has been seen with the electron microscope. However, in the heterotrich ciliate *Stentor*, where the contractile fibres shorten by a factor of seven rather than of three, contraction seems to involve a helical coiling of initially straight filaments (Bannister & Tatchell, 1968; Huang & Pitelka, 1973; Kristensen *et al.*, 1974). A similar change in the filaments in vorticellids may occur.

Our knowledge of the supramolecular structure of the spasmoneme is slight. It is not even certain that the filaments contain spasmin. Our attempts to understand the nature of the submolecular process, that is, the change in conformation, are equally hampered by lack of information.

The spasmoneme shows a much higher degree of passive extensibility than a muscle. This rubber or spring-like property is presumably involved in the contraction mechanism. Hawkes & Holberton (1975) have examined the contraction of the ciliate *Spirostomum* against different viscous loads, and find that the force of contraction does not vary with the velocity of shortening. Thus the contractile fibres in this ciliate behave like a spring or a rubber band that has been stretched and is released. In contrast, a muscle produces very little force at high shortening velocities and this has been interpreted as being due to the improbability of cross-bridge attachment at the start of the active stroke under these conditions (Huxley, 1957). Internal damping would not prevent the shortening of an elastic strand at the rate observed: this can be shown by comparison with the properties of the protein rubber resilin (Weis-Fogh, 1975).

The length/tension and birefringence measurements on the isolated spasmoneme show a rubbery behaviour at high calcium levels. Since the organelle is largely protein the molecular explanation of this result may be that polypeptide chains with a high degree of kinetic freedom are present. The effect of thermal agitation is to crumple such a chain so that its effective length is likely to be much smaller than its extended length. The chain can be made straighter by applying a force, but buffeting by solvent molecules tends to bring it back to the more disordered conformation and so produces a restoring force which is proportional to the absolute temperature. From simple assumptions, such as that the individual links in the chain are of constant length and are uniaxial elements of constant birefringence, length/tension and birefringence/stress relations can be deduced rigorously

(Treloar, 1967), and *in the high calcium state* the spasmoneme obeys them (Weis-Fogh & Amos, 1972). However, the proof of the interpretation requires thermal measurements.

The theory of rubbers suffices to explain the high calcium (contracted) state, but seems not quite compatible with the appearance of optical and mechanical anisotropy when calcium is removed from the structure. There are some differences between the spasmoneme and the lightly tanned collagen fibre, with which it has been compared. The collagen fibre becomes rubbery and contracts in the presence of concentrated salt solutions and extends, becoming birefringent, when these are washed out (Pryor, 1950; Katchalsky, Lifson, Michaeli & Zwick, 1960). But the origin of the anisotropy in the collagen is a process of crystallisation, in which hydrogen bonds unite the polypeptides in a precisely ordered three-dimensional structure. The extended collagen is extremely stiff, losing all trace of its previous mechanical behaviour. It has been suggested that the extended spasmoneme is similarly held in an ordered configuration by bonds which can be disrupted by calcium ions (Weis-Fogh & Amos, 1972). However, the extended spasmoneme is only slightly stiffer than it is in the contracted, high calcium, state. If there is considerable kinetic freedom in both states of the spasmin molecule, it is not clear how the bonds could be inserted in an ordered fashion to produce the anisotropy. Also, it is clear that the spasmin molecule must be very different from the collagen polypeptides, since it responds specifically to an ion at a concentration a million times lower than is needed to make collagen contract, and appears to be quite resistant to unspecific alterations, for instance in pH and ionic strength.

Another difficulty for the rubber theory is the recent finding (Fig. 8) that the birefringence of the isolated spasmoneme is largely form birefringence. This type of birefringence is generated by oriented structures suspended in a medium differing from them in refractive index, and it vanishes if a fluid of matching refractive index is infiltrated amongst them. The birefringence of oriented polypeptide chains would be expected to be intrinsic, that is, due to the polarisability of chemical bonds in the polypeptide backbone. This type of birefringence is unaffected by infiltration, provided media are used which do not alter the structure of the polypeptide. In the experiment of Fig. 8, a glycerinated spasmoneme was immersed in glycerol/water mixtures in which the calcium-induced contraction can occur, even in 70 % glycerol. In these relatively innocuous imbibition media, the birefringence of the spasmoneme was greatly reduced, with complete reversibility. Similar results were obtained with sucrose solutions. It is clear from this that the birefringence of the spasmoneme may not be regarded as an

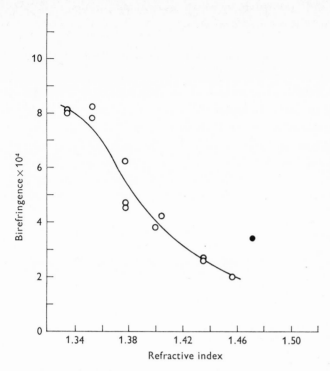

Fig. 8. Variation of birefringence of an isolated, glycerinated spasmoneme of *Zoothamnium* with the refractive index of the solution in which it is placed. The solutions contained o.1 M KCl and a Ca–EGTA buffer at pCa 8, together with various concentrations of glycerol up to 86 % (v/v). The specimen was clamped at a constant, moderately stretched length and the width remained almost constant. The filled circle indicates a measurement made in pure glycerol in which the specimen shrank to 68 % of its normal width (W. B. Amos, unpublished).

indication of the conformation of polypeptide chains and so the agreement of the optical results with rubber theory must be treated with caution.

As an extreme alternative, it may be proposed that the submolecular structure of the spasmoneme is not rubber-like in either state, but instead each subunit has a highly ordered conformation. In this case, extensibility could still exist if the spasmin molecules are linked together to form a helical or folded filament. When the spasmoneme is stretched passively, the bonds within the molecule could be stretched by a small amount, with the effect that the angle of bonding of each molecule to the next is altered slightly. As in a metal spring, small internal energy changes could correspond to a large extension of the structure as a whole. Directional properties are obviously more easily incorporated into this model.

Further analysis of the fine structure of the spasmoneme by electron microscopy and, if possible, by X-ray diffraction is needed to decide which of the two models is nearer the truth.

We thank the Science Research Council for supporting work on the spasmoneme since 1972, and also for a grant to Professor T. Weis-Fogh and Drs P. Echlin, B. L. Gupta, T. A. Hall and R. B. Moreton for the development of the Biological Microprobe Laboratory, in which the microprobe measurements were made.

REFERENCES

ALLEN, RICHARD D. (1973a). Structures linking the myonemes, endoplasmic reticulum and surface membranes in the contractile ciliate *Vorticella*. *J. Cell Biol.*, **56**, 559–579.

(1973b). Contractility and its control in peritrich ciliates. *J. Protozool.*, **20**, 25–36.

AMOS, W. B. (1971). A reversible mechanochemical cycle in the contraction of *Vorticella*. *Nature, Lond.*, **229**, 127–128.

(1972). Structure and coiling of the stalk in the peritrich ciliates *Vorticella* and *Carchesium*. *J. Cell Sci.*, **10**, 95–122.

(1975a). Structure and protein composition of the spasmoneme. In *Comparative Physiology. Functional Aspects of Structural Materials*, ed. L. Bolis, S. H. P. Maddrell & K. Schmidt-Nielsen, pp. 99–104. North-Holland: Amsterdam.

(1975b). Contraction and calcium binding in the vorticellid ciliates. In *Molecules and Cell Movement*, Symposium at the 28th annual meeting of the Society for General Physiology, ed. S. Inoué & R. E. Stephens, pp. 411–436. Raven Press: New York.

(1976). An apparatus for microelectrophoresis in polyacrylamide slab gels. *Analyt. Biochem.*, in press.

AMOS, W. B., ROUTLEDGE, L. M. & YEW, F. F. (1975). Calcium-binding proteins in a vorticellid contractile organelle. *J. Cell Sci.*, **19**, 203–213.

BAKER, J. R. (1952). *Abraham Trembley of Geneva*. Edward Arnold & Co.: London.

BANNISTER, L. H. & TATCHELL, E. C. (1968). Contractility and the fibre systems of *Stentor coeruleus*. *J. Cell Sci.*, **3**, 295–308.

CARASSO, N. & FAVARD, P. (1966). Mise en évidence du calcium dans les myonèmes pédonculaires de ciliés péritriches. *J. Microscopie*, **5**, 759–770.

CLOSE, R. (1965). The relation between intrinsic speed of shortening and duration of the active state in muscle. *J. Physiol.*, *Lond.*, **180**, 542–559.

COSTANTIN, L. L., FRANZINI-ARMSTRONG, C. & PODOLSKY, R. J. (1965). Localization of calcium-accumulating structures in striated muscle fibers. *Science, Wash.*, **147**, 158–160.

ENGELMANN, T. W. (1875). Contractilität und Doppelbrechung. *Pflügers Arch. ges. Physiol.*, **11**, 432–464.

ENTZ, G. (1892). Die elastichen und contractilen Elemente der Vorticellinen. *Math. Natur. ber. Ungarn*, **10**, 1–48.

FAURÉ-FREMIET, E. (1905). La structure de l'appareil fixateur chez les Vorticellidae. *Arch. Protistenk.*, **6**, 207–226.

FAURÉ-FREMIET, E., FAVARD, P. & CARASSO, N. (1962). Étude au microscope électronique des ultrastructures d'*Epistylis anastatica*. *J. Microscopie*, **1**, 287–312.

PLATE I

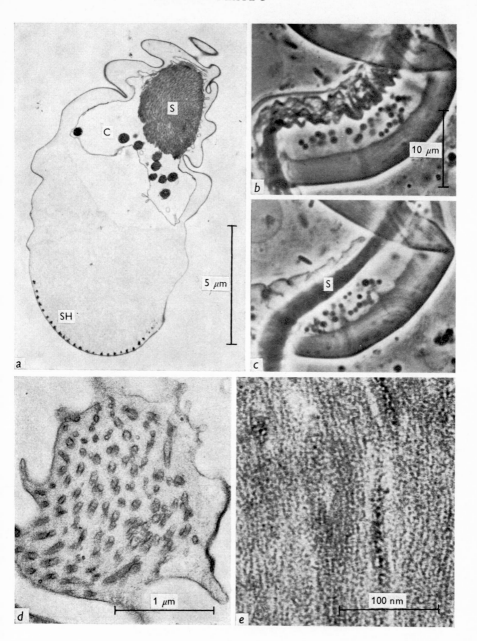

For explanation see p. 300

PLATE 2

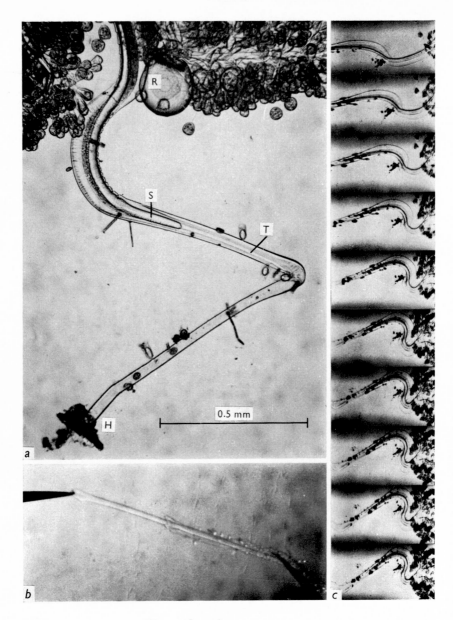

a

R

S

T

0.5 mm

H

b

c

For explanation see p. 300

PLATE 3

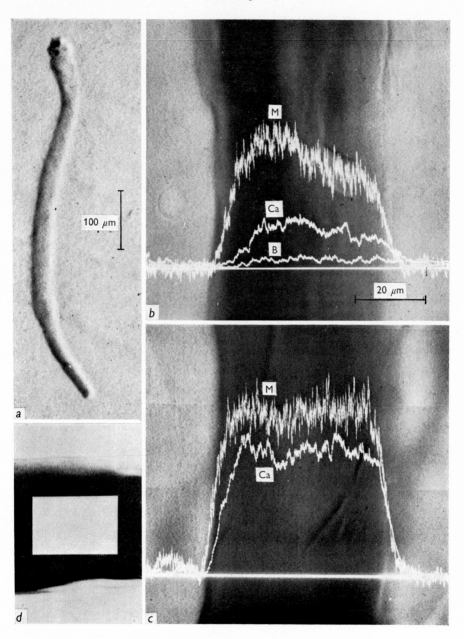

For explanation see pp. 300–1

PLATE 4

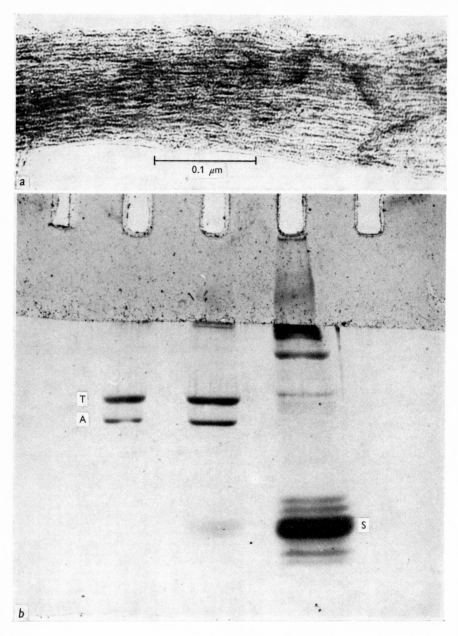

For explanation see p. 301

FAVARD, P. & CARASSO, N. (1965). Mise en évidence d'un réticulum endoplasmique dans le spasmonème de ciliés péritriches. *J. Microscopie*, **4**, 567–572.

FURSSENKO, A. W. (1929). Lebenscyclus und Morphologie von *Zoothamnium arbuscula* Ehrenberg. *Arch. Protistenk.*, **67**, 376–500.

GRAY, J. (1955). The movement of sea urchin spermatozoa. *J. exp. Biol.*, **32**, 775–801.

HALL, T. A. (1971). The microprobe assay of chemical elements. In *Physical Techniques in Biological Research*, 2nd edn, ed. G. Oster, Chap. 3, pp. 158–275. *1* Part A. Optical Techniques. Academic Press: New York.

HARTSHORNE, D. J. & PYUN, H. Y. (1971). Calcium binding by the troponin complex and the purification and properties of troponin A. *Biochim. biophys. Acta*, **229**, 698–711.

HASSELBACH, W., MAKINOSE, M. & FIEHN, W. (1970). Activation and inhibition of the sarcoplasmic calcium transport. In *Symposium on Calcium and Cellular Function*, ed. A. W. Cuthbert, pp. 74–84. Macmillan: London.

HAWKES, R. B. & HOLBERTON, D. V. (1975). Myonemal contraction of *Spirostomum*. II. Some mechanical properties of the contractile apparatus. *J. cell. Physiol.*, **85**, 595–602.

HILL, A. V. (1948). On the time required for diffusion and its relation to processes in muscle. *Proc. R. Soc. Lond.* B, **135**, 446–453.

HOFFMANN-BERLING, H. (1958). Der Mechanismus eines neuen, von der Muskelkontraktion verschiedenen Kontraktionszyklus. *Biochim. biophys. Acta*, **27**, 247–255.

HUANG, B. & PITELKA, D. R. (1973). The contractile process in the ciliate *Stentor coeruleus*. I. The role of microtubules and filaments. *J. Cell Biol.*, **57**, 704–728.

HUXLEY, A. F. (1957). Muscle structure and theories of contraction. *Prog. Biophys.*, **7**, 255–318.

JONES, A. R., JAHN, T. L. & FONSECA, J. R. (1970). Contraction of protoplasm. IV. Cinematographic analysis of the contraction of some peritrichs. *J. cell. Physiol.*, **75**, 9–20.

JONES, A. R. & MORLEY, N. (1969). Relaxation of the stalk in *Vorticella*. *J. Protozool.* (*suppl.*), **16**, 21.

KATCHALSKY, A., LIFSON, S., MICHAELI, I. & ZWICK, M. (1960). Elementary mechanochemical processes. In *Contractile Polymers*, pp. 1–40. Pergamon: London.

KRETSINGER, R. H. (1975). Hypothesis – calcium modulated proteins contain EF hands. In *Current Topics in Intracellular Regulation*, 1st international symposium on calcium transport in contraction and secretion, pp. 469–478. North-Holland: Amsterdam.

KRETSINGER, R. H. & NOCKOLDS, C. E. (1973). Carp muscle calcium binding protein. II. Structure determination and general description. *J. biol. Chem.*, **248**, 3313–3326.

KRISTENSEN, B. I., NIELSEN, L. E. & ROSTGAARD, J. (1974). Variations in myoneme birefringence in relation to length changes in *Stentor coeruleus*. *Expl Cell Res.*, **85**, 127–135.

LAEMMLI, U. K. & FAVRE, M. (1973). Maturation of the head of bacteriophage T4. I. DNA packaging events. *J. molec. Biol.*, **80**, 575–599.

LAKE, J. A. & LEONARD, K. R. (1974). Structure and protein distribution for the capsid of *Caulobacter crescentus* bacteriophage φCbK. *J. molec. Biol.*, **86**, 499–518.

LEVINE, L. (1956). Contractility of glycerinated vorticellae. *Biol. Bull.*, **111**, 319.

NOLL, D. & WEBER, H. H. (1934). Polarisationsoptik und molekularer Feinbau der Q Abschnitte des Froschmuskels. *Pflügers Arch. ges. Physiol.*, **235**, 234–246.

PROSSER, C. L. & BROWN, F. A. (1962). *Comparative Animal Physiology*, 2nd edn, p. 433. W. B. Saunders Co.: Philadelphia & London.

PRYOR, M. G. M. (1950). Mechanical properties of fibres and muscles. *Prog. Biophys.*, 1, 216–268.

RAHAT, M., PRI-PAZ, Y. & PARNAS, I. (1973). Properties of stalk 'muscle' contractions of *Carchesium* sp. *J. exp. Biol.*, 58, 463–471.

RANDALL, J. T. & HOPKINS, J. M. (1962). On the stalks of certain peritrichs. *Phil. Trans. R. Soc. Series B*, 245, 59–79.

ROUTLEDGE, L. M., AMOS, W. B., GUPTA, B. L., HALL, T. A. & WEIS-FOGH, T. (1975). Microprobe measurements of calcium-binding in the contractile spasmoneme of a vorticellid. *J. Cell Sci.*, 19, 195–201.

ROUTLEDGE, L. M., AMOS, W. B., YEW, F. F. & WEIS-FOGH, T. (1976). New calcium-binding contractile proteins. *Cold Spring Harbor Conferences in Cell Proliferation*, 3 *Cell Motility*, in press.

SCHMIDT, W. J. (1940). Die Doppelbrechung des stieles von *Carchesium*, inbesondere die optische-negative Schwankung seines Myonemes bei der Kontraktion. *Protoplasma*, 35, 1–14.

SOTELO, J. R. & TRUJILLO-CENÓZ, O. (1959). The fine structure of an elementary contractile system. *J. biophys. biochem. Cytol.*, 6, 126–127.

TRELOAR, L. R. G. (1967). *The Physics of Rubber Elasticity*, 2nd edn. Oxford University Press: Oxford.

TREMBLEY, A. (1748). Observations on several species of small water insects of the polypus kind. *Phil. Trans. R. Soc. Lond.*, 44, 627.

UEDA, K. (1954). Electrical stimulation of the stalk muscle of *Carchesium*. II. *Zool. Mag., Tokyo*, 63, 9–14.

VAN EERD, J. & KAWASAKI, Y. (1972). Calcium-induced conformational changes in troponin C. *Biochem. biophys. Res. Commun.*, 47, 859.

WEIS-FOGH, T. (1975). Principles of contraction in the spasmoneme of vorticellids. A new contraction system. In *Comparative Physiology. Functional Aspects of Structural Materials*, ed. L. Bolis, S. H. P. Maddrell & K. Schmidt-Nielsen, pp. 83–98. North-Holland: Amsterdam.

WEIS-FOGH, T. & AMOS, W. B. (1972). Evidence for a new mechanism of cell motility. *Nature, Lond.*, 236, 301–304.

WESENBERG-LUND, C. (1925). Contributions to the biology of *Zoothamnium geniculatum* Ayrton. *K. danske vidensk. Selsk. Naturvidensk og malheim. Afd.* 8, *Raekke X*, 1–53.

APPENDIX

Since protons displace calcium from Ca–EGTA and Ca–EDTA, it is necessary to take the pH into account when using these substances as calcium buffers. This can be done by using the 'apparent association constant' (k_{app}) appropriate to each pH value. The method of calculation is given by Caldwell (1970). Values of k_{app} were computed from the pH titration data of G. Schwarzenbach, H. Senn, G. Anderegg, R. Gut and H. Ackerman, quoted by Sillén & Martell (1964), and are tabulated here in logarithmic form. The original data from which these values were calculated were obrained at 20 °C and in the presence of 0·1 M KCl.

Portzehl, Caldwell & Ruegg (1964) gave a method for calculating the effect of Mg^{2+} on Ca–EGTA buffers. The effects of components of physiological solutions other than H^+ and Mg^{2+} can usually be neglected, but Perrin & Dempsey (1974) provide references to computer methods by which the Ca^{2+} concentration may be calculated in the presence of a large number of different cations and ligands.

Ogawa (1968) has measured k_{app} for Ca–EGTA directly in the presence of various physiological pH buffers and has obtained values lower than those of Schwarzenbach *et al.* by a factor of approximately 5 and pH 6.8.

REFERENCES

CALDWELL, P. C. (1970). Calcium chelation and buffers. In *A Symposium on Calcium and Cellular Function*, ed. A. W. Cuthbert, pp. 10–16. Macmillan: London.

OGAWA, Y. (1968). The apparent binding constant of glycolether-diaminetetraacetic acid for calcium at neutral pH. *J. Biochem.*, **64**, 255–257.

PERRIN, D. D. & DEMPSEY, B. (1974). *Buffers for pH and Metal Ion Control.* Chapman & Hall: London.

PORTZEHL, H., CALDWELL, P. C. & RUEGG, J. C. (1964). The dependence of contraction and relaxation of muscle fibres from the crab *Maia squinado* on the internal concentration of free calcium ions. *Biochim. biophys. Acta*, **79**, 581–591.

SILLÉN, L. G. & MARTELL, A. E. (ed.) (1964). *Stability Constants of Metal–Ion Complexes. Special Publication No. 17.* The Chemical Society: London.

Appendix Table 1

$$\log_{10} k_{app}$$

	EDTA		EGTA	
pH	Ca^{2+}	Mg^{2+}	Ca^{2+}	Mg^{2+}
5.00	4.15	2.27	2.71	−0.48
5.20	4.53	2.64	3.11	−0.28
5.40	4.90	3.01	3.50	−0.08
5.60	5.27	3.37	3.90	0.12
5.80	5.61	3.72	4.29	0.33
6.00	5.94	4.05	4.69	0.53
6.20	6.25	4.35	5.09	0.74
6.40	6.53	4.63	5.49	0.94
6.60	6.80	4.90	5.89	1.16
6.80	7.04	5.14	6.29	1.38
7.00	7.27	5.37	6.68	1.61
7.20	7.49	5.59	7.08	1.85
7.40	7.71	5.81	7.48	2.11
7.60	7.91	6.01	7.87	2.39
7.80	8.12	6.22	8.25	2.68
8.00	8.32	6.42	8.63	2.99
8.20	8.52	6.62	9.00	3.31
8.40	8.72	6.82	9.35	3.63
8.60	8.92	7.02	9.68	3.93
8.80	9.11	7.21	9.97	4.21
9.00	9.31	7.41	10.23	4.46

EXPLANATION OF PLATES

PLATE I

(a) Electron micrograph of a transverse section of the stalk in *Carchesium*, showing the cytoplasm (C) containing the spasmoneme (S) and enveloped in extracellular sheathing material (SH). The round, densely stained objects are mitochondria.

(b), (c) Light micrographs of the stalk of *Carchesium*, showing the spasmoneme (S), cytoplasm and sheath. The mitochondria appear as dark spheres. The stalk sheath is immobilised, but the spasmoneme shows cyclic length changes. A slow phase of extension, with splitting into strands (b) alternates with a rapid contraction (c). (Phase contrast, electronic flash; from Amos, 1971.)

(d) Electron micrograph of a nearly-transverse thick section of the spasmoneme of *Vorticella*, showing the membranous tubules within it.

(e) Electron micrograph of a longitudinal section of the spasmoneme of *Zoothamnium*, showing longitudinal filaments which have a beaded appearance.

PLATE 2

(a) The main stalk of *Zoothamnium geniculatum* showing the spasmoneme (S) fixed in a partially contracted state with 2 % (v/v) glutaraldehyde. The knee-joint, to the right of the picture, is flexed as a result of tension transmitted to it by the extracellular tendon (T), which appears light in the photograph. A large reproductive cell (R) is visible amongst the ordinary zooids of the colony. The holdfast (H) has been torn from a water lily leaf (from Weis-Fogh & Amos, 1972).

(b) Glycerinated spasmoneme being drawn out of the central portion of the stalk sheath by means of tungsten needles.

(c) Frames taken from a high-speed film of the contraction of a *Zoothamnium* colony after an electrical stimulus. The interval between the pictures is 0.42 msec. Graphite particles have been attached to the stalk sheath as markers. The rate of contraction obtained by measuring along the curve of the spasmoneme was 172 lengths per second (from Amos, 1975b).

PLATE 3

Images obtained in the JEOL JXA-50A microanalyser (from Routledge *et al.*, 1975).

(a) Secondary electron image of an extended spasmoneme dried down from a solution containing 10^{-8} M free calcium ions.

(b) Counting rates recorded during a linear scan across an extended spasmoneme and the supporting film. The straight line is the line of scan and is also the base line for the signals from the spasmoneme with the background signal from the supporting film subtracted in advance. Trace M is the X-ray continuum recorded in an energy selective Si(Li) detector and represents the total mass per unit area. Trace Ca is the signal from a diffracting spectrometer adjusted to detect Ca K_α radiation. Trace B is the signal with the spectrometer offset from the Ca K_α peak to record the background component. The difference between Ca and B represents the genuine calcium signal and the ratio of this signal to the mass signal M is proportional to the calcium mass fraction.

(c) As above, but from a contracted spasmoneme. The trace B is not included in this micrograph but it was not significantly different from that in Plate 3 (b). It is evident that the calcium mass fraction is much greater in this case.

(d) Transmission image of a spasmoneme oriented horizontally with a superimposed raster covering a rectangular median area of the specimen. The area is 40×26 μm^2.

PLATE 4

(a) Fragment of a glycerinated spasmoneme of *Zoothamnium* negatively stained with uranyl acetate. The lighter areas probably represent filaments with a beaded structure (from Amos, 1975a).

(b) A 15 % (w/v) polyacrylamide micro-slab gel, in which electrophoresis has been carried out on three samples. The right-hand one is a solution obtained by dissolving 10 *Zoothamnium* spasmonemes in SDS; on the left is a sample containing 0.2 μg each of actin (A) and tubulin (T); in the centre is a mixture containing actin, tubulin and a small quantity of the spasmoneme sample. No band from the spasmoneme corresponds precisely to either tubulin or actin. The spasmin band (S; mol. wt 20000) is prominent (from Amos et al., 1975).

STRUCTURE AND EVOLUTION OF
TROPONIN C AND RELATED PROTEINS

By J. H. COLLINS

Department of Muscle Research, Boston Biomedical Research Institute,
Boston, Mass. 02114, USA

There are many biological information transfer and control processes in which changes in Ca^{2+} concentration play a critical role in coupling extracellular neural or hormonal stimuli to intracellular events. These processes involve Ca^{2+} control proteins, i.e. proteins that can reversibly bind and release Ca^{2+}, undergoing conformational changes that initiate the Ca^{2+}-mediated processes. The best understood Ca^{2+} control proteins are those involved in contractility. The most thoroughly studied contractile system is that of the highly differentiated vertebrate striated muscle which has developed very highly organized arrays of overlapping thick and thin filaments (Hanson & Huxley, 1953) that slide past one another during contraction and relaxation. As a result of the work of many investigators, it is now clear that myosin is the major component of the thick filaments, while the thin filaments consist mainly of actin, tropomyosin and troponin. Rabbit white (fast-twitch) skeletal muscle may be considered a model system, since proteins isolated from this source have been biochemically very well characterized. The same, or very similar, proteins have been found in other species and muscle types, as well as in other tissues and cells such as the central nervous system (particularly in nerve endings), platelets, fibroblasts, granulocytes, and amoebae.

Actin is a globular, nearly spherical protein (about 5.5 nm diameter) of molecular weight 41 800. In muscle it exists as a double-helical polymer (half-pitch distance about 37 nm) forming the structural backbone of the thin filaments (Moore, Huxley & DeRosier, 1970). The complete amino acid sequence of rabbit white skeletal muscle actin has been determined (Elzinga, Collins, Kuehl & Adelstein, 1973; Collins & Elzinga, 1975), and preliminary sequence studies on actins from other sources indicate that the protein varies little throughout nature (Pollard & Weihing, 1974). In addition to actin, the thin filaments contain tropomyosin and troponin (for review, see Ebashi & Endo, 1968; Ebashi *et al.*, this volume). Tropomyosin is a long, thin molecule which has a high affinity for actin and binds all along the thin filament with a repeat of about 38 nm, or one tropomyosin for every seven actin monomers. The complete amino acid sequence of

rabbit white skeletal muscle tropomyosin shows a continuous 4, 3, 4...
spacing of hydrophobic residues (Stone, Sodek & Smillie, 1974). This was
predicted by Crick (1953) since tropomyosin is a two-chain, α-helical
coiled-coil. Troponin is bound to the thin filaments at regular intervals of
38 nm; each molecule is associated with one tropomyosin and thus with
seven actins. Troponin, first discovered by Ebashi & Kodama (1965), is a
complex consisting of a Ca^{2+} binding protein (TnC), an inhibitory protein
(TnI) and a protein (TnT) which interacts strongly with tropomyosin
(Greaser & Gergely, 1971, 1973). The complete amino acid sequences of
TnC (Collins et al., 1973, 1974; Collins, 1974) and TnI (Wilkinson &
Grand, 1975) from rabbit white skeletal muscle have been determined, and
studies on the sequence of TnT are well advanced (Pearlstone, Carpenter &
Smillie, 1975; Collins, 1975a).

Myosin, the major component of the thick filaments, is a large, asymmetric molecule consisting of two very similar heavy chains (molecular
weight about 200 000) and four light chains (molecular weight about 20 000;
Lowey, Slayter, Weeds & Baker, 1969; Weeds & Lowey, 1971). Each heavy
chain extends from one of two globular 'heads' (about 7 nm diameter) to
the C-terminus at the end of a long (about 140 nm) 'tail' (Lowey et al.,
1969; Starr & Offer, 1973). Interactions involving the tail, which is a two-
stranded α-helical coiled-coil, are responsible for the assembly and stability
of the thick filaments. A portion of the tail (about 50 nm long) does not
interact strongly with the thick filament and may contain 'hinges' at either
end which permit flexibility in the movement of the heads. The myosin
heads, which have ATPase activity, interact with actin in such a way that
their ATPase activity is increased about 20-fold and the tension necessary
for contraction is generated (for more details, see reviews by Huxley, 1972,
1973). Myosin light chains, whose role in muscle contraction is not yet well
understood, will be discussed in detail below.

In resting muscle, the intracellular Ca^{2+} concentration is 0.1 μM or less
and the interaction of actin and myosin is sterically blocked by tropomyosin
(Huxley, 1972). Contraction is initiated by the release of Ca^{2+} stored in the
sarcoplasmic reticulum membrane, and the concentration increases to
about 10 μM (for reviews, see Ebashi & Endo, 1968; Ebashi, Endo &
Ohtsuki, 1969; Weber & Murray, 1973). Ca^{2+} is then bound to TnC,
causing a change in the conformation of the troponin complex which results
in tropomyosin moving away from the myosin combining sites on the actin
helix (Hitchcock, Huxley & Szent-Györgyi, 1973; Margossian & Cohen,
1973; Potter & Gergely, 1974a; Hitchcock, 1975). This is the case in
vertebrate striated muscle, but in some invertebrates (notably molluscs)
troponin appears to be absent and actin–myosin interaction is modulated

by the combination of Ca^{2+} with myosin (Kendrick-Jones, Lehman & Szent-Györgyi, 1970; Kendrick-Jones, Szentkiralyi & Szent-Györgyi, 1972; Szent-Györgyi, Szentkiralyi & Kendrick-Jones, 1973). Many other invertebrates appear to possess both types of Ca^{2+} control (Lehman, Kendrick-Jones & Szent-Györgyi, 1972; Szent-Györgyi, this volume).

Muscle cells also contain many proteins which are not bound to the filaments. Among these soluble proteins, Ca^{2+} binding parvalbumin (CBP) is the major, perhaps the only (Briggs, 1975), protein which binds Ca^{2+} specifically and with high affinity. Although CBP is a chemically very well-characterized protein, its biological role has not yet been established (for reviews, see Pechère, Capony & Demaille, 1973; Demaille et al., 1974a; Gosselin-Rey, 1974; Pechère et al., 1975).

In summary, muscle cells contain three proteins which are, or could be, Ca^{2+} control proteins: myosin (thick filaments), troponin C (thin filaments) and CBP (soluble). Before discussing these proteins in more detail, it is worthwhile at this point to review briefly the structural aspects of protein Ca^{2+} binding (for more detailed reviews, see Friedberg, 1974; Kretsinger, 1974; Liljas & Rossmann, 1974). There are at present four Ca^{2+} binding proteins of known three-dimensional structure: staphylococcal nuclease (Cotton et al., 1971), thermolysin (Matthews et al., 1972), concanavalin A (Becker et al., 1975) and carp CBP (Kretsinger & Nockolds, 1973). In all cases, Ca^{2+} is coordinated to oxygen atoms. The oxygen ligands may be contributed by water, main-chain peptide bond oxygen, or side chains of aspartic acid, asparagine, glutamic acid, serine or threonine (glutamine is another possibility, but is not involved in any of these proteins). The number of carboxyl groups in a given site ranges from one to four. All of the proteins tend to use octahedral symmetry, but the ways in which the polypeptide chains are folded to produce the binding sites are all different, and there is no evolutionary or functional relationship among these proteins. With the (probable, but not proven) exception of CBP, the function of Ca^{2+} in these proteins is mainly to stabilize the tertiary structures.

CALCIUM BINDING PARVALBUMIN (CBP)

CBPs are a group of water-soluble, acidic, low molecular weight (about 12000) Ca^{2+} binding proteins first observed by Deuticke (1934). A carp CBP was crystallized by Henrotte (1955), who was the first to note a characteristic common to all CBPs: a high content of phenylalanine relative to tyrosine and tryptophan. For many years it was believed that CBP occurs only in the muscles of aquatic lower vertebrates (see e.g. Demaille et al., 1974b), but it has recently been found in higher vertebrates as well (Lehky, Blum, Stein & Fischer, 1974; Pechère, 1974). Although present in highest amounts in white skeletal muscle, CBP has also been found in red skeletal, cardiac and smooth muscles (Lehky *et al.*, 1974; Briggs, 1975; Gosselin-Rey, 1974). Recently, Blum, Pocinwong & Fischer (1974) found in dogfish a phosphate-acceptor protein which they speculate may be the true physiologically active form of CBP.

While the function of the CBPs is still a matter of speculation (see e.g. Demaille *et al.*, 1974*a*; Pechère *et al.*, 1975) their structure has been very well characterized. The following CBP amino acid sequences have been reported: three polymorphic forms from carp (Coffee & Bradshaw, 1973; Kretsinger, 1972), two from pike (Frankenne, Joassin & Gerday, 1973; Gerday, 1974), two from coelecanth (Pechère *et al.*, 1973; Demaille *et al.*, 1974*a*), and one each from hake (Capony, Ryden, Demaille & Pechère, 1973), whiting (Joassin, 1974), frog (Pechère *et al.*, 1973; Demaille *et al.*, 1974*a*) and rabbit (Enfield *et al.*, 1975). There is great variability among the sequences of these 11 CBPs, especially in the N-terminal region, and only one-fourth of the amino acid residues is invariant. However, this variability does not seem so great when considered in light of the three-dimensional structure of carp 'component B' CBP (Kretsinger & Nockolds, 1973). The residues most important for maintaining the structure, i.e. those involved in Ca^{2+} binding and in forming the internal hydrophobic core, are highly conserved. Most of the highly variable residues are those whose side chains are at the surface of the molecule. This high variability in the surface is tolerated because CBP does not interact specifically with other proteins in the muscle cell (Pechère *et al.*, 1975).

Although there are no obvious internal repeating amino acid sequences in CBP (Pechère *et al.*, 1973), Kretsinger (1972) noticed a similarity in the three-dimensional structures of residues 39–69 and 78–108. These segments are actually nearly superimposable in structure and are related to one another by an approximate two-fold axis of symmetry. Each segment contains a Ca^{2+} binding loop, flanked on either side by two helices, and the structural repeats correspond very well with weak sequence repeats. Although the N-terminal segment (residues 7–33) of CBP does not contain a third Ca^{2+} binding site, and is quite different in structure from the other two segments, it does contain two helices. Kretsinger (1972) proposed that the three segments are related and that CBP arose from a small precursor protein by means of gene triplication. The strongest internal sequence repeats are between the N-terminal and C-terminal segments, while the strongest structural repeats are between the middle and C-terminal segments (Fig. 1). McLachlan (1972) also examined the sequence and

(1–33)	Ac-ala	phe	ala	gly	val	LEU*	asn	ASP	ALA	ASP*	ile	ala	ala	LEU	glu	ALA	cys
		ser				lys	SER	ALA	glu	val	LYS	lys		PHE	LYS	ALA	val
									asp						ALA		
									lys						gly		
(34–71)| —	gly	leu	thr	ser	lys	SER	ALA	asp	ASP*	val	LYS	lys	ala	PHE	ALA	ile ile
				lys	ala	MET	THR	ASP	ALA		ile					LYS	ALA
									glu							glu	
									asn							gly	
(72–108)| ala	asp	ala	arg	ala	LEU*	THR	ASP	gly	GLU	thr	LYS*	thr	phe	LEU	LYS	ALA gly
		gly			val		SER	ALA	ALA				ala	ala		ALA	
		ser							asn								
									lys								

lys	ala	ala	asp	ser	phe*	asn	—	—	his	lys	ala	PHE*	phe	ala	lys	VAL* — — —
ASP		GLU	gly			lys			glu		GLU		thr	LEU		
ASP	gln	ASP	lys	gly	ile	phe	glu	glu	asp	GLU	LEU*	lys	leu	PHE*	LEU	gln asn phe lys
	ala			ser												lys gly
ASP	ser	ASP'	gly	asp	gly''	lys'	ile	gly	val	asp	GLU	PHE*	thr	ala	LEU*	VAL lys ala-OH
	gln					met										gly

Fig. 1. The amino acid sequence of carp 'component B' CBP, aligned as three homologous regions (Kretsinger, 1972). The alignment of the three regions is the same as that proposed by Kretsinger & Nockolds (1973). Helical segments are underlined. Hydrophobic core residues are indicated by (*) and residues involved in Ca²⁺ binding by (') if contributing one ligand, ('') if two. Substitutions which occur in one or more of the other CBPs of known sequence are shown only in cases where they result in an increased sequence similarity among the three regions. Residues identical or very similar in all three regions are capitalized.

structure of carp CBP and found no convincing evidence from the sequence alone for internal duplication. However, the correlation between weak sequence repeats and structural repeats in the Ca^{2+} binding regions he considered good evidence for gene duplication in residues 39–69 and 78–108. He also found that the 'evidence for a further repeat in residues 7–35 is not compelling'.

Considering the structure of CBP in more detail (Kretsinger & Nockolds, 1973), it is noteworthy that all hydrophilic side-chains are on the surface, except those involved in Ca^{2+} binding and in an internal salt bridge between arginine 75 and glutamic acid 81; 40% of the hydrophobic side-chains are exposed to solvent, and the remaining 60% form a compact hydrophobic core. The oxygen ligands for each of the two Ca^{2+} binding sites are arranged in approximate octahedra and each is confined to a 12-residue segment in the polypeptide chain (51–62 and 90–101). The 51–62 Ca^{2+} is 6-coordinate, corresponding to the six octahedral vertices; the 90–101 Ca^{2+} is 8-coordinate (aspartic acid 92 and glutamic acid 101 contribute both side-chain oxygen atoms), and because glycine 98 has no side-chain, the 90–101 Ca^{2+} is exposed to solvent and coordinated to a water molecule. Phenylalanine 57 and lysine 96 coordinate *via* their peptide bond oxygens and all other ligands come from side-chains of aspartic acid, glutamic acid and serine. Isoleucines 58 and 97 are close together in the structure; they are connected by a main-chain hydrogen bond and their side-chains are in contact with each other in the hydrophobic core. The N-terminal segment completes the hydrophobic core of the molecule and covers the arginine-75–glutamic acid-81 internal salt bridge. Noting that the 90–101 Ca^{2+} is more exposed to solvent, Kretsinger (1974) has postulated a series of structural alterations in CBP which could be initiated by the release of this Ca^{2+}. In support of this view, Moews & Kretsinger (1975) showed that the 90–101 Ca^{2+} can be replaced by terbium ion under conditions in which the 51–62 Ca^{2+} is not replaced.

In the first published report that CBP binds Ca^{2+}, Pechère, Capony & Ryden (1971) noted that CBP and TnC are similar and might have evolved from a common ancestor. Besides having the ability to bind Ca^{2+} specifically and with high affinity, both proteins are acidic, soluble in water, have high phenylalanine content (relative to tyrosine and tryptophan) and low cysteine content, occur abundantly in muscle, and are highly antigenic. Demaille *et al.* (1974*b*) isolated a peptide from *Varanus* (an African lizard) TnC which is similar in sequence to residues 92–96 of one of the coelecanth CBPs (note: this evidence is often incorrectly cited as proving that TnC and CBP are homologous). Malencik, Heizmann & Fischer (1975; see also Heizmann, Malencik & Fischer, 1974) found that TnC (or whole troponin) from dogfish does not cross-react with antibodies to dogfish CBP. However, a breakdown product of molecular weight about 11000 (produced in the absence of Ca^{2+} from TnC or whole troponin, presumably by contaminating proteases) showed strong cross-reactivity in the presence of Ca^{2+}. Neither dogfish CBP nor the TnC breakdown product could replace TnC in conferring Ca^{2+} sensitivity to actin–myosin interaction or reverse the inhibition induced by TnI. It should be emphasized at this point that CBP (molecular weight about 12000) is *not* a proteolytic breakdown product of the larger (molecular weight about 18000) TnC. This is conclusively proved by comparing the amino acid sequences of rabbit white skeletal muscle CBP (Enfield *et al.*, 1975) and TnC isolated from the same source (Collins *et al.*, 1973). The two proteins are coded for by different genes.

TROPONIN C (TnC)

Comparison of CBP sequences with that of rabbit white skeletal muscle TnC (Fig. 2) shows clearly that the two proteins are homologous (Collins *et al.*, 1973). The sequence of TnC can be divided into two halves (residues 16–83 and 92–159) which are similar in sequence to each other (Collins *et al.*, 1974) and to the Ca^{2+} binding region (residues 40–108) of CBP (Collins, 1974), as shown in Fig. 3. There are 34% identical residues between the two halves of TnC; this strongly implies that they have similar three-dimensional structures, since it is well known that sequences evolve more rapidly than structures (McLachlan, 1972). The similarities of each half of TnC with the C-terminal two-thirds of CBP (especially around the Ca^{2+} binding sites), and the fact that hydrophobic residues (valine, methionine, isoleucine, leucine, phenylalanine) occur in TnC in nearly all (14 out of 15 in each half) positions where hydrophobic core residues occur in CBP, suggests that the three-dimensional structure of TnC probably consists essentially of a duplication of the C-terminal region of CBP. TnC would then contain eight α-helical regions and four Ca^{2+} binding sites. This is confirmed by the analysis of Weeds &McLachlan (1974), who used a computer programme which gives high similarity scores for amino acid replacements that fit readily into the same three-dimensional structure. In fact, by making certain further assumptions (Kretsinger, 1974) it has been possible to construct a plausible model (Kretsinger & Barry, 1975) of the predicted three-dimensional structure of TnC.

At this point I will introduce a uniform residue numbering system for comparing the sequences of TnC, CBP and myosin light chains which is based on the predicted structure of TnC. The eight helical regions (each

Sequence alignment (rotated figure). Two aligned blocks are shown; alternate residues appearing beneath a position are given in parentheses.

Upper block

LC-1	X-pro lys — asn val lys lys pro ala ala ala ala ala ala pro ala lys ala pro ala pro ala
LC-2	X-pro lys — ala lys arg arg — ala ala ala — — — — — — — — —
TnC	Ac-asp (Ac-met) thr asp gln ile ala (tyr) lys glu (ala) ile lys ala ser arg ser (val glu) ser tyr (gln) LEU ser (thr) glu
LC-1	pro ala pro ala pro glu glu lys ile asp leu ser ala lys ile lys glu PHE ser (ala) lys (ala)
LC-2	— — — — glu gly gly ser ser asn val phe ser met PHE asp gln

Lower block (numbered positions A: 1–11; ab: 1–8)

Row	pre	1	2	3	4	5	6	7	8	9	10	11	1	2	3	4	5	6	7	8
TnC	glu met gln	ILE (gln/lys)	ala (asn)	glu	PHE	LYS	ala	ALA	PHE	asp	MET (ILE)	PHE (val leu)	ASP	ala	asp (gly)	gly (glu)	gly (asp)	gly	asp (cys)	ILE
LC-1	glu gln asp	gln (ILE)	ala (asp/ala)	glu	PHE	LYS	glu	ALA	PHE	leu	LEU	TYR	— ASP	arg	thr	gly	asp	gly	asp (ser)	lys (ILE)
LC-2	thr glu	glu	ILE	gln	PHE	LYS	glu	ALA	PHE	thr	VAL	ILE	— ASP	glx	arg	asn	asp	arg (gly)	asn (ile)	ILE

Fig. 2. (p. 310–313)

Protein sequence alignment (residues given in three-letter code; UPPER case = major sequence, lower case = variant/alternative residues; — = deletion/gap; * = noted residue).

Upper block — helices B, bc, C, D

Protein	B 9	1	2	3	4	5	6	7	8	9	10	11	bc 1	2	3	4	5	6	7	C 1	2	3	4	5	8	9	D 1	2	3	4
TnC	ser	val (thr)	lys	glu	LEU	gly	thr (lys)	val	MET	ARG	met	LEU	GLY	gln	—	—	pro	THR	lys	GLU	GLU	LEU								
LC-1	thr	leu	ser	gln	VAL	gly	asp	val	LEU	ARG	ALA	LEU	GLY	thr	—	ASN	pro	THR	asn	ala	GLU	VAL			ILE	ASP	ILE	GLU	PHE	glu (asp)	gln	PHE
LC-2	asp	lys	glu	glu	LEU	arg	asp	thr	PHE	ALA	MET	GLY	GLY	arg	leu	ASN	val	lys	leu	GLU	ASP	LEU			ILE		ILE	PHE	thr	val	PHE
CBP	Ac-ALA / Ac-thr	PHE / LYS	ala / met	gly / MET	—	val	leu	asn	asp	ala	ASP	ILE*	SER / lys	glu	GLU	lys	VAL								PHE	asn	his (asn)	lys	ala	PHE*	ASP (ile/glu)	TYR / GLY

Lower block — loop cd, helix D

Protein	5	6	7	8	9	10	11	cd 1	2	3	4	5	6	7	8	9
TnC	ASP	ala (gln/glu)	ILE	MET	ILE	glu (asp)	val	ASP	GLU	ASP	gly	ser	GLY	thr	—	ILE VAL
LC-1	lys	VAL	LEU	gly	asn	pro	ser	ASP	GLU	—	met	asn	ala	lys	lys	ILE
LC-2	ASP	ALA	MET	—	—	—	—	lys	GLU	ala	ser	GLY	pro	—	—	ILE
CBP	ala	ALA	ala	LEU	glu (asp/asn)	ala	cys (val)	lys	GLU / GLU	ala / glu	GLU	ASP (asn)	GLY	—	—	PHE / ASP

Upper block

	5	6	7	8	9	10	11	de 1	2	3	4	5	6	7	8	1	2	3	4	5	6
TnC	LEU / val	MET	MET	MET	val	arg	gln / cys	met / lys	glu / asp	asp	ala / ser	lys	gly	lys	SER	GLU	GLU	GLU	LEU	ala / ser	GLU / ASP
LC-1	LEU / pro	MET	LEU	gln	ala	ile	—	—	ser	asn	lys	asp	gln	gly	THR	tyr	GLU	ASP	PHE	val	GLU
LC-2	LEU / thr	MET	MET	phe	gly	glu	—	—	lys	leu	lys	lys	ala	asx	GLX	ASX	val	ILE		thr	gly
CBP	*PHE / ala	lys	VAL					—	gly		*leu	thr / lys	ser / gly / ala	lys	SER / ASP	ala / ASP	ASP / GLU	ASP / LEU	*VAL / LEU / ILE	lys	lys / GLU

thr — MET (underline, TnC) ; ASP GLU LEU ILE GLU (underline, CBP)

Lower block

	7	8	9	10	11	ef 1	2	3	4	5	6	7	8	F 1	2	3	4	5	6	7	8
TnC	cys / leu	PHE	ARG	ILE / MET	PHE	ASP	ARG / LYS	asn	asp	gly	GLY	tyr	ILE	asp	ala / leu	GLU	LEU	ala / LYS	glu / ile	ILE	PHE / LEU
LC-1	gly	LEU	ARG	VAL	PHE	ASP	LYS	GLU	asp / gly	gly	thr	val	gly	met	gly	ala	GLU	LEU	ARG	his	VAL / LEU
LC-2	ala	PHE	LYS	VAL	LEU	ASX	pro	GLU	gly	lys	GLY	thr	ILE	lys	lys	gln / phe	LEU	glx	LEU	glx	LEU / LEU
CBP	*ala / val	PHE	ala	ILE / LYS	ILE / VAL / LEU	*ASP	gln / ARG / LYS	ASP	lys	ser	GLY	phe / tyr	*ILE	glu	asp	GLU / LEU	LEU / LYS	leu / PHE		*PHE / ILE	LEU / ILE / VAL

val — LYS VAL LEU (underline, CBP)

Fig. 2 — Alignment of the amino acid sequences of TnC, LCs and CBP (sequence alignment table)

```
            fg                         G
            9   10  11  1   2   3   4   5    1    2    3    4   5    6    7    8    9   10  11
TnC        arg ala ser  —   —  GLY GLU HIS VAL THR  ASP  GLU  GLU  ILE glu  ser  LEU  MET lys asp GLY
           gln     thr                 thr ILE ile  GLU  ASP  ASP                ala          asp
LC-1       ala THR leu  —   —  GLY GLU LYS MET  —   GLU  GLU  VAL  glu ala  LEU  MET   —  ALA GLY
LC-2       thr THR gln  —   —  cys ASP ARG PHE SER  GLX  GLX  ILE  lys asn  MET  trp  ala ALA phe
                                                *
CBP        gln asn phe lys ala asp ala arg ala LEU  THR  ASP  gly GLU thr  lys  thr  PHE LEU lys ALA GLY
               lys SER              GLY                 SER               ala  LEU  MET ala asp
```

```
            gh                                          H
            1   2   3   4   5   6   7   8   9   1    2    3    4    5    6    7    8    9   10  11
TnC        asp lys asn asn asp GLY arg ILE asp PHE  ASP  glu  PHE  LEU  lys  MET  MET  glu gly val gln-OH
                                               TYR                  PHE                          glu-OH
LC-1       gln glu ASP ser asn GLY cys ILE asn TYR  GLU  ala  PHE  VAL  lys  his  ILE  met ser ile-OH
LC-2       pro pro ASX val gly GLY asn ILE VAL asp  TYR  lys  ILE  cys  tyr  VAL  ILE  thr his gly asx ala
                                *                         *              *
CBP        asp ser ASP gly asp GLY lys ILE gly VAL  ASP  glu  PHE  thr  ala  LEU  VAL  lys ala-OH
             lys                                    GLU                 MET  ILE  ser
```

```
LC-2       lys asp glu gln-OH
```

Fig. 2. Alignment of the amino acid sequences of TnC, LCs and CBP with the uniform residue numbering system based on the predicted structure of TnC. Helices are designated A to H, going from the N- to the C-terminus; interhelical loops are designated ab, bc, ..., gh. TnC: rabbit white skeletal muscle TnC (Collins, 1974) with substitutions occurring in bovine cardiac muscle TnC (Van Eerd & Takahashi, 1975). LC-1: rabbit white skeletal muscle LC-1, with substitutions that occur in rabbit white skeletal muscle LC-3 (Frank & Weeds, 1974). LC-2: rabbit white skeletal muscle LC-2. CBP: carp 'component B' CBP (Coffee & Bradshaw, 1973) with structural features (Kretsinger & Nockolds, 1973) symbolized as in Fig. 1, and substitutions selected from one or more of the other CBP sequences. Identical or very similar residues occurring in three or four proteins are capitalized.

containing 11 amino residues) are designated A to H, going from the amino to the carboxyl end of the polypeptide chain; the non-helical segments, or loops, are designated as ab, bc, etc. This system is analogous to that used for the globins (see e.g. Perutz & Ten Eyck, 1971). The sequences of the various proteins, aligned with this numbering system, are shown in Fig. 2. The sequence of all except residues 1–15 and 159 of TnC can be written as

$$AabBbcCcdDdeEefFfgGghH.$$

(An extra methionine residue is at the end of the D helix.) The best sequence alignment (Collins *et al.*, 1973, 1974) of TnC and CBP is obtained by assuming that, in CBP, 3 residues have been inserted at the beginning of the fg loop and that the following residues have been deleted: all through B6, cd1, cd7, D9–de2 (Fig. 2). The two halves of the TnC sequence can be aligned, with no deletions or insertions (Fig. 3), as:

$$AabBbcCcd\,D,$$
$$Eef\,Ffg\,GghH.$$

The de loop connects the two halves to give, in effect, a covalently linked dimer. The C-terminal half of TnC is 54% identical to E1–H10 in one or more CBPs, while the N-terminal half is 40% identical. Alignments of the whole CBP and TnC sequences (residues B7–H10) yields 47% identical residues with one or more CBPs (Fig. 2). The sequence of TnC may be further divided (Collins *et al.*, 1973, 1974) into four homologous segments, aligned as follows (Fig. 4):

$$AabB,$$
$$CcdD,$$
$$EefF,$$
$$GghH.$$

These segments correspond very well with the three homologous segments in CBP first observed by Kretsinger (1972).

These observations led to the conclusions (Collins *et al.*, 1973, 1974; Collins, 1974) that: (1) TnC has four Ca^{2+} binding sites, located at residues ab1–B3, cd1–D3; ef1–F3, gh1–H3; (2) the four homologous regions of TnC have similar three-dimensional structures to the two Ca^{2+} binding regions (E1–F11 and G1–H10) of CBP; (3) TnC and CBP are homologous, each having evolved by replication of a small (about 35 residues), ancestral Ca^{2+} binding protein; (4) TnC is a product of two gene duplications to produce a protein four times the length of the small ancestral protein; (5) CBP evolved from a larger, TnC-like precursor. Detailed analyses by computer programmes designed to detect internal

```
TnC   (A1–D11)  IAEFKAAFDMF DADGGGDISVKELGTVMRML      GQ TPTKEELDAII EEVDEDGSGT IDFEEFLVMMVRQ
                KN          I VLGAED C   T   K            N P     QEM D        V D          C

      (E1–H11)  EEELAECFRIF DRNADGYIDAEELAEIFRAS      GE HVTDEEIESLM KDGDKNNDGR IDFDEFLKMMEGV
                            SDL M   K    L   KIMLQ T        TI EDD E            Y    EF K

LC–1  (A1–D11)  QDEFKEAFLLY DRIGDSKITLSQVGDVLRAL      GT NPTNAEVKKVLGNPSDEQMNAKKIEEQFLPMLQAI
                IA

      (E1–H11)  YEDFVGGLRVF DKEDGTVGMGAELRHVIATL      GE KMKEEVEALM AGQEDSNGC INVEAFVKHIMSI

LC–2  (A1–D11)  IZZFKEAFTVI DZNRDGIINKEDLRDTFAAM      GRLNVKEEDLDAM    KEASGP INTVFLTMMFGE

      (E1–H11)  ZBVITGAFKVL BPEGKGTIKKQFLZZLLTTQ      CD RFSZZZIKNMW AAFPPBVGGN VDYKNICYVITHG

CBP   (B7–D8)              AF                           *                       *      **
                          TKKSL                       AG VLNDADIAAAL EAC KAADS FNHKAFFAKV
                          S                            D  VSEE VKK I K V EEEGG   DY E TLI
                          M                            E  K D D    D F              I  M
                                                                      N
                                                                      A

      (E1–H10)  *        ´*  ´*´;´*;´*´* *    ´*´*     *           *        ´;´*´    *´   *   **
                ADDVKKAFAII DQDKSGFIEEDELKLFLQNFKAD   AR ALITDGETKTFL KAGDSDGDGK IGVDEFTALVKA
                DA L EV KVL A A Y   E G V KS          G  V S A      AALM AD   K       E   VTMISS
                LN I        R        I   V
                E           K
```

Fig. 3. Internal sequence duplications in TnC, LCs and CBP. Segments A1–D11 and E1–H11 (Fig. 2) of each protein are aligned with each other. Alignments among the different proteins and structural features of CBP are the same as in Fig. 2. To conserve space, the single-letter code for amino acid residues is used. (A, ala; B, asx; C, cys; D, asp; E, glu; F, phe; G, gly; H, his; I, ile; K, lys; L, leu; M, met; N, asn; P, pro; Q, gln; R, arg; S, ser; T, thr; V, val; W, trp; Y, tyr; Z, glx.)

```
TnC  (A1-B11)   IAEFKAAF DMF DADGGGD ISVKELGTVMRML
                KN       I  VLGAED C  T    K

     (C1-D11)   KEELDAII EEV DEDGSGT IDFEEFLVMMVRQ
                P   QEM  D           V D       C

     (E1-F11)   EEELAECF RIF DRNADGY IDAEELAEIFRAS
                   SDL   M   K        L  KIMLQ T

     (G1-H11)   DEEIESLM KDG DKNNDGR IDFDEFLKMMEGV
                EDD  E                Y    EF K

LC-1 (A1-B11)   QDEFKEAF LLY DRTGDSK ITLSQVGDVLRAL
                IA

     (C1-D11)   NAEVKKVLGNPS DEQMNAKKIEFEQFLPMLQAI

     (E1-F11)   YEDFVGGL RVF DKEDGTV GMGAELRHVLATL

     (G1-H11)   EEEVEALM  AG QEDSNGC INYEAFVKHIMSI

LC-2 (A1-B11)   IZZFKEAF TVI DZNRDGI INKEDLRDTFAAM

     (C1-D11)   EEDLDAM      KEASGP INFTVFLTMMFGE

     (E1-F11)   ZBVITGAF KVL BPEGKGT IKKQFLZZLLTTQ

     (G1-H11)   ZZZIKNMW AAF PPBVGGN VDYKNICYVITHG
                   *                *    **
CBP  (C1-D8)    DADIAAAL EAC KAADS FNHKAFFAKV
                AE VKK I D V DEEGG DY E  TMI
                ED  D   N F A     KE      L
                    T   K   E     S
                    A
                   *    *    *  / " / /  */  /* ** *
     (E1-F11)   ADDVKKAF AII DQDKSGF IEEDELKLFLQNF
                PE I EV  EVL A A DY V E  G V KS
                DN L     KA  K         I G
                A        R
                     **    / " / /  *   /"* **
     (G1-H10)   DGETKTFL KAG DSDGDGK IGVDEFTALVKA
                AA  AALM AD  K   D   AE  ETMISGS
                VN  S        Q       A
                                     V
```

Fig. 4. Alignments of the four mutually homologous regions in TnC and LCs with the three regions in CBP. For details, see Figs. 2 and 3.

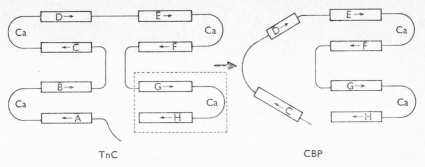

Fig. 5. Schematic drawing of the structural and evolutionary relationships of TnC and CBP. TnC is predicted to contain eight α-helices (symbolized by rectangles), designated A to H going from the amino to the carboxyl termini of the polypeptide chain. Both TnC and CBP evolved after two successive gene duplications produced a protein four times the size of a small (about 35 amino acid residues), ancestral protein with a single Ca^{2+} binding site (shown enclosed by dashed line). CBP arose from an incomplete copy of a gene for a larger, TnC-like protein. Each half (A through D and E through H) of TnC is similar in three-dimensional structure to the C-terminal part (E through H) of CBP.

repeats and similarities between protein sequences (Weeds & McLachlan, 1974; Barker & Dayhoff, 1975; Pechère et al., 1975) have amply confirmed and extended these conclusions which are summarized in Fig. 5.

Fig. 6 shows the sequences in the 12-residue segments predicted to be the four Ca^{2+} binding sites of TnC, and compares them with the known sites in CBP. It is noteworthy that aspartic acid always occurs in the X position and glutamic acid in the $-Z$ position. The residue contributing the $-Y$ ligand can be anything, since it is the main-chain oxygen, not the side-chain, which coordinates Ca^{2+}. It is conceivable, however, that the side-chain of aspartic acid ab7 of TnC may also be involved; this is the only site where an acidic side-chain occurs at this position. Of the five residues whose side-chains would be involved, at least four in each site contain oxygen atoms and at least three are acidic. A glycine always (except for an aspartic acid at ef6 in hake CBP and at gh6 in coelecanth CBP) occurs in position 6 of the loop, where a sharp bend in the polypeptide chain is required for proper formation of the site. An isoleucine (occasionally valine) is always present at position 8 of the loop, where it forms part of the hydrophobic core. As mentioned above, isoleucines ef8 and gh8 are close together in carp CBP; their side chains are in contact, and they are also connected by a main-chain hydrogen bond. An analogous situation may occur in TnC (Kretsinger, 1974) where isoleucines ab8 and cd8 would form one pair, ef8 and gh8 the other.

In the past few years there has been much disagreement on the number

	X		Y		Z		−Y		−X			−Z
CBP ef1–F3	asp	gln ala lys arg	asp	lys ala	ser	gly asp	phe tyr	ile val	glu	glu gln	asp glu	glu
gh1–H3	asp	ser lys gln	asp	gly	asp	gly asp	lys met gly	ile	gly	val ile ala	asp glu	glu
TnC ab1–B3	asp leu	ala gly	asp ala	gly glu	gly asp	gly	asp cys	ile	ser	val thr	lys	glu
cd1–D3	asp	glu	asp	gly	ser	gly	thr	ile val	asp	phe	glu asp	glu
ef1–F3	asp	arg lys	asn	ala	asp	gly	tyr	ile	asp	ala leu	glu	glu
gh1–H3	asp	lys	asn	asn	asp	gly	arg	ile	asp	phe tyr	asp	glu
LC-2 ab1–B3	asp	glx	asn	arg	asn	gly	ile	ile	asp	lys	glu	asp
ef1–F3	asx	pro	glu	gly	lys	gly	thr	ile	lys	lys	gln	phe
gh1–H3	pro	pro	asx	val	gly	gly	asn	val	asp	tyr	lys	asn
LC-1 ab1–B3	asp	arg	thr	gly	asp	ser	lys	ile	thr	leu	ser	gln
ef1–F3	asp	lys	glu	asp	gly	thr	val	gly	met	gly	ala	glu
gh1–H3	gln	glu	asp	ser	asn	gly	cys	ile	asn	tyr	glu	ala

Fig. 6. Sequences of the Ca^{2+}-binding sites of CBP and corresponding regions in TnC and LCs. Residue numbering is the same as in Fig. 2, and the sequences are aligned as in Fig. 4. Site cd1–D3 of LC-1, LC-3, LC-2 and CBP are not included because of insertions and deletions which disrupt the sites. The six octahedral vertices assigned to the Ca^{2+} ligands in CBP (Kretsinger & Nockolds, 1973) are shown at the top. All known substitutions in CBP are shown.

of experimentally determined Ca^{2+} binding sites in TnC (see discussion by Potter & Gergely, 1975). However, using an improved method, Potter & Gergely (1975; also reported by Potter et al., 1974; Potter & Gergely, 1974b) have now clearly established that rabbit white skeletal muscle TnC contains four Ca^{2+} binding sites, in agreement with the number deduced from the sequence. They further find that TnC contains two high-affinity and two low-affinity sites. Mg^{2+} appears to compete for the high-affinity sites and these are now referred to as the Ca^{2+}–Mg^{2+} sites (Potter & Gergely, 1975). The two low-affinity sites are specific for Ca^{2+} and are thus called the Ca^{2+}-specific sites. There appear to be two additional sites which bind Mg^{2+} and not Ca^{2+}, but nothing can be said about the physiological significance, if any, of these sites, nor can they be located in the sequence of TnC. The Ca^{2+}–Mg^{2+} sites are always occupied by either Ca^{2+} or Mg^{2+} and seem to be involved mainly in stabilizing the structure of TnC. The Ca^{2+}-specific sites are involved in the regulation of muscle contraction. For an understanding of the molecular mechanism of muscle contraction, it is thus vital to be able to distinguish the Ca^{2+}-specific from the Ca^{2+}–Mg^{2+} sites in the sequence of TnC. Kretsinger (1974) has suggested that this might be accomplished by analogy with CBP. As mentioned above, the gh1–H3 Ca^{2+} of CBP is more exposed to solvent than is the ef1–F3 Ca^{2+}; this is due to glycine gh9, which has no side-chain and thus does not co-ordinate (a water molecule substitutes for the equivalent glutamic acid ef8 in the other site). The presence in TnC of glycines at positions cd4, ab4 and especially ab5 may cause the ab1–B3 and cd1–D3 sites to bind Ca^{2+} with lower affinity (and thus correspond to the Ca^{2+}-specific sites). Weeds & McLachlan (1974) agree that this is the most obvious interpretation, but also point out the 'more interesting possibility' that the glycines could allow greater flexibility of polypeptide chain folding, resulting in tighter binding sites. One may also argue that increased flexibility could allow binding of the smaller Mg^{2+} to the same sites. Physical-chemical studies (reviewed by Potter *et al.*, 1975) also suggest that the ef1–F3 and gh1–H3 sites are the Ca^{2+}–Mg^{2+} sites, but the results are not entirely conclusive and the question must be regarded as still undecided.

Recently, Van Eerd & Takahashi (1975) reported the amino acid sequence of TnC from bovine cardiac muscle, which contains 161 amino acid residues (compared with 159 for rabbit skeletal TnC). Bovine cardiac TnC has two insertions: a methionine at the N-terminus and a valine at the end of helix A. There are 10 substitutions concentrated in the N-terminal segment which precedes helix A; otherwise the two TnC sequences are 68% identical, with the substitutions distributed fairly evenly throughout the sequence (Fig. 2). Two of these substitutions (aspartic acids ab1 and ab3 in rabbit skeletal TnC are changed to leucine and alanine in bovine cardiac TnC) probably cause the loss of Ca^{2+} binding in the ab1–B3 site. The other three assumed Ca^{2+} binding sites and the putative hydrophobic core residues are very highly conserved. The conclusion by Van Eerd & Takahashi (1975) that bovine cardiac TnC probably binds three Ca^{2+} is 'in rather good agreement' with the value of 2.4 Ca^{2+} per molecule deter-mined experimentally by Ebashi (unpublished, cited by Van Eerd & Takahashi, 1975). As shown in Fig. 3, internal repeats between half sequences are somewhat weaker in bovine cardiac TnC (28%) than in rabbit skeletal TnC (34%), indicating that the skeletal form is more similar to their common ancestor.

TnCs that have so far been isolated from other sources are similar in size (molecular weight 17000–21000) and amino acid composition to rabbit skeletal TnC, and all are able to sensitize rabbit skeletal actin–myosin interaction. Interestingly, the number of Ca^{2+} binding sites reported is quite variable. Demaille et al. (1974b) report that hake and python TnC bind three Ca^{2+} and lizard TnC only two. Malencik et al. (1975) find that of two Ca^{2+} bound by dogfish TnC, one is very difficult to remove in the absence of Mg^{2+}. They conclude that one Ca^{2+} may be required for the structural integrity of the protein, while the other confers Ca^{2+} sensitivity. These two sites are probably analogous to the $Ca^{2+}–Mg^{2+}$ and Ca^{2+}-specific sites of rabbit skeletal TnC. Regenstein & Szent-Györgyi (1975) report that lobster TnC binds only one Ca^{2+}. Vanaman, Harrelson & Watterson (1975) have found TnC-like Ca^{2+} binding proteins in mammalian, avian and amphibian brain, as well as in the central and peripheral nerves of rabbit. They isolated the bovine brain protein, which binds two Ca^{2+}; preliminary sequence studies on this protein show extensive similarity with rabbit skeletal TnC.

MYOSIN LIGHT CHAINS (LCs)

It is now generally assumed, though not yet proved, that all myosins contain two molecules each (one per heavy chain) of two classes of LCs. Despite the fact that the existence of the LCs has been known for a long time (Tsao, 1953; Locker, 1956; Kominz, Carroll, Smith & Mitchell, 1959), their physiological function is still not clearly understood. The LCs of rabbit white skeletal muscle have been intensively studied (for reviews see Weeds & Frank, 1972; Lowey & Holt, 1972; Dreizen & Richards, 1972). There are three types of LCs, called (Sarkar, 1972) LC-1 (molecular weight 20700), LC-2 (molecular weight 19000) and LC-3 (molecular weight 16500). Studies on thiol peptides (Weeds, 1967) showed that LC-1 and LC-3 are closely related, since they have identical sequences (residues fg4–H6 of LC-1 in Fig. 2) around their single cysteine residue. LC-2 is apparently unrelated by this criterion (and also by tryptic peptide mapping), since the sequences around its two cysteines (residues F9–fg4 and H2–H6 of LC-2 in Fig. 2) are very different from those of LC-1 and LC-3 (Weeds, 1969). Differences between the two classes of LCs are also observed in the conditions required for their dissociation from the whole molecule (Weeds & Lowey, 1971; Sarkar, 1972). LC-2 can be dissociated by treatment of myosin with DTNB (5,5′-dithiobis-2-nitrobenzoic acid) under conditions that apparently do not affect the ATPase activity of myosin; hence the commonly used name 'DTNB light chain' was coined. LC-1 and LC-3 (also called the 'alkali 1' and 'alkali 2' light chains) can only be dissociated under relatively harsh conditions (such as high pH) that usually lead to irreversible loss of ATPase activity. The stoichiometry of the LCs has been well established (Sarkar, 1972; Weeds & Frank, 1972): there are two molecules of LC-2 and a total of two molecules of LC-1 plus LC-3, with about twice as much LC-1 as LC-3. It is generally believed that each myosin 'head' has associated with it one LC-2 and one LC-1 or LC-3. This heterogeneity of the 'alkali' LCs extends even to single fibres of rabbit white skeletal muscle (Weeds, Hall & Spurway, 1975). The physiological significance, if any, of the heterogeneity, remains to be established.

LC-1 and LC-3 may be somehow involved in the binding of ATP to myosin (Dreizen & Richards, 1972; Wagner & Yount, 1975). They do not bind Ca^{2+} under physiological conditions, i.e. in the range 0.1 to 10 μM in the presence of 1 mM Mg^{2+} (J. Kendrick-Jones, unpublished, cited by Weeds & McLachlan, 1974), and no role in the Ca^{2+} regulation of muscle contraction has been found for them. The structural relationship between LC-1 and LC-3 has been established from their amino acid sequences (Frank & Weeds, 1974; see also Weeds & Frank, 1972). LC-3 is *not* a proteolytic breakdown product of LC-1, and the two proteins are coded for by different genes. Five of the first eight residues of LC-3 are different from the corresponding residues of LC-1 (see Fig. 2); the C-terminal 141 residues of both proteins (residues A3–H11) are identical. LC-1 has an additional 41-residue segment, unusually rich in alanine and proline, at its N-terminus, which accounts for the difference in molecular weight between LC-1 and LC-3. This unusual segment may be due to an 'error' in which a segment of DNA that does not normally code for a protein has been translated (Frank & Weeds, 1974; Barker & Dayhoff, 1975).

Since LC-1 and LC-3 neither bind Ca^{2+} nor seem to be involved in the regulation of actin–myosin interaction, it was very surprising to discover (Collins, 1974) that they are homologous with TnC and CBP. The most favourable sequence alignment of LC-1 with either TnC or CBP (Fig. 2) is fully consistent with the alignment of TnC and CBP with each other. Over the span of residues A1–H11 there are 38% identical residues between LC-1 (or LC-3) and one or both TnC sequences. This assumes that in LC-1 and LC-3 there are single-residue insertions following residues C8 and cd7 and a deletion at G9. LC-1 (or LC-3) is less similar to CBP than is TnC (32% versus 47% identical residues when compared with one or more CBPs), especially around the Ca^{2+}-binding sites. Internal repeats in LC-1 and LC-3 are recognizable, though weak, and follow the same pattern found in TnC and CBP. Alignment of A1–D11 with E1–H11

(Fig. 3) produces 21% identical residues in LC-1 (or LC-3), versus 34% for the same alignment in rabbit skeletal TnC. Conservation of sequence suggests that LC-1, LC-3 and TnC may have similar three-dimensional structures, even though the LCs have lost their Ca^{2+} binding sites. Indeed, of the 30 positions where hydrophobic core residues would be expected in the predicted structure of TnC, 27 are hydrophobic residues in LC-1 and LC-3 (the exceptions are: proline bc4, which also occurs in TnC; glycine ef8, and histidine H7). This raises the question of whether or not Ca^{2+} binding is really necessary to stabilize the structure of TnC. Of course it is extremely difficult to draw valid conclusions when comparing protein structures which are only predicted, not solved. No one knows to what extent the structure of LC-1 and LC-3 may be stabilized by Mg^{2+}, ATP or interaction with the myosin heavy chains.

The four homologous regions (A1–B11, C1–D11, E1–F11, G1–H11) can be recognized in LC-1 and LC-3 (Fig. 4), although the sequence repeats are so weak (for example, compare E1–F11 with G1–H11) that they might not have been detected (in the absence of a sensitive computer test) without reference to the corresponding segments in TnC. Since it is known experimentally that LC-1 and LC-3 do not bind Ca^{2+}, a study of the sequences of these four regions should provide further clues as to the sequence requirements for Ca^{2+} binding. It is easy to see why the cd1–D3 site of LC-1 does not bind Ca^{2+} (Weeds & McLachlan, 1974; Tufty & Kretsinger, 1975). Helix C is probably disrupted by proline C10, and helix D similarly distorted by proline D6. Also, the insertion of glycine (following C8) and lysine (following cd7) destroys the continuity of the whole region. Such obvious disruptions do not occur in the other three regions of LC-1 and one must look to the 12-residue segments corresponding to the Ca^{2+} binding sites of CBP and (supposedly) TnC (Fig. 6). In the ab1–B3 segment, it is difficult to understand why Ca^{2+} is not bound, since there are five residues potentially capable of donating side-chain oxygen ligands. Important considerations may be that the net charge of the segment is zero, and that glutamine (rather than the glutamic acid which always occurs in TnC and CBP) is at the $-Z$ position. The ef1–F3 site has only three potential liganding side chains and, moreover, glycine ef8 occurs where an internal isoleucine (or valine) is present in all other segments. The gh1–H3 site lacks the apparently essential aspartic acid in the X position, and has no potential ligand in the $-Z$ position.

Computer analysis (Weeds & McLachlan, 1974; Barker & Dayhoff, 1975; Pechère et al., 1975) of the sequence relationships among LC-1 (or LC-3), TnC and CBP, confirms further the suggestion (Collins, 1974) that CBP evolved from a TnC-like precursor which was produced from a smaller protein via two successive gene duplications. Barker & Dayhoff (1975) have estimated that the divergence of LC-3, TnC and CBP occurred over two billion years ago, 'long enough to account for the presence of these proteins in all eukaryotic cells'.

For several years no function could be demonstrated for rabbit skeletal LC-2. However, the following discoveries have prompted investigations into a possible regulatory role for LC-2 in rabbit white skeletal muscle myosin: (1) in molluscan muscle the Ca^{2+} sensitivity is associated with myosin, rather than troponin (the latter appears to be absent) (Kendrick-Jones et al., 1970); (2) one of the molluscan myosin light chains is somehow involved in the regulation (Kendrick-Jones et al., 1972; Szent-Györgyi et al., 1973); (3) many invertebrates possess both types of regulatory systems (Lehman et al., 1972; Szent-Györgyi, this volume). It is now established that rabbit white skeletal muscle myosin binds Ca^{2+} (Bremel & Weber, 1975; Potter, 1975; Balint et al., 1975) and that the Ca^{2+} binding sites are located on the LC-2 subunit (Werber, Gaffin & Oplatka, 1972; Gaffin & Oplatka, 1974; Morimoto & Harrington, 1974). LC-2 contains one high-affinity Ca^{2+} binding site which is strongly competed for by Mg^{2+} (Werber et al., 1972). Potter (1975) has cautioned that significant binding of Ca^{2+} to myosin under physiological conditions would require a free Mg^{2+} concentration of 0.1 mM or less. However, using a free Mg^{2+} concentration of 0.3 mM, Morimoto & Harrington (1974) found that native thick filaments or synthetic myosin filaments (but not filaments from which myosin 'heads' have been removed by papain digestion) show a reversible increase in sedimentation coefficient and decrease in relative viscosity when the Ca^{2+} concentration is increased from 0.1 μM to 10 μM. Werber & Oplatka (1974) found that removal of LC-2 from myosin by DTNB treatment is accompanied by a decrease in Ca^{2+} sensitivity of actin-activated myosin ATPase in the presence of troponin and tropomyosin. Perrie, Smillie & Perry (1973) found that LC-2 can be phosphorylated in vivo, although none of the functional properties of myosin appear to be affected by this phosphorylation. They isolated a peptide

Ala-Ala-Ala-Glu-Gly-Gly-Ser-Ser-Asn-Val-Phe

in which one of the two serine side chains was the site of phosphorylation. Pires, Perry & Thomas (1974) found good evidence that myosin is phosphorylated in resting muscle, and that similar phosphorylated light chains may be present in other muscles. The enzyme 'myosin light-chain kinase', which specifically phosphorylates LC-2, is rather tightly bound to myosin, and its activity is Ca^{2+} sensitive.

With these studies as background, it was suspected that LC-2 might be related to TnC, and thus also to LC-1, LC-3 and CBP. This would prove for the first time that the two classes of myosin light chains, so different in their biological properties, are nevertheless descended from a common ancestor and might have structures which are at least partially similar. It would also suggest that Ca^{2+} control proteins involved in the two types of muscle regulatory systems are related in structure and therefore might share some common functional characteristics. It was also hoped that the single Ca^{2+} binding site of LC-2 could be located in the sequence by comparison with CBP and TnC. Preliminary sequence studies on LC-2 quickly confirmed its relationship to the other proteins (Collins, 1975*b*). The nearly complete sequence, shown in Fig. 2, aligns with LC-1, LC-3, TnC and CBP in a manner fully consistent with the most favourable alignments deduced from individual comparisons of pairs of protein sequences. LC-2 has 30 % identical residues when compared to either TnC sequence, 26 % with either LC-1 or LC-3, and 37 % with one or more of the CBPs. In comparing the two classes of LCs with TnC and CBP, it appears that LC-1 and LC-3 are more similar to TnC (38 % versus 30 % identical residues), while LC-2 is more similar to CBP (37 % versus 32 % identical residues). Over the span of residues A1–H11, LC-2 is assumed to have a leucine inserted following bc2 and a five-residue deletion at C8–cd1. (This deletion may be due to an as yet 'missing' small peptide with an N-terminal lysine residue.) LC-2 has in addition 25 residues preceding A1 and six residues following H11. The phosphoserine-containing peptide of Perrie *et al.* (1973) is located in the N-terminal segment. As in LC-1 and LC-3, internal sequence repeats are weak in LC-2; alignment of A1–D11 with E1–H11 produces only 19 % identical residues (Fig. 3). Expected hydrophobic core residues (based on the predicted structure of TnC) are preserved in all positions except threonine B7 and a deletion at residue C8. Thus the major portion of LC-2 may also be similar in three-dimensional structure to TnC, with a shorter C helix and cd loop.

As expected, LC-2 also contains four mutually homologous regions (Fig. 4) with weak sequence repeats. It is almost certain that the single Ca^{2+} binding site is located at residues ab1–B3 (Fig. 6). There are five potential side-chain ligands (aspartic acids ab1, ab9, B3 and asparagines ab3, ab5), and aspartic acid occurs in the X position (and also in the −Z position, where glutamic acid is always found in TnC and CBP). The C1–D11 region of LC-2 cannot contain a Ca^{2+} binding site, since residues C8–d1 have been deleted (furthermore, there is no potential ligand at residue D3). The segment ef1–F3 contains only two side-chains with potential ligands. The gh1–H3 segment lacks potential ligands in the X and Z positions, and a neutral amino acid (asparagine H3) occurs at the

−Z position. A closer comparison of the (putative) Ca^{2+} binding site of LC-2 with those of rabbit skeletal TnC provides a further clue to the important question of which sites in TnC are the two Ca^{2+}–Mg^{2+} structural sites and which are the functionally important Ca^{2+}-specific sites. Fig. 6 shows that the LC-2 site resembles the ef1–F3 and gh1–H3 sites of TnC more closely than the ab1–B3 and cd1–D3 sites. There are two noteworthy features of the LC-2 site which occur in the ef1–F3 and gh1–H3 sites of TnC but not in the ab1–B3 and cd1–D3 sites: (1) there is only a single glycine residue; (2) asparagine is at the Y position. Since it is known (Werber *et al.*, 1972; Potter, 1975) that the LC-2 site is a Ca^{2+}–Mg^{2+} site, one may conclude that the ef1–F3 and gh1–H3 sites of TnC are also Ca^{2+}–Mg^{2+} sites. The two functionally important Ca^{2+}-specific sites of TnC would then be the ab1–B3 and cd1–D3 sites. It is interesting to note that the ab1–B3 site of bovine cardiac TnC has apparently lost its Ca^{2+} binding ability. This leads to the speculation that *all* TnCs might contain a functionally 'essential' Ca^{2+} binding site at cd1–D3, and that other sites which may be present are merely evolutionary relics.

A number of comparative studies on LCs from fast (white skeletal) and slow (red skeletal, cardiac) striated muscles of various higher vertebrate species have been carried out (Lowey & Risby, 1971; Weeds & Pope, 1971; Sarkar, Sreter & Gergely, 1971; Weeds & Frank, 1972; Frank & Weeds, 1974). Cardiac and red skeletal myosins each contain two molecules each of two types of LC (molecular weights about 21 000 and 18 000) which are very similar to LC-1 and LC-2 of white skeletal muscle, but there is no LC-3 present. Three thiol peptides have been isolated from LC-1 of bovine cardiac muscle, and the same peptides were found in cat red skeletal muscle (Weeds & Pope, 1971). One of these is identical to the single thiol peptide of white skeletal muscle LC-1 (residues fg4–H6). Another is very similar to residues ab8–B9:

red skeletal, cardiac   Ile-Thr-Tyr-Gly-Gln-Cys-Gly-Asp-Val-Leu-Arg,
white skeletal   Ile-Thr-Leu-Ser-Gln-Val-Gly-Asp-Val-Leu-Arg.

Bovine cardiac LC-2 contains no cysteine residues to compare with the two present in white skeletal LC-2, but the amino acid compositions of the two LC-2s are otherwise nearly identical (see Weeds & Frank, 1972). Further evidence for the correlation of the fast and slow muscle light chains comes from preliminary sequence studies on the cyanogen bromide peptides of bovine cardiac LC-1 (Frank & Weeds, 1974) and LC-2 (J. J. Leger & M. Elzinga, unpublished, cited with permission). It thus seems that, at least in higher vertebrates, the presence of LC-3 is characteristic of fast muscles with relatively high myosin ATPase activity.

Vertebrate smooth muscles also contain two classes of LCs (molecular weights about 20000 and 17000) which are somewhat different from those of striated muscle, and it is not clear whether these correspond to LC-1 (or LC-3) and LC-2 (Kendrick-Jones, 1973; Leger & Focant, 1973). The smaller light chain of chicken gizzard myosin appears to be 'essential' for myosin ATPase activity (Kendrick-Jones, 1973), and the larger LC can be phosphorylated (Conti & Adelstein, 1975). While it is well established that vertebrate striated muscles are controlled via troponin (see Ebashi, and also Szent-Györgyi, this volume) the case is not so clear in smooth muscle. While Ebashi *et al.* (1966) found that regulation is by troponin alone, Bremel (1974) found that control is exerted through myosin and that there is no troponin. Driska & Hartshorne (1975) also did not find troponin, but report that Ca^{2+} sensitivity is nevertheless associated with the thin filaments. They found a 'new' thin filament protein, of molecular weight about 130000, which they speculate may somehow be involved in the regulation.

The most thoroughly characterized invertebrate muscle LCs are those from scallop myosin (Kendrick-Jones *et al.*, 1972; Szent-Györgyi *et al.*, 1973). There are two molecules each of two types of LC, with apparently identical molecular weights (about 18000). As mentioned previously, troponin appears to be absent in scallop muscle, and the Ca^{2+} sensitivity is associated with myosin. When scallop myosin is treated with EDTA, about half of one of the two types of LCs is removed and this abolishes the Ca^{2+} sensitivity. If this 'EDTA light chain' is then recombined with the myosin, Ca^{2+} sensitivity is restored. Recently, Kendrick-Jones (1974) found that rabbit white skeletal LC-2 (but not LC-1 or LC-3) binds and restores Ca^{2+} sensitivity to EDTA-treated scallop myosin. The scallop EDTA LC also binds to DTNB-treated rabbit white skeletal muscle myosin (from which LC-2 has been partially removed), but no Ca^{2+} sensitivity was found in this hybrid. The same type of sensitizing LC has been found in lobster and frog striated muscles and vertebrate smooth and cardiac muscles (J. Kendrick-Jones, E. M. Szentkiralyi & A. G. Szent-Györgyi, unpublished, cited by Kendrick-Jones, 1974). These results suggest very strongly that LC-2 and the EDTA LC must be similar in structure, at least at their heavy chain combining sites. However, there is one possibly significant difference between these two LCs. Scallop myosin binds Ca^{2+} and removal of the EDTA LC is accompanied by a decrease in the amount of Ca^{2+} bound. However, unlike LC-2, the isolated EDTA light chain does not bind Ca^{2+}. The most obvious interpretation of this is that binding to the heavy chain stabilizes the structure of a Ca^{2+}-binding site on the EDTA LC. However, a different type of Ca^{2+}-binding site, in which ligands are contributed by both heavy and light chains, cannot be ruled out.

A number of non-muscle myosins also appear to contain two types of light chains (for a review, see Pollard & Weihing, 1974). The best characterized of these is human platelet myosin, which contains LCs of molecular weights about 20000 and 17000, with amino acid compositions distinctly different from those of striated muscle myosin LCs (Adelstein, Conti & Anderson, 1973). The larger of these can be selectively phosphorylated by an endogenous protein kinase, and in this respect seems to be analogous to rabbit white skeletal muscle LC-2. Unlike the case with LC-2, where phosphorylation does not seem to affect the functional properties of myosin, phosphorylation of the platelet LC is accompanied by a 10-fold increase in actin-activated myosin ATPase activity (Conti & Adelstein, 1975). The Ca^{2+} sensitivity of platelet actin-activated ATPase is associated with the thin filaments, and bands possibly corresponding to the troponin components have been observed on SDS–polyacrylamide gel electrophoresis (Cohen, Kaminski & DeVries, 1973).

It thus seems that LCs from all myosins so far studied fall into two functionally distinct classes. It is tempting, in using higher vertebrate white skeletal muscle as a model, to think of LC-1 and LC-3 as 'catalytic' LCs and LC-2 as a 'regulatory' LC. How then is one to recognize these two classes of LCs on a structural basis in other myosins? They cannot be classified by size, since 'regulatory' LCs may be either smaller (e.g. cardiac muscle) or larger (e.g. smooth muscle) than, or even very similar (e.g. scallops) to, 'regulatory' LCs. Comparisons of amino acid compositions, tryptic peptide maps or sequences of thiol peptides are helpful in only the few cases described above. The sequences of the LCs would appear to vary considerably in different myosins, yet similarities also exist between the two classes of LCs because of their common ancestry. It is obvious that knowledge of extensive portions of the amino acid sequences of LCs from diverse sources is necessary before common structural features which distinguish the two classes of LCs can be recognized.

On a functional level, it is not always possible to recognize the 'catalytic' LCs as being 'essential' for myosin ATPase activity. For example, lobster striated muscle has two types of LCs, with molecular weights about 17000 and 19000. The larger of these can be selectively dissociated by DTNB treatment, but the ATPase activity of myosin is then abolished (Regenstein & Lowey, 1974). 'Regulatory' LCs, at least when isolated from the whole myosin molecule, do not always bind Ca^{2+}. It remains to be seen whether all 'regulatory' LCs, and no 'catalytic' LCs, have the ability to resensitize EDTA-treated scallop myosin or to be phosphorylated by specific kinases.

CONCLUSIONS AND SPECULATIONS

In summary, then, muscle cells contain a unique family of proteins possessing four different sets of biological properties yet derived from a common ancestor. The divergence of these proteins from their common ancestor preceded the divergences which produced the various actomyosin-like contractile systems which exist today. It is tempting to speculate that other proteins, particularly Ca^{2+} control proteins (Kretsinger, 1974) may also belong to this family. However, computer searches of all available protein sequences (Tufty & Kretsinger, 1975; Barker & Dayhoff, 1975) have so far failed to find any new members. On the other hand, the fact that these different muscle proteins possess evolutionary, and probably structural, relationships suggests that at least some of their biological properties may also be similar. Also of interest is the question of possible relatedness to other myofilament proteins, particularly the heavy chains of myosin.

The idea that relationships among the muscle proteins may exist is not a new one. From a comparison of their amino acid compositions, Bailey (1948) suggested that tropomyosin may be a precursor or subunit of myosin. The 'tail' portion of myosin is an α-helical coiled-coil formed from the C-terminal part of two heavy chains. Tropomyosin has the same structure, and it would hardly be surprising to find that it is very similar in sequence, if not actually homologous, to the myosin 'tail'. Laki (1957; see also Laki, 1971) compared the amino acid compositions of actin, tropomyosin, myosin, and myosin fragments, and proposed that myosin could be considered as a complex of tropomyosin and actin. A discrepancy in the amino acid compositions of these proteins led to a prediction that myosin also contains a 'third protein' with relatively high phenylalanine and low tyrosine content which might be similar to a carp muscle protein (now known to be CBP) which had recently been crystallized by Henrotte (1955). This prediction of similarity was soon confirmed when Kominz et al. (1959) first isolated the myosin light chains and determined their amino acid composition. More recently, Weltman & Dowben (1973) speculated that the major proteins of the thick and thin filaments may have a common ancestor, and the myosin molecule may include parts which are similar to actin, tropomyosin and the whole troponin complex. In the thin filaments, TnC binds to the other two troponin components, TnI and TnT, which are rather similar in amino acid composition. Noting the symmetry between the two halves of TnC, Kretsinger (1974) suggested that TnI and TnT may bind to TnC at similar sites and thus may be similar to each other. This reasoning may be extended to the LCs, which bind to myosin heavy chains in the thick filaments. Since LCs and TnC are related, the myosin heavy chains might contain two LC binding sites similar to the TnC binding sites on TnI and TnT. The globular 'heads' of myosin contain binding sites for actin and nucleotide which may be similar to sites on actin which bind nucleotide and other actin monomers. The validity of these speculations can of course be tested by comparing the amino sequences of the various proteins. As mentioned previously, the complete sequences of actin, tropomyosin and TnI from rabbit white skeletal muscle have been determined. These proteins are not related to one another or to the TnC family. The sequence of TnT is not yet available for comparison. Although only a few short (20 residues or less) peptide sequences from the huge (about 1700 residues) myosin heavy chain have so far been determined, considerable progress is expected within the next few years (M. Elzinga, personal communication).

The sequence work on LC-2 described here was done with the very capable assistance of Erlinda Capuno and Nora Jackman, and was supported by grants from the American Heart Association, Massachusetts Affiliate, Inc. (no. 1296), the National Institutes of Health (HL-17464) and the National Science Foundation (GB-35070). The continued encouragement and support of Drs Marshall Elzinga and John Gergely are gratefully acknowledged. Sequence work on rabbit skeletal TnC was done in collaboration with Drs M. L. Greaser, J. D. Potter and M. J. Horn, and a manuscript describing the experimental details is now in preparation.

REFERENCES

ADELSTEIN, R. S., CONTI, M. A. & ANDERSON, W., JR (1973). Phosphorylation of human platelet myosin. *Proc. natn. Acad. Sci. USA*, **70**, 3115–3119.

BAILEY, K. (1948). Tropomyosin: A new asymmetric protein component of the muscle fibril. *Biochem. J.*, **43**, 271–279.

BALINT, M., SRETER, F. A., WOLF, I., NAGY, B. & GERGELY, J. (1975). The substructure of heavy meromyosin. The effect of Ca^{2+} and Mg^{2+} on the tryptic fragmentation of heavy meromyosin. *J. biol. Chem.*, **250**, 6168–6177.

BARKER, W. C. & DAYHOFF, M. O. (1975). Gene duplications in the evolution of muscle proteins. *Biophys. J.*, **15**, 121a.

BECKER, J. W., REEKE, G. N., JR, WANG, J. A., CUNNINGHAM, B. A. & EDELMAN, G. M. (1975). The covalent and three-dimensional structure of concanavalin A. III. Structure of the monomer and its interactions with metals and saccharides. *J. biol. Chem.*, **250**, 1513–1524.

BLUM, H. E., POCINWONG, S. & FISCHER, E. H. (1974). A phosphate-acceptor protein related to parvalbumins in dogfish skeletal muscle. *Proc. natn. Acad. Sci. USA*, **71**, 2198–2202.

BREMEL, R. D. (1974). Myosin-linked calcium regulation in vertebrate smooth muscle. *Nature, Lond.*, **252**, 405–407.

BREMEL, R. D. & WEBER, A. (1975). Calcium binding to rabbit skeletal myosin under physiological conditions. *Biochim. biophys. Acta*, **376**, 366–374.

BRIGGS, N. (1975). Identification of the soluble relaxing factor as a parvalbumin. *Fedn Proc.*, **34**, 540.

CAPONY, J.-P., RYDEN, L., DEMAILLE, J. & PECHÈRE, J.-F. (1973). The primary structure of the major parvalbumin from hake muscle. *Eur. J. Biochem.*, **32**, 97–108.

COFFEE, C. J. & BRADSHAW, R. A. (1973). Carp muscle calcium-binding protein. I. Characterization of the tryptic peptides and the complete amino acid sequence of component B. *J. biol. Chem.*, **248**, 3305–3312.

COHEN, I., KAMINSKI, E. & DeVRIES, A. (1973). Actin-linked regulation of the human platelet contractile system. *FEBS Lett.*, **34**, 315–317.

COLLINS, J. H. (1974). Homology of myosin light chains, troponin C and parvalbumins deduced from comparison of their amino acid sequences. *Biochem. biophys. Res. Commun.*, **58**, 301–308.

(1975*a*). Purification and analysis of the cyanogen bromide peptides of troponin T from rabbit skeletal muscle. *Biochem. biophys. Res. Commun.*, **65**, 604–610.

(1975*b*). Myosin DTNB light chain: amino acid sequence and relationship to other muscle proteins. *Fedn Proc.*, **34**, 539.

COLLINS, J. H. & ELZINGA, M. (1975). The primary structure of actin from rabbit skeletal muscle. Completion and analysis of the amino acid sequence. *J. biol. Chem.*, **250**, 5915–5920.

COLLINS, J. H., POTTER, J. D., HORN, M. J., WILSHIRE, G. & JACKMAN, N. (1973). The amino acid sequence of rabbit skeletal muscle troponin C: gene replication and homology with calcium binding proteins from carp and hake muscle. *FEBS Lett.*, **36**, 268–272.

(1974). Structural studies on rabbit skeletal muscle troponin C: evidence for gene replication and homology with calcium binding proteins from carp and hake muscle. In *Calcium Binding Proteins*, ed. W. Drabikowski, H. Strezelecka-Golaszewska & E. Carafoli, pp. 51–64. Elsevier: Amsterdam.

CONTI, M. A. & ADELSTEIN, R. S. (1975). Platelet myosin phosphorylation controls actin activation of myosin ATPase activity. *Fedn Proc.*, **34**, 670.

COTTON, F. A., BIER, C. J., DAY, V. W., HAZEN, E. E., JR & LARSEN, S. (1971). Some aspects of the structure of staphylococcal nuclease. *Cold Spring Harb. Symp. quant. Biol.*, **36**, 243–249.

CRICK, F. H. C. (1953). The packing of α-helices: simple coiled coils. *Acta Cryst.*, **6**, 689–697.

DEMAILLE, J., DUTRUGE, E., CAPONY, J.-P. & PECHÈRE, J.-F. (1974*a*). Muscular parvalbumins: a family of homologous calcium-binding proteins. Their relation to the calcium-binding troponin component. In *Calcium Binding Proteins*, ed. W. Drabikowski, H. Strezelecka-Golaszewska & E. Carafoli, pp. 643–677. Elsevier: Amsterdam.

DEMAILLE, J., DUTRUGE, E., EISENBERG, E., CAPONY, J.-P. & PECHÈRE, J.-F. (1974*b*). Troponin C from reptile and fish muscles and their relation to muscular parvalbumins. *FEBS Lett.*, **42**, 173–178.

DEUTICKE, H. J. (1934). The sedimentation constants of muscle proteins. *Z. physiol. Chem.*, **224**, 216–228.

DREIZEN, P. & RICHARDS, D. H. (1972). Studies on the role of light and heavy chains in myosin adenosine triphosphatase. *Cold Spring Harb. Symp. quant. Biol.*, **37**, 29–45.

DRISKA, S. & HARTSHORNE, D. J. (1975). The contractile proteins of smooth muscle. Properties and components of a Ca^{2+}-sensitive actomyosin from chicken gizzard. *Archs. Biochem. Biophys.*, **167**, 203–212.

EBASHI, S. & ENDO, M. (1968). Calcium ion and muscle contraction. *Prog. Biophys. molec. Biol.*, **18**, 123–183.

EBASHI, S., ENDO, M. & OHTSUKI, I. (1969). Control of muscle contraction. *Quart. Rev. Biophys.*, **2**, 351–384.

EBASHI, S., IWAKURA, H., NAKAJIMA, H., NAKAMURA, R. & OOI, Y. (1966). New structural proteins from dog heart and chicken gizzard. *Biochem. Z.*, **345**, 201–211.

EBASHI, S. & KODAMA, A. (1965). A new factor promoting aggregation of tropomyosin. *J. Biochem., Tokyo*, **58**, 107–108.

ELZINGA, M., COLLINS, J. H., KUEHL, W. M. & ADELSTEIN, R. S. (1973). Complete amino acid sequence of actin of rabbit skeletal muscle. *Proc. natn. Acad. Sci. USA*, **70**, 2687–2691.

ENFIELD, D. L., ERICSON, L. H., BLUM, H. E., FISCHER, E. H. & NEURATH, H. (1975). Amino acid sequence of parvalbumin from rabbit skeletal muscle. *Proc. natn. Acad. Sci. USA*, **72**, 1309–1313.

FRANK, G. & WEEDS, A. G. (1974). The amino acid sequence of the alkali light chains of rabbit skeletal muscle myosin. *Eur. J. Biochem.*, **44**, 317–334.

FRANKENNE, F., JOASSIN, L. & GERDAY, CH. (1973). The amino acid sequence of the pike (*Esox lucius*) parvalbumin III. *FEBS Lett.*, **35**, 145–147.

FRIEDBERG, F. (1974). Effects of metal binding on protein structure. *Quart. Rev. Biophys.*, **7**, 1–33.

GAFFIN, L. & OPLATKA, A. (1974). Physico-chemical studies on the light chains of myosin. II. Effect of metal ions on the electrophoretic behavior of myosin subunits. *J. Biochem., Tokyo*, **75**, 277–281.

GERDAY, CH. (1974). The amino acid sequence of the parvalbumin II of pike (*Esox lucius*). *Abstr. Commun. 9th Mtg Fedn Eur. Biochem. Soc.*, Abstr. S1D2.

GOSSELIN-REY, C. (1974). Fish parvalbumins: immunochemical reactivity and biological distribution. In *Calcium Binding Proteins*, ed. W. Drabikowski, H. Strezelecka-Golaszewska & E. Carafoli, pp. 679–701. Elsevier: Amsterdam.

GREASER, M. L. & GERGELY, J. (1971). Reconstitution of troponin activity from three protein components *J. biol. Chem.*, **246**, 4226–4233.

(1973). Purification and properties of the components from troponin. *J. biol. Chem.*, **248**, 2125–2133.

HANSON, J. & HUXLEY, H. E. (1953). Structural basis of the cross striations in muscle. *Nature, Lond.*, **172**, 530–532.

HEIZMANN, C. W., MALENCIK, D. A. & FISCHER, E. H. (1974). Generation of parvalbumin-like proteins from troponin. *Biochem. biophys. Res. Commun.*, **57**, 162–168.

HENROTTE, J. G. (1955). A crystalline component of carp myogen precipitating at high ionic strength. *Nature, Lond.*, **176**, 1221.

HITCHCOCK, S. E. (1975). Regulation of muscle contraction: binding of troponin and its components to actin and tropomyosin. *Eur. J. Biochem.*, **52**, 255–263.

HITCHCOCK, S. E., HUXLEY, H. E. & SZENT-GYÖRGYI, A. G. (1973). Calcium-sensitive binding of troponin to actin-tropomyosin: a two site model for troponin action. *J. molec. Biol.*, **80**, 825–836.

HUXLEY, H. E. (1972). Structural changes in the actin- and myosin-containing filaments during contraction. *Cold Spring Harb. Symp. quant. Biol.*, **37**, 361–376.

(1973). Muscle contraction and cell motility. *Nature, Lond.*, **243**, 445–449.

JOASSIN, L. (1974). The amino acid sequence of the whiting (*Gadus merlangus*) parvalbumin III. *Abstr. Commun. 9th Mtg Fedn Eur. Biochem. Soc.*, Abstr. S1D1.

KENDRICK-JONES, J. (1973). The subunit of gizzard myosin. *Phil. Trans. R. Soc. Ser. B*, **265**, 183–189.

(1974). Role of myosin light chains in calcium regulation. *Nature, Lond.*, **249**, 631–634.

KENDRICK-JONES, J., LEHMAN, W. & SZENT-GYÖRGYI, A. G. (1970). Regulation in molluscan muscles. *J. molec. Biol.*, **54**, 313–326.

KENDRICK-JONES, J., SZENTKIRALYI, E. M. & SZENT-GYÖRGYI, A. G. (1972). Myosin-linked regulatory systems: the role of the light chains. *Cold Spring Harb. Symp. quant. Biol.*, **37**, 47–53.

KOMINZ, D. R., CARROLL, W. R., SMITH, E. N. & MITCHELL, E. R. (1959). A subunit of myosin. *Archs. Biochem. Biophys.*, **79**, 191–199.

KRETSINGER, R. H. (1972). Gene triplication deduced from the tertiary structure of a muscle calcium-binding protein. *Nature New Biol.*, **240**, 85–88.

(1974). Calcium binding proteins and natural membranes. In *Perspectives in Membrane Biology*, ed. S. Estrada-O. & C. Gitler, pp. 229–262. Academic Press: New York, San Francisco, London.

KRETSINGER, R. H. & BARRY, C. D. (1975). The predicted structure of the calcium-binding component of troponin. *Biochim. biophys. Acta*, **405**, 40–52.

KRETSINGER, R. H. & NOCKOLDS, C. E. (1973). Carp muscle calcium-binding protein. II. Structure determination and general description. *J. biol. Chem.*, **248**, 3313–3326.

LAKI, K. (1957). The composition of contractile muscle proteins. *J. cell. comp. Physiol.*, **49**, 249–265.

(1971). Size and shape of the myosin molecule. In *Contractile Proteins and Muscle*, ed. K. Laki, pp. 179–217. Marcel Dekker: New York.

LEGER, J. J. & FOCANT, B. (1973). Low molecular weight components of cow smooth muscle myosins: characterization and comparison with those of striated muscle. *Biochim. biophys. Acta*, **328**, 166–172.

LEHKY, P., BLUM, H. E., STEIN, E. A. & FISCHER, E. H. (1974). Isolation and characterization of parvalbumins from the skeletal muscle of higher vertebrates. *J. biol. Chem.*, **249**, 4332–4334.

LEHMAN, W., KENDRICK-JONES, J. & SZENT-GYÖRGYI, A. G. (1972). Myosin-linked regulatory systems: comparative studies. *Cold Spring Harb. Symp. quant. Biol.*, **37**, 319–330.

LILJAS, A. & ROSSMANN, M. G. (1974). X-ray studies of protein interactions. *A. Rev. Biochem.*, **43**, 475–507.

LOCKER, R. H. (1956). The dissociation of myosin by heat coagulation. *Biochim. biophys. Acta*, **20**, 514–521.

LOWEY, S. & HOLT, J. C. (1972). An immunochemical approach to the interaction of light and heavy chains in myosin. *Cold Spring Harb. Symp. quant. Biol.*, **37**, 19–28.

LOWEY, S. & RISBY, D. (1971). Light chains from fast and slow muscle myosins. *Nature, Lond.*, **234**, 81–84.

LOWEY, S., SLAYTER, H. S., WEEDS, A. G. & BAKER, H. (1969). Substructure of the myosin molecule. I. Subfragments of myosin by enzymatic degradation. *J. molec. Biol.*, **42**, 1–29.

MALENCIK, D. A., HEIZMANN, C. W. & FISCHER, E. H. (1975). Structural proteins of dogfish skeletal muscle. *Biochemistry, N.Y.*, **14**, 715–721.

MARGOSSIAN, S. S. & COHEN, C. (1973). Troponin subunit interactions. *J. molec. Biol.*, **81**, 409–413.

MATTHEWS, B. W., COLEMAN, P. M., JANSONIUS, J. N., TITANI, K., WALSH, K. & NEURATH, H. (1972). Structure of thermolysin. *Nature New Biol.*, **238**, 41–43.

McLACHLAN, A. D. (1972). Gene duplication in carp muscle calcium-binding proteins. *Nature New Biol.*, **240**, 83–85.

MERCOLA, D., BULLARD, B. & PRIEST, J. (1975). Crystallization of troponin C. *Nature, Lond.*, **254**, 634–635.

MOEWS, P. C. & KRETSINGER, R. H. (1975). Terbium replacement of calcium in carp muscle calcium-binding parvalbumin: an X-ray crystallographic study. *J. molec. Biol.*, **91**, 229–232.

MOORE, P. B., HUXLEY, H. E. & DeROSIER, D. J. (1970). Three-dimensional reconstruction of F-actin, thin filaments and decorated thin filaments. *J. molec. Biol.*, **50**, 279–295.

MORIMOTO, K. & HARRINGTON, W. F. (1974). Evidence for structural changes in vertebrate thick filaments induced by Ca^{2+}. *J. molec. Biol.*, **88**, 693–709.

PEARLSTONE, J. R., CARPENTER, M. R. & SMILLIE, L. B. (1975). The primary structure of troponin T: cyanogen bromide fragments. *Fedn Proc.*, **34**, 539.

PECHÈRE, J.-F. (1974). Isolation of a parvalbumin from rabbit muscle. *C.R. Acad. Sci. Paris*, **278**, 2577–2579.

PECHÈRE, J.-F., CAPONY, J.-P. & DEMAILLE, J. (1973). Evolutionary aspects of the structure of muscular parvalbumins. *Syst. Zool.*, **22**, 533–548.

PECHÈRE, J.-F., CAPONY, J.-P. & RYDEN, L. (1971). The primary structure of the major parvalbumin from hake muscle. Isolation and general properties of the protein. *Eur. J. Biochem.*, **23**, 421–428.

PECHÈRE, J.-F., DEMAILLE, J., CAPONY, J.-P., DUTRUGE, E., BARON, G. & PINA, C. (1975). Muscular parvalbumin. Some explorations into their possible biological significance. In *Calcium Transport in Contraction and Secretion*, ed. E. Carafoli, F. Clementi, W. Drabikowski & A. Margreth, pp. 459–468. North-Holland: Amsterdam & New York.

PERRIE, W. T., SMILLIE, L. B. & PERRY, S. V. (1973). A phosphorylated light chain component of myosin from skeletal myosin. *Biochem. J.*, **135**, 151–164.

PERUTZ, M. F. & TEN EYCK, L. F. (1971). Stereochemistry of cooperative effect in hemoglobin. *Cold Spring Harb. Symp. quant. Biol.*, **36**, 295–310.

PIRES, E., PERRY, S. V. & THOMAS, M. A. W. (1974). Myosin light chain kinase, a new enzyme from striated muscle. *FEBS Lett.*, **41**, 292–296.

POLLARD, T. D. & WEIHING, R. R. (1974). Actin and myosin and cell movement. *CRC crit. Rev. Biochem.*, **2**, 1–65.

POTTER, J. D. (1975). Effect of Mg^{2+} on Ca^{2+} binding to myosin. *Fedn Proc.*, **34**, 671.

POTTER, J. D. & GERGELY, J. (1974a). Troponin, tropomyosin, and actin interactions in the Ca^{2+} regulation of muscle contraction. *Biochemistry, N.Y.*, **13**, 2697–2703.

(1974b). Ca^{2+} and Mg^{2+} binding to troponin and the regulation of muscle contraction. *Fedn Proc.*, **33**, 1465.

(1975). The calcium and magnesium binding sites on troponin and their role in the regulation of myofibrillar adenosine triphosphatase. *J. biol. Chem.*, **250**, 4628–4633.

POTTER, J., LEAVIS, P., SEIDEL, J., LEHRER, S. & GERGELY, J. (1975). Interaction of divalent cations with troponin and myosin. In *Calcium Transport in Contraction and Secretion*, ed. E. Carafoli, F. Clementi, W. Drabikowski & A. Margreth, pp. 415–425. North-Holland: Amsterdam & New York.

POTTER, J. D., SEIDEL, J. C., LEAVIS, P. C., LEHRER, S. S. & GERGELY, J. (1974). Interaction of Ca^{2+} with troponin. In *Calcium Binding Proteins*, ed. W. Drabikowski, H. Strzelecka-Golaszewska & E. Carafoli, pp. 129–152. Elsevier: Amsterdam.

REGENSTEIN, J. M. & LOWEY, S. (1974). Light chains of lobster myosin. *Fedn Proc.*, **33**, 1581.

REGENSTEIN, J. M. & SZENT-GYÖRGYI, A. G. (1975). Regulatory proteins of lobster striated muscle. *Biochemistry, N.Y.*, **14**, 917–925.

SARKAR, S. (1972). Stoichiometry and sequential removal of the light chains of myosin. *Cold Spring Harb. Symp. quant. Biol.*, **37**, 14–17.

SARKAR, S., SRETER, F. A. & GERGELY, J. (1971). Light chains of myosins from white, red and cardiac muscles. *Proc. natn. Acad. Sci. USA*, **68**, 946–950.

STARR, R. & OFFER, G. (1973). Polarity of the myosin molecule. *J. molec. Biol.*, **81**, 17–31.

STONE, D., SODEK, J. & SMILLIE, L. B. (1974). Relationships between the structure of tropomyosin and its amino acid sequence. *Fedn Proc.*, **33**, 1582.

SZENT-GYÖRGYI, A. G., SZENTKIRALYI, E. M. & KENDRICK-JONES, J. (1973). The light chains of scallop myosin as regulatory subunits. *J. molec. Biol.*, **74**, 179–203.

TSAO, T. C. (1953). Fragmentation of the myosin molecule. *Biochim. biophys. Acta*, **11**, 368–382.

TUFTY, R. M. & KRETSINGER, R. H. (1975). Troponin and parvalbumin calcium binding regions predicted in myosin light chain and T4 lysozyme. *Science, Wash.*, **187**, 167–169.

VAN EERD, J. P. & TAKAHASHI, K. (1975). The amino acid sequence of bovine cardiac troponin C. Comparison with rabbit skeletal troponin C. *Biochem biophys. Res. Commun.*, **64**, 122–127.

VANAMAN, T. C., HARRELSON, W. G., JR & WATTERSON, D. M. (1975). Studies on a troponin C like Ca²⁺ binding protein from brain. *Fedn Proc.*, **34**, 307.

WAGNER, P. D. & YOUNT, R. G. (1975). Subunit location of sulfhydryl groups of myosin labeled with a purine disulfide analog of adenosine triphosphate. *Biochemistry, N.Y.*, **14**, 1908–1914.

WEBER, A. & MURRAY, J. M. (1973). Molecular control mechanisms in muscle contraction. *Physiol. Rev.*, **53**, 612–673.

WEEDS, A. G. (1967). Small subunits of myosin. *Biochem. J.*, **105**, 25C–27C. (1969). Light chains of myosin. *Nature, Lond.*, **223**, 1362–1364.

WEEDS, A. G. & FRANK, G. (1972). Structural studies on the light chains of myosin. *Cold Spring Harb. Symp. quant. Biol.*, **37**, 9–14.

WEEDS, A. G., HALL, R. & SPURWAY, N. C. S. (1975). Characterization of myosin light chains from histochemically identified fibres of rabbit psoas muscles. *FEBS Lett.*, **49**, 320–324.

WEEDS, A. G. & LOWEY, S. (1971). Substructure of the myosin molecule. II. The light chains of myosin. *J. molec. Biol.*, **61**, 701–725.

WEEDS, A. G. & MCLACHLAN, A. D. (1974). Structural homology of myosin alkali light chains, troponin C and carp calcium binding protein. *Nature, Lond.*, **252**, 646–649.

WEEDS, A. G. & POPE, B. (1971). Chemical studies on light chains from cardiac and skeletal muscle myosins. *Nature, Lond.*, **234**, 85–88.

WELTMAN, J. K. & DOWBEN, R. M. (1973). Relatedness among contractile and membrane proteins: evidence for evolution from common ancestral genes. *Proc. natn. Acad. Sci. USA*, **70**, 3230–3234.

WERBER, M. M., GAFFIN, S. L. & OPLATKA, A. (1972). Physico-chemical studies on the light chains of myosin. I. Effect of metal ions on the fluorescence of a light chain (LMP-II) from rabbit myosin. *J. Mechanochem. Cell Motil.*, **1**, 91–96.

WERBER, M. M. & OPLATKA, A. (1974). Physico-chemical studies on the light chains of myosin. III. Evidence for a regulatory role of a rabbit myosin light chain. *Biochem. biophys. Res. Commun.*, **57**, 823–830.

WILKINSON, J. M. & GRAND, R. J. A. (1975). The amino acid sequence of troponin I from rabbit skeletal muscle. *Biochem. J.*, **149**, 493–496.

COMPARATIVE SURVEY OF
THE REGULATORY ROLE OF
CALCIUM IN MUSCLE

By A. G. SZENT-GYÖRGYI

Department of Biology, Brandeis University,
Waltham, Massachusetts 02154, USA

Contraction of all muscles so far studied is triggered and controlled by calcium ions. At rest, the free calcium concentration in the sarcoplasm is low and cross-links between actin and myosin cannot form. Stimulation releases calcium ions from membranous compartments. In the presence of calcium ions, actin and myosin interact and the muscle contracts as long as calcium is available for the contractile system. In conditions of rest, in the presence of magnesium ions, the ATPase activity of myosin is low and is strongly accelerated by actin. The actin activation of the ATPase activity is the result of the acceleration of the release of the product ADP from the myosin intermediate, a rate-limiting step in the chemical cycle associated with the mechanical events (Lymn & Taylor, 1971; Bagshaw & Trentham, 1974). The actin-activated ATPase activity is thus related to the contractile events and may be taken as an in-vitro measure of contractile activity. The calcium dependence of this ATPase is an in-vitro measure of regulation and forms the experimental basis of this discussion.

Pure rabbit myosin and rabbit actin do not require calcium ions for interaction, and the ATPase activity is high even in the absence of calcium. Actomyosin preparations are calcium sensitive only in the presence of control proteins. Control proteins are also needed for calcium binding. These proteins function by preventing the interaction between actin and myosin in the absence but not in the presence of calcium.

Control proteins have been discovered in vertebrate muscles, where they consist of troponin and tropomyosin (Ebashi, 1963; Ebashi & Kodama, 1965). These proteins are associated with actin and modify actin behaviour. Not all regulation, however, is actin-linked. In molluscan muscles, for example, troponin is absent; the regulatory component is a particular light chain of myosin that modifies behaviour of myosin (Szent-Györgyi, Szentkiralyi & Kendrick-Jones, 1973). This discussion will be focused on how these two regulatory systems are distributed in the animal kingdom, and I will discuss some of the problems that arise from this distribution.

[335]

Table 1. *Properties of regulatory systems*

	Actin control	Myosin control
Components	Three different troponin subunits and tropomyosin	Regulatory light chain
Interaction	Only with actin	Only with myosin
Presence of tropomyosin	Obligatory	Not needed
Calcium binder	Troponin C	Myosin (light and heavy chains are both required)
Effects in the absence of calcium	Block sites on actin	Block sites on myosin
Result	No cross-link formed	No cross-link formed
ATP requirement	Yes	Yes
Species of regulated myosin	[My** (ADP, P_i)]	[My** (ADP, P_i)]
Rigor links	Not prevented	Not prevented

For My** see text.

This discussion is based on joint experiments with W. Lehman on the comparative aspects of regulation and with J. Kendrick-Jones and E. M. Szentkiralyi on the role of the regulatory light chains of myosin.

COMPARISON OF ACTIN-
AND MYOSIN-LINKED REGULATION

The significant aspects of actin and myosin control are summarized on Table 1. The major features of actin control are as follows: tropomyosin and three subunits of troponin are all required for calcium sensitivity; the three subunits of troponin have different functions; there is a calcium binding subunit, an inhibitory subunit and a subunit responsible for the binding to tropomyosin (Greaser & Gergely, 1971); actin displays a co-operative behaviour (Bremel & Weber, 1972); a movement of tropomyosin occurs which is controlled by calcium binding on troponin (Huxley, 1972). Troponin is distributed periodically along the thin filaments and cannot be in direct contact with all the actin monomers that it controls (Ohtsuki, Masaki, Nonomura & Ebashi, 1967). The changes that take place on troponin as a result of calcium binding by troponin C are relayed by tropomyosin to the individual actin monomers. The structural studies suggest that there is a calcium-dependent movement of tropomyosin (O'Brien, Bennet & Hanson, 1971; Haselgrove, 1972; Huxley, 1972; Lowy & Vibert, 1972; Vibert, Haselgrove, Lowy & Paulsen, 1972). In the absence of calcium, tropomyosin sterically hinders combination with myosin; in

Table 2. *Calcium sensitivity, calcium binding and the regulatory light chains of scallop myofibrils*[a]

	Calcium sensitivity[b] (%)	Ca bound[c] mole myosin	Mole regulatory light chain[d] mole myosin
Myofibrils	98	1.9	2.0
Desensitized myofibrils	20	1.0	0.92
Desensitized myofibrils reconstituted with scallop light chains	93	1.7	1.9
Desensitized myofibrils reconstituted with rabbit DTNB-light chains	96	1.05	1.8
Scallop regulatory light chain	No activity	No binding	

[a] Data compiled from Szent-Györgyi *et al.* (1973), Kendrick-Jones (1974) and Kendrick-Jones, Szentkiralyi & Szent-Györgyi (1976).

[b] Calcium sensitivity: $\left(1 - \dfrac{\text{ATPase}_{EGTA}}{\text{ATPase}_{Ca^{2+}}}\right)$ 100.

[c] From Scatchard plots.

[d] From densitometry of 10 % urea acrylamide gel electrophoresis stained with Acid Fast Green.

the presence of calcium it moves away from its inhibitory position towards the centre of the actin groove where it no longer interferes with actin sites required for combination with myosin (Huxley, 1972; Parry & Squire, 1973). In contrast, myosin control does not require the presence of tropomyosin and actin monomers do not seem to be regulated in a cooperative manner (Kendrick-Jones, Lehman & Szent-Györgyi, 1970). Myosin control depends on the presence of a single regulatory subunit (Szent-Györgyi *et al.*, 1973). This regulatory light chain is required for both calcium binding and for calcium regulation. The isolated regulatory light chain, however, does not bind calcium in the presence of excess magnesium (Table 2). The location of the high-affinity calcium binding sites has not yet been clearly established. These may be present on the light chain but would require the presence of heavy chains to retain the configuration required for calcium binding; alternatively, the binding site may be on the heavy chain provided its conformation is maintained by the light chain. There is a possibility, however, that both the regulatory light chain and the heavy chain directly contribute to the calcium binding site.

Although the components of the two regulatory systems are clearly different, there are common features in their overall functions. Both regulations work by preventing bond formation between actin and myosin in the absence of calcium. Both regulations require the presence of ATP for their inhibitory function and interfere with cross-link formation only between actin and the myosin intermediate that is formed after the hydrolytic step but that still retains the products ADP and P_i [My** (ADP, P_i)] (Bagshaw & Trentham, 1974; Marston & Lehman, 1974). Rigor links, i.e. interaction between actin and myosin in the absence of ATP, are not prevented by the regulatory systems. The light chains and the components of troponin are different in composition and properties. Light chains are not bound on thin filaments and troponin or its components are not bound by myosin. Regulatory light chains do not substitute for troponin C or troponin I in actin control, and none of the troponin components substitute for the regulatory light chains in myosin control.

DISTRIBUTION OF REGULATION

Experimental approaches

Pure rabbit myosin and rabbit actin do not form a regulated actomyosin since regulatory proteins are absent or are not functional. By hybridizing unknown myosin and thin filament preparations with the respective rabbit proteins, one may probe for the presence of a particular control. Molluscan myosin preparations when combined with pure rabbit actin form a calcium-sensitive actomyosin even in the absence of tropomyosin. Correspondingly, these myosins bind calcium with a high affinity in the presence of a great excess of magnesium (Kendrick-Jones et al., 1970). In contrast, unregulated myosins from vertebrates form a calcium-insensitive actomyosin complex with pure rabbit actin. Regulated thin filaments when combined with rabbit myosin show calcium sensitivity, while similar complexes made from molluscan thin filaments are not regulated. Complementary experiments further indicate that regulated thin filaments bind calcium and contain small molecular weight components corresponding to the troponin subunits. Unregulated thin filaments bind no calcium and, in general, are lacking in low molecular weight components (Lehman, Kendrick-Jones & Szent-Györgyi, 1972; Lehman & Szent-Györgyi, 1975).

A demonstration of regulated behaviour of isolated thin filaments or myosin preparations is the most direct evidence for the presence of a particular regulatory system. There are problems, however. The procedures involved in preparing myosins or thin filaments may lead in some instances to the loss or inactivation of regulatory components (Lehman, Bullard & Hammond, 1974; Lehman & Szent-Györgyi, 1975). Isolated myosin preparations of a number of invertebrates are very labile and difficult to work with. Tropomyosin is lost easily from some thin filaments. Only limited amounts of muscle tissue can be obtained from many organisms. To avoid some of these problems, the fastest and simplest methods were used to prepare myosin and thin filaments. The aim was to avoid a significant contamination of the thin filaments with myosin and to reduce the amount of actin in the myosin without losses in regulatory function. Other impurities, particularly paramyosin in the myosin preparations, were not entirely removed.

We have been able to show myosin-linked regulation in 17 different species with the aid of purified myosins, including several different molluscan myosins and myosins from locust, tarantula, horseshoe crab, *Golfingia*, *Glottidia*, mantis shrimp, *Urechis*, and *Priapulus*. Thin filaments were obtained from 62 different animals; 39 of these thin filaments were regulated (Lehman & Szent-Györgyi, 1975).

The different properties of actin-linked and myosin-linked control systems make it possible to analyse myofibrils and unfractionated actomyosin extracts for regulation (Lehman *et al.*, 1972; Lehman & Szent-Györgyi, 1975). One may test for myosin-control in actomyosin by flooding the preparation with excess pure actin. Since troponin acts by reducing the affinity of actin to myosin (Eisenberg & Kielley, 1970; Hartshorne & Pyun, 1971; Koretz, Hunt & Taylor, 1972), the presence of regulated thin filaments does not interfere with the reaction between myosin and the pure actin added. Activation of the ATPase activity by excess pure actin in the absence of calcium indicates that the myosin lacks control since it does not require calcium to combine with actin. If the ATPase activity remains unaltered by excess pure actin, then the myosin is regulated since it requires calcium for combination with pure actin. This competitive actin activation assay can be complemented by a competitive myosin activation test (Lehman & Szent-Györgyi, 1975). Here the regulatory properties of thin filaments are tested with the aid of pure rabbit myosin that is added to actomyosin or myofibrils. Calcium sensitivity of the incremental ATPase activity due to rabbit myosin indicates the presence of actin control in the actomyosin; in the absence of such control the incremental ATPase activity does not depend on calcium. The competitive actin activation test has been used with all muscles studied. The competitive myosin activation test, however, is restricted to muscles that have relatively low ATPase activities, otherwise the contribution of the added myosin is difficult to evaluate. The muscles of 39 species have been tested by the competitive myosin activation test; 17 of these showed actin control. The competitive activation tests are suitable to test small amounts of crude muscle extracts. The interpretation of the results depends on the assumption that there is little or no species specificity between the interaction of various myosins and actins. Since thin filaments from 62 different organisms combine with rabbit myosin and isolated myosin preparations from 17 different organisms

Table 3. *Distribution of regulation*

Both actin and myosin control	Only actin control	Only myosin control
Insecta (12)		Mollusca
		Amphineura (1)
		Gastropoda (5)
Chelicerata (2)		Pelecypoda (17)
		Cephalopoda (1)
Crustacea		
Cirripedia (4)		Brachiopoda (1)
Amphipoda (7)		
Isopoda (6)		Echiuroida (1)
Stomatopoda (1)	Mysidacea (2)	Nemertina (2)
Decapoda, slow (2)	Decapoda, fast (14)	Echinoderma (2)
Annelida (4)		
Priapulida (1)		
Sipunculida (2)	Sipunculida? (1)	
Nematoda (1)		
Physarum (?)[a]		
Vertebrate	Vertebrate (13)	
smooth (?)[b]	striated	

Numbers indicate different species studied.
After Lehman & Szent-Györgyi (1975).

[a] Nachmias & Asch (1974) showed the presence of actin control. Kato & Tonomura (1975) isolated a troponin–tropomyosin complex from *Physarum* that conferred calcium sensitivity to rabbit actomyosin reconstituted from pure actin and myosin.
[b] Bremel (1974) showed the presence of myosin control.

react with rabbit actin, there appears to be little difficulty in forming hybrids between the actins and myosins of different species.

Distribution of regulatory systems

The distribution of control systems in the animal kingdom is relatively simple (Table 3). Many invertebrate muscles are doubly regulated and utilize both actin control and myosin control simultaneously. Muscles having single regulation are more of an exception. Single actin control is restricted to the striated muscles of vertebrates and to the fast muscles of some crustaceans, such as the decapods and mysids. Single myosin control characterizes molluscan muscles. Single myosin control was found also in echinoderms, brachiopods, echiuroids; however, only one or two species were examined from each of these phyla. There is a consistency in the regulation of the muscles among the diverse members of a phylum. All vertebrate striated muscles are regulated similarly, so are all molluscan muscles. The differences among crustaceans follow major groupings. On the other hand, different muscle types may be controlled differently in the same animal. The slow muscles of lobster and crayfish, for instance, are doubly controlled while the fast muscles of the same animals show only an actin control (Lehman & Szent-Györgyi, 1975). In contrast to the vertebrate striated muscles, the vertebrate smooth muscles exhibit a myosin control (Bremel, 1974). Nevertheless, the presence of a particular control system is not associated with a particular muscle structure or with a particular physiological function. Both myosin control and actin control are widely distributed, exist in the muscles of primitive organisms, and the data do not suggest that one regulatory system preceded the other.

The existence of double control makes teleonomical sense. It offers a surer way to prevent occasional cross-link formation in the resting state. The accuracy of regulation is enhanced by having two control systems operating, and the degree of accuracy may be gleaned from A. V. Hill's studies that show that heat production in contracting muscle exceeds by several orders of magnitude the heat produced at rest. There is no obvious explanation, on the other hand, why a single control would be an advantage for any muscle. Therefore, the evidence for such assignments needs further critical examination since the assignment is based on functional tests, and inactivation or loss of regulatory components during preparation could lead to errors in classification.

Muscles with a single actin control

Sodium dodecyl sulphate (SDS) acrylamide gel electrophoresis indicates qualitatively that the thin filaments and washed myofibrils of molluscs are deficient in troponin and actin control is not found, since the components of actin control are not present in sufficient amounts for regulation (Lehman et al., 1972; Lehman & Szent-Györgyi, 1975). Low molecular weight components in small amounts, however, can be detected on many molluscan thin filaments and washed myofibrils by SDS acrylamide gel electrophoresis, especially if the gels are overloaded. From the molar ratios of these components to tropomyosin, one may estimate the maximum amounts of troponin that could be associated with the thin filaments of these muscles. Thin filaments from three different species were analysed by densitometry, the striated adductors of *Aequipecten irradians*, *Placopecten magellanicus* and the anterior byssus retractor and the adductor muscles of *Mytilus edulis* (Table 4). These muscles were selected since they are particularly well suited for, and have been the subject of, physiological and structural studies. The molar ratio, actin to tropomyosin, of all these thin filament preparations is rather similar to the ratios obtained from vertebrate striated muscles. There are two low molecular weight components: one around

Table 4. *Minor components of thin filaments*

Relative molar ratio to tropomyosin

	Actin	Tropomyosin	25 000	20 000
Aequipecten				
Striated adductor	4.5	1	0.2	0.4
Placopecten				
Striated adductor	4.6	1	Traces	0.15
Mytilus				
ABRM	5.9	1	0.22	0.17
Adductor	6.1	1	0.22	0.18

Molar concentrations have been obtained from densitometry of the bands obtained by 10 % SDS acrylamide gel electrophoresis assuming an identical staining of all the components. Staining was related to tropomyosin at different thin filament concentrations. Values were obtained for each component at at least four different loading concentrations at which optical density was proportional to protein concentration.

ABRM = anterior byssus retractor muscle.

15 000–20 000 daltons is in the size range of troponin C, and another about 25 000–30 000 daltons that is in the size range of troponin I. There is about only one mole of the 25 000 chain weight component for five moles of tropomyosin in *Aequipecten* and in the byssus or adductor muscles of *Mytilus*. There is only one mole or less of this low molecular weight component for every 10 moles of tropomyosin in *Placopecten*. These figures represent the highest presumed troponin concentrations that may be present in these muscles. These figures are in line with the functional tests that failed to reveal actin regulation in molluscan muscles.

It is possible that molluscan muscles produce deficient troponin components that cannot bind to the thin filaments and remain in the sarcoplasm. A water-soluble calcium binding protein is present in scallop representing about 0.5 % of the total proteins (Lehman & Szent-Györgyi, 1975). This protein, however, does not hybridize with any of the components of rabbit or lobster troponin and does not replace any of the troponin components in regulatory activity. Although the calcium binding protein may be analogous to parvalbumins present in vertebrate muscles, it has a significantly higher molecular weight, 22 000 daltons, and also contains tryptophan and tyrosine residues. Similar calcium binding protein is also found in *Mercenaria* and in *Mytilus*. Although molluscan muscles lack troponin components, it is possible that some components of troponin, possibly in an altered form, are synthesized in these muscles.

Muscles with a single myosin control

The absence of myosin control in vertebrate striated muscles and in the fast muscles of decapods cannot be explained by the lack of regulatory light chains in these muscles. It has been shown by Kendrick-Jones (1974) that the DTNB-light chain obtained from rabbit myosin can restore regulatory function to scallop myofibrils that have been desensitized by the removal of a regulatory light chain. Subsequent studies showed that different myosins all contain light chains that hybridize with desensitized scallop myofibrils with regulatory functions regained (Kendrick-Jones et al. 1976). The possibility thus exists that in-vitro experiments do not reflect in-vivo behaviour and that the lack of myosin control is an artifact caused by a possible greater sensitivity of vertebrate striated and fast decapod myosins to experimental manipulations required to test for myosin-linked regulation. Several reports in the literature indicate that calcium, at concentrations of about 10 times greater than required for regulation, can interact with rabbit myosin. Calcium ions alter the fluorescence of the DTNB-light chains (Werber, Gaffin & Oplatka, 1972), they bind to the DTNB-light chains, induce changes in the viscosity and in the sedimentation of isolated or reconstituted rabbit myosin filaments (Morimoto & Harrington, 1974), bind to rabbit myosin and inhibit somewhat the ATPase activity of rabbit actomyosin reconstituted from pure actin and myosin (Bremel & Weber, 1975). Calcium ions also decrease the affinity between rabbit subfragment-1 or heavy meromyosin and actin, provided the DTNB-light chains remained intact (Margossian, Lowey & Barshop, 1975). X-ray diffraction studies indicate that stimulation can directly affect the thick filaments, since reflections arising from the thick filaments may disappear or change from stimulated frog sartorius and semitendinosus muscles that have been previously stretched beyond overlap to prevent an interaction between actin and myosin (Haselgrove, 1975). Although these results suggest a possible interaction between vertebrate striated myosins and calcium, it has not been possible as yet to demonstrate in-vitro myosin control in vertebrate muscles. Myofibrils, actomyosins or myosins from vertebrates do not show a calcium sensitive ATPase activity in the presence of pure actin.

A closer examination of the effect of molluscan and vertebrate regulatory light chains indicates differences between these light chains. To appreciate these differences, a somewhat more detailed description of myosin-linked regulation is required.

There are two regulatory light chains in a myosin molecule, very likely one associated with each myosin head. Only one of the two regulatory light chains is removed by EDTA, and the removal of one regulatory light chain completely desensitizes myosin. The second regulatory light chain can be selectively removed by subsequent DTNB treatment; such treatment, however, irreversibly denatures scallop myosin. The regulatory light chains obtained by EDTA and DTNB treatments are chemically the same, and both restore calcium sensitivity to desensitized scallop myosin preparations with the same efficiency. For prevention of cross-link formation with actin, the presence of both regulatory light chains of myosin is required and desensitized myosin retains one of the regulatory light chains. The hybrid scallop myosins therefore contain one scallop regulatory light chain and a light chain obtained from another species (Kendrick-Jones *et al.*, 1976).

In contrast to calcium sensitivity, calcium binding is directly proportional to light chain content. Desensitized preparations, from which one of the two light chains have been removed, lose half of their calcium binding sites. These calcium binding sites are restored when stripped scallop myosin is recombined with scallop regulatory light chains (Table 2).

Regulatory light chains are readily removed and re-added to scallop myofibrils or to scallop actomyosin preparations. There must be binding sites on the scallop heavy chains that are not blocked by actin and are dependent on divalent cations, most likely magnesium ions. Desensitized scallop myofibrils combine with a great affinity with the regulatory light chains of frog striated, bovine cardiac, chicken gizzard, lobster, squid, *Spisula* and *Mercenaria* myosins. All these light chains retain similar binding sites to the heavy chains of scallop myosin, indicating that these sites on both the heavy chains and light chains have been conserved during evolution. In addition to these invariant sites, there are calcium dependent interactions between heavy chains and the regulatory light chains that correspond to the 'on' and to the 'off' state of myosin. Little is known at present how this variable link may operate. In addition to the calcium and magnesium dependent linkages there are additional electrostatic and/or hydrophobic interactions. The presence of multiple binding sites may explain the fact that, from many animals tested, only scallop myosin is fully desensitized by EDTA treatment, and it is only in scallop that a mole of light chain dissociates off in the absence of divalent cations (Kendrick-Jones *et al.*, 1976). Although light chain recombination is not prevented by actin, the results do not mean that rigor links are not influenced by regulatory light chains.

DTNB-light chains of rabbit striated myosin restore sensitivity to desensitized scallop myofibrils when added in a mole to mole ratio. However, calcium binding remains low and, in contrast with myofibrils, the rabbit DTNB light chains only rarely resensitize scallop myosins (Kendrick-Jones, 1974). Scallop regulatory light chains restore calcium binding fully and are able to resensitize scallop myosins consistently (Table 2). Regulatory light chains from molluscs, e.g. *Spisula*, *Mercenaria*, squid, and also from chicken gizzard, behave like the regulatory light chains of scallop in restoring functions to myosin fully, while regulatory light chains from vertebrate striated, cardiac and lobster muscles behave like the rabbit DTNB-light chains, having no effect on calcium binding and resensitizing only myofibrils or actomyosin preparations (Kendrick-Jones *et al.*, 1976). These results suggest that regulatory light chains from vertebrate striated myosin and lobster are not fully competent, and that their action in conferring calcium sensitivity may be an indirect one. One may recall that the hybrids always contain a mole of scallop regulatory light chain. It is conceivable that the calcium dependent response is mediated in the hybrid myofibril by the residual scallop regulatory light chain, and that for this action the presence of another regulatory light chain that has been used for hybridization is required. Under certain conditions, i.e. if the recombination takes place on myofibrils or on actomyosin, this supporting role may not require a fully competent light chain preparation. If the light chain is a competent one, it restores calcium binding fully, and presumably can react directly with calcium once complexed with the heavy chains, and resensitization is not dependent on the presence of actin.

Such an interpretation differentiates regulatory light chains from muscles showing in-vitro myosin control from the regulatory light chains of muscles where in-vitro myosin control cannot be demonstrated. The interpretation tends to support the validity of the functional tests indicating that vertebrates have a single actin control. Myosin control may also be lost as a result of mutations on the heavy chains. A clear-cut demonstration of mutation at a critical region of the heavy chain is not an easy task considering its large size. It may not be overly optimistic to expect that a more detailed understanding of the way regulatory light chains function will help to clarify the role of myosin control in vertebrate muscles.

This research was supported by a grant from the United States Public Health Service, AM 15963. A travel grant from the National Science Foundation BMS 75–23546 contributed towards the expenses in attending the Symposium. I thank Mr Larry Schibuk for the experiments summarized in Table 4.

REFERENCES

BAGSHAW, C. R. & TRENTHAM, D. R. (1974). The characterization of myosin-product complexes and of product-release steps during the magnesium ion-dependent adenosine triphosphatase reaction. *Biochem. J.*, **141**, 331–349.

BREMEL, R. D. (1974). Myosin-linked calcium regulation in vertebrate smooth muscle. *Nature, Lond.*, **252**, 405–407.

BREMEL, R. D. & WEBER, A. (1972). Cooperation within actin filament in vertebrate skeletal muscle. *Nature New Biol.*, **238**, 97–101.

(1975). Calcium binding to rabbit skeletal myosin under physiological conditions. *Biochim. biophys. Acta*, **376**, 366–374.

EBASHI, S. (1963). Third component participating in the superprecipitation of 'natural actomyosin'. *Nature, Lond.*, **200**, 1010–1012.

EBASHI, S. & KODAMA, A. (1965). A new protein factor promoting aggregation of tropomyosin. *J. Biochem., Tokyo*, **58**, 107–108.

EISENBERG, E. & KIELLEY, W. W. (1970). Native tropomyosin: effect on the interaction of actin with heavy meromyosin and subfragment-1. *Biochem. biophys. Res. Commun.*, **40**, 50–56.

GREASER, M. L. & GERGELY, J. (1971). Reconstitution of troponin activity from three protein components. *J. biol. Chem.*, **246**, 4226–4233.

HARTSHORNE, D. J. & PYUN, H. Y. (1971). Calcium binding by the troponin complex and the purification and properties of troponin A. *Biochim. biophys. Acta*, **229**, 698–711.

HASELGROVE, J. C. (1972). X-ray evidence for a conformational change in the actin-containing filaments of vertebrate striated muscle. *Cold Spring Harb. Symp. quant. Biol.*, **37**, 341–352.

(1975). X-ray evidence for conformational changes in the myosin filaments of vertebrate striated muscle. *J. molec. Biol.*, **92**, 113–143.

HUXLEY, H. E. (1972). Structural changes in the actin- and myosin- containing filaments during contraction. *Cold Spring Harb. Symp. quant. Biol.*, **37**, 361–376.

KATO, T. & TONOMURA, Y. (1975). *Physarum* tropomyosin–troponin complex. Isolation and properties. *J. Biochem., Tokyo*, **78**, 583–588.

KENDRICK-JONES, J. (1974). Role of myosin light chains in calcium regulation. *Nature, Lond.*, **249**, 631–634.

KENDRICK-JONES, J., LEHMAN, W. & SZENT-GYÖRGYI, A. G. (1970). Regulation in molluscan muscles. *J. molec. Biol.*, **54**, 313–326.

KENDRICK-JONES, J., SZENTKIRALYI, E. M. & SZENT-GYÖRGYI, A. G. (1976). Regulatory light chains in myosins. *J. molec. Biol.*, in press.

KORETZ, J. F., HUNT, T. & TAYLOR, E. W. (1972). Studies on the mechanism of myosin and actomyosin ATPase. *Cold Spring Harb. Symp. quant. Biol.*, **37**, 179–184.

LEHMAN, W., BULLARD, B. & HAMMOND, K. (1974). Calcium-dependent myosin from insect flight muscles. *J. gen. Physiol.*, **63**, 553–563.

LEHMAN, W., KENDRICK-JONES, J. & SZENT-GYÖRGYI, A. G. (1972). Myosin-linked regulatory systems: Comparative studies. *Cold Spring Harb. Symp. quant. Biol.*, **37**, 319–330.

LEHMAN, W. & SZENT-GYÖRGYI, A. G. (1975). Regulation of muscle contraction. Distribution of actin control and myosin control in the animal kingdom. *J. gen. Physiol.*, **66**, 1–30.

LOWY, J. & VIBERT, P. J. (1972). Studies of the low-angle X-ray pattern of a molluscan smooth muscle during tonic contraction and rigor. *Cold Spring Harb. Symp. quant. Biol.*, **37**, 353–359.

LYMN, R. W. & TAYLOR, E. W. (1971). Mechanism of adenosine triphosphate hydrolysis by actomyosin. *Biochemistry, N.Y.*, **10**, 4617–4624.

MARGOSSIAN, S. S., LOWEY, S. & BARSHOP, B. (1975). Effect of DTNB light chain on the interaction of vertebrate skeletal myosin with actin. *Nature, Lond.*, **258**, 163–166.

MARSTON, S. & LEHMAN, W. (1974). ADP binding to relaxed scallop myofibrils. *Nature, Lond.*, **252**, 38–39.

MORIMOTO, K. & HARRINGTON, W. F. (1974). Evidence for structural changes in vertebrate thick filaments induced by Ca^{2+}. *J. molec. Biol.*, **83**, 83–97.

NACHMIAS, V. & ASCH, A. (1974). Actin mediated calcium dependency of actomyosin in a myxomycete. *Biochem. biophys. Res. Commun.*, **60**, 656–664.

O'BRIEN, E. J., BENNET, P. M. & HANSON, J. (1971). Optical diffraction studies of myofibrillar structure. *Phil. Trans. R. Soc.*, Series B, **261**, 201–208.

OHTSUKI, I., MASAKI, T., NONOMURA, Y. & EBASHI, S. (1967). Periodic distribution of troponin along the thin filament. *J. Biochem., Tokyo*, **61**, 817–819.

PARRY, D. A. D. & SQUIRE, M. M. (1973). Structural role of tropomyosin in muscle regulation. Analysis of the X-ray diffraction patterns from relaxed and contracting muscles. *J. molec. Biol.*, **75**, 33–55.

SZENT-GYÖRGYI, A. G., SZENTKIRALYI, E. M. & KENDRICK-JONES, J. (1973). The light chains of scallop myosin as regulatory subunits. *J. molec. Biol.*, **74**, 179–203.

VIBERT, P. J., HASELGROVE, J. C., LOWY, J. & PAULSEN, F. R. (1972). Structural changes in actin containing filaments of muscle. *J. molec. Biol.*, **71**, 757–767.

WERBER, M. M., GAFFIN, S. L. & OPLATKA, A. (1972). Physico-chemical studies on the light chains of myosin. I. Effect of metal ions on the fluorescence of a light chain (LMP II) from rabbit myosin. *J. mechanochem. Cell Motility*, **1**, 91–95.

REGULATION OF
MUSCLE CONTRACTION BY THE
CALCIUM–TROPONIN–TROPOMYOSIN
SYSTEM

By S. EBASHI, Y. NONOMURA, T. TOYO-OKA and
E. KATAYAMA

Department of Pharmacology, Faculty of Medicine,
University of Tokyo, Tokyo, Japan

Since the introduction of the 'third factor', or 'native tropomyosin', i.e. the complex of tropomyosin and troponin, into the study of the contractile system of muscle (Ebashi, 1963; Ebashi & Ebashi, 1964), abundant experimental results as regards the regulatory mechanism of contraction in skeletal muscle have been accumulated (cf. Ebashi & Endo, 1968; Ebashi, Endo & Ohtsuki, 1969; Weber & Murray, 1973). Troponin is now shown to be composed of three subunits, troponin C (TN-C), troponin I (TN-I) and troponin T (TN-T) (Greaser & Gergely, 1972); although the final mechanism has not yet been clarified, several kinds of hypothesis as to the interaction of these subunits have been proposed (cf. Drabikowski, Strzelecka-Golaszewska & Carafoli, 1974; Ebashi, 1974a).

In spite of abundance of information from skeletal muscle, fewer experimental results have been reported on the regulatory mechanism in cardiac muscle, and extremely few concerning smooth muscle. The reason for this might be partially because of the fact that the smooth muscle proteins are not an easy material to deal with from the biochemical point of view. It might be also because of the thought, tacitly held by many muscle scientists, that the events in skeletal muscle would represent the mechanism common to all kinds of muscle and that all that remains is to confirm them in smooth muscle, although there may be quantitative differences.

As will be referred to later, however, the comparative studies on skeletal and cardiac troponins have revealed that a remarkable phenomenon observed with skeletal troponin, which was once considered to be a key phenomenon to the understanding of the troponin mechanism, is not distinctly found in cardiac troponin and therefore that the phenomenon is not involved in the essential mechanism of the troponin regulation (Ebashi, 1974a, b). Thus the observation made with skeletal muscle cannot be generalized even to cardiac muscle.

This has promoted a re-examination of the regulatory system of smooth muscle (Ebashi, Nonomura, Kitazawa & Toyo-oka, 1975; Ebashi, Toyo-oka & Nonomura, 1975). As a result, the previous view (Ebashi *et al.*, 1966) that its regulation is dependent on the troponin–tropomyosin system was confirmed. However, the mode of action of the troponin–tropomyosin system in smooth muscle is distinctly different from that in skeletal muscle. In addition to this, the interaction of myosin and actin of smooth muscle in the presence of ATP is also considerably different from that of skeletal muscle.

METHODS AND MATERIALS

The preparation methods of various muscle proteins and experimental procedures were essentially the same as described in previous papers (Ebashi & Ebashi, 1964; Ebashi *et al.*, 1966; Ebashi, Toyo-oka & Nonomura, 1975). As a source of muscle proteins, rabbit skeletal muscle, chicken gizzard and bovine stomach were used.

RESULTS

Separation of troponin from gizzard native tropomyosin and its effect on desensitized myosin B

Ebashi *et al.* (1966) have shown that essentially the same procedure used for the preparation of native tropomyosin can extract a similar fraction from gizzard, and this fraction can resensitize to Ca^{2+} the gizzard myosin B preparation which had been previously desensitized by trypsin treatment. However, since then no attempt to separate native tropomyosin into troponin and tropomyosin has been made. One of the reasons for this may be that the useful desensitization method is not yet established; the desensitization by trypsin, which is fairly effective in the skeletal system (Ebashi & Ebashi, 1964), does not give consistent results in gizzard muscle.

In view of this, the first effort was directed to establishing the method of preparing desensitized myosin B. This was rather easily achieved, following the method of desensitization used for skeletal myosin B with a slight modification. Using this preparation, the Ca^{2+}-sensitizing effect of native tropomyosin was confirmed (Fig. 1).

Separation of native tropomyosin into troponin and tropomyosin was also successfully effected by essentially the same method as that used for skeletal tropomyosin (Ebashi, Nonomura, Kitazawa & Toyo-oka, 1975; Ebashi, Toyo-oka & Nonomura, 1975). If combined, the troponin and tropomyosin thus separated showed essentially the same effect as that of native tropomyosin (Fig. 1).

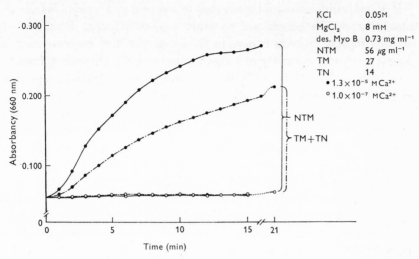

Fig. 1. Reconstitution of gizzard myosin B from desensitized myosin B (des. Myo B) and native tropomyosin (NTM) or tropomyosin (TM) plus troponin (TN). In addition to the amounts shown on the figure, the reaction mixture also contained 0.02 M Tris-maleate buffer (pH 6.8) and 0.1 mM ATP. The reaction was started by adding ATP. Temperature 21 °C.

It was found that desensitized myosin B did not show significant super-precipitation even with tropomyosin. The level of superprecipitation of desensitized myosin B plus native tropomyosin, or troponin and tropomyosin, in the absence of Ca^{2+} is always the same as that with only tropomyosin. Marked superprecipitation is exhibited only in the presence of native tropomyosin, or troponin and tropomyosin, with Ca^{2+} (Ebashi, Nonomura, Kitazawa & Toyo-oka, 1975; Ebashi, Toyo-oka & Nonomura, 1975).

This finding is quite different from that expected from the experiences with striated muscle. As is well established (Ebashi *et al.*, 1969), troponin with Ca^{2+} gives rise to the same level of superprecipitation or ATPase activity of skeletal actomyosin with tropomyosin, and the removal of Ca^{2+} from troponin induces the depression of the myosin–actin interaction; this depression cannot be seen with the system consisting of actomyosin and tropomyosin.

In this respect, it is worthy of note that gizzard tropomyosin, in the regulatory process, cannot be replaced by skeletal tropomyosin and that it forms different types of paracrystal from those of skeletal tropomyosin. On the other hand, gizzard tropomyosin can be used in place of skeletal tropomyosin in regulating the skeletal actomyosin system. Thus the regulatory mechanism based on the Ca^{2+}–troponin–tropomyosin system is definitely distinguishable in striated and gizzard muscle.

It is interesting to consider whether this property is unique for gizzard muscle or can be generalized to other smooth muscles. Since bovine stomach muscle proteins behave in the same way, this property may be at least a common property of smooth muscles of the alimentary tract.

In Plate 1 the patterns of SDS (sodium dodecyl sulphate) polyacrylamide gel electrophoresis of native tropomyosin and troponin are shown. The pattern of troponin shows a distinct fast moving band, of which the molecular weight is estimated to be around 18000, according to its electrophoretic mobility. In the original gel this fast band was seen to be separated into two components, of which the molecular weights were about 18000 and 17000, respectively. In addition to this, several slowly moving bands are observed. One of them having a molecular weight around 80000 (indicated by upper bar in Plate 1) was always associated with the main lighter bands and also with the physiological activity of troponin (Ebashi, Toyo-oka & Nonomura, 1975). However, identification of the role of each subunit has not yet been accomplished.

The effect of native tropomyosin or troponin and tropomyosin on reconstituted actomyosin

Since desensitized myosin B is far from being a pure system, the results above should be confirmed with a much simpler system, i.e. reconstituted actomyosin. As is well known, it was not an easy task to prepare actin-free myosin from smooth muscles. This difficulty was overcome by a newly developed method preparing pure myosin from myosin B (Ebashi, 1976). Actin was also prepared by the method described in that paper. The pattern of SDS polyacrylamide gel electrophoresis of these preparations is shown in Plate 2.

Using the reconstituted actomyosin thus prepared, essentially the same results as those with desensitized myosin B were obtained. Fig. 2a shows the result of fresh and well prepared reconstituted actomyosin, and Fig. 2b a somewhat deteriorated preparation. Like desensitized myosin B, the actomyosin preparation of good quality did not show significant superprecipitation, even with tropomyosin, whereas the deteriorated one shows fairly marked superprecipitation with tropomyosin and definite superprecipitation even without tropomyosin. In both cases, however, addition of troponin with Ca^{2+} accelerated the superprecipitation. The level of superprecipitation of reconstituted actomyosin with tropomyosin was the same as that of reconstituted actomyosin with both tropomyosin and troponin in the absence of Ca^{2+}. Thus the activating nature of troponin with Ca^{2+} is fully confirmed in reconstituted actomyosin system.

The ATPase activity of reconstituted actomyosin well coincided with

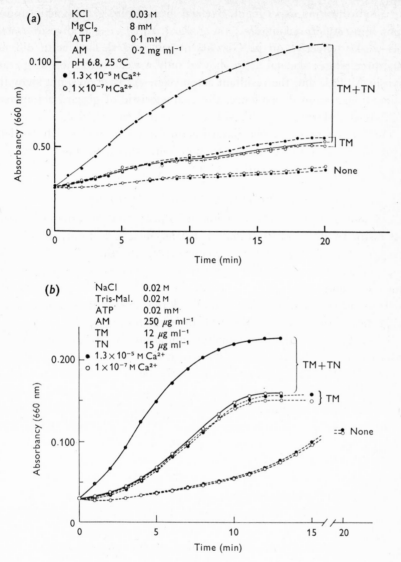

Fig. 2. Sensitization to Ca^{2+} of reconstituted gizzard actomyosin (AM) by gizzard troponin (TN) and tropomyosin (TM). (*a*) Fresh myosin; (*b*) aged myosin.

the result of superprecipitation above, i.e. Ca^{2+} enhanced the ATPase of actomyosin with native tropomyosin nearly three-fold, but did not effect the ATPase of actomyosin, without native tropomyosin, at all; the level of the ATPase in the latter was very close to that of the former in the absence of Ca^{2+} (T. Toyo-oka, T. Mikawa & M. Ikebe, personal communication).

The success of preparing actin-free myosin made it possible to carry

out a hybridization experiment. Skeletal myosin plus gizzard actin underwent some superprecipitation; the gizzard troponin–tropomyosin system
was weakly effective on it. Gizzard myosin and skeletal actin did not
superprecipitate; skeletal actin showed only a weak activation of gizzard
myosin ATPase and the resultant actomyosin–ATPase does not show the
unique magnesium dependence, the characteristic of gizzard actomyosin
ATPase (K. Maruyama & T. Mikawa, personal communication).

Thus it is now clear that gizzard actin cannot be replaced by skeletal
actin, in sharp contrast to the common belief that actin might have few
species- and tissue-specificities, if any.

The properties of gizzard actomyosin system

It was shown in the previous section that the contractile system of gizzard
was distinctly different from that of skeletal muscle. Not only myosin and
troponin, which represent the characteristics of the muscle, but also
tropomyosin and actin, which appear to be rather common among different
kinds of contractile systems, are distinguishable between gizzard and
skeletal muscle. This finding stimulated the studies inquiring into the
properties of the myosin–actin–ATP interaction of gizzard.

In Fig. 3 the extent of superprecipitation, determined by optical
absorbency (Ebashi, 1961), was plotted against Mg–ATP concentration.
The rate of superprecipitation plotted in the same way showed almost the
same relationship as shown in Fig. 3. It is worthy of mention that the
maximum superprecipitation was obtained at 10^{-5} M Mg–ATP and almost
no further change could be seen until 4×10^{-3} M. Thus there is almost no
tendency of dissociation or relaxation induced by higher concentrations of
Mg–ATP. This is in good agreement with the ATPase of gizzard actomyosin (K. Maruyama & T. Mikawa, personal communication). There is no
indication for 'substrate-inhibition' type of response; the ATPase of
gizzard actomyosin seems to follow typical Michaelis–Menten-type
kinetics.

It is well known that the contraction of smooth muscle actomyosin
system depends on free Mg^{2+} of fairly high concentrations (Filo, Bohr &
Rüegg, 1965). Formation of the filamentous structure of gizzard myosin
depends on the Mg^{2+} concentration (Plate 3), which may explain the Mg^{2+}-
dependence of contractility of smooth muscle to a certain extent. According
to preliminary work of Maruyama & Mikawa (personal communication),
however, the function of actin seems to be also dependent on the Mg^{2+}
concentration.

It was recently found that the activating effect of gizzard troponin with
Ca^{2+} mentioned above did not become distinct, if α-actinin, the 6 S com-

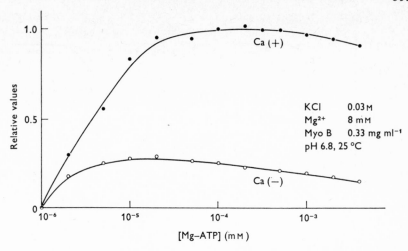

Fig. 3. Dependence of superprecipitation of gizzard myosin B (Myo B) on Mg–ATP concentrations. Relative absorbencies 20 min after addition of ATP were plotted against Mg–ATP concentrations. Other conditions were the same as those for Fig. 1 except that 0.03 M KCl was used instead of 0.05 M KCl.

ponent (Nonomura, 1967) of the crude α-actinin preparation (Ebashi & Ebashi, 1965), was not included in the protein system (the proteins used in previous experiments were more or less contaminated by α-actinin). α-Actinin, which was also found in gizzard (Ebashi *et al.*, 1966), was first noted as an accelerator of the myosin–actin–ATP interaction of skeletal muscle (Ebashi, Ebashi & Maruyama, 1964). This accelerating action, however, was later considered to be merely an in-vitro phenomenon, because it was largely abolished by native tropomyosin (Ebashi, 1967) and the localization of α-actinin was confined to the Z-band (Masaki, Endo & Ebashi, 1967).

However, the effect of α-actinin on the gizzard actomyosin system becomes appreciable only in the presence of troponin, tropomyosin and Ca^{2+}. Although further evidence is required, it is attractive to assume that α-actinin would play an indispensable role in the contractile mechanism of gizzard. In this connection, it is of interest that the 'co-factor' (Pollard & Korn, 1973) which promotes the actomyosin ATPase of *Acanthamoeba* has a similar molecular weight to α actinin, about 100000.

DISCUSSION

The results presented in this article have clearly indicated that the contraction of gizzard muscle (and bovine stomach also) is regulated by the troponin–tropomyosin system as in skeletal muscle. In this respect,

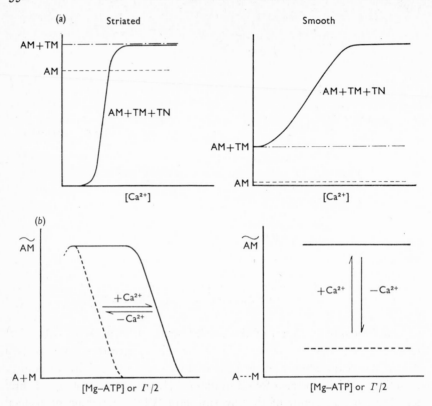

Fig. 4. Schematic illustration of the myosin–actin–ATP interaction: (*a*) as a function of the Ca^{2+} concentration (from Ebashi, Nonomura, Kitazawa & Toyo-oka, 1975); (*b*) as a function of the Mg–ATP concentration or ionic strength (Γ). $\widetilde{AM}$ = the state in which actin and myosin are actively interacting with each other; $A + M$ = the state in which actin and myosin are dissociated; $A - - - M$ = the state of actin and myosin of smooth muscle which corresponds to '$A + M$'.

gizzard smooth muscle is not fundamentally different from vertebrate striated muscle. However, the mode of action of troponin in gizzard and stomach muscle is entirely different from that of skeletal muscle. The main role of troponin in skeletal or striated muscle is to induce the depression of the interaction of actin and myosin with tropomyosin in the absence of Ca^{2+}. In contrast with this, gizzard troponin with Ca^{2+} induces a real activation of the myosin–tropomyosin–actin system. This relationship is schematically illustrated in Fig. 4*a*, *b*.

Thus the role of Ca^{2+} in skeletal and gizzard muscle, although it appears to be similar in both kinds of muscle, is substantially different in its fine mechanism. While Ca^{2+} is considered as a kind of derepressor, it is a real activator in gizzard muscle.

The clear distinction between smooth and striated muscle is not only true of the troponin–tropomyosin system but also of the actomyosin system. As is shown in this article, the interaction of gizzard myosin and actin in the presence of Mg–ATP is also distinctly different from that of skeletal muscle. The depression of the interaction by a high concentration of Mg–ATP, giving rise to a biphasic type of response to Mg–ATP, or the 'substrate-inhibition' type depression, is characteristic of the striated muscle system and has been thought to be the fundamental basis of muscle contraction, particularly from the regulatory point of view. However, this important property of the contractile system of striated muscle is not operative in the smooth muscle system. The fact that the interaction of myosin and actin in smooth muscle shows a simple response to the Mg–ATP concentration, apparently following Michaelis–Menten-type simple kinetics, might be deeply connected with the characteristic of the regulatory mechanism in smooth muscle.

All the facts and considerations mentioned above clearly indicate that the mechanisms operating in skeletal muscle do not always represent those common for every kind of muscle.

This reminds us of the facts described in previous papers (Ebashi, 1974a, b). The action of cardiac troponin is fundamentally the same as that of skeletal troponin; it can induce the depression of the actomyosin system of skeletal muscle as well as of cardiac muscle, in cooperation with either cardiac or skeletal tropomyosin.

However, the strong calcium-dependent interaction of TN-T and TN-C shown in the subunits of skeletal troponin, revealed by a spin-label study (Ebashi, Ohnishi, Abe & Maruyama, 1974; Ebashi, 1974b) as well as by a solubility experiment (Ebashi, 1974a, b), is not found in those of cardiac troponin and, therefore, does not seem to be playing an essential role in the troponin function. This is further verified by the fact that TN-T from cardiac troponin and TN-C from skeletal troponin have a weaker interaction with each other and, even so, both subunits together with skeletal TN-I form a hybrid troponin, which has a better regulatory function than that of parent troponins, whether cardiac or skeletal. This puzzling fact has naturally raised the question as to what kind of physiological role the calcium-dependent interaction of TN-T and TN-C of skeletal troponin would play.

We were apt to assume that the effect of Ca^{2+} might be exerted only on the physiological process of primary importance. This common belief must be reconsidered in view of the observation above. It is quite possible that Ca^{2+} would exert its regulatory effect on a subsidiary process of muscle contraction.

In view of this situation, it may be convenient to classify the action of Ca^{2+} to activate the contractile system as 'primary regulation'. In this category, we already know a different mechanism from the troponin-dependent mechanisms of vertebrate muscles, i.e. the myosin-linked regulation by Ca^{2+} found in molluscan muscle (Kendrick-Jones, Lehman & Szent-Györgyi, 1970). In contrast with this, the role of Ca^{2+}, not fundamental for contraction but still important from the fine regulation of the contractile process, may be called 'secondary' regulation.

It is suggested that the strong TN-C–TN-T interaction found in skeletal muscle may protect the contractile elements against excess contraction which might be induced by a high level of Ca^{2+} in the activated state of skeletal muscle (Ebashi, Nonomura, Kitazawa & Toyo-oka, 1975). Our present concern is not to discuss whether this suggestion is appropriate or not, but is to think over the possibility of whether this kind of consideration might be helpful in analysing the various kinds of roles of Ca^{2+} in muscle contraction. For instance, the Ca^{2+} binding of myosin (cf. Bremel & Weber, 1975), which certainly takes place under the high concentration of Ca^{2+} in the active state of skeletal muscle, has often been connected with the primary activation, but it is also possible that the binding would play some role which may be classified as the secondary regulation.

It is now generally accepted that myosin and actin, and probably other regulatory proteins, may also exist in almost every non-muscle tissue (cf. Pollard & Weihing, 1974). It is an interesting question whether the contractile proteins in these tissues are vertebrate striated muscle type, smooth muscle type or molluscan type.

Our experiences are, however, derived from only a few kinds of vertebrate smooth muscle. Therefore it is quite possible that in other vertebrate smooth muscles there are various kinds of contractile mechanisms, particularly regulatory mechanisms different from gizzard or bovine stomach muscle. Perhaps we should not use a simple term 'smooth muscle type'.

CONCLUSION

(1). Contraction of vertebrate smooth muscle (gizzard and bovine stomach) is also regulated by the troponin–tropomyosin system. However, the mode of regulation is entirely different from that of vertebrate striated muscle. While striated muscle troponin induces the depression of the interaction of myosin, actin, tropomyosin and ATP in the absence of Ca^{2+}, smooth muscle troponin produces the activation of the interaction in the presence of Ca^{2+}. Thus Ca^{2+} acts as a derepressor in striated muscle, but as a real activator in smooth muscle.

(2). Unlike skeletal muscle, the actomyosin system of smooth muscle does not show a biphasic response to Mg–ATP concentration. Thus, the dissociation of actomyosin into myosin and actin by a relatively high concentration of ATP is not involved in the regulation of smooth muscle contraction by the troponin–tropomyosin system.

(3). The above findings, together with some observations discussed in this article, suggest that the facts observed with vertebrate skeletal muscle do not necessarily represent the contractile mechanism common to all kinds of muscle.

We are grateful to Professor K. Maruyama, Kyoto University, for his permission to quote his unpublished data. This work was supported in part by research grants from the Muscular Dystrophy Associations of America, Inc., the Ministry of Education, Japan, the Ministry of Health and Welfare, Japan (No. 216), the Iatrochemical Foundation, Toray Science Foundation, and the Mitsubishi Foundation.

REFERENCES

BREMEL, R. D. & WEBER, A. (1975). Calcium binding to rabbit skeletal myosin under physiological conditions. *Biochim. biophys. Acta*, **376**, 366–374.

DRABIKOWSKI, W., STRZELECKA-GOLASZEWSKA, H. & CARAFOLI, E. (1974). *Calcium Binding Proteins*. PWN-Polish Scientific, Warszawa, and Elsevier Scientific: Amsterdam.

EBASHI, S. (1961). Ca-binding activity of vesicular relaxing factor. *J. Biochem., Tokyo*, **50**, 236–244.

(1963). Third component participating in the superprecipitation of "natural actomyosin". *Nature, Lond.*, **200**, 1010–1012.

(1967). Structural proteins and their interaction. In *Symposium on Muscle*, ed. E. Ernst & F. B. Straub, pp. 77–87. Academiai Kiado: Budapest.

(1974a). Regulatory mechanism of muscle contraction with special reference to the Ca-troponin-tropomyosin system. In *Essays in Biochemistry*, vol. 10, ed. P. N. Campbell & F. Dickens, pp. 1–36. Academic Press: London.

(1974b). Interactions of troponin subunits underlying regulation of muscle contraction by Ca-ion. In *Lipmann Symposium: Energy, Biosynthesis, and Regulation in Molecular Biology*, ed. D. Richter, pp. 165–178. Walter de Gruyter: Berlin & New York.

(1976). A simple method of preparing actin-free myosin from chicken gizzard. *J. Biochem., Tokyo*, **79**, 229–231.

EBASHI, S. & EBASHI, F. (1964). A new protein component participating in the superprecipitation of myosin B. *J. Biochem., Tokyo*, **55**, 604–613.

(1965). α-Actinin, a new structural protein from striated muscle. I. Preparation and action on actomyosin-ATP interaction. *J. Biochem., Tokyo*, **58**, 7–12.

EBASHI, S., EBASHI, F. & MARUYAMA, K. (1964). A new protein factor promoting contraction of actomyosin. *Nature, Lond.*, **203**, 645–646.

EBASHI, S. & ENDO, M. (1968). Calcium ion and muscle contraction. *Prog. Biophys. molec. Biol.*, **18**, 123–183.

EBASHI, S., ENDO, M. & OHTSUKI, I. (1969). Control of muscle contraction *Q. Rev. Biophys.*, **2**, 351–384.

EBASHI, S., IWAKURA, H., NAKAJIMA, H., NAKAMURA, R. & OOI, Y. (1966). New structural proteins from dog heart and chicken gizzard. *Biochem. Z.*, **345**, 201–211.

EBASHI, S., OHNISHI, S., ABE, S. & MARUYAMA, K. (1974). A spin-label study on calcium-induced conformational changes of troponin components. *J. Biochem.*, *Tokyo*, **75**, 211–213.

EBASHI, S., NONOMURA, Y., KITAZAWA, T. & TOYO-OKA, T. (1975). Troponin in tissues other than skeletal muscle. In *Calcium Transport in Contraction and Secretion*, ed. E. Carafoli, F. Clementi, W. Drabikowski & A. Margreth, pp. 405–414. North-Holland: Amsterdam & New York.

EBASHI, S., TOYO-OKA, T. & NONOMURA, Y. (1975). Gizzard troponin. *J. Biochem.*, *Tokyo*, **78**, 859–861.

FILO, R. S., BOHR, D. F. & RÜEGG, J. C. (1965). Glycerinated skeletal and smooth muscle: calcium and magnesium dependence. *Science, Wash.*, **147**, 1581–1583.

GREASER, M. L. & GERGELY, J. (1972). Reconstitution of troponin activity from three protein components. *J. biol. Chem.*, **246**, 4226–4233.

KENDRICK-JONES, J., LEHMAN, W. & SZENT-GYÖRGYI, A. G. (1970). Regulation in molluscan muscles. *J. molec. Biol.*, **54**, 313–326.

MASAKI, T., ENDO, M. & EBASHI, S. (1967). Localization of 6 S component of actinin at Z-band. *J. Biochem.*, *Tokyo*, **62**, 630–632.

NONOMURA, Y. (1967). A study on the physico-chemical properties of α-actinin. *J. Biochem.*, *Tokyo*, **61**, 796–802.

POLLARD, T. D. & KORN, E. D. (1973). *Acanthamoeba* myosin. II. Interaction with actin and with a new cofactor protein required for actin activation of Mg^{2+} adenosine triphosphatase activity. *J. biol. Chem.*, **248**, 4691–4697.

POLLARD, T. D. & WEIHING, R. R. (1974). Actin and myosin and cell movement. *CRC crit. Rev. Biochem.*, **2**, 1–65.

WEBER, A. & MURRAY, J. M. (1973). Molecular control mechanism in muscle contraction. *Physiol. Rev.*, **53**, 612–673.

PLATE I

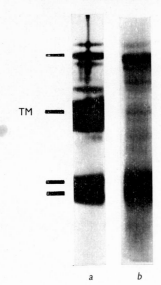

TM

a   b

SDS polyacrylamide gel electrophoresis of native tropomyosin and troponin of gizzard: (*a*) native tropomyosin; (*b*) troponin. TM: tropomyosin. Bars without denotation indicate the principal components of troponin. 10 % polyacrylamide gel was used.

PLATE 2

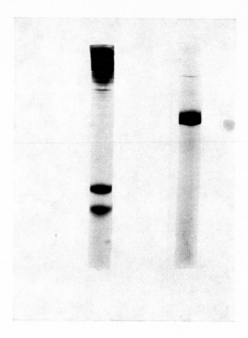

SDS polyacrylamide gel electrophoresis of gizzard myosin and actin. *Left:* myosin. *Right:* actin. 10 % polyacrylamide gel was used.

PLATE 3

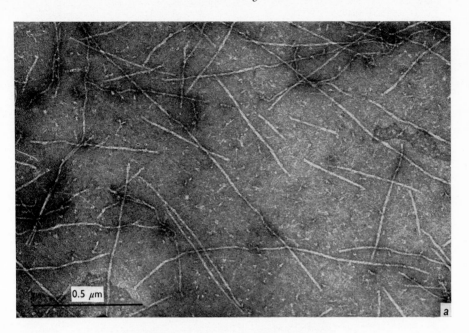

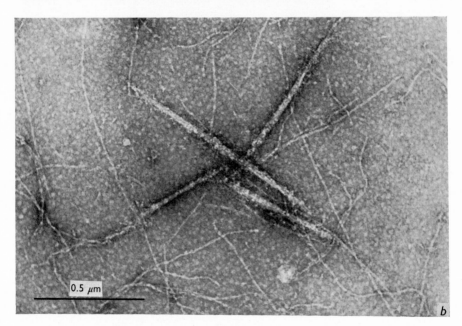

Electron micrographs of gizzard myosin B in the presence of ATP without Ca^{2+}. (a) 1 mM $MgCl_2$ and 1 mM ATP (free Mg^{2+}, ∼ 0.1 mM). (b) 8 mM $MgCl_2$ and 1 mM ATP (free Mg^{2+}, 7 mM). Since Ca^{2+} is absent, practically no interaction between myosin and actin can be found. As a result, dependence of filament formation of myosin molecules on Mg^{2+} is illustrated.

CALCIUM RELEASE AND CONTRACTION IN VERTEBRATE SKELETAL MUSCLE

By S. R. TAYLOR and R. E. GODT

Department of Pharmacology, Mayo Medical School,
Rochester, Minnesota 55901, USA

The initial event in a single contraction of a vertebrate skeletal muscle cell is the electrical excitation of the surface membrane. The final event is the development of force and/or shortening, and is believed to result from an increase of calcium in the myofilament space. The point that remains in greatest doubt is the mechanism for coupling these two events. Sandow, in 1952, introduced the term excitation–contraction (EC) coupling to describe the various steps in this process, and focused attention on many of the issues involved in determining the role of calcium. Just as it was more than 20 years ago, our current understanding of EC coupling in vertebrate skeletal muscle is based largely on experiments with frog muscle. We, among many others, may be too facile in presuming to generalize about vertebrate muscle on the basis of studies with a few species of amphibians (*R. pipiens*, *R. temporaria*, *X. laevis*). However, several characteristics of these muscles have made them well suited for critical experiments which cannot, or have not yet been performed on mammalian muscle. For example: (1) frog muscle cells are attached to easily dissected tendons, and can be isolated intact and oriented with the vector for force generation coincident with the direction of force measurement, which facilitate analyses of their mechanical properties; (2) frog muscle has a relatively low rate of metabolism at room temperature, can be adequately oxygenated by diffusion when isolated from the body, and thus, may remain functional for many days to weeks under appropriate conditions of temperature and attention to asepsis; (3) the cells are relatively large and sturdy, compared with those from other vertebrates, and survive multiple impalements with microelectrodes and microinjection of foreign substances, which facilitate studies of their electrical properties and chemical composition. Major advances in understanding the role of calcium in contraction have resulted from mechanical, chemical, and electrical studies on single cells from these muscles, and we shall discuss and evaluate some of these experiments along with selected references to experiments on non-amphibian species.

Ionized calcium is the only physiological substance that can induce

[361]

contraction when injected into a frog muscle cell. Several workers have demonstrated this in intact cells (Heilbrunn & Wiercinski, 1947; Falk & Gerard, 1954; Niedergerke, 1955), and others have extended this observation using cells from which the surface membrane has been removed (Natori, 1954; Podolsky, 1968). Accordingly, substantial effort has been devoted to studies of the change in calcium concentration in the myofilament space, and it is now known that the physiological event that induces and controls an individual contraction is an increase in calcium bound to the myofilaments (Ebashi & Endo, 1968).

Many muscle studies have been performed on 'model' systems that have had one or more components of the muscle cell selectively destroyed or modified, which makes it easier to study the somewhat simpler system that remains. We shall summarize some of the recent information obtained from studies of model systems as well as intact single fibres, and compare the relative importance of two of the possible mechanisms responsible for the intracellular release and regulation of calcium during a contraction.

The contractile filaments and their arrangement in a highly ordered lattice are the most thoroughly described structural features of frog striated muscle (Huxley, 1971). The complex internal membrane systems, on the other hand, which were discussed in a previous Society for Experimental Biology symposium on cell motility (Podolsky, 1968), have just begun to be characterized with a high degree of precision and accuracy (Franzini-Armstrong, 1975; Mobley & Eisenberg, 1975; Eisenberg & Peachey, 1975). However, for the purpose of our discussion, these intracellular membranes can be thought of in a general way; they consist of two structurally separate systems. In the frog, the membranes of the transverse tubular (T) system pass radially into the cell near the Z line at the end of each sarcomere. These tubules are believed to be in direct communication with the cell surface, and to function as the pathway for the inward spread of activation. The propagation of excitation from the cell surface along the T system seems clearly due to a sodium-dependent regenerative signal (Costantin, 1975), which has been analysed in detail (Adrian & Peachey, 1973). The sarcoplasmic reticulum (SR), on the other hand, runs parallel to the contractile filaments, does not appear to freely open either to the T system or to the cell surface, and evidently contains the store of activator* calcium that is directly responsible for contraction. At regular intervals near the Z line the SR and T tubules are separated by a small gap which appears to be bridged by rows of projections from the SR called

---

* Since calcium may function in separate steps of the EC coupling process it is often described with a qualifier that signifies a specific action, for example, activator calcium, trigger calcium, stabilizer calcium (Bianchi, 1969).

'feet' (Franzini-Armstrong, 1975). These feet may be the locus of steps in EC coupling that are yet to be understood. One of the ideas we will consider is that the membrane potential difference across the walls of the T system is the control for the release of calcium from the SR. The other idea is that small amounts of calcium are released from the T system, diffuse across the SR–T gap, and through some chemical effect induce the release of a much larger amount of calcium, enough to activate the contractile filaments fully.

CALCIUM RELEASE MECHANISMS IN SARCOPLASMIC RETICULUM OF SKINNED FIBRES

The skinned fibre preparation

A mechanically skinned skeletal muscle preparation was first used by Natori in 1954, and has proved its worth in studies of muscle ever since. A muscle fibre is dissected from a whole muscle and, after the calcium-containing liquid has been gently blotted away, the fibre is placed either under oil or in a very low-calcium medium containing magnesium and ATP that is supposed to mimic the intracellular environment. The sarcolemma may then be mechanically stripped away with little damage to the internal structures of the fibre (cf. Podolsky, 1968). Alternatively, skinned fibres can be prepared by literally ripping an intact single fibre down the midline (Endo & Nakajima, 1973). This latter procedure preserves the integrity of the surface membrane on the outside of each fibre half. Some of the transverse tubules remain patent when the cell is ripped in half. However, when the entire membrane is stripped away à la Natori, the T tubules are sheared off at their mouths and may well seal over. The skinned fibre technique is not limited to skeletal muscle. For example, Fabiato & Fabiato (1975a, b) have prepared single mechanically skinned cardiac muscle cells from homogenized heart tissue.

Once there is no longer a sarcolemma blocking diffusion, the medium bathing the skinned fibre is under experimental control, and the influence of various substances on the contractile apparatus can be studied directly. Not only is the contractile protein array in its native state but the sarcoplasmic reticulum is still functional in freshly skinned fibres. Although in frog skinned fibres the SR is greatly swollen in relaxing medium (Plate 1), especially near the SR–T gap region, the SR can still rapidly take up large amounts of calcium (Ford & Podolsky, 1972a). Ford & Podolsky used radioactive ^{45}Ca tracer techniques to show that the SR of skinned fibres could accumulate calcium to a maximum of 2 to 3 mM from bathing solutions with an EGTA-buffered calcium level of $10^{-6.2}$ M. Furthermore, they

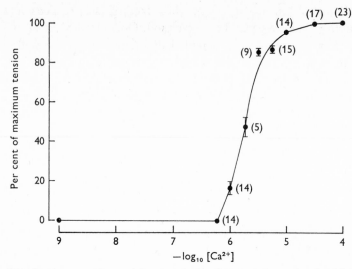

Fig. 1. Relation between free calcium concentration and the maintained isometric tension of mechanically skinned fibres from human rectus abdominis muscle. These fibres were skinned in the same way already described for frogs (Godt, 1974). Solutions contained: 2 mM Mg–ATP; 0.5 mM Mg^{2+}; 7 mM EGTA; 20 mM imidazole; 14.5 mM creatine phosphate; creatine phosphokinase 1 mg ml^{-1}; potassium propionate so that ionic strength was 0.14 M; pH 7.0. Temperature 20–23 °C. Each point is the average of several observations on separate fibres. The number of observations and the standard error of the mean are indicated with each point. The muscle fibres were obtained from tissue removed in the course of abdominal surgery for the following conditions: Wilson's disease, hernia, Crohn's disease, gallstones, and twisted bowel.

found that the maximal rate of calcium uptake by SR in skinned fibres is fast enough to account for the rapid relaxation of the twitch of intact skeletal muscle fibres.

In solutions where $[Ca^{2+}]$ is well buffered with EGTA, skinned fibres respond to increases in $[Ca^{2+}]$ by developing *tonic* contractions. Fig. 1 demonstrates the relation between free calcium and steady state isometric tension for human muscle. The relation is steep with full tension activation occurring over a range of about 1.5 decades of free $[Ca^{2+}]$. This is similar to the relationship previously determined for frog muscle (e.g. Godt, 1974), and in this regard at least, frogs do seem comparable with higher vertebrates. Avid sequestration of calcium by the SR slows the rate of tension rise in this type of contraction, but does not affect the eventual plateau level of tension.

Under certain conditions in solutions weakly buffered for calcium (low EGTA concentrations), skinned fibres develop *phasic* or *cyclic* contractions. Studies of these phasic contractions and the conditions necessary

PLATE I

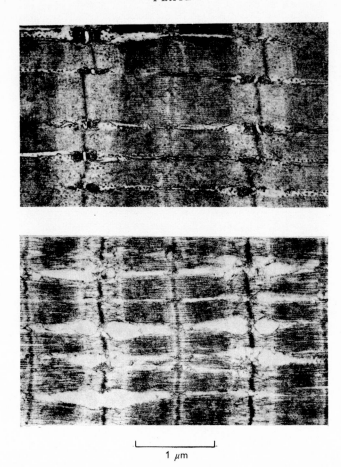

1 μm

Electron micrographs of intact and mechanically skinned muscle fibres from the frog, *Rana temporaria*. The upper photograph shows an intact fibre, and the lower photograph shows a skinned fibre with its SR swollen. The intact fibre was isolated in Ringer, and fixed in Ringer with glutaraldehyde (6·2 %). On the other hand, the skinned fibre was fixed in a relaxing solution (Maughan & Godt, 1974) containing the same amount of glutaraldehyde. The fibres were post-fixed in Ringer or relaxing solutions, respectively, with osmium tetroxide (1 %), dehydrated in an ethanol series, and embedded in Luft's Araldite.

to elicit them form the basis for comparing the two hypotheses for SR calcium release, namely, the chemical versus the electrical mechanism.

Chemical and electrical release mechanisms

According to the chemical release hypothesis an increase of 'trigger' calcium results from the action potential, and the trigger calcium, in turn, causes the release of large amounts of activator calcium from the SR. The influx of calcium is relatively small, less than 10^{-12} mole for each square centimetre of cell surface during a single action potential (Bianchi, 1969; Armstrong, Bezanilla & Horowicz, 1972). But small amounts of calcium can cause phasic contractions of skinned fibres, apparently because calcium itself can release more calcium in a seemingly 'all-or-none' fashion. It has been proposed that this is a chemically specific effect on the SR, and that the calcium induced release of calcium (CROC) may be the primary mechanism for initiating release in intact skeletal and heart muscle. The calcium that enters a frog skeletal muscle fibre with each action potential is too little by a factor of about 100 to cause contraction directly (Sandow, 1965). Therefore, the CROC may be a mechanism to amplify trigger calcium by at least this factor.

The CROC has been described by several words. For example, it has been called 'regenerative', 'autocatalytic', 'all-or-none' and 'triggered'. In our opinion these terms are not synonymous. 'Regenerative' and 'autocatalytic' mean that the stimulus and response may be proportional; that is, they may have the same or a constant ratio. In biology, regeneration usually means the renewal of something; for example, one calcium ion may stimulate the release of one other. In electronics, regeneration means that energy from the output is fed back to the input and thereby amplifies the overall signal; for example, one calcium ion may stimulate the release of several others.

On the other hand, 'all-or-none' or 'triggered' mean that the rate and amplitude of a response are independent of the strength of a stimulus. The velocity and distance travelled by a bullet are independent of the speed and strength of the squeeze that fired the gun. To our minds, an adequate stimulus should cause the SR to release activator calcium in an analogous manner if the CROC hypothesis is tenable for initiating EC coupling in intact skeletal muscle.

The electrical release hypothesis, on the other hand, involves no specific chemical intermediary. The action potential spreading along the surface membrane and into the interior of the cell via the transverse tubules induces depolarization of the SR and release of its calcium by a mechanism that may involve a non-ionic current (Chandler, Rakowski & Schneider,

1976). The interior of the SR is presumably polarized with respect to the sarcoplasm. It is not yet technically feasible to measure this potential difference directly. But relative changes in birefringence (Baylor & Oetliker, 1975) or fluorescence intensity of dye-stained cells (Bezanilla & Horowicz, 1975) are seen during muscle activation, and may be caused by SR depolarization. Depolarization of the SR membrane supposedly increases its calcium permeability and allows calcium to flow out of the SR down its concentration gradient to activate the myofibrils. This depolarization induced release of calcium (DROC) from SR may be continuously graded with the electrical stimulus. Under normal conditions in the intact muscle calcium release may appear 'all-or-none' only because the action potential is essentially all-or-none. The underlying mechanism of release is, however, not necessarily all-or-none.

Since both calcium-induced and depolarization-induced calcium release have been demonstrated in skinned fibres, a comparison of the two may suggest something about the relative importance of the CROC versus the DROC in initiating EC coupling of intact skeletal muscle. This requires a detailed examination of the properties of the two release mechanisms and the conditions under which they can be demonstrated in the skinned fibre preparation.

Calcium-induced release of calcium

Endo, Tanaka & Ogawa (1970), and Ford & Podolsky (1970), first reported findings which suggested that calcium itself could induce the further release of calcium from SR of skinned skeletal muscle fibres. They noted that skinned fibres immersed in solutions with [Ca^{2+}] weakly buffered (i.e. low EGTA concentrations) tended to give rapidly rising cyclic contractions whose magnitude could nearly equal that of an isometric tetanus. These rapid transient contractions were enhanced by low concentrations of caffeine (Endo et al., 1970) and by prior loading of the SR with calcium (Ford & Podolsky, 1970). These contractions could be abolished by destruction of the SR (Endo et al., 1970). The pattern of force development suggested that the addition of calcium was causing a net release of large amounts of calcium from the SR in an 'all-or-none' fashion. Direct measurements of calcium release by radioactive tracer methods (Ford & Podolsky, 1972b), and by the calcium-sensitive bioluminescent protein aequorin (Endo & Blinks, 1973), demonstrated that a relatively large net release of calcium does occur during these cyclic contractions.

In order to produce a CROC at room temperature (20-23 °C) it is necessary to load the SR with calcium. Evidently the physiological level of calcium in the SR is insufficient to produce a CROC since fibres skinned and transferred directly to the 'trigger' calcium solution without loading do not show this phenomenon. A CROC can, however, be induced without preloading by lowering the temperature to 5–8 °C (Ford & Podolsky, 1972b).

There seems to be an antagonism between magnesium and the calcium necessary to initiate a CROC. A low free $[Ca^{2+}]$ can initiate a CROC when $[Mg^{2+}]$ is low, but not when $[Mg^{2+}]$ is high (Ford & Podolsky, 1972b). Confirming this, Endo (1975a) found that with $[Mg^{2+}]$ at 0.9 mM, $[Ca^{2+}]$ of $10^{-3.5}$ M or greater is needed for a CROC. But with $[Mg^{2+}]$ at 0.05 mM, only 10^{-5} M calcium is sufficient. Even at the lower levels of magnesium, however, the critical level of calcium for phasic contractions is substantially higher than that necessary to activate tonic contractions (Ford & Podolsky, 1972b; Endo, 1975a, b).

The CROC is potentiated by low concentrations of caffeine (Endo et al., 1970). This action is similar to that caused by a decrease of $[Mg^{2+}]$. In solutions containing 0.9 mM magnesium, a $[Ca^{2+}]$ of about $10^{-3.5}$ M is required to initiate a CROC. Upon addition of 0.5 mM caffeine the 'trigger' calcium drops to $10^{-4.6}$ M and with 2 mM caffeine only $10^{-7.0}$ M calcium is required (Endo, 1975b). For the CROC-potentiating action of caffeine to occur the SR must be preloaded above a certain level, and since magnesium antagonizes the caffeine effect, Endo suggested that caffeine initiates contracture of both intact and skinned muscle by the CROC mechanism. If enough caffeine is applied the CROC is enhanced so strongly that the normally low calcium level in the sarcoplasm of resting muscle may be enough to release large amounts of calcium.

Depolarization-induced release of calcium

Skinned fibres can also be made to contract by rapidly changing the major monovalent anion or cation in the bathing solution. If the major monovalent anion in the relaxing medium is methanesulphonate or propionate, transfer of the fibre to a predominantly chloride-containing medium causes a rapid contraction (Ford & Podolsky, 1970; Endo & Nakajima, 1973; Stephenson & Podolsky, 1974; Thorens & Endo, 1975). Similarly, contractions can be induced in skinned fibres heavily preloaded with calcium by replacing the potassium with sodium, lithium or Tris (hydroxymethyl)-aminomethane (Endo & Nakajima, 1973). Endo & Blinks (1973), using the calcium-sensitive photoprotein aequorin, and Stephenson (1975), using ^{45}Ca as a tracer, have demonstrated a net efflux of calcium from skinned fibres during the chloride-induced contractions.

The explanation behind these observations rests on the presumption that the SR is electrically polarized in skinned fibres with the inside of the SR positive. A replacement of an impermeant anion with a more permeant one (e.g. propionate by chloride) or replacement of a permeant cation by a less permeant one (e.g. K^+ by Na^+) would thereby cause depolarization of the polarized SR membrane. The site of this action is probably the SR rather than the T-tubules since depolarization contractions can be induced either in completely skinned fibres where T-tubules may be sealed off (Costantin & Podolsky, 1967), or in partially skinned fibres where the T-tubules appear to remain open (Endo & Nakajima, 1973).

Several years ago it was reported (Lee, Ladinsky, Choi & Kasuya, 1966) that isolated vesicles of SR could respond to an electric current by releasing calcium. It was suggested at the time that this particular effect was due to something other than SR depolarization because the size of an isolated vesicle and the potential gradient across its membrane were probably quite small (Costantin & Podolsky, 1967). Recent work (Miyamoto & Kasai, 1973) has supported this suggestion by showing that the effect of an electric current on isolated SR is probably due to electrolysis at the stimulating electrodes and irreversible denaturation of the SR fragments. Nevertheless, depolarization by a permeant anion such as chloride does produce a net release of calcium from isolated SR that is graded with chloride concentration (Kasai & Miyamoto, 1973). Thus, isolated SR evidently behaves similarly to SR in skinned fibres, although the former accumulates calcium much more slowly than the latter (Ebashi & Endo, 1968).

The DROC in skinned fibres differs from the CROC contractions in a number of ways, and consideration of these differences supports the physiological importance of the DROC. First, unlike the 'all-or-none' CROC, the amount of calcium released is graded with chloride; more calcium is released as the fraction of chloride-replacement increases (Thorens & Endo, 1975). Second, there seems to be no minimal loading requirement for the DROC. The amount of calcium released, as measured by observing a test contracture induced by a high concentration of caffeine after exposure to chloride, seems to be an essentially constant fraction of the calcium in the SR before depolarization (Thorens & Endo, 1975). Next, depolarization-induced release, unlike the CROC, is not abolished by either high $[Mg^{2+}]$ or procaine (Thorens & Endo, 1975). At low temperatures, when the SR has not been loaded with extra calcium, high $[Mg^{2+}]$ has no effect on chloride-induced contractions (Stephenson & Podolsky, 1974; Thorens & Endo, 1975). On the other hand, at high temperatures, if the SR has been loaded with extra calcium, 10 to 20 times more chloride is then needed to induce contractions in high $[Mg^{2+}]$ (Stephenson & Podolsky, 1974). Finally, depolarization-induced release

Table 1. *Characteristics of calcium release mechanisms in skinned muscle fibres*

	Graded with stimulus?	Preloading of SR with calcium at room temperature?	Increasing free magnesium	Adding procaine	Adding sucrose
Chemical (Calcium-induced release of calcium)	No	Required	Diminished	Diminished	No effect
Electrical (Depolarization-induced release of calcium)	Yes	Not required	No effect	No effect	Diminished

is completely inhibited by addition of a sufficient amount of sugar (sucrose, glucose, fructose, or xylose to 40 mM or greater) to the bathing medium, but these have little effect upon the CROC contractions (Thorens & Endo, 1975). These observations are summarized in Table 1. Since the two mechanisms for calcium release respond differently under these different conditions it appears that they may be qualitatively independent processes.

The DROC mechanism and a model for its molecular basis will be discussed further in the next section. An explanation for the CROC, however, is more speculative. It has been suggested that the CROC might be related to the associated morphological changes observed in the SR of skinned fibres (Podolsky, 1975). Electron micrographs of skinned skeletal muscle cells (Plate 1) show that the SR becomes greatly dilated after skinning. However, the addition of large, biologically inert particles like polyvinylpyrrolidone to the bathing solution causes the SR to shrink back to near normal size without diminishing the ability to evoke a CROC in either skinned skeletal (R. E. Godt, unpublished observations) or cardiac (Fabiato & Fabiato, 1975a) cells. There remains the possibility, nevertheless, that the CROC may be related to availability of calcium-sensitive receptors as many SR–T junctions become separated after skinning (see Discussion in Costantin & Taylor, 1973).

CALCIUM RELEASE MECHANISMS IN
SARCOPLASMIC RETICULUM OF INTACT FIBRES

Historically, the major evidence for a graded calcium release in activation of frog skeletal muscle came from studies of potassium-induced depolarization contractures (Hodgkin & Horowicz, 1960). Direct measurements with intracellular microelectrodes showed that changes in extracellular $[K^+]$ were followed by changes in transmembrane potential, and the steady state relationship between membrane potential and the log of the external $[K^+]$ was linear over the range from the mechanical threshold to the maximum developed tension. Rüdel, Taylor & Blinks (1976) have directly observed the relative change in intracellular calcium during potassium contractures using isolated single fibres injected with the bioluminescent protein aequorin. When aequorin combines with calcium it emits light, and the amplitude of the light signal is graded with the concentration of calcium (Blinks, Prendergast & Allen, 1976). Rüdel *et al.* (1976) found that when potassium reached the level that produced detectable force, force was always associated with light emission from the fibre. Peak force development increases in a step-wise fashion with increasing $[K^+]$, and so did light emission. However, the slope of the relationship between peak force and $[K^+]$ progressively decreased and eventually became zero at high $[K^+]$, whereas the slope of the relationship between peak light and $[K^+]$ progressively increased over the same range. These studies, therefore, support the suggestion that SR calcium release and tension are graded, continuous functions of membrane potential over a steeply rising range of membrane potentials from the mechanical threshold to the potential producing maximum tension.

Other workers extended the implications of the information from the potassium contracture studies of Hodgkin & Horowicz to contractions initiated by action potentials (Sandow, 1973). The rate of calcium release was inferred from the rate of tension development during the initial part of a twitch contraction, and it was found that the apparent release of calcium could be modified by a variety of changes in the composition of the solution bathing a muscle. For example, completely replacing the chloride of normal Ringer solution with other monovalent anions did not significantly change the waveform of an action potential. However, these substitutions shifted the mechanical threshold towards the resting membrane potential and increased the initial rate of tension development and maximum amplitude of a twitch. Caffeine added to a normal Ringer produced similar effects. Low concentrations of certain divalent cations (e.g. zinc, uranyl) produced qualitatively similar effects on the amplitude of a twitch, but apparently did so by a different mechanism, by markedly slowing the repolarizing phase of the action potential. High concentrations of the cations slowed the rise as well as the fall of the action potential, and all these effects could be graded by varying the concentration of the added agent. Furthermore, these agents could produce qualitatively opposite effects in the apparent release of calcium during the initial part of the same twitch contraction. For example, high concentrations of zinc raised the mechanical threshold and slowed the rate of depolarization by an action potential, and this was reflected in the initial part of the associated mechanical response: the increased rate of tension development after stimulation was delayed (Taylor, Preiser & Sandow, 1972). Nevertheless, the same action potential was prolonged in its repolarization phase, and this was manifested in the twitch by the fact that tension development continued at a faster rate for a longer time (Fig. 6 of Taylor et al., 1972). Therefore, the action potential can evidently be manipulated in a graded manner to produce both a decrease and an increase in the rate of calcium release within a single twitch contraction. The action potential can apparently mediate graded changes in the time course of the initial events in EC coupling. These initial events as a whole do not occur in an 'all-or-none' manner, and this argues against the possibility that a CROC is the primary mechanism initiating a twitch (Sandow, 1973).

Information of a more quantitative nature became available when voltage clamp techniques were developed for studies of frog muscle(Adrian, Costantin & Peachey, 1969). One of the drawbacks to potassium contracture studies is that changes in surface potential are probably not transmitted along the T system very rapidly. If the outermost part of a fibre rapidly relaxes while the centre is being activated by the slow diffusion of potassium into the T system, it is difficult to determine the nature of calcium release from measuring only tension (Costantin, 1971). Voltage clamp studies, however, confirmed the previous observations that contractile force of frog muscle is graded with membrane potential (Bezanilla, Caputo & Horowicz, 1971).

Not only could the membrane potential in the T system be changed more rapidly with the voltage clamp technique, but the segment of the cell under voltage control could be simultaneously observed with the light microscope (Adrian et al., 1969). Direct observations of changes in striation spacing and myofibrillar orientation were used as indicators of contractile activation (Fig. 2), and permitted the presumed electrical changes in the T system to be more closely correlated with the inferred release of calcium in the cell (Costantin & Taylor, 1973). In most of the experiments previously described (e.g. Sandow, 1973) the length of the muscle was held

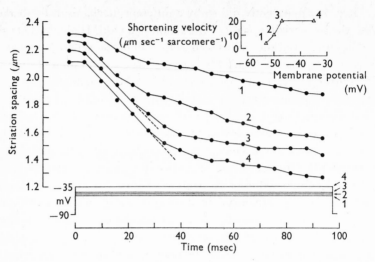

Fig. 2. The time-course of shortening of the innermost myofibrils of a muscle fibre with increasing depolarization (after Fig. 2 of Costantin & Taylor, 1973). The magnitude of each depolarizing step associated with a particular shortening curve is shown above the abscissa. Since there probably is a delay as well as a decrement in the spread of depolarization from the surface to the centre of the fibre (Adrian *et al.*, 1969) and the myofibrils are evidently coupled to one another mechanically, the initial velocity may be predominantly determined by shortening of the outer myofibrils. Therefore, the only steps plotted are those that produced active shortening of the entire fibre cross-section. The inset shows the shortening velocity measured from the slope of each shortening curve in the range of striation spacings from 2.0 to 1.9 μm. This range is the first part of each curve in which one can distinguish between actively contracting and passively shortened myofibrils. See Costantin & Taylor (1973) for further details.

constant while graded changes in the membrane potential and graded changes in the rate and amplitude of tension development were measured. However, in the experiments summarized in Fig. 2 the surface membrane potential was held constant, while the relation between graded changes in depolarization and graded changes in the rate of shortening were measured. Shortening of the entire fibre cross-section to striation spacings below 1.95 μm without the appearance of myofibrillar folding was taken to indicate that all the myofibrils were contracting actively. Nevertheless, although all the myofibrils across the entire cross-section were activated, a more rapid contraction was produced by a larger step (Fig. 2). There was no sign of an 'all-or-none' contractile response at some threshold voltage. So both the rate of tension development (Sandow, 1973) and the rate of shortening (Costantin & Taylor, 1973) are graded with the membrane potential in intact muscle, presumably because the calcium released from the SR at any moment during the initial part of a contraction is regulated by the level of depolarization in the T tubules. We believe that the weight of evidence seems to be against a CROC mechanism initiating EC coupling in contractions of intact skeletal muscle. Even with the entry of a small amount of calcium during activation (Bianchi, 1969), it evidently does not follow automatically that enough additional calcium is released to activate the myofilaments fully.

The following observations provide evidence that the initial calcium release mechanism in vertebrate skeletal muscle does not even depend upon an influx of calcium. Costantin (1971) observed a fast and slow component to tension development during potassium contractures. When the driving force for inward calcium movement was greatly reduced by bathing a fibre in a calcium-free solution, neither component was affected. Furthermore, when the relationship between depolarization and contractile activation was examined with high resolution light microscopy and a voltage clamp technique, a 10-fold decrease in extracellular calcium did not alter the pattern of contractile activation for at least the first 100 msec of a de-polarizing step (Costantin, 1974). Similarly, Armstrong et al. (1972) found that bathing single fibres in solutions with a very low calcium concentration did not eliminate twitch contractions for at least 20 min. They estimated that the calcium influx with each action potential was diminished by a factor of a million compared to the influx in normal saline solution. Nevertheless, this presumed reduction in 'trigger' calcium did not abolish contraction. Some of these experiments were done in low calcium solutions that were well buffered for calcium, and contained magnesium to replace the divalent cation that might be retained in or near the surface membrane by surface negative charges, but the twitch experiments were done in solutions containing only the calcium buffer. Therefore, Rüdel et al. (1976) repeated the twitch experiment in solutions with both EGTA (3 mM) and Mg^{2+} (3 mM), using fibres isolated from R. temporaria and injected with the photoprotein aequorin to monitor calcium transients in the myofilament space.

They found that removing extracellular calcium had no effect on the peak force in the twitch or the associated intracellular calcium transient for at least 20 to 30 min. These results support the conclusion that external calcium need not function as a 'trigger' (Armstrong et al., 1972), and that replenishment of stored calcium from the external solution is unnecessary for initiating normal twitch contractions.

Fabiato & Fabiato (1975a) have suggested that there might be a complex system with several feedbacks involved in EC coupling that would allow calcium to amplify its own release, yet can also be graded. For example, a CROC might induce transient changes in ADP or magnesium concentrations which themselves might further influence the CROC. As discussed above, this is unlikely to be the mechanism for the initiation of a normal contraction in skeletal muscle. However, there are significant differences between cardiac and skeletal muscle. For example, extracellular calcium is of greater importance in maintaining the function of cardiac muscle, and a CROC may well be the physiological mechanism initiating normal myo-cardial contraction. Moreover, even in skeletal muscle, once EC coupling is initiated by a DROC the effects of a CROC may be additive. For ex-ample, in the presence of caffeine or other agents that potentiate contrac-tion, a CROC might occur after the initial DROC (Sandow, 1973; Thorens & Endo, 1975).

The foregoing considerations do not give us any information about how depolarization of the T system is transmitted across the SR–T junc-tion to depolarize the SR and release calcium. Indeed the discussion so far suggests only what the mechanism of transmission is not: it apparently is not the chemically specific result of a net movement of calcium across the junction. However, a clue to the molecular basis for transmission has come from the work of Schneider & Chandler (1973). They discovered that when all of the time- and voltage-dependent changes in sodium and potassium currents are eliminated by appropriate manipulations of the bathing solution, they could detect a voltage-dependent movement of charge that may carry current across the T system membrane during EC coupling. It is unlikely that the current is ionic, and they suggest that the current is probably caused by charged particles that move when the electric field across the membrane is changed. Similar charge movements have been found in squid axons and are associated with sodium gating currents. However, these currents found in frog muscle are too slow to gate sodium channels, and are probably associated with the mechanism coupling T tubule depolarization to SR depolarization and the release of activator calcium. Chandler, Rakowski & Schneider (1976) have speculated that there may be some sort of mechanical linkage which extends from the charge in the T tubular membrane to a site in the SR membrane (e.g. see Fig. 11 of their paper). The SR membrane site is at a channel through which calcium can flow from the SR into the myoplasm. Their model suggests that the channel is plugged when the charge is in the resting position and the plug is removed when the charged particles move during activation. The struc-tures most likely to be associated with this mechanism are the feet across the SR–T junction (Franzini-Armstrong, 1975). However, the correlation between particles that may be sites of the observed charge movement and the feet which may allow interaction between the two membranes is not yet known in sufficient detail to implicate firmly these structures as the site for gating currents. Nevertheless, this model and future attempts to test it promise to reveal much about the molecular mechanism involved in the initiation and control of EC coupling.

CONCLUSION

The CROC phenomenon is probably not the primary mechanism by which calcium release is initiated during normal EC coupling in skeletal muscle. First, the SR of skinned fibres must be preloaded with calcium before one can induce a CROC at room temperature. Preloading is not necessary at low temperatures. The ability to initiate a CROC, therefore, decreases as temperature increases. On the other hand, the release of calcium into the myofilament space is apparently affected by temperature in the opposite manner in intact fibres. The peak of the aequorin-light transient associated with the twitch of an intact fibre increases with temperature (Taylor, Rüdel & Blinks, 1975; Rüdel et al., 1976). This point is contingent upon the elucidation of the intrinsic temperature dependence of the calcium–aequorin reaction (Blinks et al., 1976). Second, a very high level of calcium is necessary to initiate a CROC. The concentrations of calcium required would, in fact, give rise to maximal tonic contractions if applied in solutions strongly buffered for calcium with EGTA (e.g. Fig. 1). One could, however, postulate that the calcium influx during the action potential may be confined to a very small region of the cell adjoining some sensitive portion of the SR. Third, however, Armstrong et al. (1972) showed that single skeletal muscle fibres continue twitching for up to 20 min when immersed in low calcium solutions where 'trigger' calcium was probably depleted, and this result was confirmed by Rüdel et al. (1976). Armstrong et al. (1972) estimate that less than one Ca^{2+} per sarcomere would enter under these conditions. This would seem to rule out the possibility that transmembrane fluxes of calcium initiate the EC coupling process. Fourth, the CROC supposedly releases enough calcium to activate fully the contractile filaments, whereas the EC coupling process is evidently graded by the action potential (Sandow, 1973). Last, there are a number of indirect indications that intracellular $[Mg^{2+}]$ may be too high to permit a CROC process. Intracellular magnesium levels in frog skeletal muscle have not been measured, but it is known that total magnesium is about 8 to 9 mM (Bianchi, 1968). Magnesium binds strongly to ATP but the total ATP of skeletal muscle is only 2 mM (Kushmerick & Davies, 1969). Magnesium probably also binds to a number of other intracellular constituents and proteins but to an unknown degree. Polimeni & Page (1973), on the basis of a number of indirect indicators, estimate that intracellular [Mg^{2+}] in cardiac muscle is probably 10^{-3} M or less. In this regard, Endo (1975b) has utilized the magnesium-dependence of the action of caffeine on skinned fibres to estimate the physiological level of magnesium in intact skeletal muscle cells. Since caffeine rapidly penetrates the sarcolemma of intact muscle cells (Bianchi, 1962) the membrane presents no barrier to this drug. With this in mind, Endo found that when the skinned fibre bathing medium contained 0.9 mM magnesium the caffeine dose–response curves of skinned fibres were similar to those obtained for intact fibres. If, as seems likely, intracellular free [Mg^{2+}] is this high, the CROC mechanism would require an improbably high ($\sim 10^{-3.5}$ M) 'trigger' level of calcium (Endo, 1975b). For these reasons it is unlikely that the CROC plays a role in initiating the normal EC coupling process of skeletal muscle cells.

Ford & Podolsky (1972b) hypothesized that at low [Mg^{2+}] a CROC can propagate completely across a skinned fibre, whereas at high [Mg^{2+}] the CROC may attenuate with distance from the fibre surface. In the latter case, contracting regions of the fibre might be adjacent to compliant non-contracting regions and little or no tension would be monitored at the ends of the fibre. However, Fabiato & Fabiato (1975b) found that magnesium also antagonized CROC contractions in skinned skeletal muscle fibres that were about one-tenth the thickness of those used by Ford & Podolsky (1972b). They concluded that the CROC was not modified by variations in size of the preparation, and that the potentiating effect of decreased [Mg^{2+}] on the CROC is not mainly due to propagation for a greater distance, but to a decrease in the capacity and rate of uptake of calcium by the SR. Their conclusion was subsequently supported by direct measurements with ^{45}Ca (Stephenson, 1975), and is consistent with the effects of low [Mg^{2+}] on isolated SR (see Fabiato & Fabiato, 1975b for references).

On the other hand, there is convincing evidence that a CROC may play some role in cardiac muscle. Kerrick & Best (1974), and Fabiato & Fabiato (1975a, b) found that CROC contractions of single skinned cardiac cells were initiated by very low concentrations of calcium, concentrations lower than those that directly activate the myofibrils. Even if [Mg^{2+}] is in the millimolar range, CROC contractions in cardiac cells can still be initiated by low calcium levels. At these magnesium levels the CROC mechanism in skeletal muscle is completely inhibited. Contrary to the case with skeletal muscle, CROC contractions can be evoked at room temperature in single, skinned, cardiac cells with the SR loaded at the physiological levels of calcium.

While the case for a simple CROC mechanism in cardiac EC coupling seems rather strong, we conclude that EC coupling in skeletal muscle is initiated and regulated by a transmembrane potential difference in the T system and an electrical signal across the SR–T junction.

SUMMARY

A release of calcium from intracellular stores into the myofilament space is the result of the physiological events leading to contraction of vertebrate skeletal muscle. These events, which follow excitation of the muscle cell surface, have been termed excitation–contraction (EC) coupling. Many of the physical, chemical, and morphological features of EC coupling have been studied at the cellular and subcellular levels, and current ideas about the mechanism of EC coupling centre around two possibilities. One possibility is that surface membrane depolarization is transmitted to the vicinity of the intracellular calcium store, the sarcoplasmic reticulum (SR), via the transverse tubular (T) system, and electrical transmission across the SR–T junction controls the release of calcium from the SR. The evidence from studies of intact skeletal muscle cells tends to support this idea. The other possibility is that junctional transmission is chemical, specifically that exposure of the SR to a small amount of calcium causes an explosive release of a large amount of calcium, enough to activate fully the contractile filaments. Most of the evidence for this possibility has come from studies on muscle cells that have had their surface membrane removed ('skinned'). A comparison of the results in support of these two possibilities suggests that the mechanisms underlying electrically and chemically induced calcium release are independent, that the conditions required to permit chemically induced release are not likely to exist normally, and that the purely electrical changes initiate the physiological mechanism for EC coupling in intact skeletal muscle.

We wish to thank J. A. Olsen for the electron micrographs of the fibres in Plate 1, and O. E. Akwari for providing the tissue and M. J. Hahn for performing the experiments presented in Fig. 1. We are grateful to W. K. Chandler, M. Endo, and A. Fabiato for preprints of their manuscripts, and to J. R. Blinks and M. D. Thames for their criticisms of our manuscript. This work was supported in part by a grant-in-aid from the Muscular Dystrophy Association, by USPHS grants NS 10327 and AM 17828, during the tenure of an Established Investigatorship of the American Heart Association.

REFERENCES

ADRIAN, R. H., COSTANTIN, L. L. & PEACHEY, L. D. (1969). Radial spread of contraction in frog muscle fibres. *J. Physiol., Lond.*, **204**, 231–257.

ADRIAN, R. H. & PEACHEY, L. D. (1973). Reconstruction of the action potential of frog sartorius muscle. *J. Physiol., Lond.*, **235**, 103–131.

ARMSTRONG, C. M., BEZANILLA, F. M. & HOROWICZ, P. (1972). Twitches in the presence of ethylene glycol bis(β-aminoethyl ether)-N,N'-tetraacetic acid. *Biochim. biophys. Acta*, **267**, 605–608.

BAYLOR, S. M. & OETLIKER, H. (1975). Birefringence experiments on isolated skeletal muscle fibres suggest a possible signal from the sarcoplasmic reticulum. *Nature, Lond.*, **253**, 97–101.

BEZANILLA, F., CAPUTO, C. & HOROWICZ, P. (1971). Voltage clamp activation of contraction in short striated muscle fibres of the frog. *Acta Cient. Venez.*, **22**, 72–74.

BEZANILLA, F. & HOROWICZ, P. (1975). Fluorescence intensity changes associated with contractile activation in frog muscle stained with Nile Blue. *J. Physiol., Lond.*, **246**, 709–735.

BIANCHI, C. P. (1962). Kinetics of radiocaffeine uptake and release in frog sartorius. *J. Pharmac. exp. Therap.*, **138**, 41–47.

(1968). *Cell Calcium*, ed. E. E. Bittar, Appleton-Century-Crofts: New York.

(1969). Pharmacology of excitation–contraction coupling in muscle. *Fedn Proc.*, **28**, 1624–1628.

BLINKS, J. R., PRENDERGAST, F. G. & ALLEN, D. G. (1976). Photoproteins as biological calcium indicators. *Pharmac. Rev.*, **28**, 1–102.

CHANDLER, W. K., RAKOWSKI, R. F. & SCHNEIDER, M. F. (1976). Effects of glycerol treatment and maintained depolarization on charge movements in skeletal muscle. *J. Physiol., Lond.*, **254**, 285–316.

COSTANTIN, L. L. (1971). Biphasic potassium contractures in frog muscle fibers. *J. gen. Physiol.*, **58**, 117–130.

(1974). Contractile activation in frog skeletal muscle. *J. gen. Physiol.*, **63**, 657–674.

(1975). Electrical properties of the transverse tubular system. *Fedn Proc.*, **34**, 1390–1394.

COSTANTIN, L. L. & PODOLSKY, R. J. (1967). Depolarization of the internal membrane system in the activation of frog skeletal muscle. *J. gen. Physiol.*, **50**, 1101–1124.

COSTANTIN, L. L. & TAYLOR, S. R. (1973). Graded activation in frog muscle fibers. *J. gen. Physiol.*, **61**, 424–443.

EBASHI, S. & ENDO, M. (1968). Calcium ion and muscle contraction. *Progr. Biophys. molec. Biol.*, **18**, 123–183.

EISENBERG, B. R. & PEACHEY, L. D. (1975). The network parameters of the T-system in frog muscle measured with the high voltage electron microscope. *Ann. Proc. Electron Microsc. Soc. Amer.*, **33**, 550–551.

ENDO, M. (1975a). Conditions required for calcium-induced release of calcium from the sarcoplasmic reticulum. *Proc. Japan Acad.*, **51**, 467–472.

(1975b). Mechanism of action of caffeine on the sarcoplasmic reticulum of skeletal muscle. *Proc. Japan Acad.*, **51**, 479–484.

ENDO, M. & BLINKS, J. R. (1973). Inconstant association of aequorin luminescence with tension during calcium release in skinned muscle fibres. *Nature New Biol.*, **246**, 218–221.

ENDO, M. & NAKAJIMA, Y. (1973). Release of calcium induced by 'depolarization' of the sarcoplasmic reticulum membrane. *Nature New Biol.*, **246**, 216–218.

ENDO, M., TANAKA, M. & OGAWA, Y. (1970). Calcium induced release of calcium from the sarcoplasmic reticulum of skinned skeletal muscle fibres. *Nature, Lond.*, **228**, 34–36.

FABIATO, A. & FABIATO, F. (1975a). Contractions induced by a calcium-triggered release of calcium from the sarcoplasmic reticulum of single skinned cardiac cells. *J. Physiol., Lond.*, **249**, 469–496.

(1975b). Effects of magnesium on contractile activation of skinned cardiac cells. *J. Physiol., Lond.*, **249**, 497–517.

FALK, G. & GERARD, R. W. (1954). Effect of micro-injected salts and ATP on the membrane potential and mechanical response of muscle. *J. cell. comp. Physiol.*, **43**, 393–403.

FORD, L. E. & PODOLSKY, R. J. (1970). Regenerative calcium release within muscle cells. *Science, Wash.*, **167**, 58–59.

(1972a). Calcium uptake and force development by skinned muscle fibres in EGTA buffered solutions. *J. Physiol., Lond.*, **223**, 1–19.

(1972b). Intracellular calcium movements in skinned muscle fibres. *J. Physiol., Lond.*, **223**, 21–33.

FRANZINI-ARMSTRONG, C. (1975). Membrane particles and transmission at the triad. *Fedn Proc.*, **34**, 1382–1389.

GODT, R. E. (1974). Calcium-activated tension of skinned muscle fibers of the frog. Dependence on magnesium adenosine triphosphate concentration. *J. gen. Physiol.*, **63**, 722–739.

HEILBRUNN, L. V. & WIERCINSKI, F. J. (1947). The action of various cations on muscle protoplasm. *J. cell. comp. Physiol.*, **29**, 15–32.

HODGKIN, A. L. & HOROWICZ, P. (1960). Potassium contractures in single muscle fibres. *J. Physiol., Lond.*, **153**, 386–403.

HUXLEY, H. E. (1971). The structural basis of muscular contraction. *Proc. R. Soc. Lond.* B, **178**, 131–149.

KASAI, M. & MIYAMOTO, H. (1973). Depolarization induced calcium release from sarcoplasmic reticulum membrane fragments by changing ionic environment. *FEBS Lett.*, **34**, 299–301.

KERRICK, W. G. L. & BEST, P. M. (1974). Calcium ion release in mechanically disrupted heart cells. *Science, Wash.*, **183**, 435–437.

KUSHMERICK, M. J. & DAVIES, R. E. (1969). The chemical energetics of muscle contraction. II. The chemistry, efficiency, and power of maximally working sartorius muscles. *Proc. R. Soc. Lond.* B, **174**, 315–353.

LEE, K. S., LADINSKY, H., CHOI, S. J. & KASUYA, Y. (1966). Studies on the in vitro interaction of electrical stimulation and calcium movement in sarcoplasmic reticulum. *J. gen. Physiol.*, **49**, 689–715.

MAUGHAN, D. W. & GODT, R. E. (1974). Role of Donnan-osmotic forces in skinned muscle fibers. *Fedn Proc.*, **33**, 401.

MIYAMOTO, H. & KASAI, M. (1973). Reexamination of electrical stimulation on sarcoplasmic reticulum fragments in vitro. *J. gen. Physiol.*, **62**, 773–786.

MOBLEY, B. A. & EISENBERG, B. R. (1975). Size of components in frog skeletal muscle measured by methods of stereology. *J. gen. Physiol.*, **66**, 31–45.

NATORI, R. (1954). Property and contraction process of isolated myofibrils. *Jikeikai Med. J.*, **1**, 119–126.

NIEDERGERKE, R. (1955). Local muscular shortening by intracellularly applied calcium. *J. Physiol., Lond.*, **128**, 12P.

PODOLSKY, R. J. (1968). Membrane systems in muscle cells. *Symp. Soc. exp. Biol.*, **22**, 87–100.

(1975). Muscle activation: the current status. *Fedn Proc.*, **34**, 1374–1378.

POLIMENI, P. I. & PAGE, E. (1973). Magnesium in heart muscle. *Circ. Res.*, **53**, 367–374.

RÜDEL, R., TAYLOR, S. R. & BLINKS, J. R. (1976). Changes of intracellular calcium ion concentration in isolated amphibian skeletal muscle fibers: detection with aequorin. In preparation.

SANDOW, A. (1952). Excitation–contraction coupling in muscular response. *Yale J. Biol. Med.*, **25**, 176–201.

(1965). Excitation–contraction coupling in skeletal muscle. *Pharmac. Rev.*, **17**, 265–320.

(1973). Electromechanical transforms and the mechanism of excitation–contraction coupling. *J. Mechanochem. Cell Motil.*, **2**, 193–207.

SCHNEIDER, M. F. & CHANDLER, W. K. (1973). Voltage dependent charge movement in skeletal muscle: a possible step in excitation–contraction coupling. *Nature, Lond.*, **242**, 244–246.

STEPHENSON, E. W. (1975). Release and reaccumulation of Ca-45 in skinned muscle fibers. *The Physiologist*, **18**, 407.

STEPHENSON, E. W. & PODOLSKY, R. J. (1974). Chloride-induced Ca release in skinned muscle fibers. *Fedn Proc.*, **33**, 1260.

TAYLOR, S. R., PREISER, H. & SANDOW, A. (1972). Action potential parameters affecting excitation–contraction coupling. *J. gen. Physiol.*, **59**, 421–436.

TAYLOR, S. R., RÜDEL, R. & BLINKS, J. R. (1975). Calcium transients in amphibian muscle. *Fedn Proc.*, **34**, 1379–1381.

THORENS, S. & ENDO, M. (1975). Calcium-induced calcium release and 'depolarization'-induced calcium release: their physiological significance. *Proc. Japan Acad.*, **51**, 473–478.

CONTRACTILE ACTIVATION AND CALCIUM MOVEMENTS IN HEART CELLS

By R. NIEDERGERKE, D. C. OGDEN
AND SALLY PAGE

Biophysics Department, University College London,
Gower Street, London WC1E 6BT

As we have heard (Szent-Györgyi, this volume), contractile activity in different muscle cells can differ in considerable detail, yet in every case calcium ions seem to play a key role. A characteristic feature, long recognised, which distinguishes the activity of heart from fast skeletal muscle cells is that heart twitches, in contrast to those of the skeletal fibres, depend critically on the level of calcium ions in the fluids surrounding the cells. The explanation for this observation has emerged in the past 10 to 20 years and can be summarised as follows: the excitable membrane of heart cells contains special channels, or carriers, apparently missing in skeletal muscle fibres, which allow the inward passage of calcium ions into the cells during each action potential. Since, as is well known, calcium ions also serve as the activator of the contractile proteins, these ions provide a direct link between excitation and contraction in heart cells, participating, indeed, in both these events. In fast skeletal fibres, on the other hand, influx of calcium ions through the cell membrane plays only a minor role, if any, during excitation; while activator calcium for contraction is contained within the cells and recycled between the myofibrils and the sarcoplasmic reticulum.

An important consequence of the dual role of calcium in heart cells is illustrated by events occurring during regulation of the strength of the heart beat in the living organism. For example, the sympathetic transmitter noradrenaline (or, in certain animals, adrenaline), which helps to convey the signal demanding an accelerated blood supply from the heart, causes an increase of the calcium inward current during the cardiac action potential. As a result, the heart beat strengthens and this, together with other mechanisms brought into action, increases the volume of blood which is being pumped.

Calcium movements of this kind have now been observed in heart cells of different animals, e.g. mammals and amphibia, and the mechanisms regulating these movements seem similar in these cases, but again, there are interesting differences, though probably mainly of quantitative nature. For example, much evidence (summarised below) suggests that, in the frog heart, calcium inward current carries a sufficient quantity of these ions into the cells to activate the contractile proteins; yet in the mammalian heart this does not appear to be the case. Here the main function of this current is thought to be that of replenishing a cellular store of calcium, the sarcoplasmic reticulum, which is able to release the ion for contractile activation by a mechanism similar to that also present in skeletal fibres (as described in the preceding paper, Taylor & Godt, this volume; cf. also Reuter, 1973). One of the most striking indications of the different properties of the amphibian and mammalian hearts lies in the amount of sarcoplasmic reticulum detectable by electron microscopy in the cells of these two hearts. Thus, in frog heart ventricle the sarcoplasmic reticulum, expressed as its volume relative to that occupied by myofibrils, amounts to no more than 0.5 %, compared with 7.3 %, the corresponding figure for rat heart ventricle (Page & Niedergerke, 1972; Page, McCallister & Power, 1971). While this difference is clearly indicative of a greater functional role of the sarcoplasmic reticulum in the mammalian heart, it is nevertheless possible that the sarcoplasmic reticulum also plays a part during contractile activity of the frog heart, e.g. in speeding up the process of relaxation.

A considerable amount of information has accumulated in the past 15 years on the cellular fluxes and currents associated with contractile activity of heart cells, both amphibian and mammalian (for recent reviews see Langer, 1973, and Reuter, 1973), yet a detailed hypothesis incorporating the data obtained into a description of the events leading to muscular contraction has not emerged. One of the reasons for this resides in the uncertainty surrounding the magnitude of the calcium fluxes and currents which enter the cells during each individual action potential. To illustrate this point, the results of our own tracer calcium flux determinations in frog heart tissue may first be recalled (Niedergerke, 1963; Niedergerke, Page & Talbot, 1969). In this work, the extra influx of calcium crossing the cell surface per action potential was found to be no more than 0.3 to 0.4 pmole cm^{-2} of membrane surface, giving an average current density $< 0.3 \mu A cm^{-2}$ (for an action potential's duration of 300 msec), even under conditions in which the strength of the heart twitches was maximal. Fluxes of this size will cause a net uptake of calcium of 2–3 μmoles l^{-1} heart cells per action potential, a figure calculated by taking account of the cell dimensions. In the case of mammalian heart cells, calcium fluxes associated with action potentials have been estimated on the basis of voltage clamp data, and these, somewhat surprisingly, yielded values of similar magnitude, varying from 0.3 to 1.2 pmoles cm^{-2} per impulse (Reuter, 1973). As can be shown, calcium ions in these quantities, transferred into heart cells, would hardly suffice to occupy, in activating a twitch, more than 10 % of the free binding sites on the troponin molecules (in addition to sites already occupied in a state of rest; estimated from Ca binding sites on troponin (see Weber & Murray, 1973) and taking into account the myofibrillar content of frog heart cells). Further, the current density obtained is probably too small to contribute appreciably to the membrane currents during the action potential. Both these points are in conflict with various experimental findings (some of which are to be reported below) which seem to suggest a significant contribution by calcium movements to the action potential and the subsequent twitch. The origin of these discrepancies is not yet known with certainty, but, as has already been argued (Niedergerke & Orkand, 1966; Chapman & Niedergerke, 1970a), there are good reasons for believing that all the values obtained are underestimates of the true fluxes and currents. Two likely contributary factors may be mentioned. (1) In tracer calcium experiments the time resolution for the determination of cellular fluxes is, at present, not better than 10–20 sec, chiefly because of impeded diffusion of ions in the tissue to and from the myocardial cells. Consequently, any transients occurring in time intervals shorter than this, for example during an action potential, would largely be missed. (2) In the voltage clamp experiments, estimates of the size of calcium currents and fluxes rest on the validity of assumptions made about the concurrent potassium and sodium currents. For example, if, as it seems, these potassium and sodium currents do not remain constant when either external calcium is withdrawn or an inhibitor of calcium influx is applied, then the quantitative assessment of calcium inward current would be subject to uncertainty also in this case (for a recent discussion on this point, see Kass & Tsien, 1975).

In view of these and other uncertainties it is perhaps understandable that the interpretation of many results in this field is controversial and that several different mechanisms are currently invoked in explaining contractile activation and its modification by 'inotropic' agents (cf. a recent review by Langer, 1973). Thus, it is being proposed, both for the amphibian and mammalian heart, that activator calcium is derived not only from the external medium surrounding the heart cells and the sarcoplasmic reticulum within these cells but also from other cellular structures capable of storing and discharging calcium ions, the structures envisaged including the cell membrane, a specialised region adjacent to it, the mitochondria and others.

The experiments reported below were made in an attempt to clarify this situation for the case of the frog heart. First, an analysis is presented of three different experimental conditions which induce rapid changes in strength of the heart beat ($t_{\frac{1}{2}} \leqslant$ 10–20 sec) and which are explicable in terms of mechanisms controlling the magnitude of calcium inward current across the cell surface. Later, an additional mechanism which seems to modify contractile tension by processes occurring in the cell interior will also be discussed. Extensive use is made of histological information obtained in frog heart tissue, on the diffusion pathways determining the rate of ion exchange within the fluid spaces of individual heart trabeculae (cf. Page & Niedergerke, 1972; Lammel, Niedergerke & Page, 1975).

RECENT EXPERIMENTS WITH FROG HEART TISSUE

In the first experiment, the most simple to be discussed (illustrated in Fig. 1a), a single atrial trabecula was continuously superfused with Ringer's fluid whose calcium concentration was altered in a stepwise fashion (completion of fluid change within 100 msec), while twitch tensions and associated intracellular action potentials were recorded throughout. For simplicity, the discussion is confined here (but see further below under discussion of slow tension changes) to the responses obtained when, after brief exposure to high [Ca^{2+}], the external calcium concentration was lowered again, in Fig. 1a from 4 to 1 mM. As is seen, both tension and the action potential's 'overshoot' rapidly declined in the low calcium medium, in an almost parallel fashion, to the lower steady state levels which were attained within a few beats. In particular, the fall in twitch tension occurring during the initial 8 sec of exposure to the lower [Ca^{2+}], i.e. the difference in the peak of the last twitch in 4 mM Ca^{2+} and that of the first in 1 mM Ca^{2+}, amounted to some 75 % of the overall tension decline, corresponding to a tension change with a $t_{\frac{1}{2}}$ of 4 sec. It is instructive to compare this result with the theoretical $t_{\frac{1}{2}}$, of between 1.5 and 5 sec, with which the average calcium concentration at heart cell surfaces can be expected to have altered under these conditions (from data obtained by Page & Niedergerke, 1972, and Lammel et al., 1975). Thus, the fact that the experimental value for the half-time of tension decline lies within this range of theoretical values clearly suggests that the observed fall was synchronous with the decline of [Ca^{2+}] at the cell surfaces and, also, that it was the immediate response to the reduced calcium inward current for which the fall in overshoot is a measure (cf. Hagiwara & Nakajima, 1966; Niedergerke & Orkand, 1966; Rougier et al., 1969).

Another relevant point emerged from an extension of this type of experiment in which, instead of a single step, a series of step changes of [Ca^{2+}] were imposed on the preparation, each step being taken from a constant high level (3 mM) to various lower levels of calcium (ranging

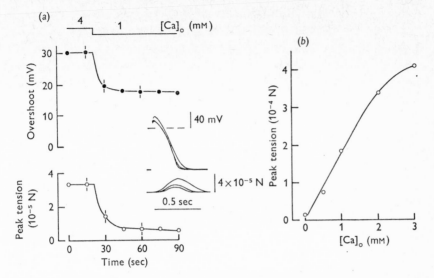

Fig. 1. Response of twitches and action potentials to reduction of external $[Ca^{2+}]$. Single atrial trabecule used in this as in all subsequent experiments (except in that of Fig. 6); continuous superfusion with Ringer's fluid, stepwise change of $[Ca^{2+}]$ in this fluid. (a) *Top trace*: single $[Ca^{2+}]$ step imposed on trabecula driven at regular rate of 4 min^{-1}. *Graphs below*: plots of overshoots of action potentials and peak twitch tensions of 7 successive twitches. *Inset*: superimposed tracings of potential and tension records at points indicated by vertical bars in main graph. (b) Series of step changes of $[Ca^{2+}]_0$ (external calcium) imposed on a different preparation, each time starting from 3 mM down to 0, 0.5, 1.0 and 2.0 mM. Plot of peak tension of first twitch in low $[Ca]_0$ against this concentration (except for steady state tension in 3 mM Ca^{2+}). Stimulation at 10 min^{-1}.

from 0 to 2 mM). The results, plotted in Fig. 1*b* as the twitch tension of the first twitch at the low level of $[Ca^{2+}]$ versus this concentration shows the monotonic relationship obtained, in this case, between tension and calcium; and the curve drawn through the results is linear over much of its range, crossing the ordinate close to its point of origin. (Equilibration of $[Ca^{2+}]$ at heart cell surfaces in this experiment, most probably, was virtually complete after the 6 sec exposure period to the low $[Ca^{2+}]$ used since the $t_{\frac{1}{2}}$ with which the average $[Ca^{2+}]$ altered had a value of approximately 1.5 sec. This was determined from the rapid phase of tension changes in response to $\Delta[Ca^{2+}]$ steps at an enhanced heart rate; for description of the analysis cf. Chapman & Niedergerke, 1970*a*.) Clearly, these results too favour the explanation of a direct link between calcium inward current and contractile activation since (*a*) the exposure time to low calcium was, again, short enough for intracellular calcium to have remained practically unaltered; and (*b*) calcium inward current is expected to rise with $[Ca^{2+}]$ in

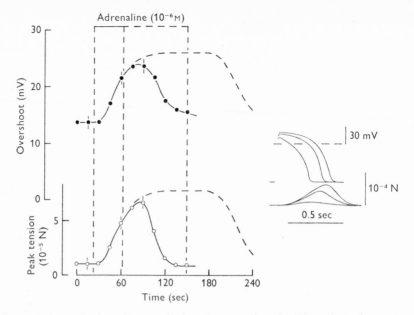

Fig. 2. Effect of adrenaline applied to heart trabecula. Plot of results, super-imposed, from two successive periods of exposure to 10^{-6} M adrenaline. For a short exposure (30 sec), points connected by solid line; for longer exposure (120 sec), results sketched in with dashed line. *Inset*: superimposed tracings of 3 potential and tension records before, during and after the short exposure period at times indicated in main graph (short vertical bars as in Fig. 1). [Ca^{2+}]: 0.75 mM through-out, stimulation at 4 min^{-1}.

a monotonic fashion, similar to, though, of course, probably not identical with, that of the tension rise.

Figs. 2 and 3 illustrate results of the second type of experiment which was concerned with the mode of action of adrenaline, the natural sympathetic transmitter in the frog heart. In the experiment of Fig. 2, an atrial trabecula had been exposed to the transmitter for two successive periods; one of these was sufficiently long (120 sec) to induce a steady tension response, the other, shorter one (30 sec) gave rise to only a transient, though nearly maximal, response. The most striking feature in both cases is the simul-taneous, again almost parallel, change in overshoot and twitch height, during both the build-up of the transmitter effect and its decline. Obviously, in this respect, the similarity with the calcium response just described is quite close, except for what is in the present context a minor detail, that the prolongation of the action potential's duration was more consistently observed in the high adrenaline than in the high calcium medium (compare inset of Fig. 2 with Fig. 1). The tentative conclusion, therefore, is that this response as well arises as the direct consequence of a change in calcium inward current, enhancement of the current by adrenaline being well documented (Vassort *et al.*, 1969; Reuter, 1973).

With regard to the time course of the response, this cannot, however, be described in terms of instantaneous effects of external concentration changes, even when the diffusion times to and from the cell surfaces are taken into account. A detailed examination of this time course will be reported elsewhere (R. Niedergerke & S. Page, unpublished), but some of the results may be outlined. On application and withdrawal of the adrenaline there was an initial delay which is reflected (*a*) by the failure of the first twitch (and overshoot) in adrenaline to be substantially, if at all, enhanced (after 3 to 5 sec exposure); (*b*) by the absence of an immediate tension and potential decline on withdrawal of the adrenaline, the twitches and potentials even continuing to rise if the preceding exposure to adrenaline had been kept short (see Fig. 2). Following this delay, potentials and tension rose, or fell, with a $t_\frac{1}{2}$ of 10 to 20 sec, i.e. much more slowly than would be expected if diffusion equilibration were the determining factor (for this, the calculated $t_\frac{1}{2}$ ranged from about 2 to 5 sec). In the light of these findings, an alternative hypothesis of adrenaline action might be considered according to which calcium inward current, enhanced by adrenaline, causes, during successive action potentials, a gradual accumulation of calcium in a cellular store which could then release increasing quantities for contractile activation. This idea was tested by comparing the two types of response evoked in a preparation which, while exposed to adrenaline, was either continuously active or in a state of rest; the supposed cumulative effects should, of course, not occur at rest. The outcome of this experiment was conclusive. As Fig. 3 shows, tension build-up in adrenaline was not critically dependent on the preparation's activity: in some experiments the response of the beating preparation was, indeed, slightly larger (Fig. 3*a*), in others no difference between the two responses under comparison was obtained (Fig. 3*b*). The same result held when external calcium was reduced to a low value (e.g. to 0.01 mM) during much of the resting period, so also eliminating the (somewhat remote) possibility that the adrenaline response is explicable in terms of an increased resting uptake by the cells. Thus, calcium accumulation inside the cells plays little part in the response to adrenaline, a conclusion which will be qualified later (p. 390) for certain well-defined conditions. As regards the time course of the adrenaline effect, a complex sequence involving at least two consecutive reaction steps seems to underlie the gradually developing and subsiding response, one of these steps being concerned, most probably, with the formation and destruction of cyclic AMP inside the cells (e.g. Tsien, Giles & Greengard, 1972; R. Niedergerke & S. Page, unpublished).

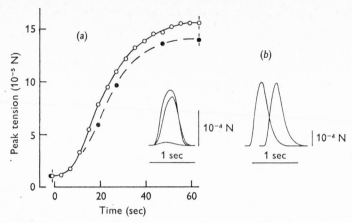

Fig. 3. Build-up of adrenaline effect with and without stimulation of trabecula. (a) 10^{-7} M adrenaline applied at time zero in both cases; ○ results with continuous stimulation, ● results after stimulation was stopped at time zero and a single test pulse applied either 19, 27, 47 or 63 sec afterwards. *Inset*: tracings of tension records of three twitches (whose tension values are marked by vertical bars in main graph), the one before, the two others 63 sec after application of adrenaline. (b) Results from a different trabecula subjected to the same procedures as that in (a). Tracings (displaced laterally for clarity) of tension records, both after 60 sec in 10^{-7} M adrenaline, and either after continuous stimulation (first record) or after 60 sec of rest (second record). [Ca^{2+}]: 1 mM and stimulation at 15 min^{-1} in both (a) and (b).

The third, and last, type of experiment in this series deals with the analysis of a phenomenon of rapid twitch facilitation in cardiac cells, already reported (see Lammel *et al.*, 1975). The phenomenon (Fig. 4), a large increase in twitch tension (up to 100-fold and more), occurs after a heart trabecula has been briefly (e.g. for 2–5 sec) exposed to a high potassium fluid ($\geqslant 20$ mM). The first twitch, termed 'facilitated twitch', after this conditioning treatment is usually the largest, and subsequent twitches decline to a low steady level with a $t_{\frac{1}{2}}$ of decline ranging from 5 to 20 sec. Again, the question was asked whether this twitch facilitation is brought about by a membrane mechanism which enhances calcium inward current or whether calcium accumulation in a cellular store plays an important part. To distinguish between these alternatives, we examined the effects of step changes of external calcium concentration applied at various times during the interval between the last twitch before high [K] exposure and the first (i.e. facilitated) twitch afterwards. Clearly, if calcium ions were to be transferred from the outside medium to a cellular store during the conditioning treatment, a response not unlikely while cells are depolarised in the high potassium medium, then the degree of facilitation should depend on the level of external calcium at that time since the magnitude of calcium

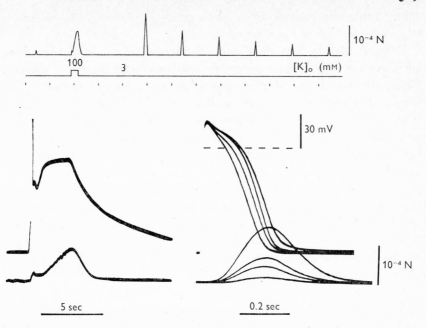

Fig. 4. Twitch facilitation after high [K$^+$] exposure. *Upper part*: semi-schematic representation of tension results to illustrate experimental procedure, showing steady state twitch before high [K$^+$] exposure, the high [K$^+$] contracture and a series of facilitated twitches afterwards; time marker: 10 sec intervals. *Lower part*: simultaneous recordings of tension and membrane potential during high [K] exposure (*left*) and twitch facilitation (*right*); exposure time to 100 mM K$^+$, 3 sec. Note abortive action potential and twitch at beginning of high [K$^+$] action (*left*). Superimposed records of steady state twitch and first, third and fourth facilitated twitches after high [K$^+$] treatment (*right*); successively shorter action potentials are associated with twitches of decreasing height. (From Lammel *et al.*, 1975; reproduced with permission of the Royal Society of London.)

uptake is determined by this level. The answer obtained (see Fig. 5) was once again unambiguous and showed that the facilitated twitch was un-affected by calcium uptake during the high [K] exposure, but was highly sensitive to changes of calcium during the period immediately preceding its appearance, a period which could be kept sufficiently short (e.g. 3 sec) to allow merely the (partial) diffusion equilibration of calcium ions at the cell surfaces. This, together with the monotonic, initially almost linear, relationship between peak tension and [Ca] (Fig. 5, cf. also Fig. 1*b*) suggests that enhanced calcium inward current is the most likely factor under-lying twitch facilitation also (see for further, more detailed discussion, Lammel *et al.*, 1975). It appears, then, that the processes initiated by the action of adrenaline on the one hand and those underlying twitch facilita-tion on the other have a final membrane event in common, as can be shown more conclusively by a closer analysis of these two responses (R. Niedergerke & S. Page, unpublished).

However, that an important point of difference also exists is clear from an inspection of the action potentials associated with the two responses. Thus, the overshoots of these potentials are enhanced by adrenaline (Fig. 2) but not, or to a much smaller extent, during twitch facilitation (Fig. 4). The most likely explanation of this difference is that the mechanisms responsible for increasing the membrane calcium current are not selective for calcium but alter the current (or currents) carried by other ions as well, so obscuring the potential shifts expected from a change in calcium current alone. If such modification of several currents were indeed to occur, then relatively small differential effects on any one of these might suffice to bring about the different levels of overshoot. It should be mentioned in this context that membrane potassium conductance is, in fact, modified under the influence of adrenaline in both mammalian and frog heart tissue (Brown & Noble, 1974; Tsien, 1974).

The discussion so far has been concerned entirely with mechanisms acting at the level of the cell membrane. In addition to these, there is evidence, in the frog heart, of another regulatory mechanism modifying contraction, which appears to operate within the cell interior. This is a conclusion derived from a study of tension changes which occur at a much slower rate than those described above (their $t_{\frac{1}{2}}$ being $\geqslant$ 1 min) and which can be observed in a variety of different experimental conditions (Chapman & Niedergerke, 1970a, b). The most interesting of these are (a) a change in heart rate, (b) a changed external calcium or sodium concentration, and (c) prolonged exposure of the tissue to adrenaline. In each case, slow tension changes are obtained, superimposed upon the more rapid ones such as those illustrated in Figs. 1 and 2, which they usually outlast for considerable periods. Most probably the common feature here is the occurrence of slow net shifts of calcium ions between heart cells and the surrounding medium; and, indeed, for the cases of a change in heart rate and of external sodium concentration, such shifts have been demonstrated. The examples shown in Fig. 6 are taken from experiments made in collaboration with Mr D. Gadsby (unpublished), using a technique which allows tissue ^{45}Ca to be followed while heart ventricles are continuously perfused with ^{45}Ca-labelled fluids of constant $[Ca^{2+}]$ and radioactivity (for details of method, see Niedergerke & Page, 1969). In the experiment of Fig. 6a which illustrates the effects of an altered heart rate, a heart ventricle was subjected to two brief periods of regular activity but was kept in a state of rest during the remaining time. As is shown, each period was associated with an increased rate of tissue ^{45}Ca uptake with respect to

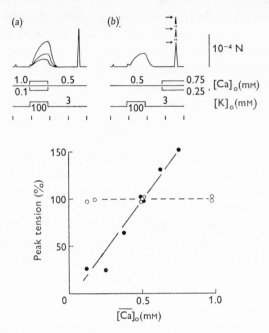

Fig. 5. Dependence of (first) facilitated twitch on [Ca] step changes induced (a) during high [K] exposure or (b) at a time immediately preceding the facilitated twitch. *Upper part*: semi-schematic representation of experimental procedures in the two conditions. Step change of $[Ca^{2+}]_0$ (external Ca) (a) simultaneous with onset (and termination) of high $[K^+]_0$ (external K) exposure, (b) induced 3.5 sec before eliciting facilitated twitch. Time marker: 5 sec. *Lower part*: peak tension of facilitated twitch (combined results from three experiments) in % of that obtained in the presence of 0.5 mM Ca throughout, plotted against [Ca] present (a) at end of high [K] treatment (open circles and dashed line), (b) at times of facilitated twitch (filled circles and solid line), in units of average [Ca] in fluid spaces of trabecula (concentrations calculated by means of values of $t_{\frac{1}{2}}$ determined from rapid twitch responses to $[Ca^{2+}]$ steps, values in three trabeculae between 1.5 and 2.2 sec; for details of procedure, see discussion in connection with Fig. 5 of Lammel *et al.*, 1975).

the uptake at rest, reflecting, in part, the accelerated tracer exchange in the cells due to the extra influx and efflux in these periods (cf. Niedergerke, 1963). In addition, a net loss of calcium ions was obtained after both periods of activity, indicative of a preceding net uptake of the ion by the cells (cf. discussion in Niedergerke *et al.*, 1969). These findings, together with the fact that the magnitude of calcium loss was identical, irrespective of whether or not tracer exchange in the tissue had been completed (four experiments), indicate that the uptake had occurred into a well-defined cellular compartment, or store.

For the demonstration of the effect of a change in $[Na^+]$ (Fig. 6b) a heart ventricle was first kept in a state of continuous activity for a prolonged period during which external sodium was switched, first from 100 % to 50 % (NaCl being replaced by choline chloride in the presence of 5×10^{-5} M atropine sulphate), and then back to 100 %. The results of main interest are (a) the increase of ^{45}Ca uptake in the low sodium fluid, a finding expected on the basis of the known enhancement of calcium movements at reduced sodium concentrations (cf. Niedergerke, 1963); and (b) a net calcium loss after the switch from the low to the high sodium concentration. After the ventricle was arrested, in the final stages of the experiment, another net release of calcium occurred analogous to that described for the after effects of activity in Fig. 6a. (The amount of calcium involved in these net shifts was normally up to about 0.06 μmole ml^{-1} heart cells.) A point of major interest resides in the time course of calcium loss which in response to both sodium enhancement and cessation of activity occurred with $t_{\frac{1}{2}} \geqslant 1$ min, similar therefore to that of the slow tension decline observed in these two conditions, or the decline in contractility in the case when the preparation had been in a state of rest.

Although these findings clearly support the idea that these net movements of calcium are linked to the slow tension changes, our views on the nature of this link and of the cellular store involved are still conjectural. In one explanation favoured by us, the cellular store in question is envisaged as having a kind of 'buffering' action to keep cytoplasmic calcium within a certain critical range. This range should be sufficiently low to prevent activation of the troponin–tropomyosin complex with the heart in a state of rest, but at the same time high enough to exert a 'priming' effect, different from that of activation, on contraction. In this context, it is important to mention that recent studies have shown the existence in muscle cells of enzyme systems, other than troponin–actomyosin, possessing high calcium sensitivity, in particular certain kinases, whose functional role in relation to the contractile proteins is under discussion at the present time (Brostrom, Hunkeler & Krebs, 1971; Ozawa, 1972; Perry, 1974; Pires, Perry & Thomas, 1974).

In conclusion, despite our lack of essential quantitative information in this area of research, certain features seem rather well established: (1) In the frog heart, contraction is induced, directly, by calcium ions which enter the heart cells during each action potential. (2) These calcium movements are controlled by one or several mechanisms, acting at the cell surface to regulate the strength of the heart beat. (3) Another regulatory mechanism of contraction, which is calcium dependent, exists in the cell interior, but its precise nature is, as yet, unknown. There is no reason to doubt that the three factors just outlined, the transmembrane move-

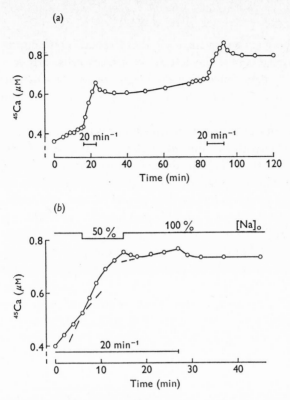

Fig. 6. Calcium movements associated (a) with periods of activity and (b) with changes of external sodium concentration. Two heart ventricles continuously perfused with ^{45}Ca-labelled Ringer's fluid; [Ca^{2+}] in this fluid either 0.8 mM (a), or 0.4 mM (b). Ordinates: ^{45}Ca-content of ventricles in μmoles ml^{-1} cell volume. Abscissae: time of perfusion with radioactive fluid (initial 30 min period of ^{45}Ca uptake omitted for simplicity). (a) Effects of short periods of stimulation, one early, the other late during calcium exchange. (b) Effects of [Na$^+$]$_0$ (external sodium) steps from 100% down to 50% and back to 100% in continuously beating ventricle (NaCl replaced by choline chloride; 5 × 10^{-5} M atropine sulphate present throughout), and of final cessation of stimulation (ventricle here, as in (a), suddenly arrested by addition of 2 × 10^{-7} g ml^{-1} tetrodotoxin to perfusate at the moment when stimulation was stopped).

ments of calcium and the two regulatory mechanisms, also participate in the function of mammalian heart cells. Most likely, however, and as already mentioned in the introductory part of this paper, an important, additional role is played, in these cells, by the sarcoplasmic reticulum which, through its capacity for storing and discharging calcium ions, contributes further to the regulation of activity.

SUMMARY

A critical short survey is made of the present knowledge of the events occurring during contractile activation of heart cells. Stress is laid on the lack of reliable data concerning the magnitude of the calcium movements across cell surfaces during the heart beat, for both the mammalian and amphibian heart.

Recent results obtained with frog heart cells are presented, from experiments made with the aim (a) of clarifying the role played by calcium ions which enter the cells at the time of a heart twitch; and (b) of elucidating the processes which regulate the strength of this twitch. The results suggest that (1) contraction is induced, directly, by calcium ions entering the cells during an action potential, and (2) two mechanisms exist which regulate contractile strength. One of these controls the magnitude of calcium inward current at the level of the cell surface, the other acts in the cell interior, involving a calcium-dependent process of an, as yet, unknown nature.

Support of this work by the Medical Research Council is gratefully acknowledged.

REFERENCES

BROSTROM, C. O., HUNKELER, F. L. & KREBS, E. G. (1971). The regulation of skeletal muscle phosphorylase kinase by Ca^{2+}. *J. biol. Chem.*, **246**, 1961–1967.

BROWN, H. F. & NOBLE, S. J. (1974). Effects of adrenaline on membrane currents underlying pacemaker activity in frog atrial muscle. *J. Physiol., Lond.*, **238**, 51–53P.

CHAPMAN, R. A. & NIEDERGERKE, R. (1970a). Effects of calcium on the contraction of the hypodynamic frog heart. *J. Physiol., Lond.*, **211**, 389–421.

(1970b). Interaction between heart rate and calcium concentration in the control of contractile strength of the frog heart. *J. Physiol., Lond.*, **211**, 423–443.

HAGIWARA, S. & NAKAJIMA, S. (1966). Differences in Na and Ca spikes as examined by application of tetrodotoxin, procaine, and manganese ions. *J. gen. Physiol.*, **49**, 793–806.

KASS, R. S. & TSIEN, R. W. (1975). Multiple effects of calcium antagonists on plateau currents in cardiac Purkinje fibres. *J. gen. Physiol.*, **66**, 169–192.

LAMMEL, E., NIEDERGERKE, R. & PAGE, S. (1975). Analysis of a rapid twitch facilitation in the frog heart. *Proc. R. Soc. Lond. B*, **189**, 577–590.

LANGER, G. A. (1973). Heart: excitation–contraction coupling. *A. Rev. Physiol.*, **35**, 55–86.

NIEDERGERKE, R. (1963). Movements of Ca in beating ventricles of the frog heart. *J. Physiol., Lond.*, **167**, 551–580.

NIEDERGERKE, R. & ORKAND, R. K. (1966). The dual effect of calcium on the action potential of the frog's heart. *J. Physiol., Lond.*, **184**, 291–311.

NIEDERGERKE, R. & PAGE, S. (1969). A new method for the determination of calcium fluxes in the frog heart by means of high precision measurement of [45]calcium concentrations. *Pflügers Arch. ges. Physiol.*, **306**, 354–356.

NIEDERGERKE, R., PAGE, S. & TALBOT, M. S. (1969). Calcium fluxes in frog heart ventricles. *Pflügers Arch. ges. Physiol.*, **306**, 357–360.

OZAWA, E. (1972). Activation of muscular phosphorylase *b* kinase by a minute amount of Ca ion. *Jap. Jl Biochem.*, **71**, 321–331.

PAGE, E., McCALLISTER, L. P. & POWER, B. (1971). Stereological measurements of cardiac ultrastructures implicated in excitation–contraction coupling. *Proc. natn. Acad. Sci. USA*, **68**, 1465–1466.

PAGE, S. G. & NIEDERGERKE, R. (1972). Structures of physiological interest in the frog heart ventricle. *J. Cell Sci.*, **11**, 179–203.

PERRY, S. V. (1974). Calcium ions and the function of the contractile proteins of muscle. *Biochem. Soc. Symp.*, **39**, 115–132.

PIRES, E., PERRY, S. V. & THOMAS, M. A. W. (1974). Myosin light-chain kinase, a new enzyme from striated muscle. *FEBS Lett.*, **41**, 292–296.

REUTER, H. (1973). Divalent cations as charge carriers in excitable membranes. *Prog. Biophys. molec. Biol.*, **26**, 1–43.

ROUGIER, O., VASSORT, G., GARNIER, D., GARGOUIL, Y. M. & CORABOEUF, E. (1969). Existence and role of a slow inward current during the frog atrial action potential. *Pflügers Arch. ges. Physiol.*, **308**, 91–110.

TSIEN, R. W. (1974). Effects of epinephrine on the pacemaker potassium current of cardiac Purkinje fibers. *J. gen. Physiol.*, **64**, 293–319.

TSIEN, R. W., GILES, W. & GREENGARD, P. (1972). Cyclic AMP mediates the effects of adrenaline on cardiac Purkinje fibres. *Nature New Biol.*, **240**, 181–183.

VASSORT, G., ROUGIER, O., GARNIER, D., SAUVIAT, M. P., CORABOEUF, E. & GARGOUIL, Y. M. (1969). Effects of adrenaline on membrane inward currents during the cardiac action potential. *Pflügers Arch. ges. Physiol.*, **309**, 70–81.

WEBER, A. & MURRAY, J. M. (1973). Molecular control mechanisms in muscle contraction. *Physiol. Rev.*, **53**, 612–673.

CALCIUM MOVEMENTS IN MUSCLE

By C. C. ASHLEY, P. C. CALDWELL, A. K. CAMPBELL,*
T. J. LEA and D. G. MOISESCU

Departments of Physiology and Zoology, University of Bristol,
Bristol BS8 1TD, and Department of Medical Biochemistry,*
Welsh National School of Medicine, Cardiff CF4 4XN

Most of the experiments described here have been performed on single striated muscle fibres. These single cells are usually 1–2 cm in length and 1–2 mm in diameter and are isolated from the crustaceans *Maia squinado* the spider crab (Caldwell & Walster, 1963) and *Balanus nubilus*, the acorn barnacle (Hoyle & Smyth, 1963). They are easily dissected and cannulated and in this state remain physiologically viable for many hours. The internal medium of these large muscle fibres can be modified by axial microinjection in a manner similar to that employed for the giant axon preparation from the squid, *Loligo* (Hodgkin & Keynes, 1956). This technique provides a convenient method for introducing radioactively labelled materials, such as ^{45}Ca, directly within these cells and this method has been used to study the efflux of ions across the surface membrane (Caldwell, 1964; Ashley & Caldwell, 1974). More recently, the single muscle fibre preparation has provided more detailed information about the internal movements of calcium within the cell during contraction by utilizing the useful indicator properties of the calcium-sensitive photoproteins, aequorin and obelin. The extent of the influx of radioactively labelled calcium from the external saline both at rest and during excitation can also be estimated by using a glass scintillator probe. This is inserted longitudinally into the muscle fibre, and detects the presence of the radioactively labelled ions within the cell.

THE USE OF THE GLASS SCINTILLATOR PROBE TO EXAMINE CALCIUM MOVEMENTS

One of the major problems of examining the influx of ions into cells is the ability to distinguish experimentally between uptake into an intracellular phase as compared to the uptake into only an external compartment. This problem is particularly important when the flux of the isotope into the intracellular phase is relatively slow. The glass scintillator probe (Caldwell & Lea, 1973) inserted intracellularly does permit this distinction

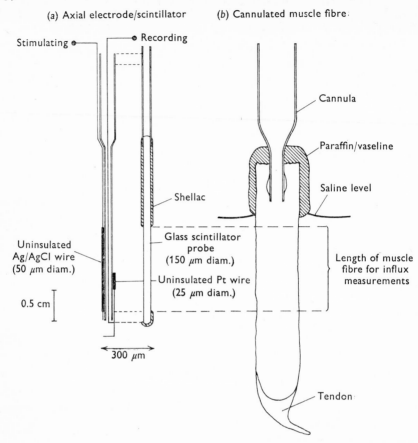

(a) Axial electrode/scintillator   (b) Cannulated muscle fibre

Stimulating ●——

●— Recording

Cannula

Paraffin/vaseline

Saline level

Shellac

Glass scintillator probe
(150 μm diam.)

Uninsulated
Ag/AgCl wire
(50 μm diam.)

Length of muscle
fibre for influx
measurements

Uninsulated Pt wire
(25 μm diam.)

0.5 cm

Tendon

←— 300 μm —→

Fig. 1. Diagrammatic representation of the apparatus for recording the glass scintillator probe responses. A dual axial stimulating-recording electrode can be attached to the probe and the assembly inserted into the muscle fibre when the effects of electrical stimulation are to be investigated. The photomultiplier tube (2 inches diameter, EMI 9324) was operated at +1600 V and the output was anode-decoupled and fed to a Panax modular counting system. The fibre was contained in a glass cuvette (volume = 2.5 ml), whose optical face was coupled (Dow–Corning coupling compound 20–057) to the face-plate of the photomultiplier tube (Ashley & Lea, 1976). The circuitry for the stimulating-recording electrode are as indicated in Ashley & Ridgway (1970).

to be made and the truly intracellular influx of at least low energy β-emitting isotopes can be readily determined. This method also has the added advantage that the measurements are made on the same single cell rather than on a population of single cells. The experimental arrangement for detecting the movements of calcium at rest and during stimulation in large muscle fibres is illustrated in Fig. 1. The β particles of ^{45}Ca ($\epsilon_{max} = 0.26$ MeV) are generally not detected from the external saline providing that the probe is inserted axially and that the fibres are at least 1 mm or greater in diameter (Caldwell & Lea, 1973; Ashley, Caldwell & Lea, 1975; Ashley & Lea, 1976). The increase in the rate of the scintillator counts with time is illustrated in Fig. 2a. There is an initial rapid phase of increase in the rate of the probe counts over the first 60–90 min which is followed by a phase where the rate of the probe counts increases approximately linearly with time. The apparent influx rate can be calculated from the rate of the probe counts as follows,

$$\text{Apparent influx rate (pmole cm}^{-2}\text{ sec}^{-1}) = \frac{M_s \cdot Y_m \cdot d \cdot 10^9}{4 \cdot Y_s \cdot t}, \qquad (1)$$

where Y_m is the count per unit time for a given length of scintillator probe in the muscle fibre; Y_s is the count per unit time for the same length of scintillator probe in the external saline containing ^{45}Ca; t is the immersion time (sec); M_s is the concentration of ion in external saline (M); and d is the diameter of the fibre (cm). The results presented in Fig. 2a are expressed as relative scintillator counts, i.e. (Y_m/Y_s) plotted versus time, so that the slope of this graph gives directly the value of $(Y_m/(Y_s.t))$ for equation (1) (see Ashley & Lea, 1976).

The slope of the linear phase of the graph predicts a calcium influx of 9–10 pmole cm^{-2} sec^{-1}, which is considerably higher than the value calculated for the same fibre by a more conventional method. In this, the total fibre ^{45}Ca content is determined by liquid scintillation assay at the end of the experiment and the influx deduced from a knowledge of the fibre dimensions (Table 1). In addition, the size of the *influx* assessed by this latter method was similar to the value of the *efflux* determined from experiments where the isotope was injected directly within the intracellular phase (Ashley, Caldwell & Lowe, 1972; Ashley, Ellory & Hainaut, 1974), that is in the range of 1–2 pmole cm^{-2} sec^{-1}. In more detailed experiments, a pre-injection of the calcium binding agent EGTA produced a rate of increase in probe counts appreciably slower than in non-injected fibres and the predicted influx (from equation (1)) was close to that determined by the other methods (Table 1). The suggested reasons for this disparity of influx values in the absence of EGTA are outlined elsewhere (Ashley & Lea, 1976), but it seems that it is in part because of the fact that EGTA internally should produce a more uniform distribution of the entering ^{45}Ca than is achieved in its absence.

The glass scintillator probe was initially used to investigate the magnitude of the extracellular space in these muscle fibres. The size of this space can be approximately assessed by extrapolation of the linear phase of increase in the rate of the probe counts to $t = 0$, the time of application of the ^{45}Ca-labelled saline. This gives a value for the extracellular space

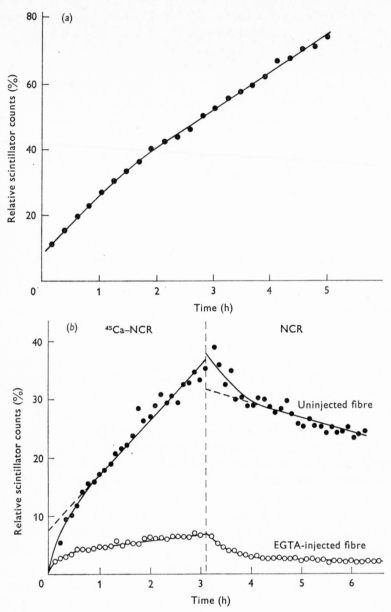

Fig. 2. Relative scintillator counts (Y_m/Y_s) versus time for a single *Balanus* muscle fibre. The external saline (Fatt & Katz, 1953) contained in addition 2 mM TES (*N*-Tris (hydroxymethyl) methyl-2-amine ethane sulphonic acid) and $^{45}CaCl_2$ (100 μCi ml^{-1}). (*a*) Fibre diameter, 1.2 mm; temperature, 20 °C. Glass scintillator, Koch-Light Laboratories, GSF 1, 150 μm diameter. (*b*) EGTA-injected fibre diameter, 1.6 mm; probe diameter, 150 μm; non-injected fibre diameter, 1.3 mm. 'Cold' (^{45}Ca-free) saline was applied at the hatched vertical line. Mean fibre resting potentials: (*a*) -45.0 mV; (*b*) -50.0 mV, -52.0 mV (from Ashley & Lea, 1976).

Table 1. *Estimated calcium influx into single barnacle muscle fibres*

Calcium influx (pmole cm^{-2} sec^{-1}) ± S.E., 20 °C

[EGTA]$_i$ (mM)	Number of fibres	Probe	Liquid scintillation assay	Liquid scintillation assay with 5 % cleft correction
0	5	13.8 ± 2.3	1.3 ± 0.3	0.6 ± 0.3
3.3–7.0 (mean 4.8)	6	1.31 ± 0.12	2.87 ± 0.41	1.53 ± 0.44

[EGTA]$_i$ = intracellular concentration of EGTA allowing for fibre dilution.

which is in the range of 4–10 % of the fibre volume (Fig. 2*b*) and is similar to that determined by the probe using ^{14}C-labelled inulin in the external saline as a more precise marker for the extracellular space (Ashley & Lea, 1976). In Fig. 2*b* are presented the results from two separate fibres, one EGTA injected and the other untreated. The initial rapid phase is observed in both and represents calcium entering mainly the extracellular space, the cleft system (Selverston, 1967) which is essentially filled in about 60–90 min with a $\tau_{\frac{1}{2}}$ of 15–20 min. When the fibres are placed in a 'cold' saline (i.e. [45]Ca-free), the rate of the probe counts decreases rapidly over the first 60–90 min and then more slowly. The rate constant for this latter phase is 1.5–1.9 × 10^{-3} min^{-1}, and is similar to the rate constant for the residual phase of calcium loss observed after axial injection of the isotope (Ashley *et al.*, 1972). It seems that the rapid phase of loss observed both with the probe and after axial injection represents loss from a mainly extracellular phase, while the slow phase of loss is from the truly intracellular compartment.

Although the calcium influx measured conventionally is somewhat lower in the absence of EGTA (Table 1), the EGTA injection method can be used in conjunction with the probe to give an estimate as to the extent of calcium entry occurring from the external saline during membrane depolarization. The effect of electrical stimulation upon the rate of the probe counts in a single muscle fibre is illustrated in Fig. 3. The fibre was initially injected with EGTA, which also prevents contraction, and the experimental arrangement is as outlined in Fig. 1 (see legend for details). The steady influx deduced from the rate of the probe counts during the linear phase was 2 pmole cm^{-2} sec^{-1}. Application of constant current pulses of 200 msec duration, at a rate of 4 per sec resulted in an increase in the *total* calcium content (ΔCa) of 2.3 pmole $pulse^{-1}$ (< 0.1 μM kg^{-1}

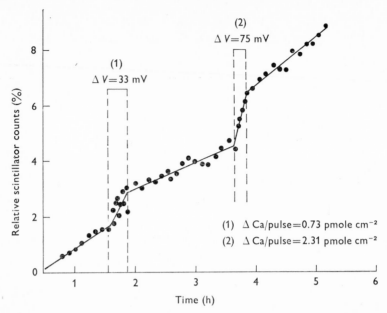

Fig. 3. The effect of electrical stimulation upon the time course of the relative probe counts in a single *Balanus* muscle fibre. Conditions as in Fig. 2. In the injected fibre, the final EGTA concentration after fibre dilution was 4.1 mM. Fibre diameter, 1.5 mm. Resting potential, −46 mV; NCR = normal crab Ringer solution (from Ashley & Lea, 1976).

pulse^{-1}) for an average membrane depolarization of $+75$ mV, when the external calcium concentration, $[Ca_o^{2+}]$, is 11.8 mM. It is known that the *free* $[Ca^{2+}]$ at rest is in the range 0.07–0.1 μM in these fibres (Ashley, 1970) and that the *free* $[Ca^{2+}]$ must rise to about 2 μM to produce 50% maximum tension (P_0) in barnacle myofibrils (Ashley & Moisescu, 1974). Although the *total* calcium bound to the myofibrils to elicit 50% P_0 is not known accurately, it must be considerably greater than *free* $[Ca^{2+}]$ of 2 μM, since the total calcium bound to crustacean troponin when fully saturated is about 0.1 mM kg^{-1} wet weight of muscle (Lehman, Kendrick-Jones & Szent-Györgyi, 1972). On a simple basis therefore at least 50 μM kg^{-1} would be required to be bound to elicit the 50% P_0 expected for this level of depolarization (Hagiwara, Takahashi & Junge, 1968), while even more is needed for two calcium ions acting cooperatively (Ashley & Moisescu, 1972a, b). It seems difficult to ascribe a direct role in myofibrillar reactions to the relatively small amount of total calcium entering per pulse observed in these experiments. Thus the light emission changes recorded with aequorin internally (Ashley & Ridgway, 1970) must reflect mainly calcium released (and reaccumulated) via internal calcium stores, such as exist in

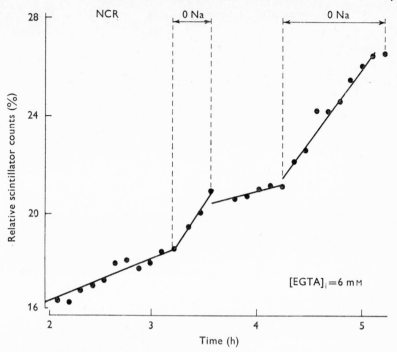

Fig. 4. The effect of sodium replaced, 0 Na (Li), salines upon the time course of the relative scintillator counts for a single *Balanus* muscle fibre. Salines as in Fig. 2. The normal and 0 Na salines were matched so that they had virtually identical specific activities. Mean fibre resting potential, −43.0 mV; fibre diameter, 1.2 mm (from Ashley & Lea, 1976).

the longitudinal elements of the sarcoplasmic reticulum (SR) (Ashley, Griffiths, Moisescu & Rose, 1975).

The glass scintillator probe can also be used to investigate the nature of transport processes occurring across the surface membrane of these muscle fibres. If the sodium content of the external saline is isosmotically replaced by lithium, the rate of the probe counts is increased by some 200–500 % compared to the resting rate (Fig. 4). Thus calcium influx in this preparation is enhanced by a reversal of the normal sodium gradient operating across the cell membrane. This finding supports the idea that calcium movements across the cell surface are at least partly dependent upon a sodium–calcium exchange system (Ashley, Ellory & Hainaut, 1974) as seems to be the case for squid axon and other systems (Baker, Blaustein, Hodgkin & Steinhardt, 1969; Baker, 1972). Certainly the calcium efflux from this preparation is sensitive to the removal of external sodium (Ashley, Ellory & Hainaut, 1974; Russell & Blaustein, 1974), as is the efflux of magnesium ions (Ashley & Ellory, 1972).

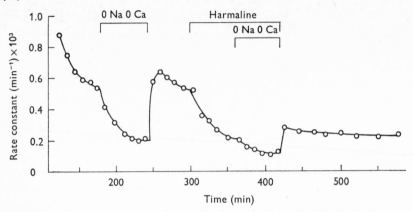

Fig. 5. The effect of 3 mM harmaline applied externally in physiological saline upon the efflux of ^{45}Ca from a single *Balanus* muscle fibre. The fibres were equilibrated for 2 h in physiological saline before the beginning of the experiment (see Ashley, Ellory & Hainaut, 1974). Subsequent removal of both sodium and calcium in the external saline produced only a small additional inhibition of the efflux. The 0 Na, 0 Ca salines were choline replaced. All experimental points were quench corrected. Mean fibre resting potential, -50 mV (from Lea, Ashley & Ellory, 1976).

In recent experiments performed in collaboration with Dr J. C. Ellory, University of Cambridge, the compound harmaline, a known inhibitor of Na^+–K^+ ATPase (Canessa, Jaimovich & Fuente, 1973) has a marked inhibitory effect not only on calcium efflux but also on the 0 Na-stimulated influx in this preparation (Lea, Ashley & Ellory, 1976) (Fig. 5). This may provide a useful method of inhibiting sodium–calcium and calcium–calcium exchange systems.

INTERNAL CALCIUM CHANGES AS DETECTED BY CALCIUM-SENSITIVE PHOTOPROTEINS

One of the major problems in relating ion movements to contraction and secretion has been the difficulty of utilizing external ionic fluxes, particularly those of calcium, as a direct index of free calcium changes occurring within the cell during excitation (Ashley & Caldwell, 1974). It has been possible in recent years to overcome this problem in many larger cells by utilizing the properties of the calcium-sensitive photoproteins, such as aequorin (Shimomura, Johnson & Saiga, 1962, 1963a, b) and more recently obelin (Morin & Hastings, 1971; Campbell, 1974). These proteins have many advantages as calcium-indicators; they do not require the presence of ATP or other co-factors, nor the presence of oxygen. Aequorin was first used biologically to detect free calcium changes occurring within large striated muscle fibres during contraction (Ridgway & Ashley, 1967;

Ashley & Ridgway, 1968, 1970). These experiments indicated that the change in free calcium concentration, as indicated by the change in aequorin light-emission, preceded the change in the isometric tension response. Recently, these initial observations using the photoprotein aequorin have been reinvestigated using obelin as an internal indicator for free calcium (see Fig. 6a). The time course of light response from the obelin essentially precedes the development of tension, as with aequorin, and the peak of the light-emission response occurs at about the maximum rate of development of tension in these fibres.

However, if a quantitative relationship between the calcium change and tension is to be investigated, it is essential to know the precise magnitude and time course of the free calcium change as indicated by the photoproteins. This has been investigated in a number of different ways as far as the relationship between free calcium and light-emission is concerned, all of which indicate that with both buffered and unbuffered calcium solutions, the light emission from aequorin increases as the square of the free calcium while that for obelin increases as the cube (Ashley, 1970, 1971; Baker, 1972; Moisescu, Ashley & Campbell, 1975). These observations as summarized in Fig. 7a. Here the lines are the theoretical prediction for two Ca^{2+} acting cooperatively in a *consecutive* reaction scheme (Ashley & Moisescu, 1972b) for aequorin and three Ca^{2+} in a consecutive scheme for obelin (Moisescu et al., 1975). However, the fluorescence binding experiments of Shimomura & Johnson (1970), suggest that aequorin may also bind a total of three Ca^{2+}, as does obelin. The theoretical predictions indicated in Fig. 7a for aequorin would suggest therefore that the first calcium with aequorin should be bound with a very much higher affinity than the square root of the product of the affinities of the other two sites, so that usually the first calcium for aequorin remains unobserved (see also Shimomura & Johnson, 1970, 1975). It was also of importance to estimate from these experiments the binding constant for the two calcium binding sites of aequorin under conditions that are as close as possible to those existing within the muscle fibre. The result from such an experiment for a simulated internal environment is presented in Fig. 7b and the line, fitted to the experimental points, represents the cooperative action of two Ca^{2+} acting in a *consecutive* scheme with binding constant $K_1 K_2 = 5 \times 10^7 \ \mathrm{M^{-2}}$, and $K_1 \ll K_2$. This value for the combined binding constant for the two sites is very similar to that deduced from the in-vivo experiments with aequorin in intact muscle fibres (Ashley & Moisescu, 1972b). An important point that was apparent from these experiments was that the binding constants of calcium to aequorin were markedly affected by the presence of monovalent cations in the medium, although the overall

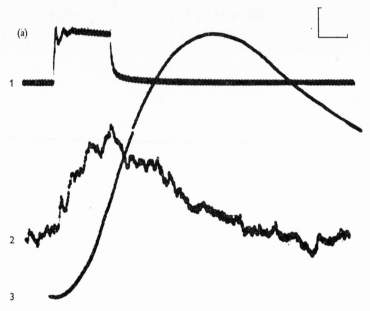

Fig. 6. (a) The effect of electrical stimulation upon the membrane response (trace 1) light-emission (trace 2) and isometric tension (trace 3) of a single *Balanus* muscle fibre injected with 0.2 μl of a saturated obelin solution dissolved in 100 mM TES buffer, pH 7.3: temperature 20 °C. Normal physiological saline (see legend to Fig. 2 for details) externally. The experimental apparatus was as described by Ashley & Ridgway (1970).
Calibration. Vertical: trace 1, 50 mV; trace 2, 20 nA; trace 3, 3 mN. Horizontal: 200 msec. The response in trace 2 has time constant of 40 msec. The details of obelin extraction and purification are given in Campbell (1974).

calcium stoichiometry for light-emission was unaffected. This implies that great care must be taken when constructing calibration curves such as those indicated in Fig. 7a, to ensure that the concentration of monovalent cations, such as H^+, does not alter during the experiment, particularly if unbuffered calcium solutions are being employed. In a more detailed analysis the effect of monovalent cations of physiological interest has been investigated at constant calcium concentrations. This permits the relative absolute binding constants for each of the ions of interest to be assessed (D. G. Moisescu & C. C. Ashley, unpublished).

$$2Ca^{2+} + Aeq \underset{k^{-I}}{\overset{k^{I}}{\rightleftharpoons}} (2Ca - Aeq) \xrightarrow{k^{II}} X \xrightarrow{k^{III}} Y* \xrightarrow{fast} Y + h\nu, \qquad (2)$$

where X and Y* are intermediates in the reaction and Y is the final product.

Rapid mixing experiments of Hastings *et al.* (1969) have indicated that the response of aequorin to rapid changes in free calcium is in the msec range and their scheme has been adapted for two Ca^{2+}. The rise time of

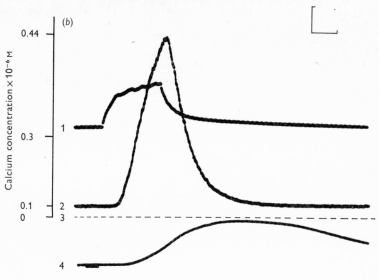

Fig. 6 (*cont.*)

(*b*) The effect of electrical stimulation upon the membrane response (trace 1), light-emission (trace 2) and isometric tension (trace 4) of a single *Balanus* muscle fibre injected with 0.2 μl of a saturated solution of aequorin in 100 mM TES buffer pH 7.3. Trace 3 represents the base line for [Ca^{2+}] and light. Temperature 10–12 °C. Normal physiological saline externally (see legend to Fig. 2). The change in the free calcium concentration is indicated on the vertical axis, since the resting free calcium concentration is close to 0.1 μM in this preparation (Ashley, 1970) (from Ashley & Ridgway, 1970; Ashley, 1971).

Calibration. Vertical: trace 1, 20 mV; trace 4, 5 g wt. Horizontal: 100 msec.

the light was *c*. 10 msec, at 20 °C, i.e. 100 sec^{-1} for k^{III} in equation (2), or 80 sec^{-1} at 10 °C (van Leeuwen & Blinks, 1969) (see Ashley & Moisescu, 1972*b*; Ashley, Moisescu & Rose, 1974*b*). It is now possible to attempt to relate the free calcium changes to relative tension in a quantitative way. It is a straightforward procedure, based upon the overall reversability of the calcium reactions, to determine both the calcium stoichiometry per functional unit for tension, as well as the number of slowly equilibrating steps, directly from the experimental records (Ashley & Ridgway, 1970; Ashley & Moisescu, 1972*a*, *b*). This analysis of the experimental records indicated that at least two Ca^{2+} per functional unit were required to produce tension and in a similar analysis that there was more than one, and most likely there were two, slowly equilibrating steps involved in tension production. The detailed calculations that enable the time course of relative isometric tension to be predicted from the time course of the free calcium change have been given in detail elsewhere (Ashley & Moisescu, 1972*b*; Ashley, Moisescu & Rose, 1974*b*; Ashley & Caldwell, 1974).

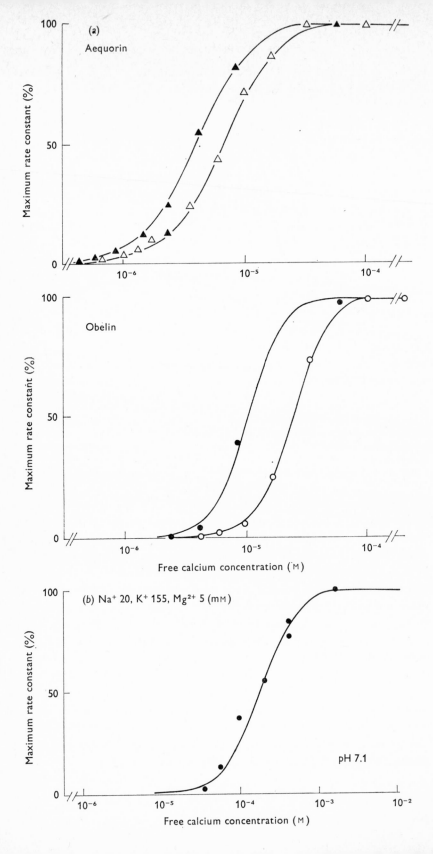

The results from such a calculation are illustrated in Fig. 8a where the calculated points for relative isometric tension provide a reasonable prediction as to the time course of tension recorded experimentally. An important point not directly considered in these initial calculations was whether the diffusion of calcium from the site of release, the SR, to the site of action on the myofilaments played an important part in the kinetics of tension development. This point was investigated theoretically by considering that the myofibrils were a series of cylinders of radius a, surrounded uniformly by the calcium release sites, the SR (see also Ashley, Moisescu & Rose, 1974b). The radius of each myofibrillar cylinder was divided into a large number of intervals, and the free calcium and tension calculated at each point. These calculations provide not only the predicted tension incorporating calcium diffusion across the myofibril, but they also indicate the time course of the free calcium change at different points across the diameter of the myofibril as a function of time. This is illustrated in Fig. 9 and indicates the result for the condition of aequorin uniformly distributed such that a gradient for calcium should exist across the myofibril. The mean tension response is, however, virtually unaffected as illustrated in Fig. 8b. In these calculations aequorin was considered to be uniformly distributed across the myofibril, and this seemed a reasonable assumption based upon

---

Fig. 7. (a) Apparent relative rate (k_r) of the photoprotein utilization as a function of pCa in a calcium buffered medium of the following composition (mM): $K_2EGTA + Ca-K_2EGTA$, 20; KCl, 40; K-TES, 10. The lines represent the theoretical predictions from a *consecutive* scheme of reaction (Ashley & Moisescu, 1972b), involving two Ca^{2+} for aequorin $(k_r^A = K_1^A K_2^A [Ca^{2+}]^2/(1 + K_1^A K_2^A [Ca^{2+}]^2))$ and three Ca^{2+} for obelin $(k_r^O = K_1^O K_2^O K_3^O [Ca^{2+}]^3/(1 + K_1^O [Ca^{2+}] + K_1^O K_2^O [Ca^{2+}]^2 + K_1^O K_2^O K_3^O [Ca^{2+}]^3))$. The superscripts A and O refer to aequorin and obelin respectively, and K_i represents the binding constant of the photoprotein for the ith Ca^{2+}.

Aequorin: $\triangle$, pH 6.8 ± 0.01; theoretical fit for $K_1^A < 1000\ K_2^A$, $K_1^A K_2^A = 2.3 \times 10^{10}$ M^{-2}. $\blacktriangle$, pH 7.1 ± 0.01; theoretical fit for $K_1^A < 1000\ K_2^A$, $K_1^A K_2^A = 6.46 \times 10^{10}$ M^{-2}.

Obelin: $\bigcirc$, pH 6.8 ± 0.01; theoretical fit for $K_1^O < 0.001\ \sqrt{(K_1^O K_2^O)}\ K_2^O\ K_3^O$ and $K_3^O > 1000\ \sqrt{(K_2^O K_3^O)}\ K_1^O K_2^O$, $K_1^O K_2^O K_3^O = 7.95 \times 10^{13}$ M^{-3}. $\bullet$, pH 7.1 ± 0.01; theoretical fit for $K_1^O < 0.001\ \sqrt{(K_2^O K_3^O)}\ K_2^O K_3^O$ and $K_3^O > 1000\ \sqrt{(K_1^O K_2^O)}\ K_1^O K_2^O$, $K_1^O K_2^O K_3^O = 1 \times 10^{15}$ M^{-3}.

The maximum overall rate of utilization, k, for aequorin was 1.5 sec^{-1} and for obelin was 3.9 sec^{-1} at 23 °C, pH 6.5–7.5 (from Moisescu, Ashley & Campbell, 1975).

(b) Apparent relative rate constant (k_r) (1.3 sec^{-1} at 20 °C), of aequorin utilization as a function of pCa in a simulated internal muscle environment. The reagents were all 'Specpure' (Johnson Matthey, London) and the concentrations used (see Figure) were chosen from published analyses of the ionic composition of *Balanus* fibres (Ashley & Ellory, 1972). The line represents the theoretical prediction for two calciums acting cooperatively in a *consecutive* reaction scheme (Ashley & Moisescu, 1972b) (from Ashley & Moisescu, 1975).

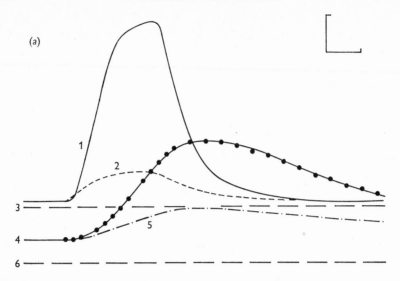

Fig. 8. (*a*) Tracings from the barnacle muscle fibre experiment: (1) aequorin light-emission; (2) [Ca^{2+}_{free}] calculated according to the scheme of Hastings *et al.* (1969), adapted for two calciums; (3) base line for light and [Ca^{2+}_{free}]; (4) isometric tension; (5) calculated Ca^{2+} bound to the tension sites; and (6) base line for [Ca^{2+}_{bound}] and tension. ●, Tension values calculated from model (see text).

For scheme (1) (Ashley & Moisescu, 1972*b*) $K_i = k_i/k_{-i} = K$; $K \times [Ca^{2+}_r] = 0.15$; $k_{-i} = 1.29$ sec^{-1}: for scheme (2) $_1K = 2K$, $_2K = \frac{1}{2}K$; $K \times [Ca^{2+}_r] = 0.15$; $_1\_k = \frac{1}{2}(_2\_k) = 1.29$ sec^{-1}. For resting calcium, [Ca^{2+}_r] is $0.7-1.0 \times 10^{-7}$ M (Ashley, 1970), K is $1.5-2.0 \times 10^6$ M^{-1}. $_1K$ and $_2K$ are the equilibrium constants for the first and second calcium binding steps, $_1\_k$ and $_2\_k$ are the respective reverse rate constants for these two steps.

Calibration: horizontally, 100 msec; vertically, (1) 3.8 nlm, (2) $6.67 \times [Ca^{2+}_r]$, (4) 2.5 g wt (2.6 % P_0), (5) 20 % calcium bound (see text). Temperature, 10–12 °C (from Ashley & Moisescu, 1972*a*).

the known apparent diffusion coefficient for aequorin (see later section) and upon the size of the aequorin molecule compared with the size of the myofilament lattice. It was instructive, however, to consider the condition where aequorin was only located at the surface of the myofibril. The predicted tension for this case is also illustrated in Fig. 8*b* and although there is a shift in the predicted relative tension response, it is sufficiently small as to be readily accommodated by a slight change in the reverse rate constant for calcium coming off the functional unit. In summary, these calculations suggest that, in this preparation, diffusion distances are not of importance in the kinetics of tension development.

It is also of importance to point out that these calculations make no prediction as to the exact nature of the functional unit involved in tension. It is likely that this involves calcium binding to the troponin system

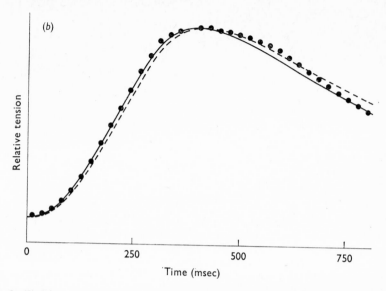

Fig. 8. (b) The time course of recorded tension (●): solid line, the time course of relative isometric tension calculated from the *independent* scheme incorporating the process of calcium diffusion for aequorin uniformly distributed; dashed line, calculated relative isometric tension aequorin considered at the surface of the myofibril (see text and Ashley & Moisescu, 1972b). The parameters for the tension calculation are as quoted in legend to Fig. 8a. $D\,((\partial^2 Ca(r,t)/\partial r^2) + 1/r\,(\partial Ca(r,t)/\partial r)) = (\partial Ca(r,t)/\partial t) + k\,Ca(r,t)\,M(r,t) - k_-\,CaM\,(r,t)$, and where tension $P(t)$ is given by $P(t) \sim \int_0^a CaM^2(r,t)\,r\,dr$, r is the distance from the axis of the myofibril, M is the concentration of free calcium binding site and CaM the concentration of occupied calcium binding site.

The boundary condition is the following for $Ca(r,t)$: (1) aequorin uniformly distributed, light intensity $I(t) = \int_0^a Ca^2(a,t)\,r\,dr$; (2) aequorin at surface of myofibril, light intensity $I(t) \sim Ca^2(a,t)$, where a, the radius of the myofibril, is 2 μm and D, the apparent diffusion coefficient for calcium, is 6×10^{-6} cm^2 sec^{-1} (from Ashley, Moisescu & Rose, 1974b).

on actin, but in addition it is also possible that it may involve sites upon the myosin (Ashley & Moisescu, 1972b), these being induced *in vivo*.

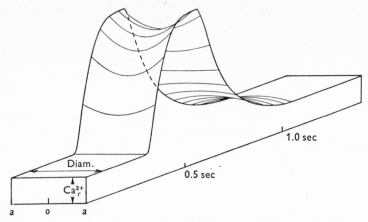

Fig. 9. Three-dimensional reconstruction of the free calcium transient across the diameter of the myofibril calculated for the condition of aequorin uniformly distributed (see legend to Fig. 8*b*). The complete transient is shown only at the surface of the myofibril. The lines joining the transients are at points of equal time and illustrate the effect of 'diffusion' upon the time course and magnitude of the free calcium: *o*, axis of myofibril; *a*, surface of myofibril (from Ashley, Moisescu & Rose, 1974*b*).

CALCIUM AND THE CONTRACTION OF ISOLATED BUNDLES OF MYOFIBRILS

In order to gain more detailed information as to the part played by calcium in contraction, the Natori preparation was used (Natori, 1954). Here the surface membrane of the muscle fibre is removed, exposing bundles of myofibrils. The bundles of isolated myofibrils may be clamped in the apparatus illustrated in Fig. 10, in order to record force development when the external ionic environment is manipulated. One end of the myofibrillar bundle is attached to a pair of fine jeweller's forceps, while the other is indirectly attached to the anode peg of an RCA mechano-electric transducer. The bundles of myofibrils used were usually in the range of 50–150 μm in diameter and some 1–3 mm in length. Myofibrils from frog as well as from the crab *Maia* and the barnacle *Balanus nubilus* have been used (Moisescu, 1974). A familiar observation with the Natori preparation in both the frog *Rana temporaria* (Hellam & Podolsky, 1969) and in barnacle is that tension can develop quite slowly even at relatively low pCa values (pCa = $-\log_{10}[Ca^{2+}]$). This observation is illustrated in Fig. 11*a*, for a bundle of frog myofibrils. This slow development of tension was observed even if agents such as deoxycholate (DOC) were included in order to inactivate the calcium-accumulating machinery of the SR (Fig. 11*c*). However, from the analysis of the relationship between free

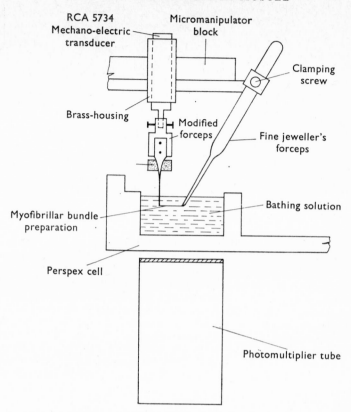

Fig. 10. Apparatus employed to record light and tension responses from isolated bundles of myofibrils. The output of the photomultiplier tube (EMI 9635A) was D.C. coupled to the input stage of a Tektronix storage oscilloscope (after Moisescu, 1974).

calcium and tension in intact single muscle fibres outlined in the previous section, it seemed that the myofibrils can make at least as equal a contribution to the rate of removal of free calcium from the sarcoplasm as the SR, at least during the early stages of the contractile response (Ashley & Moisescu, 1973a; Ashley, Moisescu & Rose, 1974b). Thus a change in the calcium buffering conditions in the supporting medium was required so that the free calcium within the bundle of myofibrils was more adequately stabilized. This was achieved by increasing the concentration of Ca–EGTA complex in the activating solution to a large value, while reducing the concentration of free EGTA in the relaxing solution to a minimum. Now the rate of development of tension in both frog (Fig. 11b) and in barnacle was much faster, and in barnacle the rate approached that observed in intact muscle fibres during a maximum isometric tetanus. A theoretical calculation taking

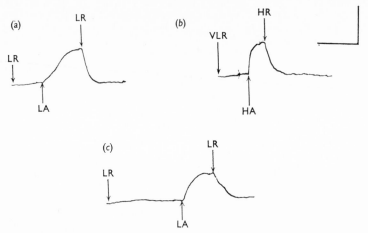

Fig. 11. Time course of tension development in bundles of frog sartorius myofibrils activated by the external application of Ca–EGTA buffer solutions (*a–c*). Tension responses measured by an RCA 5734 transducer and the output recorded on a Devices M2 pen recorder. The bundles of myofibrils were dissected in a relaxing solution ([Ca^{2+}] < 10^{-9} M) and paraffin oil. The solutions were changed by a method similar to that described by Hellam & Podolsky (1969). Bundle diameter, 82 μm; segment length, 1.5–2.0 mm; temperature, 20 °C. *Calibration*: vertical, 50 mg wt; horizontal, 25 sec. Solutions: all pH 7.10 ± 0.03; [ATP], 4 mM; free magnesium ion concentration, 0.05 mM. (Ca–EGTA + EGTA): VLR, 0.3 mM; LR, LA, 3 mM; HA, HR, 30 mM; DOC (K-deoxycholate), 1 mM throughout trace (*c*). pCa: for relaxing, 9; for activating, 6.2. Time in relaxing solutions before activation, 1–3 min. Ionic strength throughout 0.17–0.18 M. VLR = very low relaxing; LR = low relaxing; and HA = high activating; LA = low activating; HR = high relaxing (from Ashley & Moisescu, 1973*b*).

into account the inward diffusion and equilibration of the various Ca–EGTA buffer species, also supported these experimental findings and suggested that in barnacle, for the conditions outlined in Fig. 11*a*, tension should equilibrate in 8–10 sec, while with improved calcium buffering conditions tension should equilibrate within 1–2 sec (Fig. 11*b*).

In fact in a barnacle myofibrillar bundle preparation the rate of development of tension can readily be increased simply by increasing the external free calcium concentration in an ionically balanced solution. At very high calcium concentrations the rate of rise can be increased apparently faster than *in vivo* and in the range of 10–15 sec^{-1} at 20 °C (Ashley & Moisescu, 1975).

Aequorin was used in order to investigate the relationship between free calcium changes within the myofibrillar bundle and the tension response as a function of time. These experiments permitted a direct estimate to be made of the speed at which the free calcium concentration equilibrated within the bundle in response to activation by an externally applied Ca–

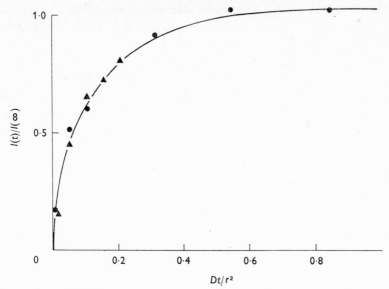

Fig. 12. Aequorin diffusion into a bundle of barnacle myofibrils. Two sets of experimental results ($\bullet$, $\blacktriangledown$), both with radii, $r = 78\ \mu$m; temperature, 20 °C. The line is predicted diffusion of aequorin with an apparent diffusion coefficient 1×10^{-7} cm^2 sec^{-1}. To convert abscissa to time units multiply by 10 (min).

The light intensity $I(t)$ is proportional to the aequorin concentration within the bundle at time t, and $I(\infty)$ is proportional to the aequorin concentration at steady state (i.e. time ∞), so that

$$I(t)/I(\infty) = 1 - 4 \sum_{n=1}^{\infty} 1/\mu_n^2 \exp\left(-\mu_n^2\left(D_{\text{Aeq}}\, t/r^2\right)\right)$$

and D_{Aeq} is the apparent diffusion coefficient for aequorin, r is the radius of the bundle and μ_n is the solution of the Bessel function of order zero ($\mathcal{J}_0(\mu) = 0$) (from Ashley & Moisescu, 1975).

EGTA buffer system. It was of importance, however, before these experiments were attempted, to have an accurate estimate of the apparent diffusion coefficient for aequorin within the preparation. This was investigated by allowing the myofibrillar bundle to remain in an ionically balanced relaxing solution containing aequorin for known periods of time, after which it was removed, rinsed briefly and then activated in a solution at a lower pCa to utilize the aequorin. The area under the light-emission curve was directly proportional to the number of aequorin molecules entering the bundle in the particular time interval. The results from this investigation are illustrated in Fig. 12 where the experimental points have been fitted to an apparent diffusion coefficient of 1×10^{-7} cm^2 sec^{-1} at 20 °C and ionic strength = 0.3 M. This result indicates that a $c.$ 80 μm diameter myofibrillar bundle will be 90 % 'loaded' with aequorin after 5 min. This apparent diffusion coefficient is some eight times slower than the diffusion of aequorin in free solution (Shimomura & Johnson, 1969) and some 40 times slower than the diffusion of EGTA within the myofibrils (calculated as 4×10^{-6} cm² sec⁻¹ from a similar type of analysis in relaxed myofibrils). These observations implied that aequorin, because of its relatively slow diffusion compared with Ca^{2+} or Ca–EGTA, could be used successfully as an indicator of intra-myofibrillar bundle calcium changes. Separate theoretical calculations suggested that over a period of 10 sec, for a 100 μm diameter bundle, at least 80% of the observed light-emission would still be from photoprotein contained within the preparation.

In the experiments described below, bundles of both frog and barnacle myofibrils were isolated under paraffin oil and were then subjected to the following routine. First, they were transferred to a relaxing solution containing a high total EGTA, pCa < 9 to remove contaminating calcium and this step also disrupts the SR in barnacle, as judged by electron microscopical observations. Secondly, the myofibrillar bundle was transferred to a relaxing solution with a low total EGTA content, pCa ~ 9; the time in the high and low relaxing solutions was dependent upon the size of the bundle. Thirdly, the preparation was transferred back to oil where 20–40 nl of aequorin contained in a low relaxing solution was applied along the length of the myofibrillar segment. The preparation remained under oil for a time sufficient to give c. 90% loading of the preparation with the photoprotein. Finally, the preparation was removed from oil, rinsed briefly (1–3 sec) in low relaxing solution and then activated in a high total EGTA solution of low pCa. The experimental light and tension responses are illustrated in Fig. 13 for a barnacle myofibrillar bundle. Similar responses were observed for bundles of frog myofibrils (Ashley & Moisescu, 1975). In both preparations the light response reached a steady value within 1–2 sec. This suggests that the free calcium equilibrates within the preparation in this period of time. The theoretical studies of diffusion into a 100 μm diameter cylinder certainly predict equilibration times of 1–2 sec (Hill, 1948, 1949). However, in contrast, the tension response equilibrates more slowly taking some 10 sec to reach a steady value at this pCa (trace 2), despite the obviously constant free calcium concentration. The slow rise in the tension response suggests that the reactions in which calcium is involved to produce force, equilibrate relatively slowly with time (Ashley & Moisescu, 1972b). At lower pCa values, tension develops more rapidly and this is to be expected if the tension reactions are calcium dependent. At very high activating calcium concentrations, tension in barnacle myofibrillar bundles rises at least as fast as in vivo during maximal activation (see Ashley & Moisescu, 1975).

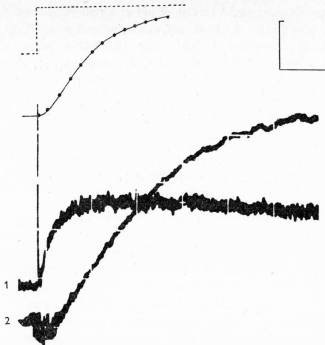

Fig. 13. Aequorin-light (trace 1) and tension (trace 2) responses from a bundle of barnacle myofibrils (diameter 175 μm) when activated with an externally applied calcium buffer solution ([Ca^{2+}] = 8.7 × 10^{-8} M). Initially, the preparation was in a high EGTA relaxing solution (10 min), followed by a low EGTA relaxing solution (5 min). Aequorin, in low relaxing, was applied uniformly along the segment while immersed under paraffin oil (3 mm length). After 6 min the bundle was washed in low relaxing (2–3 sec). The 'blip' on trace 1 indicates the moment of immersion in the activating solution. Insert, (●), are the theoretical tension points calculated from the following equation assuming a step increase in calcium (dotted line). Solid line represents the tension in trace 2. Scale reduced by half.

$$P/P_\infty = 1 + (\gamma_2/\gamma_1 - \gamma_2) \exp(-\gamma_1 t) + (\gamma_1/\gamma_2 - \gamma_1) \exp(-\gamma_2 t),$$

where P/P_∞ is the ratio between the instantaneous value of tension and its steady state value; the parameters $\gamma_1 = 0.38$ sec^{-1} and $\gamma_2 = 0.48$ sec^{-1} depend upon the rate constants and [Ca^{2+}] in the *consecutive* scheme (Ashley & Moisescu, 1972b). Composition of solutions: high EGTA relaxing ([Ca^{2+}] < 10^{-9} M), 20 mM EGTA; low EGTA relaxing ([Ca^{2+}] ~ 10^{-9} M), 0.1 mM EGTA, 20 mM K$_2$SO$_4$; activating ([Ca^{2+}] = 8.7 × 10^{-8} M), total EGTA, 20 mM. All solutions contained in addition (mM): 40 KCl, 10 TES, 0.1 Mg^{2+}, 4 ATP (total), A23187 2.5 μg ml^{-1}, buffered with KOH to 7.1 ± 0.01. Ionic strength = 0.13 M, temperature 20 °C. *Calibration*: vertical, 100 nA (1), 0.15 mN (2); horizontal, 2 sec (from Ashley, Moisescu & Rose, 1974a).

These responses can be simulated theoretically if the diffusion of a Ca–EGTA buffer system within the bundle is considered. In addition, by using the more satisfactory calcium buffering conditions, the effect of [H$^+$] and [Mg^{2+}] could be investigated upon the steady state tension versus pCa curves for barnacle. A change of pH of 0.5 alters by some five to six times the $[Ca^{2+}]$ value at which 50% of maximum tension (P_0) is attained. There was a similar effect upon the tension:pCa relationship when the free $[Mg^{2+}]$ was changed from 1 mM to 0.05 mM at constant pH. These results suggest that the affinities of the functional unit for $[H^+]$ and $[Ca^{2+}]$ are of about the same order of magnitude while that for $[Mg^{2+}]$ is some three to four orders of magnitude smaller (Ashley & Moisescu, 1974).

FREE CALCIUM CHANGES WITHIN ERYTHYROCYTE 'GHOSTS' DETECTED USING THE PHOTOPROTEIN OBELIN

Changes in the concentration of free calcium ions are not only important in controlling force development in muscle but also play an important role, together with the cyclic nucleotides, in the action and secretion of a number of hormones (Rasmussen, Goodman & Tenenhouse, 1972). Pigeon red blood cells, in common with other avian erythrocytes, produce large amounts of cyclic AMP (cAMP) when stimulated with adrenaline (Davoren & Sutherland, 1963). The calcium-sensitive photoprotein obelin (Morin & Hastings, 1971; Campbell, 1974) has been introduced into pigeon erythrocyte 'ghosts' so that the effects of hormones such as adrenaline on the permeability of the cell membrane to calcium could be investigated. In the experiments reported here, the effect of the bivalent cation ionophore A23187 on the calcium permeability of the 'ghost' cell membrane is demonstrated. The details of the preparation of the ghosts is similar to that outlined by other authors and is given in detail elsewhere (Campbell & Dormer, 1975a, b). Variations in osmolarity, temperature and pH during preparation were found to affect significantly the calcium permeability of the cell membrane of the final 'ghost' cell preparation. The relative permeability of the 'ghost' cell suspension to calcium was assessed by adding the ion externally and subsequently recording the light-emission from the obelin within the cells. 'Ghost' cells relatively impermeable to calcium could however be prepared, after initial haemolysis at pH 7.0 by resealing the cells in the presence of an ATP-regenerating system plus 150 mM KCl at 37 °C for 1 h. This preparation was used subsequently to investigate the effect of A23187 on the obelin luminescence within the 'ghosts'. Optimum effects of the ionophore, as judged by the obelin response, was obtained by preincubating the 'ghosts' with A23187 in the absence of low external calcium and magnesium (Fig. 14a, b). The addition of external calcium (final concentration 1 mM) to the 'ghost' cell suspension resulted in an immediate stimulation of the obelin luminescence,

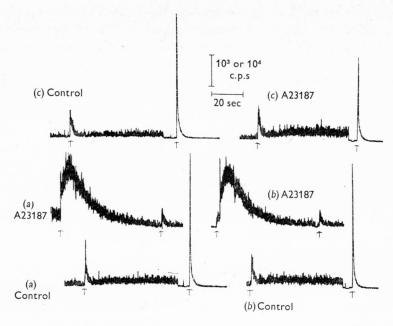

Fig. 14. Effect of ionophore A23187 on pigeon erythrocyte 'ghosts'. Pigeon erythrocytes were haemolysed at pH 7.0 in 6 mM NaCl, 3 mM MgCl₂. After washing, the 'ghosts' were resealed for 60 min at 37 °C in a medium containing obelin, 10 mM TES, 150 mM KCl, 6 mM MgCl₂, 2 mM ATP (disodium salt), phosphoenolpyruvate (potassium salt), pyruvate kinase (19 U ml⁻¹) pH 7.4. The 'ghosts' were centrifuged, washed and resuspended in medium A (10 mM TES, 140 mM NaCl, 5 mM KCl, 2 mM MgCl₂, pH 7.4). 'Ghosts' (50 μl) were suspended in 0.5 ml of medium A (initial Mg²⁺ concentration, 0.2 mM or 2 mM). This suspension was incubated for 20 sec with ionophore A23187 (16 μg ml⁻¹) or ethanol (2 μl ml⁻¹). Then 0.5 ml medium A + 2 mM CaCl₂ was added (final Mg²⁺ concentration, 0.2 mM or 2 mM). The rate of obelin luminescence was recorded on a chart recorder (Campbell, 1974; Campbell & Dormer, 1975a). Rate-meter time-constant 0.01 sec, 10 mV output equivalent to 1000 c.p.s., except after the addition of triton X-100 which occurred 60 sec after the addition of Ca²⁺, when 10 mV output was equivalent to 10 000 c.p.s.

	[Mg²⁺] during preincubation (mM)	Final [Mg²⁺] after addition of Ca²⁺ (mM)
(a) Control	0.2	0.2
A23187	0.2	0.2
(b) Control	0.2	2
A23187	0.2	2
(c) Control	2	2
A23187	2	2

(From Campbell & Dormer, 1975a.)

and more than 90% of the photoprotein was utilized within 60 sec (first arrow Fig. 14a, A23187). Subsequent addition of triton X-100 (second arrow), a non-ionic detergent, resulted in little additional luminescence.

In the absence of A23187, the addition of calcium resulted in a slight luminescence response which was either from extracellular obelin or from obelin contained within damaged 'ghost' cells (Fig. 14a, control). Here subsequent addition of triton X-100 resulted in destruction of both the cells and the intracellular obelin. It seems that the luminescence response recorded in the presence of A23187 truly represents intracellular calcium changes brought about by the ionophore. If the concentration of magnesium is raised in the preincubation medium, the ionophore response is considerably reduced (Fig. 14c, control and A23187). This finding could suggest that there is a competition between calcium, magnesium and ionophore for binding to the 'ghost' cell membrane.

Much of this work was supported by grants from the Medical and Science Research Councils. A.K.C. wishes to thank the Director, Marine Biological Laboratory, Plymouth, for facilities.

REFERENCES

ASHLEY, C. C. (1970). An estimate of calcium concentration changes during the contraction of single muscle fibres. *J. Physiol., Lond.*, **210**, 133–134P.

(1971). Calcium and the activation of skeletal muscle. *Endeavour*, **30**, 18–25.

ASHLEY, C. C. & CALDWELL, P. C. (1974). Calcium movements in relation to contraction. *Biochem. Soc. Symp.*, **39**, 29–50.

ASHLEY, C. C., CALDWELL, P. C. & LEA, T. J. (1975). Calcium influx into single crustacean muscle fibres as measured with a glass scintillator probe. *J. Physiol., Lond.*, **248**, 9–10P.

ASHLEY, C. C., CALDWELL, P. C. & LOWE, A. G. (1972). The efflux of calcium from single crab and barnacle muscle fibres. *J. Physiol., Lond.*, **223**, 733–755.

ASHLEY, C. C. & ELLORY, J. C. (1972). The efflux of magnesium from single crustacean muscle fibres. *J. Physiol., Lond.*, **226**, 653–674.

ASHLEY, C. C., ELLORY, J. C. & HAINAUT, K. (1974). Calcium movements in single crustacean muscle fibres. *J. Physiol., Lond.*, **242**, 255–272.

ASHLEY, C. C., GRIFFITHS, P. J., MOISESCU, D. G. & ROSE, R. M. (1975). The use of aequorin and the isolated myofibrillar bundle preparation to investigate the effect of SR calcium releasing agents. *J. Physiol., Lond.*, **245**, 12–14P.

ASHLEY, C. C. & LEA, T. J. (1976). Calcium fluxes in single muscle fibres using a glass scintillator probe. *J. Physiol., Lond.*, in press.

ASHLEY, C. C. & MOISESCU, D. G. (1972a). Tension changes in isolated muscle fibres as predicted by the free calcium concentration. *J. Physiol., Lond.*, **226**, 82–84P.

(1972b). Model for the action of calcium in muscle. *Nature New Biol.*, **237**, 208–211.

(1973a). The mechanism of the free calcium change in single muscle fibres during contraction. *J. Physiol., Lond.*, **231**, 23–25P.

(1973b). Tension changes in isolated bundles of frog and barnacle myofibrils in response to sudden changes in the external free calcium concentration. *J. Physiol., Lond.*, **233**, 8–9P.

(1974). The influence of Mg^{2+} and of pH upon the relationship between steady-state isometric tension and Ca^{2+} concentration in isolated bundles of barnacle myofibrils. *J. Physiol., Lond.*, **239**, 112–114*P*.

(1975). The part played by Ca^{2+} in the contraction of isolated bundles of myofibrils. In *Calcium transport in Contraction and Secretion*, ed. E. Carafoli, F. Clementi, W. Drabikowski & A. Margreth, pp. 517–525. North-Holland: Amsterdam.

ASHLEY, C. C., MOISESCU, D. G. & ROSE, R. M. (1974*a*). Aequorin light and tension responses from bundles of myofibrils following a sudden change in free calcium concentration. *J. Physiol., Lond.*, **241**, 104–106*P*.

(1974*b*). Kinetics of calcium during contraction: myofibrillar and SR fluxes during a single response of a skeletal muscle fibre. In *Calcium-binding Proteins*, ed. W. Drabikowski, H. Strzelecka-Golaszewska & E. Carafoli, pp. 609–642. Elsevier: Amsterdam.

ASHLEY, C. C. & RIDGWAY, E. B. (1968). Simultaneous recording of membrane potential, calcium transient and tension in single barnacle muscle fibres. *Nature, Lond.*, **219**, 1168–1169.

(1970). On the relationships between membrane potential, calcium transient and tension in single barnacle muscle fibres. *J. Physiol., Lond.*, **209**, 105–130.

BAKER, P. F. (1972). Transport and metabolism of calcium in nerve. *Prog. Biophys. molec. Biol.*, **24**, 177–233.

BAKER, P. F., BLAUSTEIN, M. P., HODGKIN, A. L. & STEINHARDT, R. A. (1969). The influence of sodium on calcium efflux in squid axons. *J. Physiol., Lond.*, **200**, 431–458.

CALDWELL, P. C. (1964). Calcium and the contraction of *Maia* muscle fibres. *Proc. R. Soc. Lond. B*, **160**, 512–516.

CALDWELL, P. C. & LEA, T. J. (1973). Use of an intracellular glass scintillator for the continuous measurement of the uptake of ^{14}C-labelled glycine into squid giant axon. *J. Physiol., Lond.*, **232**, 4–5*P*.

CALDWELL, P. C. & WALSTER, G. E. (1963). Studies on the micro-injection of various substances into crab muscle fibres. *J. Physiol., Lond.*, **169**, 353–372.

CAMPBELL, A. K. (1974). Extraction, partial purification and properties of obelin, the calcium-activated luminescent protein from the hydroid, *Obelia geniculata*. *Biochem. J.*, **143**, 411–418.

CAMPBELL, A. K. & DORMER, R. L. (1975*a*). The permeability to Ca^{2+} of pigeon erythrocytes studied using the calcium-activated luminescent protein obelin. *Biochem. J.*, **152**, 255–265.

(1975*b*). Studies on free calcium inside pigeon erythrocyte ghosts by using the calcium-activated luminescent protein, obelin. *Biochem. Soc. Trans.*, **3**, 709–711.

CANESSA, M., JAIMOVICH, E. & DE LA FUENTE, M. (1973). Harmaline, a competitive inhibitor of Na ion in the (Na$^+$ + K$^+$)-ATPase system. *J. Memb. Biol.*, **13**, 263–282.

DAVOREN, P. R. & SUTHERLAND, E. W. (1963). The effect of 1-epinephrine and other agents on the synthesis and release of adenosine 3′,5′-phosphate by whole pigeon erythrocytes. *J. biol. Chem.*, **238**, 3009–3015.

FATT, P. & KATZ, B. (1953). The electrical properties of crustacean muscle fibres. *J. Physiol., Lond.*, **120**, 171–204.

HAGIWARA, S., TAKAHASHI, K. & JUNGE, D. (1968). Excitation–contraction coupling in a barnacle muscle fibre as examined with voltage clamp techniques. *J. gen. Physiol.*, **51**, 157–175.

HASTINGS, J. W., MITCHELL, G., MATTINGLY, P. H., BLINKS, J. R. & VAN LEEUWEN, M. (1969). Response of aequorin bio-luminescence to rapid changes in calcium concentration. *Nature, Lond.*, **222**, 1047–1050.

HELLAM, D. C. & PODOLSKY, R. J. (1969). Force measurements in skinned muscle fibres. *J. Physiol., Lond.,* **200,** 807–819.

HILL, A. V. (1948). On the time required for diffusion and its relation to processes in muscle. *Proc. R. Soc. Lond.* B, **135,** 446–453.

—— (1949). The abrupt transition from rest to activity in muscle. *Proc. R. Soc. Lond.* B, **136,** 399–420.

HODGKIN, A. L. & KEYNES, R. D. (1956). Experiments on the injection of substances into squid giant axons by means of a micro-syringe. *J. Physiol., Lond.,* **131,** 592–616.

HOYLE, G. & SMYTH, T. (1963). Neuromuscular physiology of giant muscle fibres of a barnacle, *Balanus nubilus* Darwin. *Comp. Biochem. Physiol.,* **10,** 291–314.

LEA, T. J., ASHLEY, C. C. & ELLORY, J. C. (1976). The effect of harmaline on calcium transport in single muscle fibres from the barnacle *Balanus nubilus. Biochim. biophys. Acta,* in press.

LEHMAN, W., KENDRICK-JONES, J. & SZENT-GYÖRGYI, A. G. (1972). Myosin-linked regulatory systems: comparative studies. *Cold Spring Harb. Symp. quant. Biol.,* **37,** 319–330.

MOISESCU, D. G. (1974). The intracellular control and action of calcium in striated muscle and the forces responsible for the stability of the myofilament lattice. Ph.D. thesis, University of Bristol.

MOISESCU, D. G., ASHLEY, C. C. & CAMPBELL, A. K. (1975). Comparative aspects of the calcium-sensitive photoproteins, aequorin and obelin. *Biochim. biophys. Acta,* **393,** 133–140.

MORIN, J. G. & HASTINGS, J. W. (1971). Biochemistry of the bioluminescence of colonial hydroids and other coelenterates. *J. cell. comp. Physiol.,* **77,** 305–312.

NATORI, R. (1954). The property and contraction process of isolated myofibrils. *Jik. med. J.,* **1,** 119–126.

RASMUSSEN, H., GOODMAN, D. P. B. & TENENHOUSE, A. (1972). The role of cyclic AMP and calcium in cell activation. *Crit. Rev. Biochem.,* **1,** 95–148.

RIDGWAY, E. B. & ASHLEY, C. C. (1967). Calcium transients in single muscle fibres. *Biochem. biophys. Res. Commun.,* **29,** 229–234.

RUSSELL, J. M. & BLAUSTEIN, M. P. (1974). External cation-dependent calcium efflux from barnacle muscle fibres: evidence for Na:Ca exchange. *J. gen. Physiol.,* **63,** 144–167.

SELVERSTON, A. I. (1967). Structure and function of the transverse tubular system in crustacean muscle fibres. *Am. Zool.,* **7,** 515–525.

SHIMOMURA, O. & JOHNSON, F. H. (1969). Properties of the bioluminescent protein aequorin. *Biochemistry, N.Y.,* **8,** 3991–3997.

—— (1970). Calcium binding, quantum yield and emitting molecule in aequorin bioluminescence. *Nature, Lond.,* **227,** 1356–1357.

—— (1975). Regeneration of the photoprotein aequorin. *Nature, Lond.,* **256,** 236–238.

SHIMOMURA, O., JOHNSON, F. H. & SAIGA, Y. (1962). Extraction, purification and properties of aequorin, a bioluminescent protein from the luminous hydromedusan, *Aequorea. J. cell. comp. Physiol.,* **59,** 223–239.

—— (1963a). Further data on the bioluminescent protein aequorin. *J. cell. comp. Physiol.,* **62,** 1–8.

—— (1963b). Micro-determination of calcium by aequorin bioluminescence. *Science, Wash.,* **140,** 1339–1340.

VAN LEEUWEN, M. & BLINKS, J. R. (1969). Properties of aequorin relevant to its use as a calcium indicator in biological work. *Fedn Proc.,* **28,** abstr. 571.

INTRACELLULAR AND EXTRACELLULAR ROUTES IN BIOMINERALIZATION

By K. SIMKISS

Department of Zoology, University of Reading, Whiteknights Park,
Reading RG6 2AJ, Berkshire

At an earlier symposium of this Society, Williams (1960) made an interesting distinction between the processes of first and second order modelling in biological research. A first-order model simply asserts reality as it is thought to exist, while a second-order model attempts to explain a phenomenon in terms of simpler systems with fewer but better understood elements.

Biologists are trained to consider a wide range of phenomena so that it is perhaps one of their prime functions to describe first-order models. Almost by definition many of the models they propose are poorly understood and a current example of a biological phenomenon that has not progressed beyond the level of a first-order model is magnetic orientation in birds (Keeton, 1971). Most phenomena remain in this stage until a great deal of supporting evidence is obtained but they are then subjected to second-order modelling which is usually associated with biochemistry, biomathematics and biophysics. When a biological process reaches this level of understanding its investigation usually proceeds at a faster and more predictable rate.

If one wishes to pursue this simplistic approach and trace the origin of the current interest in cell calcium to any single source then it would probably be Heilbrunn (1940) who first produced a first-order model of the role of calcium in cell physiology. The phenomenon has since been pursued as second order models which include

(*a*) the control of intracellular calcium by membrane pumps, and

(*b*) its role as a 'second messenger', studied by its effects on a number of enzymatic and structural proteins.

These systems have been well reviewed at this symposium and it seemed, therefore, that it might be an interesting exercise for a biologist to try and restate a first-order model based on a more comparative approach to cell calcium and then attempt to speculate as to its functions in the light of existing hypotheses. As most of the evidence presented at this symposium has approached cell calcium through biochemical investigations, often of vertebrate tissues, I will start by considering anatomical studies on

[423]

representatives of other phyla. Furthermore, since over 95 % of the calcium in a vertebrate is present as mineralized deposits it seems perverse of this symposium to ignore this as if it had no relevance to cellular activities. I will therefore also turn my attention to the phenomenon of biomineralization.

THE OCCURRENCE OF CALCIUM GRANULES

In the past decade a large number of invertebrate tissues have been studied with the electron microscope and calcium deposits have been found to occur as intracellular granules in a number of them. Some examples are listed in Table 1 and it can be seen that they cover many phyla and several types of organ systems. Typically, the granules form in membrane-bound vesicles often closely associated with the Golgi apparatus. The chemical composition of the granules is very variable as are the functions ascribed to them. It is obvious, however, that in order to make any progress in characterizing the phenomena involved in granule formation some system of classifying them is necessary. There appear to be two basic types.

Type A

Granules of this type are normally sculptured, skeletal structures with a high degree of purity. They are usually formed in a vesicle which is closely associated with a Golgi complex, and contain an organic matrix. Crystal deposition occurs in close association with this material so that the resulting structure is characteristically a crystallographically well-organized calcareous structure. Examples of Type A intracellular products are the coccoliths of *Coccolithus huxleyi* (Wilbur & Watabe, 1963) and *Cricosphaera carterae* (Manton & Leedale, 1961, 1969) and the scleroblast spicules of the sea pansy *Renilla reniformis* (Dunkelberger & Watabe, 1974).

Type B

These are typically spherical structures. In the hepatopancreas of the blue crab *Callinectes sapidus*, Becker *et al.* (1974) identified three main sorts of granule according to their appearance under the electron microscope (Plate 1): Type I, large granules up to 20 μm in diameter staining like lipid but with hard properties and a pronounced layered structure (Plate 1*b*). They often appeared to be heterogeneous and to have no distinct bounding membrane. Type II were about 3.0 μm in diameter giving the appearance of apparently shrinking during tissue preparation and with hard, electron dense cores. Clusters of needle-like crystals frequently surrounded the periphery of these structures (Plate 1*d*). Type III were striking structures

PLATE I

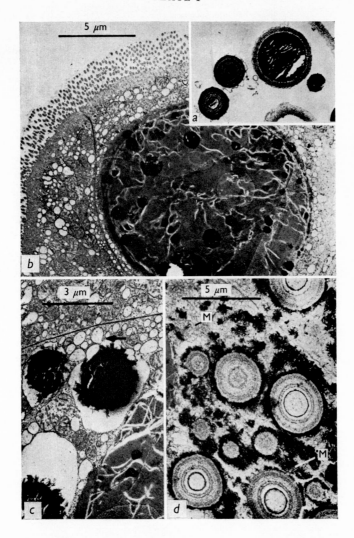

Intracellular granules from the hepatopancreas of the blue crab *Callinectes sapidus*. (*a*) Extruded granules in the lumen of the gland. (*b*) Type B I granules containing smaller dense inclusions. (*a*) and (*b*) are at the same magnification. (*c*) Type B II granules with electron-dense core and needle-like crystals on the surface. The arrow indicates coalescing vesicle that may be involved in granule secretion. (*d*) Type B III granules distinguished by concentric layers: M = mitochondrion (from Becker *et al.*, 1974).

PLATE 2

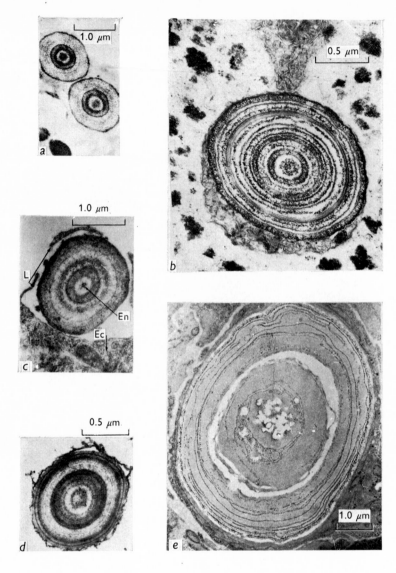

Examples of Type B III granules from various tissues. (a) Granules from the mantle of the bivalve *Anodonta cygnea* (from Istin & Masoni, 1973). (b) Granule from midgut wall of homopteran insect *Philaenus spumaris* (from Gouranton, 1968). (c) Granule from bladder system of trematode *Cyathocotyle bushiensis*: Ec = cytoplasm of excretory cells; En = nucleus of granule; L = lamella of cell (from Erasmus, 1967). (d) Zinc-containing granule from gut of barnacle *Lepas anatifera* (from Walker *et al.*, 1975). (e) Granule from cestode (Nieland & von Brand, 1969).

PLATE 3

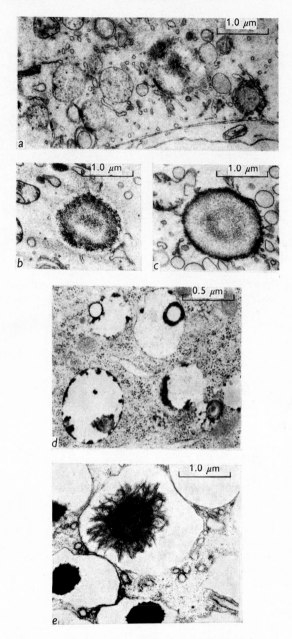

Different mechanisms of formation of intracellular granules. (*a*)–(*c*) The formation and mineralization of an organic granule found in the digestive cell of the snail *Helix pomatia*. The organic granules are formed within the vesicles (*a*) and become progressively more electron dense (*b*, *c*) (from Abolins-Krogis, 1970). (*d*) Dense deposits accumulating on the inner surface of the vesicle membrane of the silkworm *Bombyx mori* (from Waku & Sumimoto, 1974). (*e*) Needle-like clusters in an intermediate stage of granule formation in the collembolan midgut (from Humbert, 1974).

PLATE 4

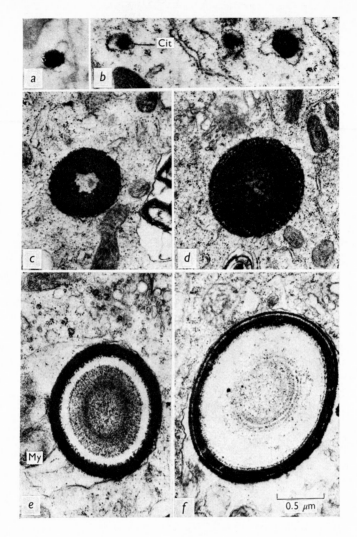

Stages in the formation of granules in the larvae of the insect *Cercopis sanguinea*. The granule originates in the cisternae (Cit.) of the endoplasmic reticulum (*b*). Note the similarity of (*c*) to the Type B II granule (Plate 1*c*) and the final concentric structure in (*e*) and (*f*) (cf. Plate 1*d*). All parts are at the same magnification (from Gouranton, 1968).

PLATE 5

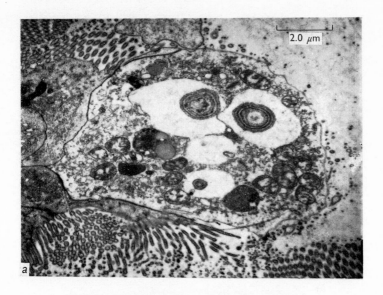

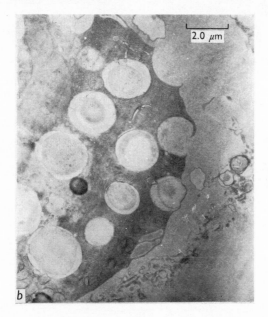

Mechanisms of granule extrusion. (*a*) Showing apocrine secretion in the collembolan *Tomocerus minor* (from Humbert, 1974). (*b*) Merocrine secretion of granules in midgut cells of the silkworm *Bombyx mori* (from Waku & Sumimoto, 1974).

PLATE 6

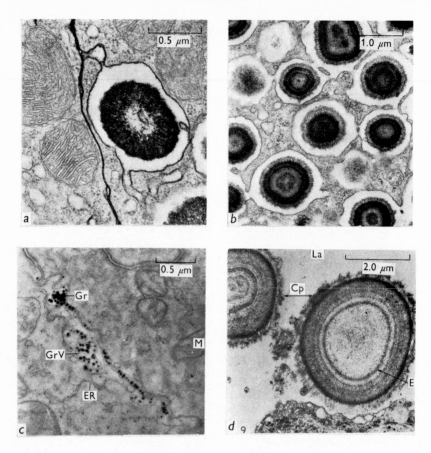

The occurrence of calcium granules in intercellular spaces. (*a*, *b*) Granules in invaginations of the basal and lateral cell membranes of the posterior caecum of the amphipod *Orchestia cavimana* (*a*, from Graf, 1971; *b*, F. Graf, unpublished). (*c*) Granules in the intercellular spaces of the mantle of the bivalve *Mercenaria mercenaria*: ER = endoplasmic reticulum; Gr = granules; GrV = vesicles containing granules; M = mitochondrion (from Neff, 1972). (*d*) Granules extruded from the cytoplasm of the trematode *Cyathocotyle bushiensis*: Cp = excretory corpuscle; E = concentric rings of granule; La = excretory lacuna (from Erasmus, 1967).

Table 1. *Some examples of intracellular 'calcium' granules*

Phylum	Species	'Organ'	Reference
Protozoa	*Proroden morgani*	—	André & Fauré-Fremiet (1967)
Coelenterata	*Renilla reniformis*	Scleroblasts	Dunkelberger & Watabe (1974)
	Aurelia aurata	Statoliths	Spangenberg (1976)
Platyhelminthes			
Trematoda	*Cyathocotyle bushiensis*	Excretory	Erasmus (1967)
Cestoda	*Taenia taeniaeformis* and numerous others	—	Nieland & von Brand (1969)
Mollusca			
Gastropoda	*Ferrisia wautieri*	Connective tissue	Richardot & Wautier (1972)
	Helix pomatia	Hepatopancreas	Abolins-Krogis (1970)
Lamellibranchia	*Anodonta cygnea*	Mantle	Istin & Masoni (1973)
	Mercenaria mercenaria	Mantle	Neff (1972)
Arthropoda			
Crustacea	*Orchestia cavimana*	Posterior caecum	Graf (1971)
	Callinectes sapidus	Hepatopancreas	Becker, Chen, Greenawalt & Lehninger (1974)
	Lepas anatifera	Intestine	Walker, Rainbow, Foster & Crisp (1975)
Onychophora	*Peripatus acacioi*	Intestine	Lavallard (1967)
Diplopoda	*Pleuroloma* sp.	Ovary	Crane & Cowden (1968)
	Polydesmus complanatus	Ovary	Petit (1970)
Insecta	*Pogonognathellus longicornus*	Intestine	Humbert (1974)
	Rhodnius prolixis	Malpighian tubule	Wigglesworth & Salpeter (1962)
	Gryllus domesticus	Malpighian tubule	Berkaloff (1958)
	Gryllotalpa gryllotalpa	Malpighian tubule	Lhonoré (1971)
	Bombyx mori	Midgut	Waku & Sumimoto (1974)
	Cercopis sanguinea	Intestine	Gouranton (1968)
	Blatella germanica	Male accessory	Ballan-Dufrançais (1970)

Table 2. *Ratios of some common ions found in intracellular granules (after Becker et al., 1974; Burton, 1972; von Brand, Scott, Nylen & Pugh, 1965; von Brand et al., 1967)*

Species	Ca^{2+}		Mg^{2+}		PO_4^{3-}		CO_3^{2-}	Crystal form
Helix pomatia	1	:	0.6	:	1.0	:	0.1	Calcite
Callinectes sapidus	1	:	0.4	:	0.8	:	0.1	—
Ligula intestinalis	1	:	1.6	:	0.1	:	1.6	Amorphous[a]
Taenia crassiceps	1	:	3.4	:	0.1	:	1.9	Amorphous[a]
Raillietina cesticellus	1	:	0	:	0.4	:	1.9	Calcite[b]
Cysticerus cellulosae	1	:	0.5	:	0.04	:	3.8	—

[a] After heating at 450 °C gives $CaCO_3$, MgO and hydroxyapatite.
[b] After heating at 450 °C gives $CaCO_3$ and hydroxyapatite.

built of concentric layers of what appeared to be dense needle-like crystals (Plate 1c). Granules of Types I and II often occurred in the same cell but Type III was usually only present in badly damaged cells. At the present time it is not clear whether the distinction between these three types is real and it appears quite likely that Types II and III represent intermediate stages in granule formation (p. 427).

Type B granules are frequently rich in magnesium and phosphate as well as calcium and carbonate ions and are thus easily distinguished from Type A. In granules from the hepatopancreas of the snail *Helix pomatia*, Burton (1972) found molar ratios of Ca^{2+}, Mg^{2+}, PO_4^{3-}, and CO_3^{2-} of 1.0:0.6:1.0:0.1 while Becker *et al.* (1974) give ratios of 1.0:0.4:0.8:0.1 for samples from the digestive gland of the blue crab. Among the cestodes, von Brand, Nylen, Martin & Churchwell (1967) give a wide variety of ratios, some of which are very rich in calcium and carbonate ions whilst others contain much magnesium and phosphate. Most of these granules are amorphous when subjected to X-ray diffraction but various crystal forms could be made to appear on heating them to about 450 °C (Table 2).

Most of the Type B granules have the following common properties:

(*a*) a similar intracellular location,

(*b*) a poorly crystalline or amorphous form,

(*c*) a sparse lipid and protein–polysaccharide base or matrix,

(*d*) trace amounts of low molecular weight compounds such as organic phosphates,

(*e*) a characteristic concentric layered structure (Plate 2).

THE BASIC PHENOMENON

A first-order model simply states reality as it is thought to exist and the reality of ultrastructural studies of invertebrate tissues appears to be that they frequently contain mineral deposits within membrane-bound vesicles (Table 1). This contrasts with the descriptions of intramitochondrial minerals which have only been described occasionally, e.g. in the earthworm calciferous glands (Crang, Holden & Hitt, 1968), the insect Malpighian tubules (Wigglesworth & Salpeter, 1962) and the blue crab hepatopancreas (Chen, Greenawalt & Lehninger, 1974). There are reasons for questioning whether at least some of these observations represent normal in-vivo situations either because the resolution of the section is not very good, or because of the way the tissue was treated before microscopy. If intramito-chondrial granules do occur normally, they are certainly not very common, and one is left, therefore, with the impression that the cytoplasmic vesicle is the most important mineral-forming structure in invertebrate cells.

The details of the actual process of secretion of the granules vary be-tween species but there is considerable similarity in the general picture. The initial concretions appear in small vesicles (Berkaloff, 1958) or in cisternae of the endoplasmic reticulum which often appear to be associated with the Golgi apparatus (Waku & Sumimoto, 1974). When initially se-creted they appear to be associated with organic material which may ac-tually be involved in their induction (Abolins-Krogis, 1970) (Plate 3a–c). The centre of the granule is usually electron opaque when formed but becomes progressively clearer as the calcospherite grows (Gouranton, 1968) (Plate 4). As the cisternae increase in size, the ribosomes, if present, become lost and additional material appears to be supplied by the fusion of smaller bodies (Humbert, 1974). Usually the granule occupies most of the volume of the vesicle but in *Bombyx* it initially appears to be only a small structure attached to one region of the vesicle wall (Waku & Sumimoto, 1974) (Plate 3d). As the granule grows it frequently passes through a stage in which it has an irregular periphery apparently formed of numerous needle-like crystals and the whole structure then resembles the B Type II granules described previously (Plate 3e). This stage is followed by the formation of the typical B Type III structure of alternating electron clear and opaque layers (Gouranton, 1968; Humbert, 1974) (Plate 4). One interpretation of this type of granule formation is shown in Fig. 1. Varia-tions of this pattern frequently occur; e.g. in *Bombyx mori*, granules of glycogen occur in large numbers within the vesicle (Waku & Sumimoto,

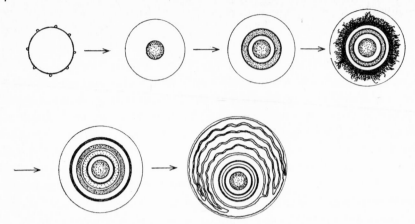

Fig. 1. Schematic representation of the mechanism of granule formation in the Collembola (after Humbert, 1974).

1974) and it has been often noted that 'myelin' bodies are associated with the vesicles (Gouranton, 1968; Ballan-Dufrançais, 1970; Humbert, 1974).

In some cases, the membrane lining the vesicle extends into the cytoplasm (Waku & Sumimoto, 1974) and in *Orchestia* it is claimed that the membranes can be traced to the extracellular spaces (Plate 6*a*, *b*) implying that the granules are in fact never truly intracellular (Graf, 1971).

This last observation raises the whole question of whether the granules remain within the cells or whether they are involved in the general physiology of the animal. Unfortunately, the evidence on this matter is relatively sparse and speculations should be viewed with much caution. The results of such ideas are sufficiently different, however, from the generally accepted models of 'cell calcium' that it is worth reviewing the evidence that has accumulated in the past decade on the possible functions of these granules.

THE FUNCTIONS OF CALCIUM GRANULES

Type A granules are usually sculptured and highly crystalline. They may be regarded as performing mechanical or skeletal functions among, for example, the Algae and Coelenterata. It is probable that coccoliths, for example, may also represent a method for the extrusion of calcium or carbonate ions but the whole structure remains as a protective layer, associated with the cell that secretes them and they may, therefore, be classified as skeletal.

Many of the Type B granules appear to have a more dynamic function. Those that remain within the secreting cell, as is the case in the parasitic cestodes, have been suggested as storing carbonate ions which neutralize the acid products of metabolism or protect the organism as it passes through the stomach of its definitive host (Nieland & von Brand, 1969). It has been stated that the phosphate present in the granule may provide an inorganic store necessary for an animal with a large carbohydrate metabolism, while the mineral deposits may also be a way of protecting an animal with a large permeability to calcium (Desser, 1963) from the danger of mineralization (Nieland & von Brand, 1969). Similar suggestions have been made in other animals and Walker *et al.* (1975) found that in the barnacles *Balanus balanoides* and *Lepas anatifera* these granules were a major route for the removal of contaminating ions such as zinc. Similar proposals have been made for a number of other animals where the granules occur in excretory organs such as the Malpighian tubules of insects, e.g. *Lucilia cuprina* larvae (Waterhouse, 1950) and *Gryllotalpa gryllotalpa* (Lhonoré, 1971); in midgut cells, as in the silkworm *Bombyx mori* (Waku & Sumimoto, 1974) and the homopteran *Cercopides* (Gouranton, 1968); and in the excretory bladder of trematodes (Erasmus, 1967). In all these cases the granules are shed into the lumen, either to be excreted or used in processes such as the formation of dwelling tubes in machaerotid larvae (Marshall & Cheung, 1973). In the case of the beetle *Ceramlynx*, the granules pass forward through the intestine and are used to form an operculum (Chapman, 1971).

In many cases the granules are recycled within the animal. This occurs in the amphipod *Orchestia cavimana* where the epithelium of the posterior caecum stores calcium in granules at the time of the moult, but releases or dissolves them again when the new exoskeleton is mineralized (Graf, 1971). A similar phenomenon occurs in the hepatopancreas of the blue crab *Callinectes sapidus* at the time of ecdysis (Becker *et al.*, 1974). In the freshwater snail *Pomacea paludosa*, granules are stored in the cells of the albumen glands of the reproductive system. In this case the granules are in the form of the mineral vaterite and they disappear as the calcareous egg capsules are formed (Meenakshi, Blackwelder & Watabe, 1974). Granules are, in fact, found in a large number of tissues such as the hepatopancreas, mantle and foot of molluscs. They have been studied under a variety of conditions by Abolins-Krogis (1965, 1968, 1970) and shown to be released from calcium cells in the hepatopancreas and transported around the body in amoebocytes when, for example, the shell is damaged. In the clam *Mercenaria mercenaria*, there are small calcium-rich granules in the intercellular spaces and in the microvilli and apical cytoplasm of the mantle cells (Neff, 1972). Calcium ions move across this tissue towards the shell, during calcification, and away from the shell during periods of anaerobic acidosis. The exact role of the granules in these processes is not clear but they appear to be transported to the mantle by subepidermal amoebocytes and are the major source of calcium in this tissue.

From this brief survey it will be apparent that intracellular granules have been implicated in a large number of physiological phenomena. In order to advance these ideas, however, we need more information on

(a) the solubility of these granules and the energy requirements for their formation,

(b) the mechanisms for granule extrusion from cells,

(c) the state of ions in extracellular fluids,

(d) the movement of ions along cellular channels and their possible relationship with granules.

Granule solubility and energy requirements for formation

Many of the functions proposed for intracellular granules are related to their availability to the animal, i.e. their solubility. This, of course, is the converse of the energy requirements for their formation. Thus, in a solution containing a simple crystal there will be a solubility product constant for the ion species represented in the mineral, and calcification will only occur when there is a tendency for this to be exceeded. Energy may therefore have to be provided to initiate mineralization by increasing the concentration of ions until they exceed the solubility product. At the cellular level it may be assumed first, that the calcium ion concentration in the cytoplasm is in the region of $1 \mu M$, and second that organelles forming mineral deposits derive their calcium ions from that source. The fluid within a mineral-containing organelle will be saturated with these ions which means that there will be roughly the same concentration gradients across the plasma membrane from the extracellular fluids to the cytoplasm as from the organelle to the cytoplasm. In the case of calcium ions this concentration gradient is several thousand-fold, and is presumably maintained by ion pumps. It is, therefore, possible to recognize the energy barriers shown in Fig. 2 as being involved in intracellular calcification. This, however, is not the only problem since there will also be a tendency for the ions to leak out of the organelle back into the cytoplasm. The greater the solubility of the mineral the greater will be the energy required to crystallize it in an organelle and the greater the energy necessary to prevent it dissolving out into the cytoplasm.

Let us first consider the possibility that intracellular granules are primarily a way of conserving energy. This would be possible if the gradient for the pump removing calcium ions from the cytoplasm was less in the

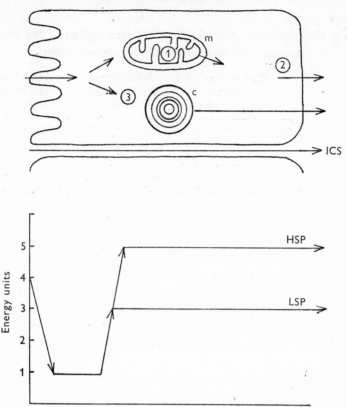

Fig. 2. Energy barriers involved in the regulation of cell calcium. Calcium enters the cell down a concentration gradient but if it enters a mitochondrion (m) it parasitizes energy which would otherwise be available to the cell and also enters a region where the electrochemical gradient requires the expenditure of energy in order to return it to the cytoplasm (c) (1). If it re-enters the cytoplasm it must then be pumped out across the plasma membrane (2). Calcium entering granules (3) must be pumped up an energy barrier that will be small for a low solubility product (LSP) granule but somewhat bigger for a high solubility product (HSP). In the case of LSP granules the energy may be less than either (1) or (2) and for both types of granule the total energy could be less than (1) + (2).

direction of the granule than it was across the plasma membrane into the body fluids. This situation might exist if the organelle produced a highly insoluble mineral.

The converse situation, namely the formation of granules which were more soluble than the normal skeletal deposits, might be necessary where the minerals were to be remobilized after storage or used to keep the body fluids supersaturated. If these functions were to be performed, they would increase the energy demands made upon the cell during intracellular mineralization.

Table 3. *Composition of granules from different regions of the snail* Helix aspersa *and their solubilities in various salines*

Type of granule	Composition of granule (molar ratios) $Ca^{2+}:Mg^{2+}:PO_4^{3-}:CO_3^{2-}$	Solubility in Kreb's saline (mM) Ca^{2+} PO_4^{3-} pH			Solubility in snail saline (mM) Ca^{2+} Mg^{2+} PO_4^{3-} pH			
None (control)	—	2.30	1.30	7.6	4.16	4.8	0.00	7.9
Foot	1 : 0.06: 0.06 : 1.04	9.40	0.61	7.7	12.50	4.5	0.15	7.8
Mantle	—	8.20	0.95	7.7	—	—	—	—
Hepatopancreas	1 : 0.96: 1.35 : 0.53	2.10	1.50	7.7	5.31	8.7	0.3	7.8

Clearly more information is necessary but virtually no work has been done on this critical problem of the solubility of the various types of intracellular granules that are found in the animal kingdom. It could in fact be argued from the sort of data produced in Table 2 that such information would be of limited value for there is an enormous diversity of chemical components present in these structures. It is, however, important to try and establish the range of values since this will define a number of problems in relation to the formation and possible physiological functions of granules.

For these reasons we have made some preliminary observations on the solubility of granules obtained from various tissues of the snail *Helix aspersa*. A difficult problem in such work is in deciding how to isolate the granules without risking either the solution of the minerals in the various tissue digests or the modification of the crystal form by strong reagents. We have used the quaternary ammonium hydroxide-containing product 'Soluene' (Packard Instruments) since it is toluene based and contains little water while being a fast-acting tissue solubilizer. Samples of hepatopancreas, mantle and foot have been mainly used although shell and epidermal mucus granules have also been studied on occasion. The tissues were digested in 'Soluene' at 60 °C for 1–2 h and the granules were then collected by centrifugation and cleaned of the solvent by repeated washings in alcohol and acetone. The dry powder so obtained was analysed for calcium, magnesium, phosphate and carbonate ions and samples were then suspended in a Kreb's saline containing 0.5 M 'Tris' buffer. The solution was gassed with 5 % CO_2 95 % O_2 and adjusted to pH 7.6 with hydrochloric acid. The suspended powder was shaken for 48 h at which time it was analysed for constituent ions and the pH measured. Some typical results are shown in Table 3. It is apparent that granules from different organs have widely different effects upon the composition of the saline. The granules from the foot and mantle are very soluble, raising the calcium concentration of the saline almost four-fold and causing the loss of phosphate ions. Granules from the hepatopancreas are rich in magnesium and phosphate ions and tend to modify the composition of the saline accordingly (Table 3). The blood of *Helix* normally contains high levels of calcium (4.5 mM) magnesium (3.6 mM) and bicarbonate (20 mM) ions but little phosphate. When granules were equilibrated with an artificial saline of this type somewhat similar results were obtained, namely the calcium levels remained normal with hepatopancreas granules but were raised by foot granules. Magnesium and phosphate levels rose with hepatopancreas but remained normal with foot granules. These results are in keeping with the composition of these granules (Table 3).

The mechanism of granule extrusion from cells

In a number of animals, e.g. the cestodes, the intracellular granules appear to remain within the cytoplasm throughout the life of the animal. In the majority of cases, however, they are extruded from the cell usually into the lumen of the excretory or intestinal system, although in some animals they may pass into the intercellular spaces of the tissue.

The mechanism of granule extrusion appears to vary. In the coccolitho-phorids it can be inhibited by cytochalasin B which apparently acts by inhibiting the fusion of the vesicle and plasma membranes so that the calcified bodies accumulate within the cytoplasm (Weiss & Wilbur, 1976). In the Metazoa both apocrine and merocrine systems of granule extrusion have been described. Apocrine secretion, in which the apical end of the cell is lost together with the granules, has been described by Humbert (1974) in the mesenteron of the collembolan *Tomocerus minor* (Plate 5a). Frequently the granules are also released by the degeneration of the whole of this tissue. In contrast, merocrine secretion is illustrated by Waku & Sumimoto (1974) in their study of the midgut of the silkworm *Bombyx mori* (Plate 5b). In this case the cell remains intact while the granule is extruded. In these and similar studies (Becker *et al.*, 1974; Erasmus, 1967; etc.) the granules can frequently be detected within the lumen of the organ (Plates 1a, 2c). In other cases, however, the granules remain within the tissues in the intercellular spaces. This is clearly seen in Neff's (1972) study of the clam mantle (Plate 6c), whilst in the amphipod *Orchestia cavimana* the posterior caecal epithelium similarly appears to be involved in storing and then releasing the calcium obtained during the moult of the exoskeleton (Graf, 1968, 1971). During this process a large number of B Type III granules are formed in a network of basal and lateral infoldings of the membranes of the epithelial cells (Plate 6a, b). The calcium appears to move in a basal to apical direction during granule formation and is resorbed in the opposite direction during the recalcification of the exoskeleton. There is evidence of a close association between the plasma membrane, and the granules, but throughout the whole process the mineral deposits appear to reside in a convoluted extension of the extracellular space (Plate 6b).

Evidence from studies such as those on *Mercenaria* and *Orchestia* suggest that the granules present in intercellular spaces could influence the composition of the extracellular fluids. We have therefore attempted to induce granule extrusion by using the ionophore Lasalocid (X-537A). In these experiments the tentacle of a snail (*Helix aspersa*) was cannulated and injected wth 200 μl of ^{45}Ca in 45 mM $CaCl_2$ and left for 5 days. An injection of this sort should almost double the calcium concentration of the body fluids and thus facilitate granule formation. Certainly at the end of this time the granules can be shown to have accumulated a considerable amount of label and the blood has returned to its normal calcium concentration. A blood sample is therefore taken for analysis at this time and then 100 μg of ionophore is injected in 20 μl of dimethyl sulphoxide. Control animals were injected with just the solvent and then left for 2 h. A second blood sample was then collected and analysed for both ^{45}Ca and total calcium. The results of such an experiment are shown in Fig. 3. It is clear that the ionophore produces a rise of over 40 % in blood calcium levels and almost doubles the specific activity of the plasma. Control animals showed no significant change in either blood calcium or blood specific activity. It is clear therefore that the ionophore causes the release of calcium from sites which were heavily labelled at the time of the initial injection. It is obviously not possible to identify these sites from this type of experiment but the results would not be in disagreement with what we know so far about intracellular granules. If this interpretation is correct then such minerals cause a rise in blood calcium levels presumably by exerting the types of effects shown in Table 3. It is, therefore, worth considering further the generally accepted ideas of the relationships of mineral deposits to body fluids.

The state of ions in extracellular fluids

The problem of the ionic composition of the body fluids and their relationship to the solubility product constants of various minerals has plagued physiologists for many years (Neuman & Neuman, 1958). The facts are that the extracellular fluids of many animals appear to be supersaturated with the relevant ions (Table 4), and this has led to two possible

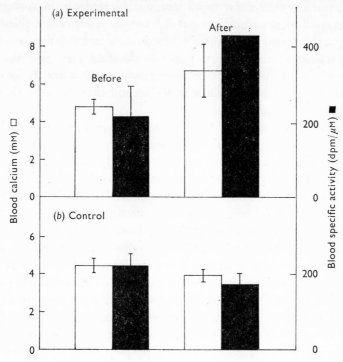

Fig. 3. The concentration of plasma calcium ($\square$) and its specific activity ($\blacksquare$) before and 2 h after the injection of the ionophore X-537A into the snail *Helix aspersa*.

Table 4. *Ionic products of calcium and carbonate in the blood of various invertebrates in relation to the solubility product of calcium carbonate under similar conditions (from Potts, 1954)*

Sample	PO_4^{3-} content (M)	Ionic product, $Ca^{2+} \times CO_3^{2-}$ (mM)^2	Solubility product, aragonite (mM)^2	State of body fluids
Mytilus blood	5×10^{-4}	2.31×10^{-6}	1.95×10^{-6}	Supersaturated
Anodonta blood	2×10^{-4}	0.33×10^{-6}	0.09×10^{-6}	Supersaturated
Helix blood	—	0.60×10^{-6}	0.55×10^{-6}	Supersaturated
Limulus blood	—	2.08×10^{-6}	1.74×10^{-6}	Supersaturated

explanations for this phenomenon. The first postulates that the energy barriers necessary to overcome the effects of nucleation and crystal poisons means that high concentrations of free ions are able to coexist within the blood without inducing crystalization (Fleisch & Neuman, 1960). The second theory proposes that the extracellular fluids are separated from the mineral phase by a cellular membrane which regulates the movement of ions between the two and, therefore, controls calcification (Neuman & Ramp, 1971). Clearly the possible functions of intracellular granules are closely related to these theories for, in the first case, granules which become extruded into the intercellular spaces could affect the electrolyte composition of the blood, whilst in the second case, the cells themselves might mobilize these minerals.

One way of approaching these problems experimentally has been to displace the normal mineral–fluid equilibrium by inducing respiratory acidosis in the animal and then studying the body fluids. In the frog, *Rana temporaria*, Simkiss (1968) used this approach to show that the calcareous deposits in the endolymphatic sacs appeared to maintain a constant $[Ca^{2+}] \times [CO_3^{2-}]$ level in the blood. No attempt was made to determine an absolute solubility product but it was shown that the animal responded as if the blood was in equilibrium with a calcareous mineral.

A somewhat similar experiment was performed by Burton & Mathie (1975) who exposed the snail *Helix pomatia* to hypercapnia. They showed that with 5 to 10 % CO_2 the calcium and bicarbonate levels of the blood rose in the ratio of 1:2. The average initial and final values of pH and ionized calcium are shown in Fig. 4 with the continuous curve corresponding to a constant ionic product, $[Ca^{2+}] \times [CO_3^{2-}]$ of 3.6 (mM)^2. This suggests that the haemolymph is in equilibrium with solid calcium carbonate. The problem which Burton & Mathie see with this interpretation is that calcite should have a solubility product constant of only 0.1 (mM)^2 under normal conditions. For this reason they reject this hypothesis and prefer instead to consider that the cell membranes, in some way, regulate the movement of calcium and carbonate ions into the blood. While this may indeed be the case there remains the alternative explanation, namely that one is dealing with the solubilities of impure deposits in complex solutions of electrolytes.

The movement of ions along cellular channels

If, as has been proposed, there is at least some evidence to suggest that intracellular calcium deposits might be released into extracellular locations where they might at least partially dissolve, then the question arises as to how that calcium might move around the body. The classical approach to calcium translocation has tended to emphasize the role of membrane pumps and cellular organelles. It has been argued, however, that these theories are not very clear and furthermore that the movement of calcium across cells involves many problems because of the obvious tendency to

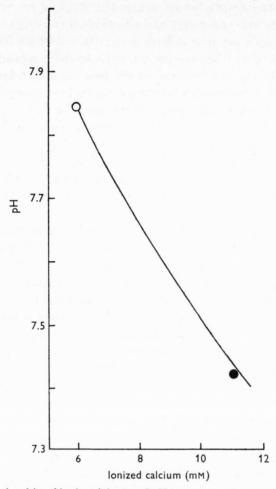

Fig. 4. The relationship of ionic calcium and pH in the blood of normal snails (○) and snails exposed to 5–10 % carbon dioxide (●). The solid line represents the relationship of calcium and pH corresponding to a constant $[Ca^{2+}] \times [CO_3^{2-}]$ product of 3.6 (mM)^2 (data from Burton & Mathie, 1975).

upset intracellular calcium levels (Simkiss, 1974). Evidence is now accumulating that in a number of tissues a 'calcium electrode' effect can be obtained. It has always been recognized in the mammalian intestine that there is a large component of the calcium movement which is concentration dependent and independent of active transport (Wasserman & Taylor, 1969). The experiments of Warner & Coleman (1975) suggest that in this tissue the calcium moves largely through the intercellular spaces. Similar conclusions were reached for the epidermis of the freshwater snail *Limnaea stagnalis*, which behaved in artificial tap water as if the surface epithelium was highly and selectively permeable to calcium ions, thus setting up a potential difference as predicted by the Nernst equation (Greenaway, 1971). The same interpretation has been applied to the results obtained with the mantle tissue of the lamellibranchs *Anodonta grandis* and *Amblema costata* (Istin & Kirschner, 1968). They showed that normally there was a transepithelial potential difference of 25 to 50 mV with the blood side negative. The potential difference varied with the external calcium concentration and was zero at about 12–16 mequiv. l^{-1}. The voltage was very sensitive to carbon dioxide and this effect was traced to calcium granules (Plate 2a) which contained carbonic anhydrase and were situated in the interstitial tissue of the mantle (Istin & Girard, 1970).

Taken at their face value these experiments suggest, first, that calcium may move through the intercellular spaces; second, that the movement is constrained by the absence of a moving anion so that a 'Nernst' potential is set up; and, third, that granules in intercellular spaces may release calcium ions into these situations. Some caution needs to be exercised with this interpretation, however, since Eddy (1975) has shown that in the goldfish (*Carassius auratus*) the gills also show a transepithelial potential that is sensitive to calcium ions. In this case the calcium effect appears to arise because in its presence the sodium flux falls from 128.3 to 18.6 μmole 100 g^{-1} h^{-1} while chloride fluxes only change from 110.3 to 28.2 μmole 100 g^{-1} h^{-1}. The result is that calcium converts a negative transmembrane potential into a positive one. The results are very similar to a calcium electrode effect, but are purely due to a differential constraint on sodium and chloride movements.

The possibility that a similar explanation might account for the calcium potential found across the snail epidermis was investigated by Simkiss & Wilbur (1976). Using the snail *Helix aspersa*, they were able to show that the potential due to calcium ions decreased markedly in the absence of a sodium and chloride ion gradient across the skin. There was, however, little change in sodium fluxes in the presence or absence of calcium. It was suggested on this evidence that there may be an important intercellular route for the movement of calcium ions across the molluscan epidermis but that there is normally a large exchange component which in the presence of sodium ions reduces the observed potential difference. It might be concluded, therefore, that in molluscs there is evidence for intercellular calcium movements and similar situations may exist in the mammalian intestine (Warner & Coleman, 1975) and the chick chorioallantois (Saleuddin, Kyriakides, Peacock & Simkiss, 1976).

Clearly more work needs to be done to investigate the effects of calcium on the transepithelial movement of other ions as well as upon the possibility of intercellular routes of calcium translocation.

SOME CONCLUSIONS...

The initial aim of this article was to provide an alternative first-order model of cell calcium. Studies on a variety of invertebrate tissues suggest that the 'reality as it exists' is as follows:

(1) Many cells have the ability to produce intracellular granules which are often rich in calcium.

(2) Several different types of granule can be recognized varying from crystallographically pure minerals to amorphous impure deposits.

(3) These granules are formed in vesicles situated in the cytoplasm and usually closely associated with either the Golgi system or the endoplasmic reticulum. Deposits in mitochondria are rare.

(4) The granules may remain within the forming cells but in many cases they can be extruded into either the lumen of organs or into the intercellular spaces where they may play an important role in the physiology of the animal.

...AND SOME SPECULATIONS

If one now tries to explain these phenomena in terms of simpler and better understood systems, i.e. if one attempts some second-order modelling, the following possibilities exist:

(5) It appears likely that the intracellular granules provide an alternative site for an intracellular system of regulating cytoplasmic calcium (cf. Borle, 1973).

(6) If the intracellular mineral which is formed has a lower solubility product than the extracellular fluids then there will be a corresponding smaller concentration gradient across the organelle membrane. For this reason it could be energetically less expensive to dispose of calcium influxes by forming calcium deposits in vesicles than by pumping the ion back into the body fluids.

(7) The use of membrane pumps in this way may have a number of advantages. Thus contaminating ions would not be returned to the body fluids (cf. zinc; Walker et al., 1975) whilst valuable metabolites such as phosphate may be stored for reuse later (cf. Table 2). The suggestion that a variety of membrane pumps may also discharge into such vesicles is an interesting one. Thus the jelly fish *Aurelia aurita* maintains neutral bouyancy by extruding sulphate ions and it is also the only known animal to produce intracellular granules of gypsum ($CaSO_4.2H_2O$) which it uses

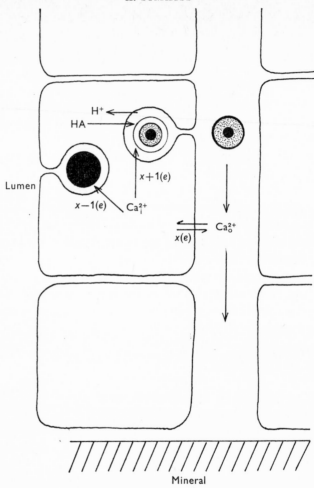

Fig. 5. Possible energy and solubility relationships of intracellular granules. A membrane pump removes calcium entering the cell so maintaining Ca^{2+}; by using $x(e)$ amount of energy. An 'insoluble' granule with a lower $[Ca^{2+}]$ inside the vesicle would save energy $(x-1(e))$ and thus provide a more economical way of removing intracellular calcium and other ions into the lumen of an organ. A 'soluble' granule would require a greater energy input $(x+1(e))$ but would maintain the extracellular calcium levels (Ca_o^{2+}) if released into the extracellular fluids. Sites of granule formation involving proton release would therefore be intracellular but the recrystallization of these deposits may appear as sites of extracellular mineralization. Calcium is shown as moving to these sites via intercellular routes.

as a statolith (Spangenberg, 1976). The mechanism of 'directing' membrane pumps into vesicles may therefore be a not uncommon phenomenon and it could provide a basis for various types of biomineralization.

(8) The process of calcification involves the extrusion of protons

(Simkiss, 1969, 1976), i.e.

$$Ca^{2+} \times HA^{-} \longrightarrow CaA + H^{+}.$$

This may be used as a definition of one type of calcification. If this reaction occurs in intracellular vesicles then the process of calcification is clearly an intracellular one. If the granules are later extruded into the extracellular spaces then, depending upon their solubility, they may dissolve and re-crystallize elsewhere in the body. Such sites of recrystallization will obviously appear as regions of mineralization, but they will not be sites of calcification as defined above. It may therefore be possible to separate sites of calcification from sites of mineralization and this may help to clarify some of the current problems in biomineralization (Anderson, 1973; Simkiss, 1976).

(9) The argument in (8) pre-supposes the movement of calcium through intercellular spaces. Some of the problems inherent in this concept have already been discussed. It provides an important alternative, however, to the current ideas on transcellular transport.

(10) The solubilities of intracellular granules will obviously determine whether they increase or decrease in size once released into the animal. If they are capable of dissolving they may, of course provide very different solubilities from the main skeletal minerals. This may explain some of the paradox of abnormal ion products for extracellular fluids (cf. Neuman & Neuman, 1958; Burton & Mathie, 1975).

Some of these speculations are summarized in Fig. 5. If *any* of them are true they will provide some new insights into cell calcium and will, there-fore, justify the further study of intracellular granules.

REFERENCES

ABOLINS-KROGIS, A. (1965). Electron microscope observations on calcium cells in the hepatopancreas of the snail *Helix pomatia*. *Ark. Zool.*, **18**, 85–92.

— (1968). Shell regeneration in *Helix pomatia* with special reference to the elementary calcifying particles. *Symp. zool. Soc. Lond.*, **22**, 75–92.

— (1970). Electron microscope studies of the intracellular origin and formation of calcifying granules and calcium spherites in the hepatopancreas of the snail *Helix pomatia*. *Z. Zellforsch. mikrosk. Anat.*, **108**, 501–515.

ANDERSON, H. C. (1973). Calcium-accumulating vesicles in the intercellular matrix of bone. Hard tissue growth, repair and remineralization. *Ciba Foundation Symposium*, **11** (new series), 213–226.

ANDRÉ, J. & FAURÉ-FREMIET, E. (1967). Formation et structure des concrétions calcaires chez *Proroden morgani* Kahl. *J. Microscopie*, **6**, 391–398.

BALLAN-DUFRANÇAIS, C. (1970). Données cytophysiologiques sur un organe ex-créteur particulier d'un insect *Blatella germanica* L. (Dictyoptère). *Z. Zell-forsch. mikrosk. Anat.*, **109**, 336–355.

BECKER, G. L., CHEN, C-H, GREENAWALT, J. W. & LEHNINGER, A. L. (1974). Calcium phosphate granules in the hepatopancreas of the Blue Crab, *Callinectes sapidus*. *J. Cell Biol.*, **61**, 316–326.

BERKALOFF, A. (1958). Les grains de sécrétion des tubes de Malpighi de *Gryllus domesticus* (Orthoptère). *C. r. hebd. Séanc. Acad. Sci. Paris*, **246**, 2807–2809.

BORLE, A. B. (1973). Calcium metabolism at the cellular level. *Fedn Proc.*, **32**, 1944.

BURTON, R. F. (1972). The storage of calcium and magnesium phosphate and of calcite in the digestive glands of the Pulmonata. *Comp. Biochem. Physiol.*, **43**, 655–663.

BURTON, R. F. & MATHIE, R. T. (1975). Calcium and pH homeostasis in the snail (*Helix pomatia*). Effect of CO_2 and $CaCl_2$ infusion. *Experientia*, **31**, 543–544.

CHAPMAN, R. F. (1971). *The Insects, Structure and Function*. English Universities Press: London.

CHEN, C-H, GREENAWALT, J. W. & LEHNINGER, A. L. (1974). Biochemical and ultrastructural aspects of Ca^{2+} transport by mitochondria of the hepatopancreas of the blue crab *Callinectes sapidus*. *J. Cell Biol.*, **61**, 301–315.

CRANE, D. F. & COWDEN, R. R. (1968). A cytochemical study of oocyte growth in four species of millipedes. *Z. Zellforsch. mikrosk. Anat.*, **90**, 414–431.

CRANG, R. E., HOLDEN, R. C. & HITT, J. B. (1968). Calcite production in mitochondria of earthworm calciferous glands. *BioScience*, **18**, 299–301.

DESSER, S. S. (1963). Calcium accumulation in larval *Echinococcus mulvilocularis*. *Can. J. Zool.*, **41**, 1055–1059.

DUNKELBERGER, D. G. & WATABE, N. (1974). An ultrastructural study on spicule formation in the pennatulid colony *Renilla reniformis*. *Tissue and Cell*, **6**, 573–586.

EDDY, F. B. (1975). The effect of calcium on gill potentials and on sodium and chloride fluxes in the goldfish *Carassius auratus*. *J. comp. Physiol.*, **96**, 131–142.

ERASMUS, D. A. (1967). Ultrastructural observations in the reserve bladder system of *Cyathocotyle bushiensis* Khan 1962 (Trematoda, Strigeoiden) with special reference to lipid excretion. *J. Parasit.*, **53**, 525–536.

FLEISCH, H. & NEUMAN, W. F. (1960). On the role of phosphatase in the nucleation of calcium phosphate by collagen. *J. Am. chem. Soc.*, **82**, 3783–3784.

GOURANTON, J. (1968). Composition, structure et mode de formation des concrétions minérales dans l'intestin moyen des Homoptères Cercopides. *J. Cell Biol.*, **37**, 316–328.

GRAF, F. (1968). Le stockage de calcium avant la mue chez les Crustacés Amphipodes *Orchestia* (Talitride) et *Niphargus* (Gommaride Lynoge). Thèse, Faculté des Sciences de L'Université de Dijon.

— (1971). Dynamique du calcium dans l'épithélium des caecums postérieurs d'*Orchestia cavimana* Heller (Crustacé Amphiode). Rôle de l'espace intercellulaire. *C. r. Acad. Sci. Paris*, **273**, 1828–1831.

GREENAWAY, P. (1971). Calcium regulation in the freshwater mollusc *Limnaea stagnalis* L. The effect of internal and external calcium concentration. *J. exp. Biol.*, **54**, 199–214.

HEILBRUNN, L. V. (1940). *An Outline of General Physiology*. Saunders: Philadelphia.

HUMBERT, W. (1974). Localisation, structure et genèse des concrétions minérales dans le mésentéron des Collemboles Tomoceridae (Insecta, Collembola). *Z. Morph. Ökol. Tiere*, **78**, 93–109.

ISTIN, M. & GIRARD, J. P. (1970). Carbonic anhydrase and mobilization of calcium reserves in the mantle of lamellibranchs. *Calc. Tiss. Res.*, **5**, 247–260.

ISTIN, M. & KIRSCHNER, L. B. (1968). On the origin of the bioelectrical potential generated by the freshwater clam mantle. *J. gen. Physiol.*, **51**, 478–496.

Istin, M. & Masoni, A. (1973). Absorption et redistribution du calcium dans le manteau des lamellibranches en relation avec la structure. *Calc. Tiss. Res.*, **11**, 151–162.

Keeton, W. T. (1971). Magnets interfere with pigeon homing. *Proc. natn. Acad. Sci. USA*, **68**, 102–106.

Lavallard, R. (1967). Ultrastructure des cellules prismatique de l'épithélium intestinal chez *Peripatus acacioi*. *C. r. Acad. Sci. Paris*, **264**, 929–932.

Lhonoré, J. (1971). Données cytophysiologiques sur les tubes de Malpighi de *Gryllotalpa gryllotalpa*. *C. r. Acad. Sci. Paris*, **272**, 2788–2790.

Manton, I. & Leedale, G. F. (1961). Further observations on the fine structure of *Chrysochromulina ericina*. *J. mar. biol. Ass. UK*, **41**, 145–155.

—— (1969). Observations on the microanatomy of *Coccolithus pelagicus* and *Cricosphaera carterae*, with special reference to the origin and nature of coccoliths and scales. *J. mar. biol. Ass. UK*, **49**, 1–16.

Marshall, A. T. & Cheung, W. W. K. (1973). Calcification in insects. The dwelling tube and midgut of machaerotid larvae. *J. Insect Physiol.*, **19**, 963–972.

Meenakshi, V. R., Blackwelder, P. L. & Watabe, N. (1974). Studies on the formation of calcified egg capsules of ampullanid snails. I. Vaterite crystals in the reproductive system and the egg capsules of *Pomacea paludosa*. *Calc. Tiss. Res.*, **16**, 283–291.

Neff, J. M. (1972). Ultrastructure of the outer epithelium of the mantle in the clam *Mercenaria mercenaria* in relation to calcification of the shell. *Tissue and Cell*, **4** (4), 591–600.

Neuman, W. F. & Neuman, M. W. (1958). *Chemical Dynamics of Bone Mineral.* University of Chicago Press: Chicago.

Neuman, W. F. & Ramp, W. K. (1971). The concept of a bone membrane: some implications. In *Cellular Mechanisms for Calcium Transfer and Homeostasis*, ed. G. Nichols & R. H. Wasserman, pp. 197–206. Academic Press: London.

Nieland, M. L. & von Brand, T. (1969). Electron microscopy of cestode calcareous corpuscle formation. *Expl Parasit.*, **24**, 279–289.

Petit, J. (1970). Sur la nature et l'accumulation de substances minérales dans les ovocytes de *Polydesmus complanatus* (Myriapode Diplopode). *C. r. Acad. Sci. Paris*, **270**, 2107–2110.

Potts, W. T. W. (1954). The inorganic composition of the blood of *Mytilus edulis* and *Anodonta cygnaea*. *J. exp. Biol.*, **31**, 376–385.

Richardot, M. & Wautier, J. (1972). Les cellules à calcium du conjonctif de *Ferrissia wautieri* (Moll. Ancylidae). Description minéralogie et variations saisonnières. *Z. Zellforsch. mikrosk. Anat.*, **134**, 227–243.

Saleuddin, A. S. M., Kyriakides, C. P. M., Peacock, A. & Simkiss, K. (1976). Physiological and ultrastructural aspects of ion movements across the chorioallantois. *Comp. Biochem. Physiol.*, **54A**, 7–12.

Simkiss, K. (1968). Calcium and carbonate metabolism in the frog, *Rana temporaria* during respiratory acidosis. *Am. J. Physiol.*, **214**, 627–634.

—— (1969). Intracellular pH during calcification. A study of the avian shell gland. *Biochem. J.*, **111**, 254–259.

—— (1974). Calcium translocation by cells. *Endeavour*, **33**, 119–123.

—— (1976). Cellular aspects of calcification. International symposium on calcification in invertebrates and plants. *Baruch Libr. Mar. Sci.*, in press.

Simkiss, K. & Wilbur, K. M. (1976). The molluscan epidermis and its secretions. *Symp. zool. Soc. Lond.*, in press.

Spangenberg, D. B. (1976). Intracellular statolith synthesis in *Aurelia aurita*. International symposium on calcification in invertebrates and plants. *Baruch Libr. Mar. Sci.*, in press.

444 K. SIMKISS

von Brand, T., Nylen, M. U., Martin, G. N. & Churchwell, K. (1967). Composition and crystallization patterns of calcareous corpuscles of cestodes grown in different classes of hosts. *J. Parasit.*, **53**, 683–687.

von Brand, T., Scott, D. B., Nylen, M. U. & Pugh, M. H. (1965). Variations in the mineralogical composition of cestode calcareous corpuscles. *Expl Parasit.*, **16**, 382–391.

Waku, Y. & Sumimoto, K. (1974). Metamorphosis of midgut epithelial cells in the silkworm (*Bombyx mori*) with special regard to the calcium salt deposits in the cytoplasm. II. Electron microscopy. *Tissue and Cell*, **6**, 127–136.

Walker, G., Rainbow, P. S., Foster, P. & Crisp, D. J. (1975). Barnacles: possible indicators of zinc pollution? *Mar. Biol.*, **30**, 57–65.

Warner, R. R. & Coleman, J. R. (1975). Electron probe analysis of calcium transport by small intestine. *J. Cell Biol.*, **64**, 54–74.

Wasserman, R. H. & Taylor, R. H. (1969). Some aspects of the intestinal absorption of calcium with special reference to vitamin D. In *Mineral Metabolism*, vol. 3, ed. C. L. Connor & F. Bronner, pp. 321–403. Academic Press: London.

Waterhouse, D. F. (1950). Studies on the physiology and toxicology of blowflies. XIV. The composition formation and fate of the granules in the Malpighian tubules of *Lucilia cuprina* larvae. *Aust. J. Sci. Res. Ser. B*, **3**, 76–112.

Weiss, R. E. & Wilbur, K. M. (1976). Effects of cytochalasin B on division and calcium carbonate extrusion in a calcifying alga. In preparation.

Wigglesworth, V. B. & Salpeter, M. M. (1962). Histology of the Malpighian tubules in *Rhodnius prolixis*. *J. Insect Physiol.*, **8**, 299–307.

Wilbur, K. M. & Watabe, N. (1963). Experimental studies on calcification in molluscs and the alga *Coccolithus huxleyi*. *Ann. N.Y. Acad. Sci.*, **109**, 82–112.

Williams, W. T. (1960). The problem of communication in biological teaching. *Symp. Soc. exp. Biol.*, **14**, 243–249.

SYNOPSIS OF THE PRESENTATIONS

By A. WEBER

Department of Biochemistry and Biophysics,
University of Pennsylvania, School of Medicine,
Philadelphia, Pennsylvania 19174, USA

I wish Heilbrunn could have been at this meeting! Twenty-five years ago his constant emphasis on the central role of calcium seemed exaggerated – now we know that even he did not think of all the reactions in which calcium is involved.

The word calcium primarily evokes the association of mineralization and messenger function. These associations are specific for calcium, magnesium does not substitute; rather magnesium serves as a constantly available co-factor for intracellular reactions, especially those involving organic phosphates. By contrast, calcium concentrations are constant only in extracellular fluids. There, calcium acts as co-factor; for instance calcium enables prothrombin factors VII, IX, X to bind the phospholipid required for their activity, and calcium stabilizes plasma membranes. Also in these activities calcium is not interchangeable with magnesium.

I remember from many years ago how chemists refused to believe these claims of discrimination by biological systems between two doubly charged positive spheres without any unfilled electron shells. This week, however, Mary Truter demonstrated graphically, what Eigen pointed out sometime ago, how a ligand can displace effectively the water of the inner coordination sphere of only either the large (0.095 nm) calcium ion or the small (0.065 nm) magnesium ion.

R. J. P. Williams suggested that the ability of calcium to fit tightly into many protein sites that bind magnesium only loosely or not at all may result from calcium's permissiveness in bond length, which may vary over more than 0·05 nm. As a result calcium is able to form more extensive cross-links which may play an important role in membrane stabilization and may be related to the low solubility of calcium phosphate and carbonate complexes.

At this meeting we considered mostly the specific role of calcium as a messenger, with a small amount of time given to mineralization. I would think that it is impossible to cover both topics adequately in one meeting – the wide range of subjects relating to mineralization is indicated by the two talks on the subject. On the one hand, K. Simkiss pointed out how

[445]

little is known about formation of intracellular mineral granules in invertebrates, and on the other hand A. R. Terepka made a novel suggestion as to
how calcium may be transported from the outside world to the site of
mineralization. Terepka and his colleagues, with the aid of an electron
probe, observed very circumscript calcium densities that are localized in
intracellular regions usually free from mitochondria and, on this basis,
proposed that calcium may be transported across cell layers while sequestered in vesicles. That is very different from calcium running into the
cell, down the concentration gradient, on one side and being pumped out
again on the other. Assuming that the cytoplasmic calcium level must always
be low to prevent inhibition of enzymes and mitochondrial calcium gorging,
it seems to me that calcium transport in vesicles probably would save
energy because it would avoid massive transport of calcium against a large
concentration gradient. One waits with interest to see whether this transport mechanism will be proved.

Messengers achieved general popularity in the sixties when, on the one
hand, cyclic AMP appeared on the scene (I believe Sutherland introduced
the term second messenger), and on the other, the messenger function of
calcium in muscle, with all its ramifications, was being elucidated. For
muscle, the need for a second messenger that would transmit the signal
given by the action potential on the membrane surface to the contractile
material in the centre of the fibre had long been recognized. In the late
forties and fifties, evidence accumulated in favour of calcium for this role,
which was proposed by Sandow in 1952. Early evidence came from the
injection of calcium into living muscle by Heilbrunn and Wiercinski (later
repeated very elegantly by Niedergerke) and the discovery, by Marsh, of
the Marsh–Bendall factor, which conferred calcium dependence on the
contraction of glycerinated muscle fibres by Mg–ATP. In the following
15 years, it became evident that in muscle calcium meets all the requirements for a second messenger. We showed myofibrils from rabbit muscle
to have regulatory calcium binding sites and S. Ebashi discovered troponin
bound to actin filaments as the calcium receptor. Later A. G. Szent-
Györgyi and his colleagues found in many invertebrate muscles another
regulatory calcium site associated with one of the small subunits of myosin.

Ebashi, as well as Hasselbach, identified the Marsh–Bendall factor as
calcium transporting vesicles (first so recognized by Hasselbach), which
can accomplish rapid removal of calcium. The vesicles are formed from the
sarcoplasmic reticulum discovered by Porter and Pallade and extensively
described by Porter, Franzini-Armstrong, Peachey, and Page. The reticulum contains much of the cellular calcium during rest, as shown by Winegrad's autoradiographic studies. One component of the reticulum, the

terminal cisterna, was identified as the putative site for calcium release because of its close association with the excitable T-tubules, and on the basis of experiments from a number of investigators, notably Podolsky and Costantin. Calcium release from cisternae no more than 2 μm apart solved the problem recognized by A. V. Hill in 1948, who pointed out that messengers could not, in sufficient amounts, reach the centres of the fibres from the periphery (a distance of 50 μm) in the short interval between action potential and contraction.

In the sixties, it became apparent that signalling by calcium is a general phenomenon. Calcium as first messenger, carrying the current during depolarization, was mentioned by several speakers who discussed unicellular organisms (see R. Eckert, R. W. Piddington, and W. B. Amos *et al.*). R. W. Meech suggested that in this role calcium may have preceded sodium during evolution. Maybe it is meaningful, with respect to a common origin of the ion pore, that sodium and calcium (also known as current carrier in some invertebrate muscles and nerve cells and the late phase of the action potential in vertebrate heart) have nearly the same ionic radius and similar variable bond lengths. Calcium as a modulator of the primary signal was discussed extensively by Meech. Internal calcium is capable of increasing potassium conductance (according to Meech even in vertebrate heart, but see Kass & Tsien (1975) for a different opinion), and of decreasing the permeabilities for sodium and calcium. Such signal modulations occur in muscles, nerve cells and sensory organs and not, according to Meech, in axons which are meant to transmit the signal precisely as received. Of special interest are the proposed modulations of sodium current in eyes. Calcium is supposed to be responsible for adaptation in the *Limulus* eye by inhibiting depolarization, and for the translation of the light signal into the code of the nervous system in the vertebrate rod by interrupting the sodium dark current. It is not difficult to picture the same ion in such diverse activities as turning on and turning off channels if the receptor for calcium is a small regulatory protein that binds calcium and responds with a conformational change that alters activity of a second protein. This mode of regulation has not yet been demonstrated for ionic channels but there is precedent for it in other calcium regulated systems.

An example of calcium as secondary messenger in secretion was given by J. L. Mongar (see Foreman *et al.*) who described histamine release stimulated by antibody binding. Calcium was identified as second messenger because, after addition of the ionophore A23187, external calcium was capable of substituting for antibody as the stimulus for histamine release. Furthermore, antibody caused increased calcium influx.

The number of cellular activities that are supposed to receive their signals through calcium is now quite large, including such diverse activities as motion, secretion, light-emission, and cell division. I think everybody would agree that the messenger role of calcium may be considered as established if all the following criteria are met: (1) raising the cytoplasmic calcium concentration replaces the primary signal (e.g. action potential, hormone, antibody binding, etc.); (2) injection of EGTA eliminates all response to the primary signal; (3) the primary stimulus is always followed by an increase in cytoplasmic calcium (as indicated by aequorin or a similar indicator); and (4), *in vitro*, the isolated active system (contractile proteins, secretory vesicles, enzymes) is dependent on calcium for its activity. Not all of these criteria are yet met by many of the systems which are supposed to receive their signals through calcium. In order to test the effect of raising cytoplasmic calcium, ionophores are frequently used because the cells are too small for injection. In view of the rising tide of papers describing such experiments, I should like to stress a few cautions mentioned by M. J. Berridge in his talk. Rapid calcium inflow through ionophores in the plasma membranes may cause gorging of mitochondria and reticulum with calcium, resulting in changes of metabolism that produce their own effects or give a positive response not due to the direct action of calcium. As far as I know, ionophore penetration into the membranes of intracellular organelles of intact cells has not yet been demonstrated but has only been inferred from positive ionophore effects in the absence of external calcium. Sometimes the intracellular calcium pumps are so fast that the cytoplasmic calcium level remains low, even with high concentrations of ionophore and external calcium, thus producing false negative results. (For an example see Hainaut & Desmedt, 1974.)

If calcium is a universal messenger the cytoplasmic calcium concentration must be low in all cells, not only in muscle. The discovery of aequorin by Shimomura and Johnson and its introduction to biological experimentation by Ashley and Ridgway and later by Blinks made it possible to demonstrate the low level of cytoplasmic calcium in various cells. Even the squid axon, where calcium has not yet been found to act as messenger, has a concentration of free calcium as low as $0.1\ \mu\text{M}$. N. B. Gilula in his elegant presentation discussing the properties of gap junctions pointed out that cell–cell communication is interrupted when intracellular calcium concentrations rise near the junctional surface, as was shown by Loewenstein and Rose using aequorin. Overall cellular calcium content must be maintained against steady influx from the outside by pumps located outside the plasma membrane, which were discussed by P. F. Baker. These pumps are difficult to investigate, and although Baker's work has greatly increased our knowledge we are not quite certain yet whether the sodium-dependent squid axon pump derives all its energy from the sodium gradient, or whether it uses very low concentrations of ATP. While the plasma membrane pumps take care of the long-term cell homeostasis, a sudden influx of calcium into the cytoplasm is dealt with by internal calcium pumps (cf. P. F. Baker and E. Carafoli). We had a lively argument with respect to the relative merits of reticulum and mitochondria, especially in heart. Nobody contested the generally held view that reticulum picks up calcium in fast skeletal muscle where massive amounts of calcium (in the order of 100 μmole g^{-1} muscle) must be removed with great speed (in less than 14 msec in the cricothyroides muscle of the bat). The reticulum pump, a transport system built so closely along the lines of the sodium–potassium pump that in these evolution-minded days one suspects a common origin, is highly specific for calcium. Its K_m for transport is well below 1 μM at physiological magnesium concentrations (assumed to be about 1 mM). Skeletal muscles that vary widely in speed of relaxation, i.e. the time allowed for collecting 100 μmole of calcium or more, released before contraction, vary correspondingly in the amount of reticulum, suggesting that the turnover number per transport channel may be similar in all these muscles.

In mammalian heart muscle the amount of reticulum is less than in frog muscle and values for calcium capacity, speed of uptake, and K_m are much more uncertain since the isolated preparations are so very unstable. On the other hand, the concentration of mitochondria is very high (80 mg g^{-1} wet weight). However, since calcium uptake by mitochondria, in contrast to that by reticulum, is considerably depressed by magnesium one can evaluate the mitochondrial contribution only from K_m values obtained in the presence of physiological magnesium concentrations of about 1 mM. This value has been reported for heart mitochondria to be as high as 40–60 μM (Scarpa & Graziotti, 1973). We agreed eventually that mitochondria probably take up calcium in the initial phase of relaxation of the heart but that reticulum restores the resting level of calcium. How the calcium distribution between mitochondria and reticulum in heart is maintained was not solved since discussion brought out, first, that change in intracellular sodium is insufficient to cause calcium release from mitochondria, and, second, that reported cyclic AMP effects have not been confirmed by other laboratories.

In cells other than muscle, for instance the squid axon, where inhibition of the mitochondrial pump results in an immediately increased efflux from the cell, mitochondria may be the main intracellular calcium pump. However, as E. Carafoli pointed out, one cannot generalize from the properties of one set of mitochondria to another since mitochondrial calcium pumps vary greatly from cell to cell type, to the point that they may be missing altogether, as in yeast cells.

The talks by C. C. Ashley, R. Niedergerke and S. R. Taylor dealt extensively with two aspects of messenger release: first, the reservoir from which calcium is released and, second, the exact stimulus that causes opening of the calcium channels of this reservoir. All three speakers used muscles for their experiments. One expects the calcium reservoir to be closely associated with the plasma membrane since calcium relays the message from the membrane to the interior. Since the days of Shanes, who kindled interest in this question, one first tends to look at the extracellular fluid. While that was ruled out as the reservoir for vertebrate skeletal muscle, which contracts without calcium in the external medium, it is still being considered for all cells that cannot fulfill their function without calcium in the extracellular fluid. Even if the external calcium is necessary for the complete execution of the primary signal (including the dispatch of the second message) or if it is required to prevent leakage from the internal compartments to the outside, it is difficult to rule out the possibility that the external calcium may, in addition, function as a second messenger. That can only be ruled out if it can be shown that calcium from the outside cannot get to the site of action fast enough, or that not enough external calcium enters to saturate the receptor sites, or that much more calcium is released from an internal site than enters from the outside. Both C. C. Ashley's and R. Niedergerke's measurements suggest that not nearly enough calcium enters from the outside to provide calcium for all of the troponin molecules on the myofibril, not to mention the competing binding sites on myosin, phosphorylase kinase, parvalbumin and other sites. One may argue that isotope exchange experiments require series of stimulations and, therefore, miss the fraction of calcium that returns immediately after each contraction to the outside, without mixing with calcium in the organelles, and that current measurements may underestimate charge influx if the outward currents are greater than assumed; nevertheless the discrepancy seems too large for *external* calcium to be the messenger that signals contraction. (R. Niedergerke's experiments, in addition, suggested that external calcium is not required in frog heart to keep any internal reservoirs filled; after calcium deprivation, readdition of calcium restored contraction in about the time required for diffusion through extracellular space, and faster than expected for the time needed to refill an intracellular compartment.) However, external calcium could be the means by which internal calcium is released from its intracellular reservoir by the mechanism of calcium-induced calcium release, as discussed by S. R. Taylor.

Whether or not *external* calcium is the second messenger in all or any of the systems other than muscle cannot yet be decided. Increased entry of calcium during stimulation can only be taken as a criterion if the amount is adequate to saturate the receptor sites as was made clear by Ashley and by Niedergerke.

The internal calcium reservoir that provides the second messenger in vertebrate skeletal muscle is presumably the terminal cisternae. They have the expected apposition to the excitable membranes of the T-tubules, and, as shown by Winegrad, it is from this region that calcium disappears after stimulation. From this reservoir, massive amounts (at least 100 μmole g^{-1} muscle) must be released in a few milliseconds. Aequorin injected into fibres in the experiments by C. C. Ashley and by S. R. Taylor and his colleagues show that calcium release precedes contraction. (Further interpretations become difficult because absolute calibrations of the luminescence as a function of free calcium require exact knowledge of free magnesium and pH, which we do not have. Therefore we do not know the concentration of free calcium during peak tension.) S. R. Taylor assembled all the evidence suggesting that in frog skeletal muscle calcium release is caused directly by depolarization and not by calcium-induced calcium release. However, he allowed that the latter may be the responsible mechanism in other cells such as heart.

A number of years ago, Rasmussen pointed out and reviewed how calcium relates with the other second messengers, the cyclic nucleotides. At this meeting M. J. Berridge gave a well-organized review of present-day opinion. Even relying only on the harder data, one can say that the effects of cyclic AMP on the calcium message are not predictable. On the one hand, cyclic AMP may shorten the duration of the calcium message either by hastening calcium removal, as reported for mammalian heart, or attenuate it by inhibiting calcium influx, as described by J. L. Mongar (see Foreman *et al.*) during this meeting. On the other hand, cyclic AMP may accentuate the message by shifting the receptor response to lower calcium concentrations as has been described by E. Krebs for phosphorylase kinase. Furthermore, not only does cyclic AMP modulate the calcium message, calcium may also control production and removal of cyclic nucleotides. For instance, in brain there exists a cyclase and a phosphodiesterase that are both dependent on calcium for functioning, an intriguing phenomenon.

Last, but not least, receptors for the calcium message were discussed, such as the cilia of *Paramecium* (R. Eckert), aequorin (O. Shimomura as well as C. C. Ashley), the *Vorticella* stalk (W. B. Amos), troponin-C (J. H. Collins as well as S. Ebashi) and the regulatory light chain of myosin (J. H. Collins as well as A. G. Szent-Györgyi). As yet, least defined are the *Paramecium* cilia: after triton treatment, added calcium determines the position of the cilia; if ATP is also added the cilia beat and calcium ion concentrations above 10^{-6} M reverse the direction of the beat. We do not know whether or not calcium combines with a small regulatory protein; nor do we know in detail what causes reversal of beat. Is the whole motor turned around or does it go into reverse gear?

The most unusual receptor is the *Vorticella* stalk, i.e. the spasmoneme. Here calcium gives a signal by being itself the agent of contraction rather than by allowing ATP to cause contraction. In the absence of ATP, calcium binding causes shortening, calcium removal re-extension. It will be very interesting to know how this is done; for instance does calcium convert the protein into a perfect rubber? (One wonders how that could be achieved by so relatively little bound calcium.)

Among the remaining receptors aequorin stands out because it is not described as having a separate regulatory protein like troponin-C (TN-C), a subunit of troponin, or the regulatory light chain of myosin, or the calcium binding subunit of phosphorylase kinase or the calcium binding protein of brain that combines with cyclase and phosphodiesterase (cf. M. J. Berridge). J. H. Collins presented the evidence that at least two of these proteins, TN-C and the myosin chain, may have a common ancestor together with parvalbumin. Since parvalbumin was crystallized by Kretsinger we have information on its calcium binding sites, e.g. that calcium is bonded to six oxygen atoms, four of them carboxyl groups, with bond lengths varying between 0.20 and 0.24 nm. This information may be applied to TN-C in spite of its greater number of calcium binding sites (four compared to two for parvalbumin) since they probably, according to Kretsinger, resulted from a gene duplication of the two parvalbumin sites.

Two aspects of calcium binding are of special physiological interest: its speed of binding and release, and the sharpness of the switch, i.e. the range of calcium concentrations required to go from complete relaxation to maximal activity. Manfred Eigen pointed out a number of years ago that calcium can be bound more rapidly than magnesium and, therefore, for a given affinity, released in a shorter time. Since in some muscles troponin must be saturated in a few milliseconds and since the dissociation constant for the last calcium is around micromolar it seems likely that the receptor proteins make use of the intrinsically high reaction rate of calcium; direct measurements, however, have not yet been made.

Steepness of the switch function of troponin or any other receptor protein, i.e. a narrow range of effective calcium concentrations, may be desirable for several reasons. With a steep switch function complete in-activation of the switch occurs after no more than a 100-fold reduction of free calcium, at relatively high calcium levels (0.1 μM). As a result the message can be terminated quite efficiently because the pump can operate in a relatively high range of calcium concentrations (from 10 to 0.1 μM) where it is not yet severely limited by the rate of calcium diffusion. It is notable that all the calcium receptors mentioned above have fairly steep switches, achieving the increase from 10 to 90 % maximal activity by an increase in calcium concentration much less than 100-fold. That requires several cooperating binding sites and can be achieved by two means: either positive cooperativity of binding, i.e. binding of one calcium creates a binding site for the next calcium that has a higher affinity than the first, or by having several calcium ions together turn the switch, without influencing each others' binding constants. The latter mechanism makes itself felt mainly in the range of calcium concentrations below K_d of the individual site, causing a very steep fall in activity from 10 to less than 0.01 %, whereas the range for an activity decrease from 90 to 10 % is only narrowed to 30–40-fold rather than a 100-fold as for a single calcium site. However, many calcium-regulated activities rise from 10 to 90 % over a narrower range of calcium concentration, e.g. tension development, myofibrillar ATPase activity and phosphodiesterase activity in brain. In these cases one must suspect some cooperativity of binding between one or the other calcium ions. The fact that that has not been demonstrated for troponin bound to actin filaments may be due to the scatter of the data in the critical region.

S. Ebashi summarized knowledge of how the calcium receptor transmits information to the active site in the only system about which we have some knowledge. This system, the regulated actin filament, is quite com-plicated. Calcium binding to the calcium binding subunit of troponin, TN-C, alters the conformation of TN-I, the second troponin subunit, causing tropomyosin to move away from a position on the actin filament where it blocks the myosin binding site in the absence of calcium, thus permitting ATP-activated myosin to interact with actin. As Ebashi pointed out, we are not yet quite sure about many of the details of these inter-actions, such as the interplay with the third troponin subunit, TN-T, or the manner in which tropomyosin is moved.

It will be very interesting to compare transmission of information in the other systems.

In conclusion, I want to speculate why calcium may have been a natural choice for a second messenger. The cell of course depends on relatively high phosphate concentrations for a large part of its energy metabolism. To avoid precipitation of phosphate it was necessary for cells to keep the high concentrations of calcium in the plasma and the extracellular space from mixing with the millimolar concentrations of phosphate in the cytoplasm. Thus, it became necessary to develop effective calcium pumps with which to maintain low levels of intracellular calcium. These, together with the readily available extracellular calcium reservoir are just the conditions necessary for an intracellular messenger. Apparently the opportunity thus presented was not overlooked during biological evolution.

SELECTED KEY REFERENCES

ASHLEY, C. C. & RIDGWAY, E. B. (1970). On the relationships between membrane potential, calcium transient and tension in single barnacle muscle fibres. *J. Physiol., Lond.*, **209**, 105–130.

BROSTROM, C. O., HUNKELER, F. L. & KREBS, E. G. (1971). The regulation of skeletal muscle phosphorylase kinase by Ca^{2+}. *J. biol. Chem.*, **246**, 1961–1967.

COSTANTIN, L. L., FRANZINI-ARMSTRONG, C. & PODOLSKY, R. J. (1965). Localization of calcium-accumulating structures in striated muscle fibres. *Science, Wash.*, **147**, 158–159.

COSTANTIN, L. L. & PODOLSKY, R. J. (1965). Calcium localization and the activation of striated muscle fibres. *Fedn Proc.*, **24**, 1141–1145.

EBASHI, S. & KODAMA, A. (1965). A new protein factor promoting aggregation of tropomyosin. *J. Biochem., Tokyo*, **58**, 107–108.

EBASHI, S. & LIPMANN, F. (1962). Adenosine triphosphate-linked concentration of calcium ions in a particulate fraction of rabbit muscle. *J. Cell Biol.*, **14**, 389–400.

EIGEN, M. & HAMMES, G. G. (1963). Elementary steps in enzyme reactions (as studied by relaxation spectronomy). *Adv. Enzymol.*, **25**, 1–38.

FORD, L. E. & PODOLSKY, R. J. (1972). Calcium uptake and force development by skinned muscle fibres in EGTA buffered solutions. *J. Physiol., Lond.*, **223**, 1–19,

FRANZINI-ARMSTRONG, C. & PORTER, K. R. (1964). Sarcolemmal invaginations and and *T*-system in fish skeletal muscle. *Nature, Lond.*, **202**, 355–357.

HAINAUT, K. & DESMEDT, J. E. (1974). Calcium ionophore A23187 potentiates twitch and intracellular calcium release in single muscle fibres. *Nature, Lond.*, **252**, 407–408.

HASSELBACH, W. & MAKINOSE, M. (1961). Die Calciumpumpe der "Erschlaffungs-grana" des Muskels und ihre Abhängigkeit von der ATP-Spaltung. *Biochem. Z.*, **333**, 518–528.

HEILBRUNN, L. V. & WIERCINSKI, F. J. (1947). The action of various cations on muscle protoplasm. *J. cell. comp. Physiol.*, **29**, 15–32.

HILL, A. V. (1948). On the time required for diffusion and its relation to processes in muscle. *Proc. R. Soc. Lond. B*, **135**, 446–453.

KASS, R. S. & TSIEN, R. W. (1975). Multiple effects of calcium agonists on plateau currents in cardiac Purkinje fibers. *J. gen. Physiol.*, **66**, 169–192.

KENDRICK-JONES, J., LEHMAN, W. & SZENT-GYÖRGYI, A. G. (1970). Regulation in molluscan muscles. *J. molec. Biol.*, **54**, 313–326.

KRETSINGER, R. H., NOCKOLDS, C. E., COFFEE, C. J. & BRADSHAW, R. A. (1971). The structure of a calcium-binding protein from carp muscle. *Cold Spring Harbor Symp. quant. Biol.*, **36**, 217–220.

LLINÁS, R., BLINKS, J. R. & NICHOLSON, C. (1972). Calcium transient in presynaptic terminal of squid giant synapse: detection with aequorin. *Science, Wash.*, **176**, 1127–1129.

MARSH, B. B. (1951). A factor modifying muscle fibre synæresis. *Nature, Lond.*, **167**, 1065–1066.

MURRAY, J. M., WEBER, A. & BREMEL, R. D. (1975). In *Calcium Transport in Contraction and Secretion*, ed. E. Carafoli, F. Clementi, W. Drabikowski & A. Margreth, pp. 489–496. North-Holland: Amsterdam & New York.

NIEDERGERKE, R. (1955). Local muscular shortening by intracellularly applied calcium. *J. Physiol., Lond.*, **128**, 12P–13P.

PAGE, S. G. (1965). A comparison of the fine structures of frog slow and twitch muscle fibres. *J. Cell Biol.*, **26**, 477–497.

PORTER, K. R. (1961). The sarcoplasmic reticulum: its recent history and present status. *J. biophys. biochem. Cytol.*, 10 no. 4 (pt 2), 219–226.

PORTER, K. R. & PALADE, G. E. (1957). Studies on the endoplasmic reticulum. III. Its form and distribution in striated muscle cells. *J. biophys. biochem. Cytol.*, **3**, 269–299.

RASMUSSEN, H. (1970). Cell communication, calcium ion, and cyclic adenosine monophosphate. *Science, Wash.*, **170**, 104–112.

RIDGWAY, E. B. & ASHLEY, C. C. (1967). Calcium transients in single muscle fibres. *Biochem. biophys. Res. Commun.*, **29**, 229–234.

ROSE, B. & LOEWENSTEIN, W. R. (1975). Permeability of cell junction depends on local cytoplasmic calcium activity. *Nature, Lond.*, **254**, 250–252.

SANDOW, A. (1952). Excitation–contraction coupling in muscular response. *Yale Jl Biol. Med.*, **25**, 176–201.

SCARPA, A. & GRAZIOTTI, P. (1973). Mechanisms for intracellular calcium regulation in heart. I. Stopped-flow measurements of Ca^{++} uptake by cardiac mitochondria. *J. gen. Physiol.*, **62**, 756–772.

SHIMOMURA, O., JOHNSON, F. H. & SAIGA, Y. (1962). Extraction, purification and properties of aequorin, a bioluminescent protein from the luminous hydromedusan, *Aequorea*. *J. cell. comp. Physiol.*, **59**, 223–239.

STENFLO, J., FERNLUND, P., EGAN, W. & ROEPSTORFF, P. (1974). Vitamin K dependent modifications of glutamic acid residues in prothrombin. *Proc. natn. Acad. Sci. USA*, **71**, 2730–2733.

WEBER, A. & HERZ, R. (1963). The binding of calcium to actomyosin systems in relation to their biological activity. *J. biol. Chem.*, **238**, 599–605.

WEBER, A., HERZ, R. & REISS, J. (1966). Study of the kinetics of calcium transport by isolated fragmented sarcoplasmic reticulum. *Biochem. Z.*, **345**, 329–369.

WINEGRAD, S. (1965). Autoradiographic studies of intracellular calcium in frog skeletal muscle. *J. gen. Physiol.*, **48**, 455–479.

WOLFF, D. J. & BROSTROM, C. O. (1974). Calcium-binding phosphoprotein from pig brain: identification as a calcium-dependent regulator of brain cyclic nucleotide phosphodiesterase. *Arch. Biochem. Biophys.*, **163**, 349–358.

AUTHOR INDEX

Figures in bold type indicate pages on which references are listed.

SUBJECT INDEX